12th International Workshop on Slow Positron Beam Techniques 2010

AA008848

(SLOPOS12)

Journal of Physics: Conference Series Volume 262

North Queensland, Australia
1 – 6 August 2010

Editors:

Stephen Buckman
James Sullivan
Ronald White

ISBN: 978-1-61782-267-4
ISSN: 1742-6588

Printed from e-media with permission by:

Curran Associates, Inc.
57 Morehouse Lane
Red Hook, NY 12571

Some format issues inherent in the e-media version may also appear in this print version.

Printed by Curran Associates, Inc. (2011)

For permission requests, please contact the Institute of Physics
at the address below.

Institute of Physics
Dirac House, Temple Back
Bristol BS1 6BE UK

Phone: 44 1 17 929 7481
Fax: 44 1 17 920 0979

techtracking@iop.org

Additional copies of this publication are available from:

Curran Associates, Inc.
57 Morehouse Lane
Red Hook, NY 12571 USA
Phone: 845-758-0400
Fax: 845-758-2634
Email: curran@proceedings.com
Web: www.proceedings.com

TABLE OF CONTENTS

These proceedings arose from the 12th International Workshop on Slow Positron Beam Techniques (SLOPOS12), which was held on Magnetic Island, North Queensland, Australia, between 1–6th August 2010. Meetings in the SLOPOS series are held (roughly) every three years and have now been held on (almost) all continents, indicating the truly international nature of the field. SLOPOS12 marked the second time that the Workshop had been held in the southern hemisphere, and the first time in Australia.

SLOPOS12 attracted 122 delegates from 16 countries. Most encouraging was the attendance of 28 student delegates, and that about half of the overall delegates were early career researchers – a good sign for the future of our field. We also enjoyed the company of more than a dozen partners and families of delegates.

In a slight departure from previous SLOPOS meetings, the International Advisory Committee approved a broader scope of scientific topics for inclusion in the program for the 2010 Workshop. This broader scope was intended to capture the applications of positrons in atomic, molecular and biomedical areas and was encapsulated in the byeline for SLOPOS–12: *The 12th International Workshop on Slow Positron Beam Techniques for Solids, Surfaces, Atoms and Molecules.*

The scientific and social program for the meeting ran over 6 days with delegates gathering on Sunday August 1st and departing on August 6th. The scientific program included plenary, invited, contributed and student lectures, the latter being the subject of a student prize. In all there were 53 oral presentations during the week. There were also two poster sessions, with 63 posters exhibited, and a prize was awarded for the best poster by a student delegate. The standard of the student presentations, both oral and posters, was outstanding, so much so that the judging panel recommended an additional number of prizes be awarded.

Topics that were the focus of invited presentations and contributed papers at SLOPOS–12 included:

- *Positron Interactions with Surfaces*
- *Positron Beam and Detector Technology*
- *Positron Interactions with Atoms and Molecules*
- *Positronium Science*
- *Defects and Vacancies in Materials*
- *Porosity and Open Volume in Materials*
- *Antimatter in Biomedical Science*
- *Anti-hydrogen Studies*
- *Positron Transport*
- *Annihilation*

On a sad note, delegates paid tribute to the contributions of one of our colleagues, Chris Beling, who tragically passed away shortly before the meeting. Chris' contributions to positron science and to the education of young scientists were noted in a number of the invited presentations. It is an honour for our community to begin these proceedings with a short tribute to Chris' life by Professor Paul Coleman.

The Workshop could not have occurred without the generous support of our sponsors: The ARC Centre for Antimatter–Matter Studies, The Australian National University, Flinders University, James Cook University, The Institute of Physics (UK) and the Australian Government's Department of Innovation, Industry, Science and Research.

It would also not have been possible without the hard work of the Local and International Organising Committees and the friendly and efficient staff at the All Seasons Resort, Magnetic Island. We are most grateful for the on-site assistance of Gillian Drew, the CAMS student and postdoc team, the financial wizardry of Chris Kalos, and the post-Workshop editorial assistance of Julia Wee and Adam Edwards.

Finally we would like to thank all of the attendees at SLOPOS12 for their scientific contributions to the Workshop, and for the warm spirit of engagement which characterised the scientific discussions and social occasions.

SLOPOS13 will be held in Germany in 2013 and we all look forward to the occasion.

Stephen Buckman, James Sullivan and Ronald White
(Guest Editors)

Local Organising Committee

Stephen Buckman (Chair, ANU, Canberra)
James Sullivan (Secretary, ANU, Canberra)
Ronald White (JCU, Townsville)
Jim Williams (UWA, Perth)
Suzanne Smith (ANSTO, Sydney)
Igor Bray (Curtin U., Perth)
Casten Makochekanwa (ANU, Canberra)
Michael Went (ANU, Canberra)
Adric Jones (ANU, Canberra)
Peter Caradonna (ANU, Canberra)
Ryan Weed (ANU, Canberra)
Jason Roberts (ANU, Canberra)
Josh Machacek (ANU, Canberra)

International Committee

G Amarendra (India)
M-F Barthe (France)
C Beling (Hong Kong)
R Brusa (Italy)
P Coleman (UK)
C Corbel (France)
M Fujinami (Japan)
R Krause-Rehberg (Germany)
K Lynn (USA)
H Schut (Netherlands)
P Simpson (Canada)
R Suzuki (Japan)
F Tuomisto (Finland)

A Weiss (USA)

SLOPOS–12 Delegates, 1–6 August 2010, Magnetic Island, Australia

SPONSORS

ARC Centre of Excellence for Antimatter-Matter Studies

Australian Government
Department of Innovation Industry, Science and Research

IOP Publishing
Journal of Physics B
Journal of Physics Conference Series

ANU
THE AUSTRALIAN NATIONAL UNIVERSITY

Flinders UNIVERSITY

JAMES COOK UNIVERSITY AUSTRALIA

Journal of Physics: Conference Series **262** (2011) 012001 doi:10.1088/1742-6596/262/1/012001

Antiparticle sources for antihydrogen production and trapping

M Charlton[1], G B Andresen[2], M D Ashkezari[3], M Baquero-Ruiz[4], W Bertsche[1], P D Bowe[2], C C Bray[4], E Butler[1], C L Cesar[5], S Chapman[4], J Fajans[4], T Friesen[6], M C Fujiwara[7], D R Gill[7], J S Hangst[2], W N Hardy[8], R S Hayano[9], M E Hayden[3], A J Humphries[1], R Hydomako[6], S Jonsell[1,10], L V Jørgensen[1], S J Kerrigan[1], L Kurchaninov[7], R Lambo[5], N Madsen[1], S Menary[11], P Nolan[12], K Olchanski[7], A Povilus[4], P Pusa[12], F Robicheaux[13], E Sarid[14], S Seif El Nasr[8], D M Silveira[9,15], C So[4], J W Storey[7], R I Thompson[6], D P van der Werf[1], D Wilding[1], J S Wurtele[4] and Y Yamazaki[15] (ALPHA Collaboration)

[1] Department of Physics, Swansea University, Swansea SA2 8PP, United Kingdom
[2] Department of Physics and Astronomy, Aarhus University, DK-8000 Aarhus C, Denmark
[3] Department of Physics, Simon Fraser University, Burnaby BC, V5A 1S6, Canada
[4] Department of Physics, University of California, Berkeley, CA 94720-7300, USA
[5] Instituto de Física, Universidade Federal do Rio de Janeiro, Rio de Janeiro 21941-972, Brazil
[6] Department of Physics and Astronomy, University of Calgary, Calgary AB, T2N 1N4, Canada
[7] TRIUMF, 4004 Wesbrook Mall, Vancouver BC, V5T 2A3, Canada
[8] Department of Physics and Astronomy, University of British Columbia, Vancouver BC, V6T 1Z4, Canada
[9] Department of Physics, University of Tokyo, Tokyo 113-0033, Japan
[10] Fysikum, Stockholm University, SE-10609, Stockholm, Sweden
[11] Department of Physics and Astronomy, York University, Toronto, ON, M3J 1P3, Canada
[12] Department of Physics, University of Liverpool, Liverpool L69 7ZE, United Kingdom
[13] Department of Physics, Auburn University, Auburn, AL 36849-5311, USA
[14] Department of Physics, NRCN-Nuclear Research Center Negev, Beer Sheva, IL-84190, Israel
[15] Atomic Physics Laboratory, RIKEN, Saitama 351-0198, Japan

E-mail: M.Charlton@Swansea.ac.uk

Abstract. Sources of positrons and antiprotons that are currently used for the formation of antihydrogen with low kinetic energies are reviewed, mostly in the context of the ALPHA collaboration and its predecessor ATHENA. The experiments were undertaken at the Antiproton Decelerator facility, which is located at CERN. Operations performed on the clouds of antiparticles to facilitate their mixing to produce antihydrogen are described. These include accumulation, cooling and manipulation. The formation of antihydrogen and some of the characteristics of the anti-atoms that are created are discussed. Prospects for trapping antihydrogen in a magnetic minimum trap, as envisaged by the ALPHA collaboration, are reviewed.

1. Introduction

The controlled formation of antihydrogen by the interaction of cold clouds of antiprotons and positrons has become routine since 2002, when the first successful experiments were reported by the ATHENA [1] and ATRAP [2] collaborations. Several investigations have been performed since which have shed light on the antihydrogen created under the conditions of these two experiments, which typically involved mixing antiprotons with positron plasmas in Penning traps, with their concomitant electric and strong magnetic fields. The work has included experiment, theory and simulation. The individual references to these studies are too numerous to list here; the interested reader is referred to the following articles, where much of the work is synthesised or reviewed [3, 4, 5, 6].

The motivation for producing antihydrogen is to eventually compare its properties with those of hydrogen. In spectroscopic terms, the 1S-2S two-photon [7], and the ground state hyperfine [8] transitions in atomic hydrogen are some of the best known quantities in physics, and it is these lines which are a natural starting point for antihydrogen spectroscopy. Furthermore, experimental studies of the gravitational interaction of antimatter with matter have never been performed, and will surely be a focus of experimental attention [9, 10]. Theoretical motivation for the latter has been summarised elsewhere [3], whilst extensions to the Standard Model of Particle Physics to incorporate effects which may produce spectral shifts between the lines of hydrogen and antihydrogen (i.e. CPT violation) have been given by Kostelecký and co-workers and others [11, 12, 13, 14].

In several of the experimental approaches to comparisons of hydrogen and antihydrogen (and in particular spectroscopy), it is envisaged that the latter needs to be held in a neutral atom trap to prolong its interaction with the interrogating fields. This is the approach being taken by the ALPHA and ATRAP collaborations that are both actively pursuing antihydrogen trapping using modified versions of the classic Ioffe-Pritchard magnetic minimum trap [15], superimposed upon the apparatus they use to produce the anti-atoms [16, 17]. Here we will describe the ALPHA apparatus and approach.

The remainder of this contribution is comprised of summaries of the techniques used by ALPHA to confine and manipulate positron and antiproton clouds and the production of antihydrogen atoms in their current neutral atom trapping apparatus. Progress towards the trapping of antihydrogen will be briefly discussed.

2. Antiproton capture and manipulation

The antiprotons used in the antihydrogen studies emanate from the Antiproton Decelerator (AD) facility [18], which is located at the European Particle Physics Laboratory, CERN. This facility is unique worldwide as it decelerates stored antiprotons to a kinetic energy of around 5.3 MeV and supplies them to the experiments in pulses around 100 ns wide, each containing just over 10^7 particles, every 100 seconds or so. A simple, but well established, foil-moderation technique [19] is used to reduce the kinetic energy of the antiprotons on entry to the ALPHA apparatus. A fraction of the antiprotons ($\sim 10^{-2}$-10^{-3}) emerge from the foil with kinetic energies which allow them to be dynamically captured in a Penning-type trap (the so-called catching trap) in which radial confinement is provided by a strong axial magnetic field (typically around 3 T), whilst axial confinement is achieved by electric fields which are suitably pulsed to allow the antiprotons to enter, but not escape. In this manner around 40,000 antiprotons can be captured per AD shot. A schematic illustration of the ALPHA apparatus is given in figure 1, with the strength of the axial component of the magnetic field shown in the lower box.

Before the antiproton pulse arrives a cloud of electrons with a density of 10^8-10^9 cm^{-3} is loaded into the catching trap. These cool rapidly [20] via the emission of synchrotron radiation in the strong magnetic field and, in the absence of any sources of heat, will reach equilibrium with the temperature of the surrounding trap electrodes, which in the case of ALPHA is ~ 7

Journal of Physics: Conference Series **262** (2011) 012001 doi:10.1088/1742-6596/262/1/012001

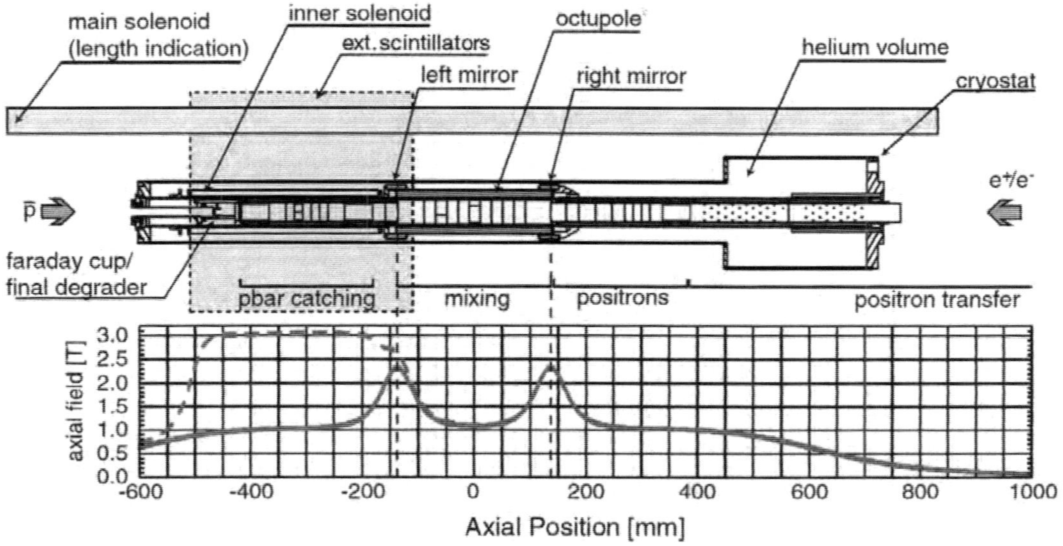

Figure 1. Schematic illustration of the antiproton capture and antihydrogen formation sections of the ALPHA apparatus, with a section through the magnetic coil system. The positron accumulator is located to the right of this apparatus. The positions of external scintillators used to monitor antiproton and antihydrogen annihilations are shown, though the inner antiproton annihilation vertex detector is omitted for clarity. The lower panel shows the behaviour of B_z, the axial magnetic field. The radial component (not shown) is provided by the octupole.

K. (We note here that it is very challenging to achieve such low electron plasma temperatures. For instance, since plasma expansion due to trap imperfections [21, 22] is difficult to eliminate, Joule heating will raise the temperature of the cloud.) The trapped antiprotons, confined to the axis of the instrument, lose kinetic energy to the colder electron cloud, and within a few seconds reach equilibrium and reside in the same potential well as the electrons [23]. The electrons can be removed, if desired, by rapidly lowering the confining potential, though this manipulation may also result in a somewhat heated antiproton ensemble. In ALPHA, the antiprotons are initially captured in the 3 T region to the left of figure 1 to enhance the efficiency [24], before being transferred to the 1 T region for mixing to form antihydrogen. As well as further electron cooling, other manipulations are performed on the antiprotons once they reside in this region, as will be described.

One of the major problems facing the antihydrogen field is to trap the anti-atoms for further study. As will be described in section 4, the neutral atom traps used in this effort, based upon the magnetic minimum principle [15], are only $\sim$ 1 K deep. Thus, the antihydrogen must be created at a temperature which will allow as much as possible to be captured into these shallow traps. The antihydrogen is typically formed via positron-antiproton interactions in a positron plasma which may have a density around 10^8 cm^{-3}. The plasma will have a self radial electric field, E_r, given by $E_r = n_e er/2\epsilon_0$, where n_e is the plasma density, r is the radius in the plasma (up to the edge at r_p) and e and ϵ_0 have their usual meanings. When this electric field is combined with the axial magnetic field, B_z, a tangential drift speed results, given by E_r/B_z, as the plasma rotates about the axis with a constant angular frequency. It is easily seen that, to leading order (e.g. [6]), the rotation speed is independent of mass such that the antiprotons will co-rotate with the positrons. Such motion will add extra kinetic energy to the antiproton which

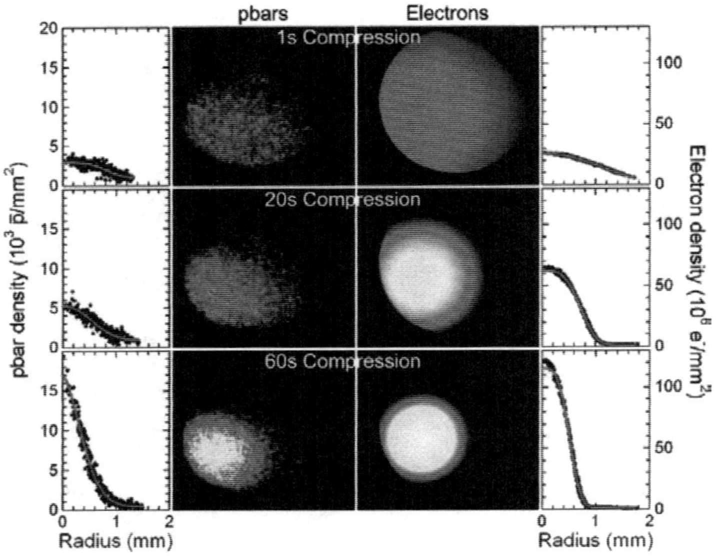

Figure 2. Examples of images of the antiproton and electron clouds for three different compression times, including the respective radial profiles. The images were obtained using an MCP/phosphor screen arrangement similar to that shown in figure 4a. The electrons were initially captured into a 136 mm long trap. Typical antiproton samples contained $\sim 3 \times 10^4$ particles.

will transfer to the antihydrogen and be dependent upon the radial position of production of the anti-atom in the plasma. For typical examples of $n_e = 10^8$ cm^{-3}, $r = 1$ mm and $B_z = 1$ T, it can be found that the extra kinetic energy is several 10s of degrees K, and thus much larger than the typical trap depths.

Thus, ALPHA has developed a technique to control the radial extent of the antiprotons using a variant of the rotating wall technique which has been used successfully over a decade or more to manipulate the radial profile of a number of plasma species, including positrons: see e.g. [25, 26, 27, 28, 29, 30, 31]. The technique involves using controlled compression of an electron plasma in which the antiprotons are embedded [32]. The rotating wall was achieved using a fixed frequency of typically 10 MHz applied to an azimuthally segmented electrode located in the 3T region of the ALPHA apparatus. The radial profiles of the antiprotons and electrons could be measured by individually ejecting each cloud onto a multi-channel plate/phosphor imaging system located upstream in a low magnetic field region.

Examples of the compression are presented in figure 2, which shows the progressive narrowing of the radial profile of the electron and antiproton clouds; see [32] for more details. Figure 3 shows the tracking of the radii of the respective clouds with compression time. Note that if the electron compression is undertaken too rapidly, the antiprotons do not track their behaviour. In this instance, it is likely that they quickly find themselves in a region of very low electron density, such that they can no longer couple via collisions. Reductions in areal density of a factor of 10 have been achieved, which translates directly into an order of magnitude decrease in the temperature due to rotational effects, with an accompanying reduction in the antihydrogen temperature. Compressed antiproton clouds are also beneficial in that they ensure complete spatial overlap with the positron plasma and help maintain the stability of the cloud against radial drift in the presence of the magnetic minimum field (see section 4).

It is also notable that manoeuvring the antiprotons from trap-to-trap (obviously using

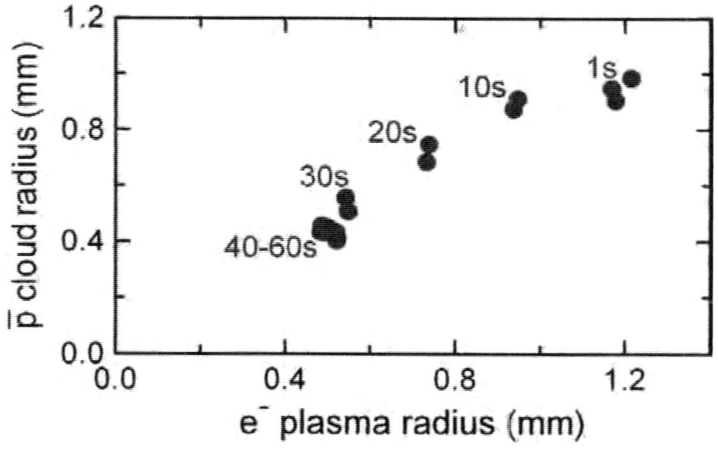

Figure 3. The tracking of the radii of the electron plasma and the antiproton cloud during sympathetic compression for various durations as indicated.

electric fields to accelerate and then decelerate and re-trap them) can lead to increases in the temperature of a cold ensemble. Re-cooling with electrons is always feasible, but ejecting these before antihydrogen experiments can take place can, as mentioned above, also heat the antiprotons. ALPHA has recently [33] developed a technique familiar in cold atom physics, namely evaporative cooling, to reduce the antiproton temperature into the cryogenic regime, and has directly measured such low antiproton temperatures for the first time.

The part of the ALPHA apparatus used for this work, and examples of some of the potential wells used, are illustrated schematically in figure 4. The potential well holding the antiprotons could be progressively lowered, and, allowing time after each step for the remaining cloud to re-thermalise via collisions, the temperature of the remaining ensemble was reduced. The temperature could be measured by slowly allowing the antiprotons to escape from the cloud, to be detected by the downstream MCP/phosphor arrangement. The first antiprotons to escape will be from the exponential tail of the Maxwell-Boltzmann distribution, such that the slope of a logarithmic plot of the integrated counts can be used to derive the temperature of the trapped cloud. Such a technique has been in use in non-neutral plasma physics for some time [34], and examples plots are shown in figure 5. It can be seen that lowering the antiproton well carefully to 10 mV produces a (9 ± 4) K ensemble, with the uncertainty mainly due to that on the voltages applied to the traps. It is notable that at this low trap depth/temperature, (6 ± 1) % of the antiprotons remained trapped.

3. Positron capture and manipulation

The positron accumulator, and most of the associated methodology, used by the ALPHA collaboration was developed for the predecessor ATHENA experiment. Much of this has been described briefly elsewhere [30, 35], though there is no major journal account of the device and its performance.

The ALPHA positron apparatus is shown schematically in figure 6. The beamline is based around a ^{22}Na β^+ source and a conical [36] solid neon moderator [37], which is attached to, though electrically isolated from, a closed cycle helium refrigerator. The latter could reach temperatures around 5 K, and with the neon optimally plated onto the source/cone arrangement, low energy ($\sim$ few eV) positrons were emitted with a yield of around 2×10^5 $e^+s^{-1}mCi^{-1}$. The low energy positrons were guided at a kinetic energy of around 80 eV using magnetic fields to

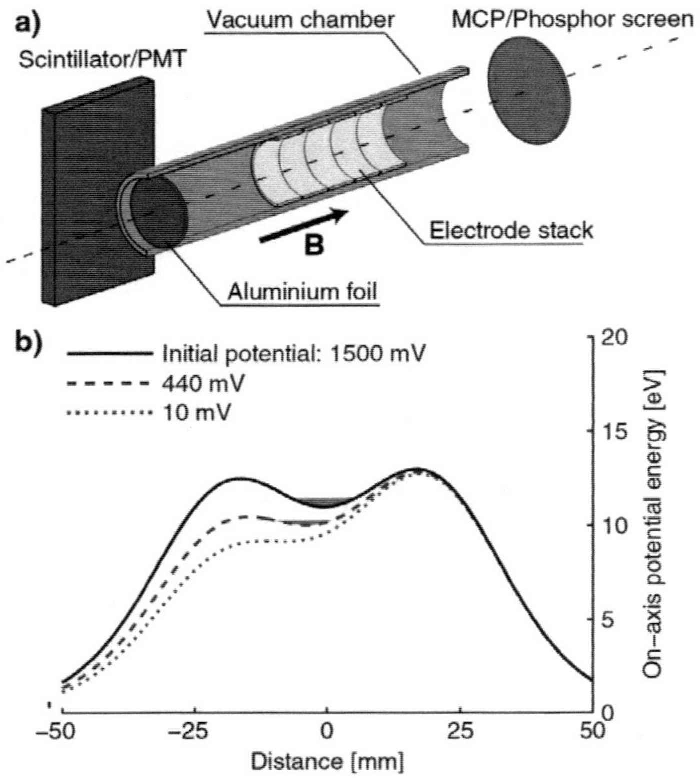

Figure 4. (a) Schematic of the part of the ALPHA apparatus used to investigate the evaporative cooling of antiprotons. (b) Three examples of the confining potential wells used in the experiments. The shallowest well (10 mV) was that which achieved the lowest antiproton temperature of ~ 9 K (see text).

the 3-stage buffer gas Penning-Malmberg trap of the type pioneered by Surko and co-workers [38, 39, 40].

Collisions with the nitrogen buffer gas cause a fraction, ϵ, of the positrons to lose kinetic energy whilst traversing the 3-stage electrode arrangement and become trapped. As a result of further collisions with the nitrogen gas in the second and third stages of the trap, the positrons eventually reside in the latter, where they accumulate. The addition of a cooling gas to the third stage, typically CF_4, SF_6, or CO_2, with a high thermalisation rate for positrons (see e.g. [41, 42]) is useful as it allows more effective rotating wall compression of the cloud. However, a cooling gas is typically not applied during antihydrogen production runs, as the pumpout time (see below) becomes extended. The accumulator rotating wall is typically applied for the full 200 s of the accumulation time at a frequency of around 600 kHz. Once accumulation is complete, the nitrogen buffer gas is pumped out [43], and after about 20-30 s the pressure is low enough for the positrons to be ballistically transported [30] to the main system (figure 1) and recaptured. The positrons cool in the strong axial magnetic field of the mixing region solenoid via the emission of synchrotron radiation and are compressed using a second rotating wall system which operates at a frequency of 700 kHz for a period of around 100 s.

In controlling the positrons, and their eventual interaction with the antiprotons, it has been found useful to reduce their overall number. This reduces the positron density, such that the

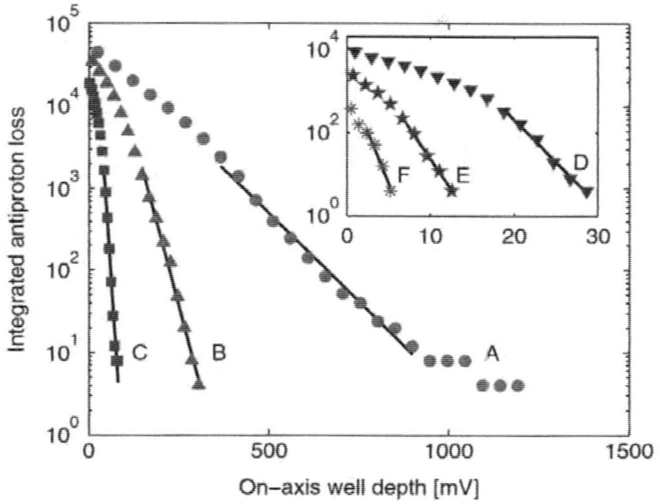

Figure 5. Integrated antiproton loss as the respective potential wells are lowered. The lines are the exponential fits used to deduce the antiproton temperature in each case. The data correspond, from A to F, to antiproton temperatures of 1040, 325, 57, 23, 19 and 9 K, respectively. Before each cooling experiment a cloud of $\sim 45,000$ antiprotons with a radius of 0.6 mm and a density of 7.5×10^6 cm^{-3} was prepared [33].

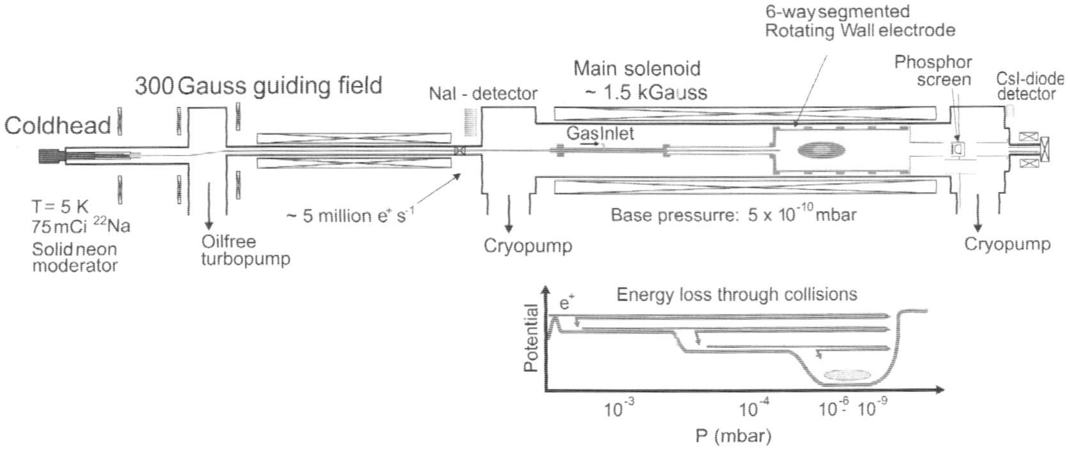

Figure 6. Schematic of the ALPHA positron accumulator. The three stages of the trap (see the lower figure) have progressively lower gas pressures due to differential pumping. The third stage electrodes are around 30 cm in diameter.

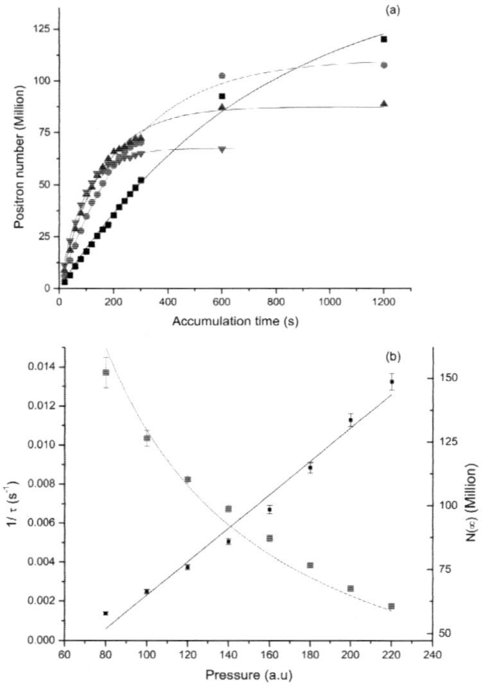

Figure 7. a) Examples of accumulation curves for pressures as measured in the gas supply line of 80 (■) , 120 (▼), 160 (▲) and 200 (•) mbar. Note that 100 corresponds to a first stage pressure of around 10^{-3} mbar. b) The variation of $1/\tau$ (black) and $N(\infty)$ (red) with gas pressure, P, together with fits to the expected forms (see text).

plasma, and hence the antiproton, rotational speed is lowered. Furthermore, simulations [6] have found that at high positron densities, the antiprotons undergo rapid antihydrogen formation and re-ionization cycles, which effectively transports them across the magnetic field lines to the edge of the plasma. Here, the weakly bound states that result from the three-body antihydrogen formation reaction (see below) are mostly field ionized by the plasma electric fields, such that the antihydrogen yield is depressed. Higher positron densities also mean higher space charge potentials. The desire to control antiproton-positron interactions on the degree-K level, in the presence of space charge measured in terms of eV, is a challenge. Thus, ALPHA has devised a method of cutting the positron ensemble into two, and discarding one of the halves. In principle, this can be carried out many times, though a reduction by a factor of 4 is typical. The sequence of positron manipulations, from the accumulator to preparation for mixing, takes around 400 s in total.

Returning to the positron accumulator, the performance of this instrument has been investigated from time-to-time over the years. The accumulation rate, R, from the low energy beam is a constant and given by $R = \epsilon I_0$, where I_0 is the intensity of the moderated beam. Thus, the time dependence of the number of accumulated positrons will be $N(t) = N(\infty)(1 - e^{-t/\tau})$, where the saturation limit $N(\infty) = R\tau$. It has been argued elsewhere [44] that this can be written as $N(\infty) = f_{ex}I_0(1 - e^{-DP})/BP$, where f_{ex} is the branching ratio for positron capture into the trap (essentially a ratio of scattering cross sections), P is the nitrogen gas pressure and D and B are constants. The factor $(1 - e^{-DP})$ accounts for the attenuation of the positron beam and $1/\tau = BP$: note that strictly the "P" in the exponential factor relates to that in the first

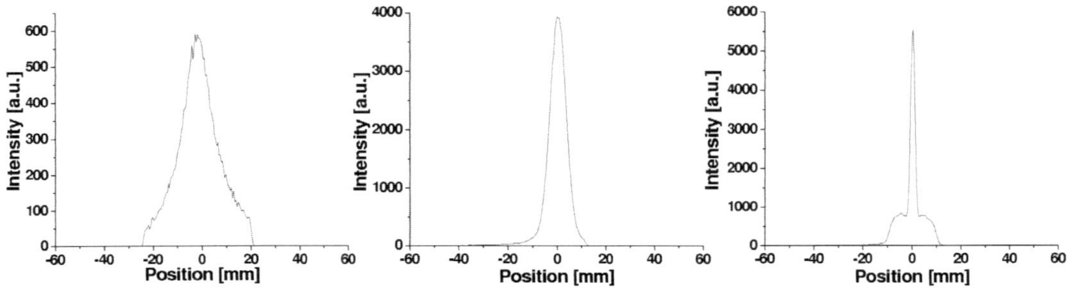

Figure 8. One dimensional projection of the phosphor screen image of the ejected positron cloud following dipolar rotating wall compression. (Note that the overall distribution is ellipsoidal.) (a) no rotating wall; (b) rotating wall with N_2 gas only and (c) rotating wall with N_2 and added CO_2 cooling gas. The integrated yields are around 29, 38 and 30 million positrons for (a), (b) and (c) respectively.

stage of the trap, where the primary scattering event takes place, whilst the "P" in BP refers to the third stage, where the particles are stored. When only nitrogen gas is used, the pressures are related, since the gas is admitted into the first stage and the pressure in the third stage is proportional, since molecular flow conditions prevail.

The expression for $N(\infty)$ tends to a constant value of $f_{ex}I_0D/B$ for $DP \ll 1$, whilst for $DP \gg 1, N(\infty) \rightarrow f_{ex}I_0/BP$. The quantity B is related to the particle loss via annihilation and possibly radial transport in the trap (the latter can be arrested by the application of a rotating wall) and can be derived from experiments measuring $N(t)$ versus P. The beam intensity, I_0, can be determined using a calibrated detector, leading to a value for f_{ex}. This was found earlier [44] to be ~ 0.4. Examples of curves of $N(t)$ versus t and $N(\infty)$ and $1/\tau$ versus P are given in figure 7, showing that the ALPHA 3-stage accumulator behaves according to expectations, at least over the main pressure range of its application. (Note the offset pressure range in Figure 7. Data at lower pressure seem to indicate that the accumulator has another mode of operation, which will be investigated in future.)

Figure 8 shows the spatial profile of the positron cloud ejected from the accumulator both without the rotating wall compression and for two cases with the rotating wall applied. Here a segmented electrode located in the third stage of the accumulator configured in a dipolar mode was used to apply the rotating wall. It is clear from the figure that the rotating wall substantially increases the central density of the plasma. This is desirable, as smaller radius positron plasmas can be created on transfer to the main mixing magnet for antihydrogen production.

4. Antihydrogen production in a magnetic minimum trap

When positrons are mixed with antiprotons under the conditions of the ALPHA experiment, it has long been supposed that two reactions may lead to antihydrogen formation: namely the spontaneous radiative and three-body (two positrons and an antiproton) reactions. It has also been assumed that the three-body reaction would dominate antihydrogen formation, however, as mentioned in the previous section, it has been realised relatively recently [6, 17] that it also results in the transport of the antiprotons to the outskirts of the positron plasma. Depending upon the density of the positrons, this may occur on a timescale as to preclude significant antihydrogen formation via spontaneous radiative recombination. Thus, most of the antihydrogen formed will be via the reaction

$$e^+ + e^+ + \overline{p} \rightarrow \overline{H} + e^+, \quad (1)$$

which will produce states bound by around $k_B T_e$, where T_e is the temperature characteristic of the positron plasma. Such weakly bound states are susceptible to field ionization, either from the electric fields associated with the plasma, or the trap fields.

Experience from ATHENA has shown that injection of antiprotons into the positron plasma with kinetic energies in the 10-20 eV range results in spatial annihilation distributions which can only be explained if antihydrogen is formed efficiently by antiprotons which have not thermalised with the positrons [45]. The implications for holding antihydrogen in traps which (for the ground state) are $\sim$ 1 K deep are obvious. There are also several mysteries from the ATHENA data, which are as yet unresolved. Despite the supposed dominance of the three-body reaction (which has a textbook dependence of $T_e^{-9/2}$), attempts to measure the temperature dependence of the antihydrogen formation rate have produced much shallower dependencies, typically $\sim T_e^{-1}$ [46, 47]. Currently this is not understood, and further work is ongoing. Taken together, what we do, and dont, know about the details of antihydrogen formation in strong field Penning-type trap arrangements means that we should have detailed control over the parameters of both the positron and antiproton clouds, including the manner in which they are mixed. The manipulations performed on the positrons and antiprotons have been summarized above.

A distinctive feature of the ALPHA antihydrogen apparatus is, as shown in figure 1, the superposition of a magnetic minimum neutral atom trap onto the charged particle traps used to hold and mix the antiprotons and positrons. As has been elaborated elsewhere [48], ALPHA has created its neutral trap using a pair of mirror coils and an octupolar arrangement to provide the radial confinement. The choice of the latter, over the more familiar quadrupole [15], was motivated by the shallower magnetic field gradient near the axis of the system in the octupolar case, which should result in plasmas with extended lifetimes and greater stability [49, 50]. The magnitude of the depth of the neutral atom trap is given by $U = \mu\Delta B$, where μ is the magnetic moment of the atom, and ΔB is the difference in magnetic field between the point of antihydrogen formation (taken to be approximately on the axis of the system) and the Penning trap electrode wall, located at a radius of around 22 mm. For the ground state of antihydrogen $\mu = \mu_B$, the Bohr magneton, such that the trap depth is $\sim$ 0.7 KT^{-1}. Using state-of-the-art superconducting technology [48], ALPHA has developed a system with around 1 T of field change across the radial extent of its trap, such that ground state atoms with kinetic energies below $\sim$ 0.7 K should be confined.

ALPHA has recently demonstrated the efficient formation of antihydrogen in its combined charged particle/neutral atom trap [17]. It has detected the antihydrogen via its annihilation on the electrode wall, as well as using the field ionization technique developed by ATRAP [2]. Good correlation between these complementary signals has been found [17]. The distributions of the radial and axial annihilation positions of the antihydrogen, as reconstructed using ALPHA's vertex detector [51], are shown in figure 9 for the cases with the neutral trap off, and fully energized. In the former case the annihilation distributions are smooth, though there is some structure in the data with the trap on. It has been argued [17] that this is due to the transport to the wall of antiprotons which have resulted from field ionization of antihydrogen at large radii in the trap. Beyond a critical radius [52], which depends upon the strength of the octupolar field, any field ionized antiprotons will be automatically swept to the wall due to the radial behaviour of the octupolar field. Such a process, as described above, also has support from recent simulations [6], though in the context of the ATHENA experiment.

ALPHA has developed a strategy to search for trapped antihydrogen atoms [53], which it is continuing to refine as the positron and antiproton manipulation techniques described in sections 2 and 3 are fully integrated into the experimental protocol. Briefly, positron-antiproton mixing is performed for about 1 s, whereupon the charged particle traps are emptied. This is followed

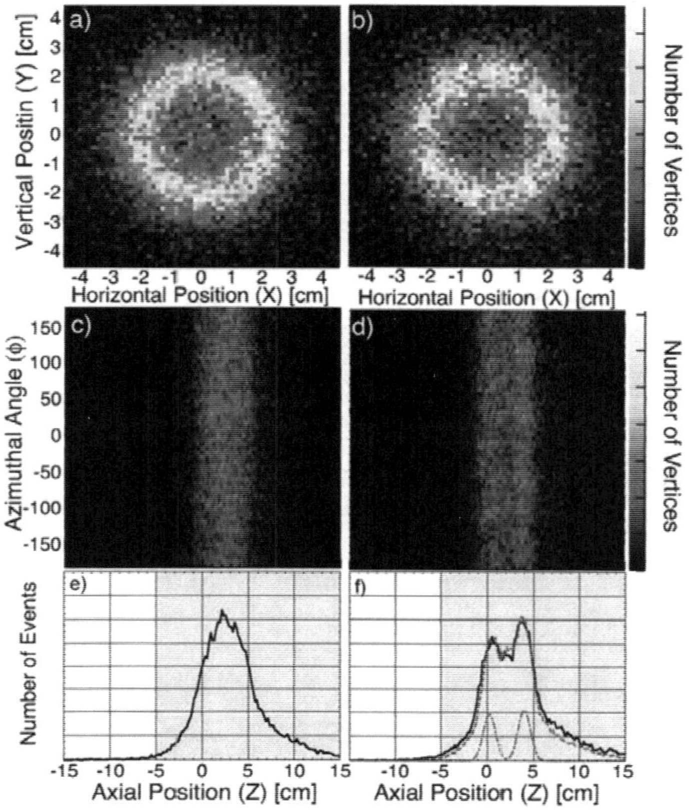

Figure 9. a),b) Azimuthal projections of antihydrogen annihilations with (b) and without (a) the neutral trap energised. (c),(d) and (e),(f) are the corresponding z-ϕ and z-distributions respectively. In (f), the red curve is a fit using a scaled version of (e) together with the two green peaks. The latter are caused by field ionized antihydrogen (see text). When these data were taken the vertex detector was incomplete and had only one layer of silicon modules for the unshaded region. This caused a lower vertex reconstruction efficiency, resulting in the slight asymmetry of the distributions evident in (e) and (f).

by a deliberate quench of the neutral atom trap. The current from the coils is dumped into a resistive load and the $1/e$ time for the decay is just under 10 ms [53]. ALPHA then searches the output of the vertex detector in a time window of 30 ms to look for annihilations which could be characteristic of antihydrogen released from the neutral atom trap. The narrow time window is an effective veto for cosmic ray events [54]. Work is ongoing.

5. Concluding remarks

We have described how antiprotons and positrons can be manipulated using a number of advanced techniques, before being combined to form antihydrogen. Current efforts to trap antihydrogen atoms have been discussed in the context of the ALPHA experiment, which is located at the unique AD facility at CERN. The motivation for attempting these difficult experiments was briefly sketched.

6. Acknowledgements

Our work has been supported by a number of national funding agencies. We are grateful to: EPSRC and the Leverhulme Trust (UK); CNPq and FINEP (Brazil); ISF (Israel); FNU (Denmark); VR (Sweden); NSERC, NRC/TRIUMF and AIF (Canada); DOE and NSF (USA). We are very appreciative of the support we receive at CERN.

7. References

[1] Amoretti M *et al* (ATHENA collaboration) 2002 *Nature* **419** 456
[2] Gabrielse G *et al* (ATRAP collaboration) 2002 *Phys. Rev. Lett.* **89** 213401
[3] Holzscheiter M H, Charlton M and Nieto M M 2004 *Phys. Rep.* **402** 1
[4] Gabrielse G 2005 *Adv. At. Mol. Opt. Phys.* **50** 155
[5] Robicheaux F 2008 *J. Phys. B: At. Mol. Opt. Phys.* **41** 192001
[6] Jonsell S, van der Werf D P, Charlton M and Robicheaux F 2009 *J. Phys. B: At. Mol. Opt. Phys.* **42** 215002
[7] Niering M *et al* 2000 *Phys. Rev. Lett.* **84** 5496
[8] Essen L, Donaldson P W, Bangham M J and Hope E G 1971 *Nature* **229** 110
[9] Kellerbauer A *et al* (AEGIS collaboration) 2008 *Nuc. Inst. Meth. in Phys. Res. B* **226** 351
[10] Pérez P *et al* in *Cold Antimatter Plasmas and Application to Fundamental Physics* (AIP Conf. Proc. **1037**) Eds Y. Kanai and Y. Yamazaki (2008) p. 35
[11] Bluhm R, Kostelecký V A and Russell N 1997 *Phys. Rev. Lett.* **79** 1432
[12] Bluhm R, Kostelecký V A and Russell N 1999 *Phys. Rev. Lett.* **82** 2254
[13] Bluhm R 2004 *Nuc. Inst. Meth. in Phys. Res. B* **221** 6
[14] Shore G M 2005 *Nuc. Phys. B* **717** 86
[15] Pritchard D E 1983 *Phys. Rev. Lett.* **51** 1336
[16] Gabrielse G *et al* (ATRAP collaboration) 2008 *Phys. Rev. Lett.* **100** 113001
[17] Andresen G B *et al* (ALPHA collaboration) 2010 *Phys. Lett. B* **685** 141
[18] Maury S 1997 *Hyperfine Interact.* **109** 43
[19] Gabrielse G *et al* 1986 *Phys. Rev. Lett.* **57** 2504
[20] Beck B R, Fajans J and Malmberg J H 1996 *Phys. Plasmas* **3** 1250
[21] Malmberg J H and Driscoll C F 1980 *Phys. Rev. Lett.* **44** 654
[22] Notte J and Fajans 1994 *Phys. Plasmas* **1** 1123
[23] Gabrielse *et al* (TRAP collaboration) 1989 *Phys. Rev. Lett.* **63** 1360
[24] see eg Andresen G B *et al* 2008 *J. Phys. B: At. Mol. Opt. Phys.* **41** 011001
[25] Huang X P, Anderegg F, Hollman E M, Driscoll C F and O'Neil T M 1997 *Phys. Rev. Lett.* **78** 875
[26] Anderegg F, Hollman E M and Driscoll C F 1998 *Phys. Rev. Lett.* **81** 4875
[27] Greaves R G and Surko C M 2000 *Phys. Rev. Lett.* **85** 1883
[28] Danielson J R and Surko C M 2005 *Phys. Rev. Lett.* **94** 035001
[29] Danielson J R, Surko C M, and O'Neil T M 2007 *Phys. Rev. Lett.* **99** 135005
[30] Jørgensen L V *et al* (ATHENA collaboration) 2005 *Phys. Rev. Lett.* **95** 025002
[31] Funakoshi R *et al* (ATHENA collaboration) 2007 *Phys. Rev. A* **76** 012713
[32] Andresen G B *et al* 2008 *Phys. Rev. Lett.* **100** 203401
[33] Andresen G B *et al* 2010 *Phys. Rev. Lett.* **105** 013003
[34] Eggleston D L, Driscoll C F, Beck B R, Wyatt A and Malmberg J H 1992 *Phys. Fluids* **4** 3432
[35] Amoretti M *et al* (ATHENA collaboration) 2004 *Nuc. Inst. Meth. in Phys. Res.* **A 518** 679
[36] Khatri R, Charlton M, Sferlazzo P, Lynn K G, Mills Jr A P and Roellig L O 1990 *Appl. Phys. Lett.* **57** 2374
[37] Mills Jr A P, and Gullikson E M 1986 *Appl. Phys.* **49** 1121
[38] Surko C M, Passner A, Leventhal M and Wysoki F J 1988 *Phys. Rev. Lett.* **61** 1831
[39] Murphy T J and Surko C M 1992 *Phys. Rev. A* **46** 5696
[40] Surko C M and Greaves R G 2004 *Phys. Plasmas* **11** 2333
[41] Greaves R G and Surko C M 2001 *Phys. Plasmas* **8** 1879
[42] Al Qaradawi I, Charlton M, Borozan I and Whitehead R 2000 *J. Phys. B: At. Mol. Opt. Phys.* **33** 2725
[43] Surko C M, Greaves R G and Charlton M 1997 *Hyperfine Interact.* **109** 181
[44] Clarke J C, van der Werf D P, Griffiths B, Beddows D C S, Charlton M, Telle H H and Watkeys P R 2006 *Rev. Sci. Inst.* **77** 063302
[45] Madsen N *et al* (ATHENA collaboration) 2005 *Phys. Rev. Lett.* **94** 033403
[46] Amoretti M *et al* (ATHENA collaboration) 2004 *Phys. Letts B* **583** 59
[47] Fujiwara M C *et al* (ATHENA collaboration) 2008 *Phys. Rev. Lett.* **101** 053401
[48] Bertsche W *et al* (ALPHA collaboration) 2006 *Nuc. Inst. Meth. in Phys. Res. A* **566** 746
[49] Fajans J, Bertsche W, Burke K, Chapman S F and van der Werf D P 2005 *Phys. Rev. Lett.* **95** 155001

[50] Andresen G B *et al* (ALPHA collaboration) 2007 *Phys. Rev. Lett.* **98** 023402
[51] Fujiwara M C *et al* in *Cold Antimatter Plasmas and Application to Fundamental Physics* (AIP Conf. Proc. **1037**) Eds Y. Kanai and Y. Yamazaki (2008) p. 208
[52] Fajans J, Madsen N and Robicheaux F 2008 *Phys. Plasmas* **15** 032108
[53] Andresen G B *et al* in *Cold Antimatter Plasmas and Application to Fundamental Physics* (AIP Conf. Proc. **1037**) Eds Y. Kanai and Y. Yamazaki (2008) p. 241
[54] Butler E *et al* 2010 to appear in *Hyperfine Interactions*

Journal of Physics: Conference Series **262** (2011) 012002 doi:10.1088/1742-6596/262/1/012002

The status of the positron beam facility at NEPOMUC

Hugenschmidt, C

FRM II and Physik Department E21, Technische Universität München, D-85747 Garching, Germany

E-mail: Christoph.Hugenschmidt@frm2.tum.de

Abstract. The NEutron induced POsitron source MUniCh NEPOMUC provides a high intensity positron beam with $9 \cdot 10^8$ moderated positrons per second with a primary beam energy of 1 keV. After remoderation, the positron beam is magnetically guided to five experimental setups: a coincident Doppler-broadening spectrometer (CDBS), a positron annihilation induced Auger-electron spectrometer (PAES), a pulsed low-energy positron system (PLEPS) as well as an interface for providing a pulsed beam with further improved brightness. An apparatus for the production of the negatively charged positronium ion Ps^- is currently in operation at the open multi-purpose beam port, where additional experiments can be realized. Within this contribution, an overview of the positron beam facility NEPOMUC with its instrumentation at the research reactor FRM II is given.

1. Introduction

At nuclear reactors, the absorption of high energy γ radiation can be used to generate positrons by pair production. Besides the γ radiation released during the fission process, one can profit from γ cascades emitted by the de-excitation of excited nuclear states after capture of thermal neutrons. At the research reactor in Delft, the γ radiation from nuclear fission is absorbed in an assembly of thin tungsten tubes for the electron-positron pair production [1]. At NEPOMUC, the positrons are generated by pair production from absorption of high-energy prompt γ-rays after thermal neutron capture in ^{113}Cd (see e.g. [2]). For mounting such an in-pile positron source close to the fuel element of a reactor, several aspects have to be taken into account, e.g. fast and thermal neutron flux, burn-up of the absorbing isotope ^{113}Cd, neutron flux depression, and γ heating. Based on the principle of the NEPOMUC source, two further positron sources have been designed at research reactors. The first one with a large positron emitting area of 900 cm^2 delivered recently about $5 \cdot 10^8$ positrons per second at the PULSTAR reactor, USA [3]. Another new project has been funded in order to install a reactor based positron source in Hamilton, Canada [4].

From the beginning in 1998, NEPOMUC was planned and built as a user-dedicated facility at the *Forschungs-Neutronenquelle Heinz Maier-Leibnitz* FRM II. The reactor is in operation for typically 240-260 days per year according to at least four fuel elements with the commissioned lifetime of 60 days each. Hence, the positron beam facility including several spectrometers is available and open for all external scientists which propose experiments with the need of a high-intensity low-energy positron beam. Besides the experiments at the multi-purpose open beam port and first high-resolution Auger studies at PAES a large variety of positron experiments have been carried out at PLEPS and CDBS. Up to now, among more than 130 submitted external proposals 95 experiments have been accepted and carried out at NEPOMUC.

2. The High Intensity Positron Beam at NEPOMUC

2.1. In-Pile Positron Source

In summer of 2004 the first setup of NEPOMUC delivered positrons with an intensity of up to $5 \cdot 10^8$ moderated positrons per second at a beam energy of 1 keV. In 2008, a new in-pile γ converter and Pt moderator was installed inside the NEPOMUC beam tube. This change lead to the unprecedented intensity of about 10^9 moderated positrons per second at a beam energy of 1 keV. For details of NEPOMUC, see [5] and references therein.

The positron source is mounted as an in-pile component of the inclined beamtube SR 11 inside the moderator tank of the reactor. An important advantage of the position of the source at FRM II is the high flux ratio of thermal to fast neutron of better than 10^4 which leads to a relatively small amount of irradiation induced defects compared to the previous source at the former reactor FRM where this ratio was close to unity.

A cadmium cap (natural abundance of ^{113}Cd: 12.2%) inside the tip of the beam tube absorbs thermal neutrons very efficiently, and hence, the binding energy of 9.05 MeV of the additional neutron in ^{114}Cd is released as γ cascade. The lifetime of the source setup is limited by the burn-up of the ^{113}Cd and amounts to 1250 days of reactor operation at the nominal power of 20 MW with the presently used thickness of 3 mm cadmium. For this reason, the beamtube SR11 with the in-pile positron source will be replaced by a new one in November 2010 The lifetime of the new positron source is extended to 25 years by using Cd enriched with 80% ^{113}Cd.

A structure of Pt foils is used for the conversion of the high-energy γ radiation into positron-electron pairs and for positron moderation. The energy of the primary positron beam is set to 1 keV. Its band width, i.e. the positron energy distribution, is given by the spectrum of the moderated positrons at the Pt surface with a width of about ΔE=1 eV and the potential differences of the Pt foils of typically 3 V. Consequently, the maximum width of the energy spectrum of the primary positron beam is assumed to be not higher than $\Delta E \approx 4$ eV.

2.2. Positron Beam Facility

The moderated positron beam is magnetically through three bends in the biological shield in order to eliminate background of fast neutrons as well as γ radiation from the reactor core and the cadmium cap in the tip of the beam tube. The beam line leads from the biological shield of the reactor pool via a remoderation unit to the experimental platform in the experimental hall of the FRM II. The energy of the remoderated beam can be adjusted between 20 and 200 eV and is presently set to 20 eV for most experiments. The total efficiency of the setup is about 5%, and the beam diameter of the remoderated beam is less than 2 mm (FWHM) in a 6 mT guiding field [6]. The beam switch on the experimental platform can be turned to five different outgoing ports. At several experiments such as CDBS and PAES, a purely electrostatic beam guidance is required. For this purpose, magnetic field terminations of μ-metal are mounted at the entrances of those devices as well as at the remoderator unit in order to release the low-energy positron beam non-adiabatically from the magnetic guiding field.

3. The Positron Instrumentation

At present, three experiments are in routine operation at NEPOMUC which use the 20 eV positron beam. In the following, the instruments are listed chronologically according to their commissioning at NEPOMUC.

3.1. Positron annihilation induced Auger electron Spectroscopy — PAES

PAES is a non-destructive technique for element selective surface studies which has several advantages over conventional AES such as suppressed secondary electron background in the range of Auger-transition energies and exceptional sensitivity to the topmost atomic layer. The main disadvantage of PAES of the very long data acquisition time in the range of typically several days per spectrum is overcome with the availability of the strong NEPOMUC positron source. The aim of our own developments was to further improve this exceptional technique in order to profit from its attractive features for high sensitive surface investigations. After upgrading the spectrometer, we succeeded not only to obtain PAES spectra within considerably reduced data acquisition times of several minutes, which is comparable to measurement times with conventional AES, but also to improve the energy resolution and the signal-to-noise ratio.

Within the last year, the surface coverage and island growth in the (sub-)monolayer region were studied in particular in the systems Cu/Pd and Cu/Fe. Moreover, the theoretically predicted segregation process of Cu in Pd was observed, which is the first *time-dependent* PAES measurement reported so far [7].

3.2. Coincident Doppler-Broadening Spectroscopy — CDBS

The CDB spectrometer enables defect studies by DBS and element specific measurements with CDBS in the near surface region and in the bulk of the specimen. A positioning device allows us to scan the sample in two dimensions with a lateral resolution of 300 μm which is currently improved to below 100 μm. Depth dependent measurements up to a few μm are performed by accelerating the positrons up to an energy of 30 keV.

With laterally resolved DBS defects can be mapped in two dimensions (see e.g. [8, 9]) and as function of positron implantation depth [10]. Depth dependent CDBS is particularly suited to study the elemental surrounding of the positron annihilation site in thin layered samples [11].

3.3. Pulsed Low-Energy Positron System — PLEPS

In order to perform depth dependent positron lifetime measurements for the investigation of defect types and concentration, the PLEPS was developed at the Universität der Bundeswehr (UniBW) München and transferred to NEPOMUC. The lifetime spectra are not only recorded within a low measurement time of a few minutes but also show an exceptional peak-to-background ratio of 10^4, which allows one to extract at least three lifetime components according to different positron states reliably (see [12]).

Examples of recent PLEPS experiments are the determination of the free volume in polymer films [13] or in bioadhesive [14]. In particular, positron lifetime experiments with PLEPS and additional DBS were applied to get a deeper insight into the nature of defects. For instance, the annealing of irradiation induced defects after He implantation in InN and GaN [15] or the vacancy defects in a thin film perovskite oxide have been studied [16].

3.4. Scanning Positron Microscope — SPM

The SPM, which allows positron lifetime measurements with a spatial resolution in the μm range, has been developed and operated at the UniBW München [17]. In order to connect the SPM to the NEPOMUC beam line, an interface including pulsing units and an additional remoderator for brightness enhancement was developed [18]. With this interface, which is required to achieve a sufficiently high phase space density to operate a positron micro-beam for lifetime measurements, a triply moderated pulsed positron beam was obtained for the first time.

3.5. Open multi-purpose beam Port — OP

At the OP, where experimental setups can easily be connected to the positron beam line, the following experiments have been performed up to now: The positron moderation by inelastic scattering in nitrogen was investigated by cooling the beam in a gas-filled drift chamber [19]. Together with the group of A. Dorn from the Max-Planck Institute for Nuclear Physics in Heidelberg first experiments on positron-He scattering were carried out. In addition, the correlated electron-electron and electron-positron emission at surfaces after slow positron impact was studied by the group of J. Kirschner from the Max-Planck Institute of Microstructure Physics in Halle using a coincidence setup for the detection of the ejected particles [20]. An apparatus for the production and lifetime measurements of the negatively charged positronium ion Ps^-, which was developed in the group of D. Schwalm at the Max-Planck Institute for Nuclear Physics in Heidelberg [21], was transferred to NEPOMUC. For the Ps^- experiments at the OP, first the primary positron beam at an energy of 500 eV and later the remoderated 20 eV beam was used. Recently, the whole setup was improved in order to determine the Ps^- decay rate with enhanced accuracy.

4. Outlook

At the CDB spectrometer, a new sample stage was installed recently in order to enable measurements in an temperature range between 110 K and 1000 K, and four additional pairs of Ge detectors will lead to a corresponding reduction in measurement time by a factor of four. In 2010, funding was granted to develop a novel a coincident setup with two segmented high-resolution Ge detectors, which will enable the detection of the electron momentum in three dimension for each annihilation event, in contrast to ACAR or (C)DBS where only projections are measured. In another project, an ACAR spectrometer is planned to be built in collaboration with the group of A. Alam, University of Bristol, which will allow high resolution electron momentum measurements. At the PAES system, a STM is mounted in order to enable the determination of the surface topology, and a X-ray source for XAES and XPS is installed to gain element information of the surface with complementary techniques. Managed by the UniBW München, additional funding was granted in order to upgrade the pulsed beam facilities, e.g. for extending the time window for the measurement of long positron lifetimes with PLEPS or for accelerating the positrons inside the SPM-interface. In 2012/2013, it is planned to move the positron instruments to a new experimental hall located at the east side of the reactor building.

4.1. Acknowledgments

The author would like to thank the positron team at NEPOMUC at the TUM as well as the collaborators partners from the UniBW for their continuous support. In addition, the author acknowledges the fruitful discussions with many external users working at NEPOMUC. Funding by the DFG (project no.: A3 within the Transregional Collaborative Research Center TRR 80) and the BMBF (project no.: 05K10WOB) is gratefully acknowledged.

References

[1] van Veen A, Schut H, de Roode J, Labohm F, Falub C, Eijt S and Mijnarends P 2001 *Mat. Sci. For.* **363-365** 415
[2] Hugenschmidt C, Schreckenbach K, Stadlbauer M and Straßer B 2005 *Nucl. Instr. Meth. A* **554**(1-3) 384
[3] Hawari A I, Gidley D W, Xu J, Moxom J, Hathaway A G, Brown B and Vallery R 2009 *AIP Conf. Proc.* **1099** 862
[4] Personal communication 2009 P. Mascher, McMaster University, Hamilton
[5] Hugenschmidt C, Löwe B, Mayer J, Piochacz C, Pikart P, Repper R, Stadlbauer M and Schreckenbach K 2008 *Nucl. Instr. Meth. A* **593** 616
[6] Piochacz C, Kögel G, Egger W, Hugenschmidt C, Mayer J, Schreckenbach K, Sperr P, Stadlbauer M and Dollinger G 2008 *Appl. Surf. Sci.* **255** 98
[7] Mayer J, Hugenschmidt C and Schreckenbach K 2010 *Surf. Sci.* **604** 1772
[8] Hugenschmidt C, Qi N, Stadlbauer M and Schreckenbach K 2009 *Phys. Rev. B* **80** 224203
[9] Hain K, Hugenschmidt C, Pikart P and Bni P 2010 *Sci. Technol. Adv. Mater.* **11** 025001
[10] Stadlbauer M, Hugenschmidt C, Schreckenbach K and Böni P 2007 *Phys. Rev. B* **76** 174104

[11] Hugenschmidt C, Pikart P, Stadlbauer M and Schreckenbach K 2008 *Phys. Rev. B* **77** 092105
[12] Sperr P, Egger W, Kögel G, Dollinger G, Hugenschmidt C, Repper R and Piochacz C 2008 *Appl. Surf. Sci.* **255** 35
[13] Egger W, Sperr P, Kgel G, Wetzel M and Gudladt H J 2008 *Appl. Surf. Sci.* **255** 209
[14] Rtzke K, Wiegemann M, Shaikh M Q, Harms S, Adelung R, Egger W and Sperr P 2010 *Acta Biomater.* **6** 2690
[15] Reurings F, Tuomisto F, Egger W, Lwe B, Ravelli L, Sojak S, Liliental-Weber Z, Jones R E, Yu K M, Walukiewicz W and Schaff W J 2010 *phys. stat. sol. (a)* **208** 1
[16] Keeble D J, Mackie R A, Egger W, Löwe B, Pikart P, Hugenschmidt C and Jackson T J 2010 *Phys. Rev. B* **81** 064102
[17] David A, Kögel G, Sperr P and Triftshäuser W 2001 *Phys. Rev. Lett.* **87** 067402
[18] Piochacz C, Egger W, Hugenschmidt C, Kgel G, Schreckenbach K, Sperr P and Dollinger G 2007 *phys. stat. sol. (c)* **4** 4028
[19] Lwe B, Schreckenbach K and Hugenschmidt C 2008 *Appl. Surf. Sci.* **255** 96
[20] van Riessen G A, Schumann F O, Birke M, Winkler C and JKirschner 2008 *J. Phys. Cond. Matt.* **20** 442001
[21] Schwalm D, Fleischer F, Lestinsky M, Degreif K, Gwinner G, Liechtenstein V, Plenge F and Scheit H 2004 *Nucl. Instr. Meth. B* **221** 185

Use of superconducting linacs for positron generation: the EPOS system at the Forschungszentrum Dresden-Rossendorf (FZD)

R Krause-Rehberg[1], M Jungmann[1], A Krille[1], B Werlich[1], A Pohl[1], W Anwand[2], G Brauer[2], M Butterling[1,2], H Büttig[2], K M Kosev[2], J Teichert[2], A Wagner[2] and T E Cowan[2]

[1]University Halle, Department of Physics, 06099 Halle, Germany
[2]Forschungszentrum Dresden-Rossendorf, 01314 Dresden, Germany

E-mail: reinhard.krause-rehberg@physik.uni-halle.de

Abstract. Intense positron sources require the pair production process for the positron generation. In case a pulsed positron source shall be constructed, a superconducting LINAC-based accelerator allows generating the required final time structure for the electron beam. This simplifies the positron beam construction. The first such setup, the EPOS system (ELBE Positron Source) at the Forschungszentrum Dresden-Rossendorf (FZD), is described.

1. Introduction

Pair production for positron sources requires electrons of high energy which can be obtained near the fuel cell of a reactor or using bremsstrahlung of an accelerator. In case of a reactor or conventional LINAC with a repetition time of about 1 ms, the time structure of the positron beam must be obtained later in the positron beamline. An alternative is the use of superconducting (sc) LINACs. Here, the required time structure can be obtained for the whole system, i.e. the electron beam has the required time structure which is retained until the sample. For positron lifetime measurements the maximum time to be measured is the vacuum lifetime of ortho-Positronium (o-Ps), $\tau_{o\text{-}Ps}$ = 142 ns. Since the decay is an exponential one, a beam repetition time which is about ten times longer is required. Thus, ideal systems should operate at 1 MHz or 500 kHz provided the bunch charge is large enough to obtain the required intensity. Due to the RF-acceleration in the GHz range, the pulse width of the electron bunch hitting the bremsstrahlung target is only a few ps and is thus short enough.

2. sc-LINACs for positron generation

A sc-LINAC has rather compact dimensions, and 10 MeV beam energy can already be obtained with a 9-cell TESLA cavity of about 1m length. At the ELBE (Electron Linac for beams with high Brilliance and low Emittance) radiation source of the Forschungszentrum Dresden-Rossendorf (FZD) four such 9-cell accelerator cavities are combined to obtain 40 MeV (figure 1). The cavities are made from niobium and operated at 1.8 K to obtain superconductivity. The basic frequency is 26 MHz in cw-mode. Any lower frequency $f = 26/2^n$ MHz (n=1,2,3...) is also possible. Moreover, a macropulse structure can be provided. A thermionic electron gun is the standard electron source so far. Electrons from a hot cathode are treated by a triode consisting of a pulse grid acting as a beam chopper and the anode system with a DC accelerating high voltage of 250 kV (figure 2). The obtained electron pulse is rather long (≈500 ps). It is bunched in two stages with resonant bunchers of 260 and 1300 MHz, respectively. The time focus is located at the entrance point of the first accelerator module. The pulse length there is < 5 ps which is preserved up to the user experiments. The thermionic source has been operated for several years. It is very robust and stable.

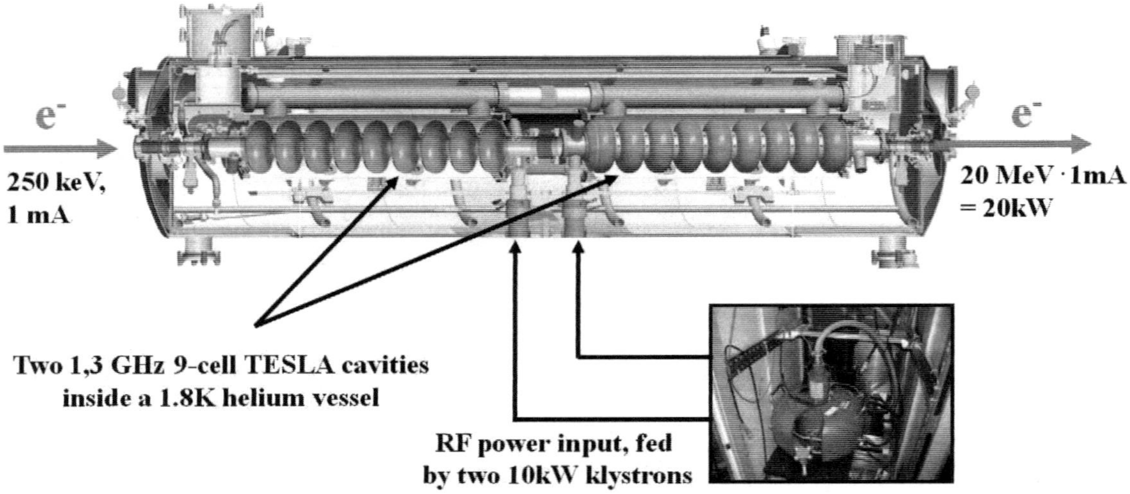

Figure 1. Two 9-cell TESLA cavity accelerator stages of the ELBE radiation source at the FZD. Four stages allow a maximum energy of 40 MeV with an average current of 1 mA (40 kW).

However, the bunch charge for an individual bunch is limited to 77 pC. In case a wider time structure of about 1 □s is required (see discussion above), the average current drops down to about 80 □A (3.2 kW). As a future alternative, the newly developed sc-photo-gun can be used (figure 3).

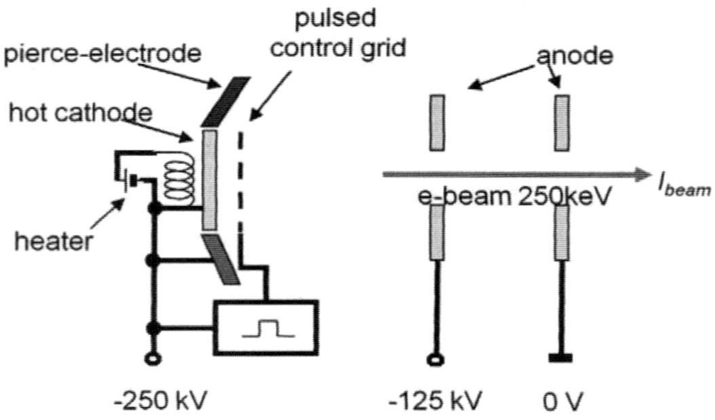

Figure 2. The thermionic gun at ELBE provides a bunched electron beam of 250 keV. The bunch length is about 500 ps and requires a two-stage bunching system to reach a 5 ps pulse width corresponding to 1 mm length at the entrance of the first sc-accelerator stage [1].

Here, 232-nm laser pulses of about 10 ps length liberate electron bunches from a photo-cathode by the photoelectric effect. This so-called SRF gun (Superconducting Radio Frequency Photoinjector) can be operated with a bunch charge of 77 pC at 13 MHz, but also in a slow mode with 500 kHz and 1 nC corresponding to an average current of 0.5 mA. This beam is then further accelerated with the four cavity accelerators of ELBE to a final energy of about 50 MeV. This would be an ideal electron source for the intense positron source EPOS [2]. However, although the SRF gun at ELBE has shown its functionality, the final intensity and energy is not obtained so far and the system needs further development.

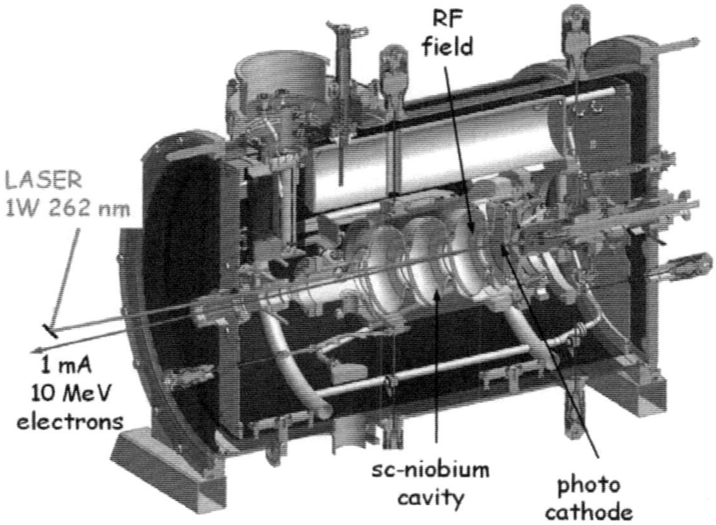

Figure 3. The SRF gun of ELBE [3]. Short pulses of a 232 nm UV-laser generate electron bunches with bunch charges up to 1 nC which are then accelerated up to 10 MeV in a very compact setup.

Such an SRF gun represents a complete table-top 10 MeV electron accelerator with ideal timing properties for positron generation. Although the positron yield will clearly be smaller compared to a 40-MeV system (see figure 4), such a system would have the advantage to prevent radioisotope activation because of the electron energy being under 10 MeV. In this case, no (□,n) processes are to be expected. Hence, after switching off the electron beam, there will be no radiation anymore. This would make its operation much easier. Moreover, electron sources of several tens MeV provide average positron energies of a few MeV, which decreases the moderation efficiency drastically. The average positron energy for a 9-MeV system is expected to be < 1 MeV. This effect will at least partly compensate for the lower positron yield.

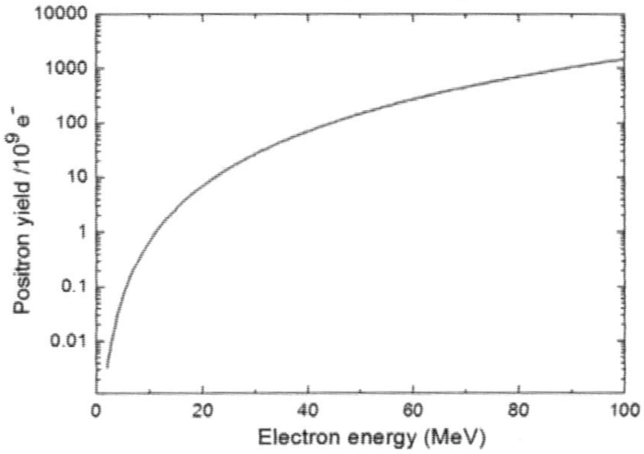

Figure 4. The positron yield calculated for a bremsstrahlung target of optimum thickness at each energy [4].

3. The EPOS system at the radiation source ELBE

The EPOS system consists now of three parts: the mono-energetic pulsed positron beam (MePS), the gamma-induced positron spectroscopy (GiPS), and two conventional, isotope-based systems (CoPS). CoPS does consist of the continuous slow-positron beam SPONSOR [5] and a lifetime/Doppler setup. The latter system is still operated with analogue electronics which will be replaced in the near future by a 4-PMT (PMT=photomultiplier tube) digital system combined with a Ge detector, equipped with an automated He refrigerator system.

The GiPS setup, which is in operation for some time now, is unique so far [6]. The bremsstrahlung impinges onto the sample and generates electron-positron pairs via pair production in the whole sample volume. A multi-detector system surrounding the sample detects the annihilation quanta that are emitted by positron annihilation directly inside the sample. Different types of detectors are used for the application of different positron spectroscopies. Fast PMTs equipped with BaF_2 scintillators are used for lifetime measurements. The GiPS setup contains four HPGe detectors and two BaF_2 detectors. Two HPGe detectors each facing an opposite BaF_2 detector act as two independent AMOC (Age-Momentum Correlation) spectrometers. Two further HPGe detectors directly above and under the sample are used as a CDBS (CDBS=coincidence Doppler broadening spectroscopy) spectrometer. The distances between sample and detectors can be adjusted. By using this detector setup, lifetime, (coincidence) Doppler broadening and age-momentum correlation measurements can be realized simultaneously.

The MePS system [2] will make use of moderated, monoenergetic positrons for near-surface studies but is still under construction. First, a 2 keV positron beam is extracted by an electrostatic lens from a Pt moderator which is placed close to the bremsstrahlung target. The positron beam is then fed into a magnetic guidance field which is guiding the beam from the production cave to the laboratory over a distance of about 10 m. The beam properties could already been tested at the entrance of the chopper/buncher system. Due to the positron moderation process in combination with an energy spread during positron acceleration to 2 keV and the magnetic beam guidance process itself, the positron bunch widens in accordance with earlier simulations from originally less than 5 ps FWHM (FWHM=full width at half maximum) to a peak of about 1.9 ns FWHM. However, an adjacent buncher will provide the time focus with a width of < 100 ps at the sample position. An accelerator will give the final positron energy range of 0.05 to 30 keV. A fully digital and remote controlled multi-detector system (8 PMTs and 2 HPGe) will ensure measurements at a high count rate.

The whole EPOS system is established as a user-dedicated facility. A web-based application system allows access for interested users [7].

4. References

[1] http://www.fzd.de/ELBE/

[2] Krause-Rehberg R, Sachert S, Brauer G, Rogov A and Noack K 2006 *Appl. Surf. Sci.* **252** 3106; Krause-Rehberg R, Brauer G, Jungmann M, Krille A, Rogov A and Noack K 2008 *Appl. Surf. Sci.* **255** 22

[3] Arnold A, Büttig H, Janssen D, Lehnert U, Michel P, Möller K, Murcek P, Schneider Ch, Schurig R, Staufenbiel F, Teichert J, Xiang R, Kamps T, Lipka D, Marhauser F, Lehmann W-D, Stephan J, Volkov V, Will I, Klemz G 2007 *Nucl. Instr. & Meth.* A **577** 440

[4] Ley R 2000 Wissenschaftlich-Technische Berichte FZR (ISSN 1437-322x) **295** 9

[5] Anwand W, Kissener H.-R. and Brauer G 1995 *Acta Phys. Pol. A* **88** 7

[6] Butterling M, Anwand W, Brauer G, Cowan T.E., Hartmann A, Jungmann M, Kosev K, Krause-Rehberg R, Krille A, Schwengner R and Wagner A 2010 *Physica Status Solidi A* **207** 334

[7] http://www.fzd.de/db/Cms?pNid=1732

Acknowledgments

Authors are grateful for permanent support from all colleagues involved in the project at University Halle and FZD, and in particular to the ELBE accelerator crew for providing excellent and stable electron beams. Furthermore, we thankfully acknowledge financial support from the German Bundesministerium für Bildung und Forschung (BMBF) under project number 05 KK7NH1.

Miscibility and crystallization behavior of poly (3-hydroxybutyrate) and poly (ethylene glycol) blends studied by positron annihilation spectroscopy

E E Abdel-Hady[1], M O Abdel-Hamed[1] and A M Hammam[1]

[1]Physics Department, Faculty of Science, Minia University,Minia , Egypt

esamhady@yahoo.com

Abstract. Positron annihilation Lifetime (PAL) spectroscopy has been used to study the effect of PEG concentrations on the free volume properties of PHB. The data revealed that the ortho-positronium (o-Ps) lifetime τ_{Ps} increases with 20% increase in concentration, decrease as the concentration increases to 40%, then rapid increase at 50% concentration of PEG. The o-Ps intensity, I_3, shows a linear dependence as the concentration increases with a discontinuity at 20% concentration of PEG. Furthermore, the results presented and discussed in this work show that the PHB and PEG are miscible up to 40% of PEG but greater than 40%, the blend is immiscible. In addition, the mechanical properties of PHB are well improved by the addition of PEG with a low concentration up to 20%, while at higher concentration the blend becomes waxy.

1. Introduction

Positron annihilation lifetime (PAL) technique turned out to be a very useful method for the investigation of different aspects of polymer properties. It has been demonstrated that PAL technique is uniquely sensitive to free volume-based structural changes (glass transition, relaxation processes) in polymers. PHB is biotechnological polyester, highly crystalline, totally biodegradable with low versatility in mechanical properties and offers a high potential for application in drug-delivery systems. PHB has disadvantage: Thermally unstable during processing, stiff and brittle and low biodegradation rate. To overcome these problems, a reactive blend with a water soluble polymer is formed. Among the water soluble polymers, PEG is the most preferred candidate. The aim of this work is to improve the mechanical properties of PHB to be used in medical applications.

2. Experimental

2.1 Materials and Sample Preparation

PHB and PEG are white powders purchased from ALDERICH with a molecular weight of 230 kg/mol and 6000 g/mol, respectively. All PHB/PEG blend films were prepared by solution-cast technique using chloroform as the co-solvent. The PHB and PEG were dissolved separately in chloroform before blending. Since PHB is insoluble in chloroform at room temperature, a magnetic stirrer was used to stir it for 150 min at 60 °C. The polymer concentration of each solution was 30 g/l. After the PHB and PEG solutions were well homogenized, they were mixed and stirred for 12 h. The mixed solution was poured on a poly Petri dish, then left at room temperature for 48 h and subsequently dried at 60 °C for further 120 h to eliminate the solvent completely [1].

2.2. *PAL measurements*

The PAL measurements were performed using a conventional fast-fast coincidence system. A positron source (^{22}Na), sealed in a thin Kapton foil, and was sandwiched between sample (10 x10 mm^2 with a thickness of 1 mm) materials. The time resolution of the system was 240 ps (full width at half maximum, FWHM) with a time calibration of 54 ps/ch. PAL spectra containing 1.5×10^6 counts were analyzed into three components (τ_1, τ_2, and τ_3) and their intensities (I_1, I_2 and I_3) determined using PALS Fit [2] and LT 9.0 programs [3] without source correction.

3. Results and discussion

The effect of concentration of PEG in PHB varying from 10% to 50% by weight on τ_3 & I_3 at room temperature is shown in (Figure 1). As shown in figure 1, the o-Ps lifetime, τ_3, increases with increasing PEG concentration up to 20% and then starts to decrease as the concentration increases up to 40% followed with a rapid increase at 50% concentration of PEG. The o-Ps intensity, I_3, shows a linear dependence as the concentration increased with a discontinuity at 20% concentration of PEG. The behavior of the lifetime parameters can be explained as follows: some PEG molecules must be retained in the PHB matrix and can also enter the free-volume holes of the PHB matrix during the transformation of blend solution into the dried blend membrane. However, the interaction between PHB and PEG can only occurs through the terminal hydroxyl groups of PEG via hydrogen bonds with the carbonyl groups of PHB. Increases in PEG increase the space between the chains of PHB, (forcing them a part) and therefore weakens the interactions between them. This effect is major at low concentrations, and hence the o-Ps lifetime, τ_3, increases as the PEG content rises till 20% [4].

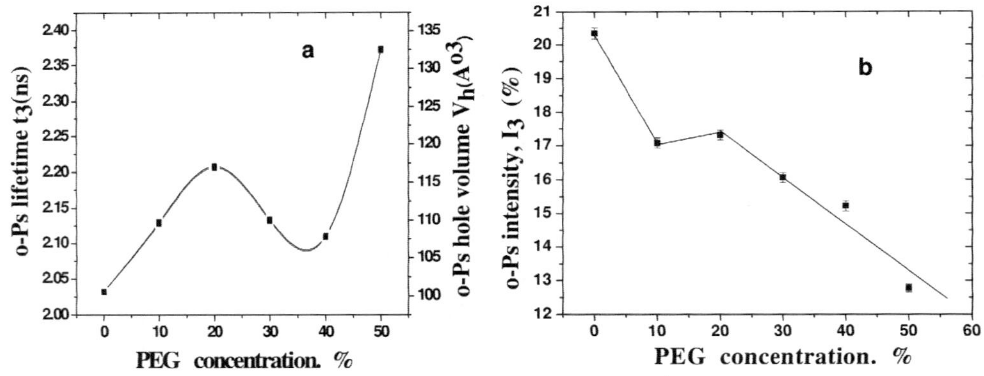

Figure 1. The effect of PEG concentration on (a)The o-Ps lifetime, τ_3, and the free volume hole size(b) The o-Ps lifetime intensity, I_3.

At the range (0 - 20 %) of PEG concentration, the interaction between PEG segments and PHB chains is slightly increased and this lead to a partition effect to the free-volume holes of the PHB matrix. With further addition of PEG until 40%, the PEG molecules will fill the free volume inside the PHB matrix. Therefore, the o-Ps lifetime τ_3 and its intensity I_3 decrease .i.e. the PHB/PEG blend is fully miscible in the range (0 - 40 %) concentration [5]. At the concentration greater than 40%, the PEG molecules apt to crystallize and form a separate phase. Their poor interaction causes phase separation, and the molecules in the interfacial zone of the PHB/PEG blend are less compact than within the PHB matrix. Therefore, the free-volume size in the interfacial zone was much larger than it is in the PHB matrix or the PEG phase. So, at higher concentration greater than 40%, the PHB/PEG blend is partially miscible [6]. The variation of the fractional free volumes with PEG concentrations shows two inflection points at 20% and 40% concentration of PEG (figure 2). It can be noticed that, the decrease in the fractional free volumes does not exceed 3.37% until 10% concentration of PEG. However, from 10 to 20% concentration of PEG, FFV increases then an appreciable decrease is observed in fractional free volume up to 40%. In addition, at 50% concentration of PEG, the FFV increases leading to the formation of much larger free volume holes which is due to the phase separation.

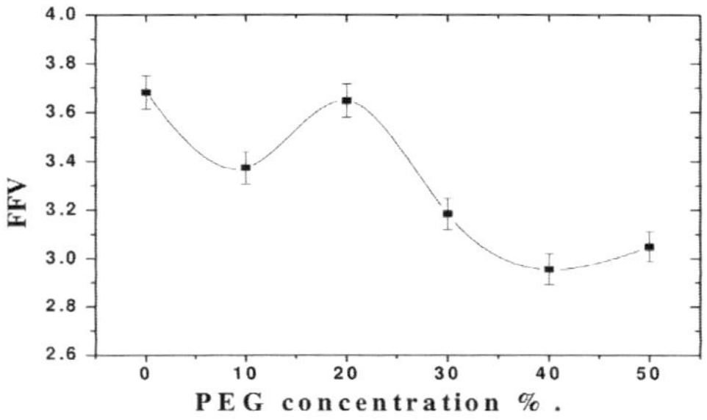

Figure 2. The Relative Fractional Free Volume hole,FFV, as a Function of PEG Concentration.

Figure 3 shows the effect of PEG concentration on the mechanical properties of PHB. As shown from the figure, the young`s modulus (E) slightly increases as the concentration of PEG increases, up to 40%, above which a rapid increase in E is observed. This is attributed to the interaction of PEG with PHB which hinders the chains from motion and reduces the force of secondary intermolecular bonds between PHB chains [7]. On the other hand, it is obvious that the tensile strength decreases with increasing the PEG concentration. It can be concluded that, the mechanical properties of PHB are approved with low concentration of PEG (20%), while at higher concentrations of PEG the flexibility decreases due to the increase in Young`s modulus.

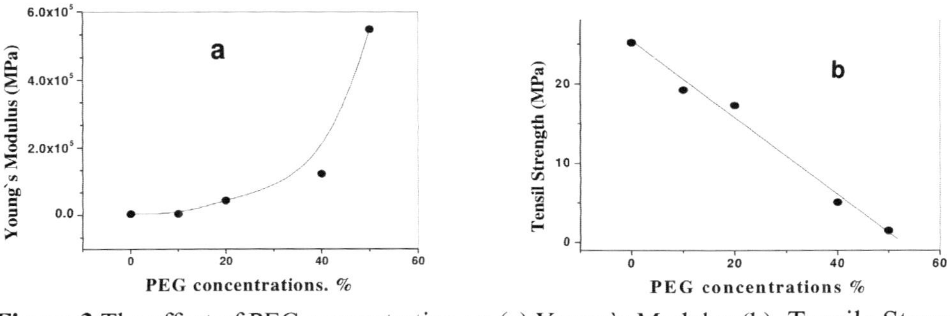

Figure 3.The effect of PEG concentration on (a) Young`s Modulus (b) Tensile Strength

Using FT-IR, the frequency of the functional group, such as carbonyl, ether, or amide, is usually determined by the force constant (K) and mass of the boned atoms [8]. If there is intermolecular hydrogen bonding between the hydroxyl group of PEG and the carbonyl group of PHB, the carbonyl bond energy of PHB may be reduced. Consequently, K of the carbonyl group will weaken and the peak may shift to a lower frequency [8]. Figures 4 & 5 show FT-IR spectra for the pure PHB and 50% concentrations of PEG, respectively. As can be seen from these figures the carbonyl peak of the

Journal of Physics: Conference Series **262** (2011) 012004 doi:10.1088/1742-6596/262/1/012004

blends shifts to lower numbers gradually as the PEG contents are increased. This indicates that with increasing PEG content, the intermolecular hydrogen bonding in the PHB/PEG blends is gradually increased.

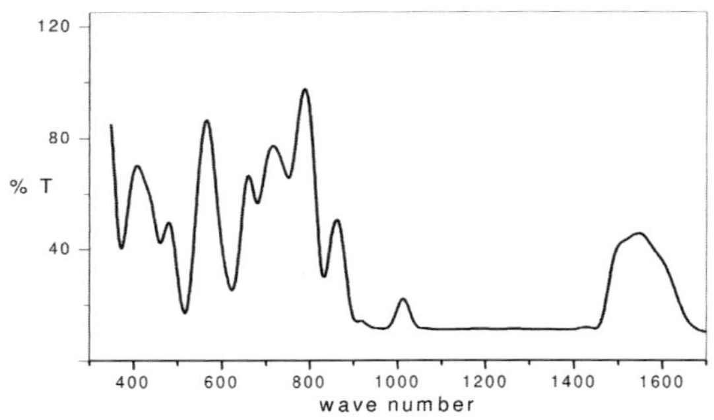

Figure 4. FT-IR spectrum for pure PHB.

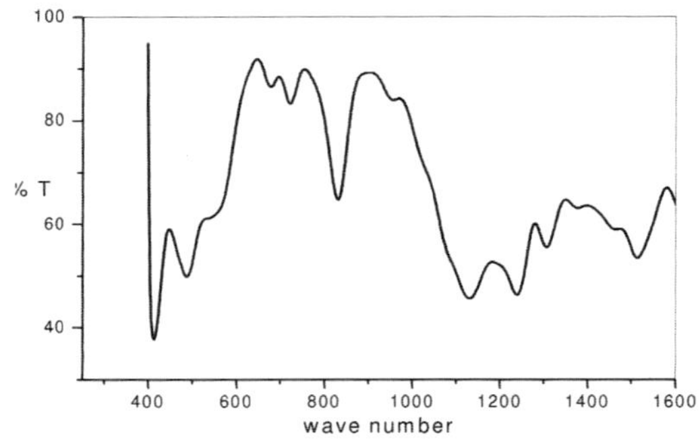

Figure 5. FT-IR spectrum for PHB/PEG (50%).

From the wave number the force constant can be calculated [8]. It was found that the force constant decreases with increasing the PEG concentration then starts to increase at 50% of PEG concentration. This may be attributed to the fact that not all PEG molecules are included in the PHB matrix, i.e. PEG molecules agglomerate. Since the region between 880 cm^{-1} and nearly 1300 cm^{-1} is a crystalline-sensitive region [9]. One can confirm that the crystallinity of the PHB in the blend was suppressed by increasing PEG contents. The number of absorption peaks decreases with PEG content and disappear at the concentration greater than 20% of PEG concentration.

3. Conclusion.

The differences in the miscibility of two blend systems in the solid state were demonstrated with PALS methods. PHB and PEG (6000) are miscible up to 40% of PEG but greater than 40% the blend is immiscible (PEG agglomeration). Mechanical properties of PHB are well improved by the addition of PEG up to 20%, while at higher concentration the blend becomes waxy. PHB interacts with PEG via the formation of a hydrogen bond but the hydrogen bond between PEG molecules is greater than that between PHB and PEG at high concentrations and consequently the phase separation (immiscibility) occurs at higher concentrations.

5. References

[1] Cheng G, Cai Z and Wang L 2003 *Journal of Materials Sscience:Materials in Medicine* **14** 1073
[2] Olsen J V, Kirkegaard P, Pedersen N J, and Eldrup M 2007 *Phys. Status Solidi C* **4** 4004
[3] Kansy J 1996 *Nucl. Instrum. Methods A* **374** 235
[4] Pintye-Hódi K , Regdon Jr G, Erős I, Süvegh K, Marekc T, Kery I and Zelko R 2006 *International Journal of Pharmaceutics* **313** 66
[5] Chang-Sik Ha and Won-Jei 2002 *Prog.Polym.Sci.* **27** 59
[6] He Y , Asakawa N and Inoue Y 2000 *Polym Int.* **49** 609
[7] Parra D F, Fusaro J, Gaboardi F and Rosa D S 2006 *Polymer Degradation and Stability* **91** 1954
[8] Lim J S, Noda I and Soon S 2006 *J. Polym. Sci. Part B: Polym. Phys.* **44** 2852
[9] Reis K C, Pereira J , Smith A C , Carvalho C W P , Wellner N and Yakimet I 2008 *Journal of Food Engineering* **89** 361

Journal of Physics: Conference Series **262** (2011) 012005 doi:10.1088/1742-6596/262/1/012005

Positron annihilation study of acryl amide/poly (metha acrylic acid) membrane

E E Abdel-Hady[1], M O Abdel-Hamed[1], M M Eltoony[2], A M Hammam[1] and M R M Elsharkawy[1]

[1]Physics Department, Faculty of Science, Minia University, Minia, Egypt
[2]Chemistry Department, Atomic Energy Authority, Cairo 11115, Egypt

esamhady@yahoo.com

Abstract. Gamma irradiation posses a serious role for casting the membranes. Acryl amide /poly (methacrylic acid) membrane was synthesized under γ-radiation effect. The structure of the membrane was characterized by FTIR, thermo-gravimetric analysis and the scanning electron microscope. The properties of the membranes were also investigated in terms of proton conductivity and positron annihilation lifetime (PAL) parameters. On the basis of the values of the long-lived components in the lifetime spectra, the size of the free volume and their intensity were calculated. The positron lifetime study on these irradiated casted membranes shows that the cross-linking and degradation within the membrane matrix affect the free volume content and hence the microstructure.

1. Introduction

Positron annihilation spectroscopy has been established as a powerful probe for the determination of free volume hole sizes in the polymer [1]. Recent investigations in membrane materials and film systems have been reported mainly in the bulk using conventional positron annihilation lifetime technique [2-4]. The automotive fuel cell represents perhaps the largest volume of all potential fuel cell applications and has also set some of most challenging requirements for component performance, durability and cost. Currently, there is much ongoing research for developing non-perfluorinated polymers with better performance and lower cost as alternative PEM materials. The aim of this work was to obtain low-cost membrane specifically for fuel cell use. The specific aim was to determine whether Acrylamide / Polymeth-acrylic Acid Membrane could be developed that would give a competitive performance to an established Nafion membrane in a fuel cell.

2. Experimental

2.1 Materials and Sample Preparation

Acrylamide of purity 98.5% was supplied by Merck, Germany. Poly (meth acrylic acid) and powdered activated carbon (PAC) was commercially purchased from local company OPTCo, Egypt. 10 % of Acrylamide was solvated in doubly distilled water with the addition of 1% of phosphoric acid, 1% poly methacrylic acid, 0.5% of phathalic anhydride (which is used as a plastisizing agent) and powdered activated carbon at different ratios 0.1, 0.2 and 0.3% to the acrylamide. The content of the sealed vial was heated at 80°C for 3 hours. The viscous solution was poured into clean and dry glass Petri dish. The Petri dish content was subjected to γ-rays from a ^{60}Co source with different doses (10 & 20 KGy). Evaporation of the irradiated soft gelatinous mixture at dried ambient temperature for 48 hours.

2.2. Positron annihilation lifetime spectroscopy
The positron annihilation lifetimes (PAL) of the membranes were determined by detecting the prompt γ-ray(1.28Mev)from the nuclear decay that accompanies the emission of a positron from the ^{22}Na radioisotope and the subsequent annihilation γ-rays (0.511Mev). A conventional fast-fast coincidence circuit of PAL spectrometer with a time resolution 240 Ps was used to record all PAL spectra. The PAL spectra containing1.5x10^6 counts were analyzed into three components (τ_1, τ_2, and τ_3) and their intensities (I_1, I_2 and I_3) using PALSfit program[5].

2.3. Characterization
FTIR spectra were obtained with Mattson 1000, Pye-Unicam, England.Fourier transform infrared spectrophotometer. Thermo-gravimetric analysis (TGA) was recorded on a thermal analysis system at a heating rate of 10 °C /min. Proton conductivity measurements were derived from AC impedance spectroscopy measurements over a frequency range of 50 to 10^6 Hz with an oscillating voltage of 50–500mV, using a system 3532 Hioiki bridge LCR hitester. Each membrane sample was cut into sections 2.5cm×2.0 cm prior to being mounted in the cell. The sample was equilibrated at 100% RH at ambient atmospheric pressure and clamped between two electrodes. The proton conductivities of the samples were measured through the membrane and were calculated from the impedance data, using the following relationship: σ=l/RS where σ is the proton conductivity (in S/cm), *l* is the membrane thickness, *S* is the area for protons to transport through the membrane (in cm^2) and *R* is derived from the intersection of the semicircle on a complex impedance plane with the Re (Z) axis.

3. Results and discussion

3.1. FT-IR characterization
Figure 1 shows strong bands at 1040–910 cm^{-1} that belong to asymmetric stretching vibrations of the P–OH group and at 1150 cm^{-1} that corresponds to P-O stretching. At higher poly phosphoric acid ammonium salt ratio, the P–O–H vibration at 930 cm^{-1} becomes stronger indicating the existence of excess acidic protons. Additionally, phosphoric acid units give rise to broad bands with medium intensity at 1700–1590 cm^{-1} and 2850–2750 cm^{-1} [figure 1 b]. The band broadening around 3100 cm^{-1} is due to the hydrogen bonding network which is necessary for proton conduction.

3.2. Thermo-gravimetric analysis
The thermogram could be divided into 5 divisions as shown in figure 2 .The first division described the membrane stability upon raising the temperature (working temperature of the membrane). The Second division explained the slow weight decrease upon raising the temperature. The third division described the convex zone regarding to temperature axe. The fourth division described the abrupt change of weight by temperature raising up, via which 0.1% PAC membrane, the temperature range was 320-370 while it expanded to a range of 320-380 °C for 0.3% PAC-membrane. The weight loss through this division showed similar loss of weight for 0.1 and 0.3% PAC membrane which was 13%. The fifth division showed a weight reduction to 52 % from the original weight for 0.1% PAC while it was 50 % for 0.3 % PAC. This loss of weight has occurred up to 550 °C.

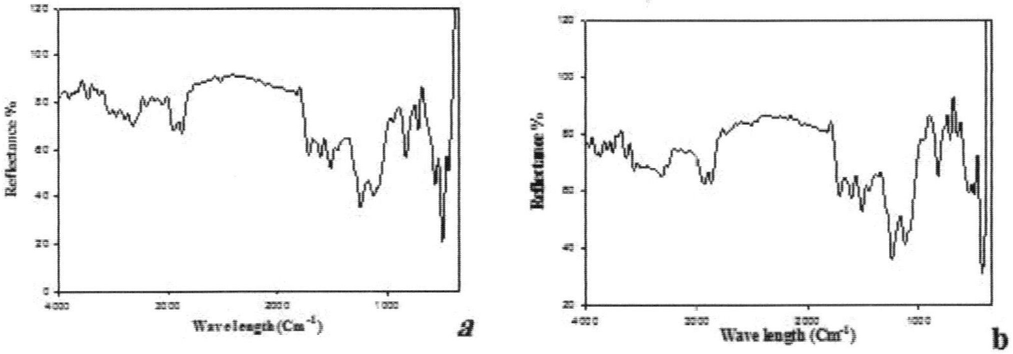

Figure 1.FTIR of casted membrane at different Gamma irradiation dose; a) 20 KGy irradiation dose and 0.1 % PAC; b) 20 KGy irradiation dose and 0.2 % PAC.

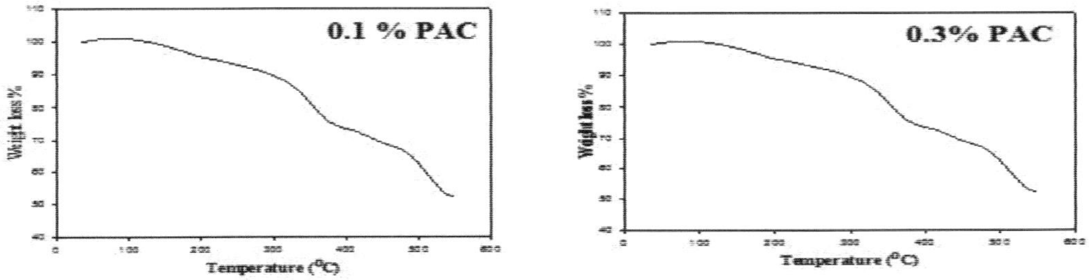

Figure 2. Thermogram of casted membrane in (0.1% & 0.3% PAC) at 20 KGy irradiation dose

3.3. Scanning electron microscope
SEM shows all the particles appeared to be tightly packed, evenly distributed carbon aggregates in figure 3 (a) which is explained by their diffusion onto deepest layer of matrix. Whitish zones have been seen and the disappearance of black and white may due to cross-linking between poly phosphoric acid ammonium salt and poly meth acrylic acid. By increasing the PAC content [figure 3 (b)], it can be seen larger content of PAC aggregate diffused into the net polymer while exceeding scattered regularly over the membrane surface.

a

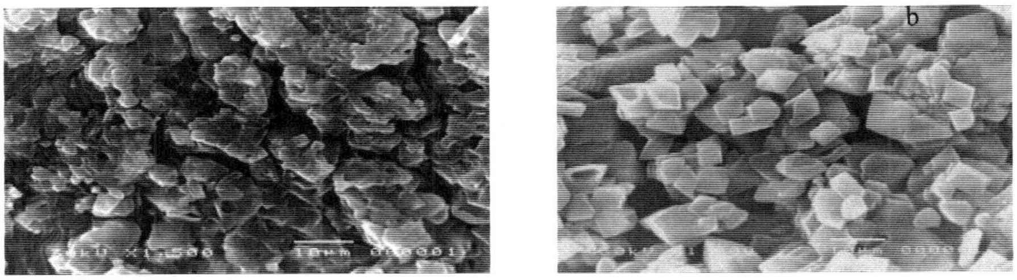

b

Figure 3: Scan Electron Microscope a) 20 KGy and 0.1 % PAC; b) 20 KGy and 0.2 % PAC.

3.4. Proton conductivity
Table 1. shows that, the casted membrane with different PAC ratio and doses has relatively good proton conductivity values ranging from 1.46 to 3.14 mS/cm compared to Nafion (70 mS/cm).

Journal of Physics: Conference Series **262** (2011) 012005 doi:10.1088/1742-6596/262/1/012005

Table 1: The proton conductivity for casted membrane with different PAC ratio and doses

Samples	PAC ratio	Dose KGy	Proton conductivity (mS/cm)
1	0.1	10	2.70
2	0.2	10	1.46
3	0.3	10	0.31
4	0.1	20	1.80
5	0.2	20	2.09
6	0.3	20	2.90

3.5. Positron lifetime studies

Table 2. shows the positron annihilation parameters for casted membrane with different PAC ratio and doses. The o-Ps lifetime τ_3, and its intensity I_3, decreased for the sample irradiated with 10 KGy with increasing PAC concentration. The decrease in τ_3 is related to the change in the free volume as a result of the formation of new bonds or cross linking. The decrease in I_3 reflects a decrease in the number of free volume holes fits well with increase in cross linking. While at 20 KGy, the increase in o-Ps lifetime due to chemical degradation and agglomeration of free volume hole size. This means that the casted membrane with different PAC concentration at 10 KGy had a good characteristics compared with the irradiated sample at 20 KGy.

Table 2: The positron annihilation parameters for casted membrane

Sample	PAC ratio	Dose KGy	Lifetime τ_3(ns)	Intensity I_3	Free volume(A^{o3})	Fractional free volume
1	0.1	10	1.70±0.022	4.90±0.121	70.27	3.44
2	0.2	10	1.68±0.027	4.21±0.128	68.33	2.88
3	0.3	10	1.61±0.024	5.03±0.156	62.20	3.13
4	0.1	20	1.62±0.015	9.20±0.167	63.09	5.80
5	0.2	20	1.62±0.028	3.94±0.134	63.51	2.50
6	0.3	20	1.68±0.015	2.11±0.122	67.86	1.44

4. Conclusion

Acryl amide /poly (methacrylic acid) membrane was synthesized under γ-radiation effect. The casted membranes with different PAC ratio and doses has relatively good proton conductivity values ranging from 1.46 to 3.14 mS/cm compared to Nafion (70 mS/cm). The positron lifetime study on these irradiated casted membranes shows that the cross-linking and degradation within the membrane matrix affect the free volume content and hence the microstructure.

Acknowledgements

The authors wish to sincerely thank the STDF of Egypt, (ID220), for financially supporting this project.

5. References

[1] Jean Y C 1990 Microchem.J.**42** 72
[2] Winberg P, Eldrup M, Pedersen N J, Van Es M A and Maurer F H J 2005 Polymer **46** 8239
[3] Satyanarayana S V, Subrahmanyam V S, Verma H C, Sharma P K and Bhatta-Charya A 2006 Polymer **47** 1300
[4] Pujari P K, Sen D, Amarendra J, Abhaya S, Pandey A K, Dutta D and Mazumder S 2007 Nucl.Instrum.Methods Phys Res.sect.B **254** 278
[5] Olsen J V, Kirkegaard P, Pedersen N J, and Eldrup M 2007 *Phys. Status Solidi C* **4** 4004

Journal of Physics: Conference Series **262** (2011) 012006 doi:10.1088/1742-6596/262/1/012006

Study of phase transformations in metal silicides using positrons

S Abhaya and G Amarendra†

Materials Physics Divison, Indira Gandhi Centre for Atomic Research, Kalpakkam- 603 102, Tamil Nadu, India

E-mail: †amar@igcar.gov.in

Abstract. Positron beam studies on the silicidation in Pd/Si, Ni/Si and Co implanted Si thin film systems are presented. In the case of Pd/Si, the evidence for the occurrence of divacancies across the Pd_2Si/Si interface consequent to Pd_2Si formation is shown. In Ni/Si system, the identification of a multitude of silicide phases of Ni_2Si, NiSi and $NiSi_2$ is obtained from depth-resolved S-parameter, which is corroborated with GIXRD and AES results. In the case of the Co implanted Si, the recrystallization of amorphized silicon and the formation of $CoSi_2$ precipitates in Si are brought out clearly.

1. Introduction

Metal silicides, a class of intermetallic compounds comprising of metal and silicon, have very rich technological applications as gates, Schottky barriers, Ohmic contacts and interconnects in semiconductor industries [1]. Diffusion of the metal and silicon atoms across the interface and point defect generation consequent to diffusion are the key factors in metal silicide formation. In general, silicon diffusion leads to vacancy injection, while metal diffusion leads to injection of Si self-interstitials [2]. Italyantsev and Kuznetsov [3] using model calculations, have shown excess vacancy production across silicide/silicon interface owing to enhanced diffusion of Si from the substrate to the silicide region beyond 870 K and the diffusion of these defects from interface to substrate regions at higher annealing temperatures. Injection of vacancy defects and interstitials induced by silicide formation is of fundamental interest for the applicability of silicides in Si technology and needs to be studied in detail using defect characterization techniques. Positron Annihilation Spectroscopy (PAS) is one such technique which has high sensitivity and selectivity to open-volume defects [4]. Fractional vacancy defect concentration ranging from 10^{-7} to 10^{-4} can be probed using PAS. While conventional PAS techniques have no depth selectivity, variable low energy positron beam enables probing of defects at specific depths [5]. The present paper covers the application of the depth-resolved positron annihilation technique in investigating various atomistic processes involved in silicide formation in Pd/Si, Ni/Si thin film systems and Co implanted Si.

2. Experimental Details

Pd (100 nm)/Si(100), Ni(130 nm)/Si(100) and Co implanted Si(100) of size 10 mm x 10 mm were annealed at various temperatures between 370 and 1070 K in steps of 100 K in a vacuum

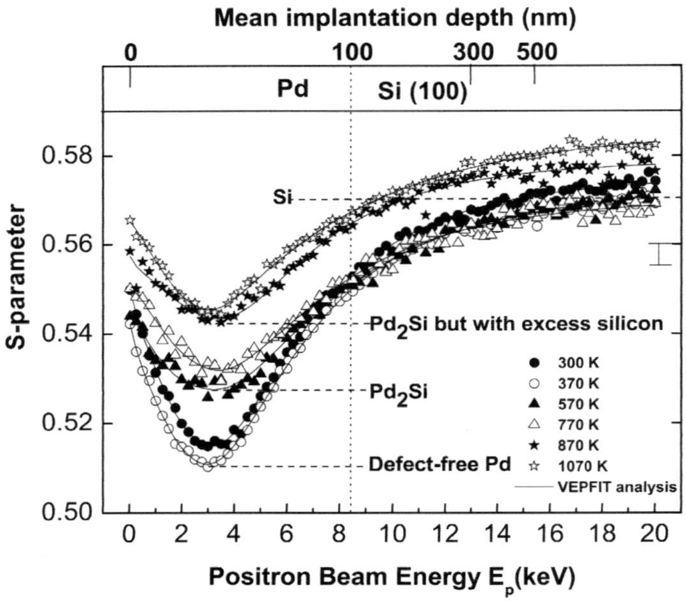

Figure 1. S vs. E_p curve for Pd (100 nm)/Si annealed at various temperatures. The mean implantation depth of the positrons is shown on the top axis calculated with Pd density up to 100 nm and with Si density beyond that. The solid line through the data points is a result of VEPFIT analysis. The horizontal dashed lines represent the different states of the Pd overlayer as a function of annealing [7].

of 1 x 10^{-6} Torr. Depth-resolved positron annihilation measurements were carried out on as-grown and annealed samples at room temperature using a compact magnetically guided positron beam system [6]. By varying the positron beam energy E_p (0-20 keV), the positron beam was implanted at various depths of the sample starting from the near-surface region up to a depth of 1.5 μm. The implanted positrons get thermalized and annihilate with the electrons of the medium giving rise to a Doppler broadened 511 keV annihilation spectrum. From the measured spectrum, a defect-sensitive lineshape S-parameter, which corresponds to the fraction of positrons annihilating with the valence electrons and complementary W-parameter which corresponds to the fraction annihilating with core electrons are deduced for each E_p. The S-parameter is obtained by taking the ratio of counts in the central portion (511 $\pm$ 1 keV) of the curve to the total photopeak counts (511 $\pm$ 10 keV) and W-parameter is deduced by taking the ratio of counts in wing regions (507 $\pm$ 1)+(515 $\pm$ 1) to the total counts (511 $\pm$ 10 keV) under the curve. S-parameter shows an increase in presence of open-volume defects and therefore from the analysis of S vs. E_p curve, depth-resolved chemical as well as defect information are obtained. As compared to the Pd/Si system, the HPGe detector used for Ni/Si and Co implanted Si was having a better energy resolution of 1.4 keV at 662 keV γ-ray peak.

3. Results and Discussion

3.1. Pd/Si system

Figure 1 shows the experimental S vs. E_p curve for as-grown and annealed samples. In as-grown Pd/Si sample, for low positron energies ($\lesssim$ 1 keV), the S-parameter is high and corresponds to annihilation in defects present at the surface ($S_{surface}$). As the positron beam energy is slowly

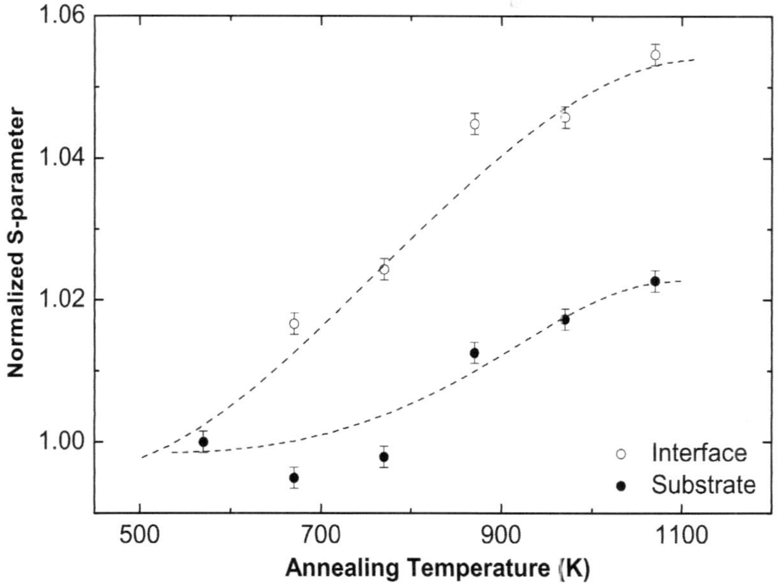

Figure 2. Normalized $S_{interface}$ (open circles) and $S_{substrate}$ (filled circles) plotted as a function of annealing temperature. The normalization for the interface and the substrate have been done with respect to the corresponding VEPFIT deduced values at 570 K, which is the formation temperature of Pd_2Si [7]. The dashed line through the data points is a guide to the eye.

increased, the S-parameter decreases, reaches a minimum around 3 keV which corresponds to annihilation of positrons in Pd over-layer (S_{Pd}). Beyond 3 keV, the S-parameter slowly increases and saturates for positron energies higher than 15 keV. The increase in the S-parameter beyond 3 keV is because the positron beam is now traversing from metallic Pd over-layer to semiconducting Si region. The saturation value of S-parameter at higher positron energies corresponds to annihilation in Si substrate (S_{Si}). At 370 K, S_{Pd} decreases indicating annealing of structural defects in the Pd over-layer. Beyond 470 K, the S-parameter increases and stays constant till 770 K. This increase in S_{Pd} above 470 K is attributed to Pd_2Si phase formation as seen from Glancing Incidence X-ray diffraction (GIXRD) results [7]. Beyond 870 K, there is a sudden increase in S vs. E_p curve i.e., the S-parameter values corresponding to surface, Pd over-layer, Pd/Si interface and Si substrate are higher and stay constant up to 1070 K. This increase is explained using VEPFIT analysis [8] of the experimental S vs. E_p curves. Makhovian implantation profile has been used in VEPFIT analysis. It should be mentioned that up to about 1070 K, Pd_2Si is the only phase present as seen from GIXRD results.

Figure 2 shows the VEPFIT normalized $S_{interface}$ and $S_{substrate}$ as a function of annealing temperature. $S_{interface}$ has been invoked beyond 570 K to account for the defect production across the interface. $S_{interface}$ gradually increases with annealing temperature and reaches a value of 1.055 $\pm$ 0.001 by 1070 K while $S_{substrate}$ stays at $\sim$ 1.00 up to about 770 K and increases to 1.023 $\pm$ 0.001 by 1070 K. These S-parameter values are compared with the ab-initio calculated S-parameter values of 1.018 for V_1, 1.045 for V_2, 1.053 for V_3 and 1.067 for V_4 in crystalline Si, to estimate the size of the vacancy defects. Since, monovacancies in Si are highly mobile, and at room temperature only divacancies are stable, the normalized S-parameter in the range of 1.022-1.045 is attributed to the presence of divacancies. It may also be noted that

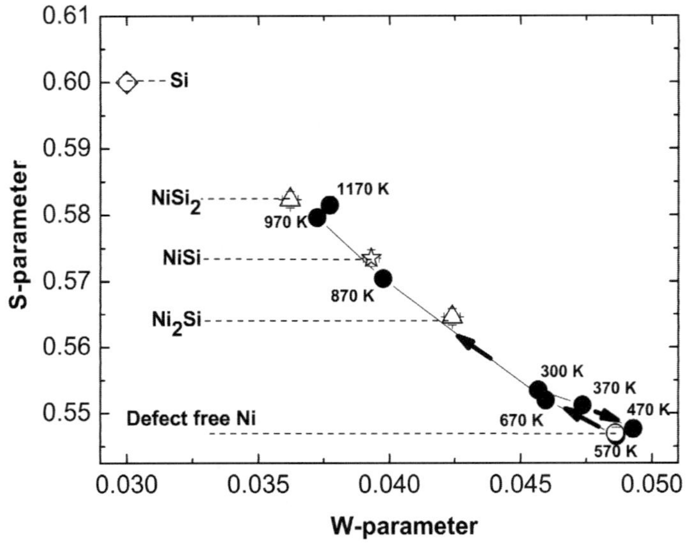

Figure 3. (S,W) correlation plot of the Ni overlayer as a function of annealing temperature [11]. Filled symbols are experimental points, while open symbols correspond to theoretical estimates of the respective silicide phases.

S-parameter values are larger at the interfacial region than at the substrate region suggesting larger concentration of divacancies in the interface than in the Si substrate region. In view of their high mobility, divacancies, produced at the interface, diffuse into the Si region beyond 870 K, thereby increasing the S-parameter value of substrate region to 1.022. Production of divacancies due to enhanced Si diffusion (as seen from Auger Electron Spectroscopy (AES) and Rutherford Backscattering (RBS) measurements [9]) from the substrate region to Pd_2Si region beyond 870 K and the diffusion of these defects from interface to substrate region at higher annealing temperatures is the first time experimental proof validating the model predictions of Italyantsev *etal* [3]. The present results highlight the capability of positrons to detect defect annealing in the Pd overlayer, Pd_2Si formation in the over-layer and the production of divacancies in the interface and substrate consequent to silicide formation.

3.2. Ni/Si system

Depth-resolved positron annihilation measurements were carried out on as-grown and annealed Ni/Si samples. Figure 3 shows the experimental (S,W) correlation plot of the Ni overlayer deduced as a function of annealing temperature. The (S,W) coordinate decreases with annealing temperature up to about 570 K and beyond 570 K, the (S,W) coordinate increases and approaches very close to that of the Si value. The (S,W) minimum at 570 K corresponds to the defect-free Ni overlayer. In order to identify the various silicide phases present in the overlayer as a function of annealing temperature, the experimental (S,W) coordinates are compared with the theoretical (S,W) coordinate of various silicide phases obtained by taking the weighted average of the S or W-parameter of the individual elements in the phase. This assumption does not take into account the crystal structure of the silicide phase and the positron affinity of the individual elements in a silicide phase [10]. For existence of two or more phases in the over-layer, the

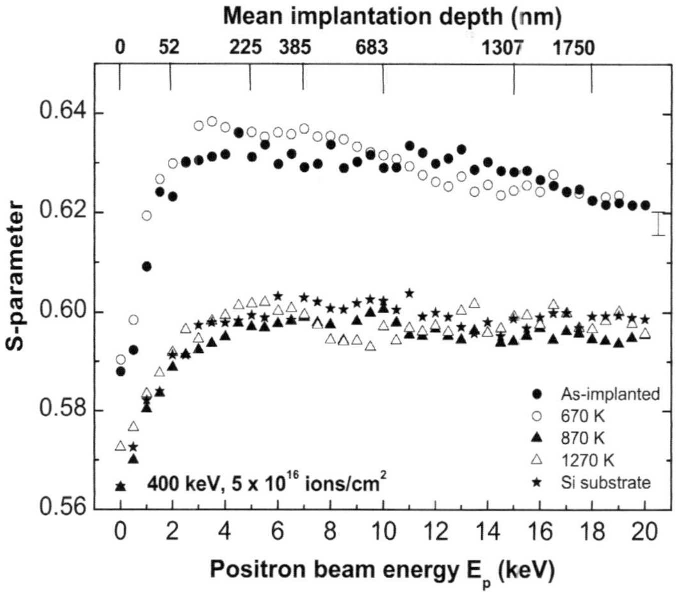

Figure 4. Variation of S-parameter with positron beam energy for Si sample implanted with 400 keV Co ions at a dose of 5 x 10^{16} ions/cm^2 and annealed at 670 K, 870 K and 1270 K. The top axis shows the depth probed by the positron beam in Si substrate [14].

(S,W) coordinate is taken to be an average of the (S,W) coordinate of the respective phases. Using this, the (S,W) coordinates of Ni_2Si, NiSi and $NiSi_2$ have been estimated and are shown as open symbols in Figure 3. At 570 K, the over-layer comprises of the defect-free Ni. At 670 K, the intermixing reaction at the interface has started. At 870 K, the over-layer comprises of NiSi and NiSi and $NiSi_2$ phases. Between 970 K and 1070 K, the over-layer consists of NiSi and $NiSi_2$ phases. At 1170 K, the over-layer consists of $NiSi_2$ phase alone. Since, the (S-W) coordinate compares well with the estimated (S,W) coordinate of $NiSi_2$ phase at 1170 K, it is concluded that the final silicide over-layer is devoid of vacancy defects. The phases identified using this calculation scheme compares well with GIXRD and Rutherford backscattering (RBS) measurements [11], confirming the sensitivity of positron annihilations to phase changes in the over-layer.

3.3. Co implanted Si

400 keV Co ions were implanted in 500 μm thick p-type Si (100) wafer at room temperature to a dose of 5 x 10^{16} ions/cm^2 using the 1.7 MV Tandetron accelerator facility. SRIM code [12] has shown that the Co ions are distributed up to a depth of 500 nm with maximum distribution around 350 nm depth range. Figure 4 shows the S vs. E_p curves for as-implanted and sample annealed at selected temperatures. When compared to the S-parameter curve of the unimplanted Si where the S-parameter slowly increases and saturates at 0.6 beyond 3 keV attributable to annihilations in Si substrate, the S vs E_p curves for as-implanted and sample annealed at 670 K are higher. Here, the S-parameter increases, reaches a maximum around 4 keV and saturates at higher value. The normalized S-parameter having a value of 1.055 at 4 keV is attributed to the presence of amorphized Si, containing vacancy clusters having five to six vacancies produced as a result of implantation [13]. From VEPFIT analysis [8], it is found that the thickness of the

amorphized region extends up to a depth of $\sim$ 450 nm from the surface and the positron diffusion length has drastically reduced to 6.25 $\pm$ 0.43 nm. GIXRD measurements at these temperatures show broad peaks attributable to amorphized silicon [14]. It may be pointed out that because of the broad implantation profile beyond 10 keV, the positrons implanted corresponding to this depth will also sense the defected region making the S-parameter value higher than the Si reference value of 0.602. Beyond 870 K, the S-parameter curve drastically reduces and matches with that of Si substrate. This suggests that recrystallization of amorphized silicon has taken place at 870 K. This is consistent with the GIXRD results which show crystalline peaks of Si beyond 870 K. The formation of $CoSi_2$ phase as shown by GIXRD is not brought out clearly by the positron beam measurements because of the formation of very small $CoSi_2$ grains distributed in Si substrate resulting in a non uniform layer of $CoSi_2$ on Si. SEM together with EDAX measurements on 870 K annealed sample confirm the presence of $CoSi_2$ precipitates of size varying between 0.2 and 1.3 μm in Si [14]. Here, positron sensitivity to recrystallization of amorphized silicon at 870 K is evident from the results.

4. Summary

Detailed investigation on the various aspects of silicidation on thin film systems of Pd/Si, Ni/Si and Co implanted Si have been carried out using the depth-resolved positron annihilation technique. From these studies, positron's sensitivity and selectivity has been exploited (a) to study defect annealing in the metal over-layer prior to the silicidation reaction, (b) interdiffusion of the metal and silicon across the interface prior to silicide formation, (c) formation of silicide phases starting from metal rich silicide to silicon rich silicide, (d) identification of the various silicide phases present in the over-layer from an analysis of the positron annihilation parameters (S,W) and (e) the existence of vacancy defects across the interface and substrate.

5. Acknowledgements

The authors would like to thank Dr. C S Sundar for his encouragement.

6. References

[1] Murarka S P 1983 *Silicides for VLSI Applications* (New York: Academic Press); Shi- Li Zhang and Mikael Ostling 2003 *Critical Reviews in Solid State and Materials Sciences B* **28** 1.

[2] Wen D S, Smith P L, Osburn C M and Rozgonyi A 1987 *Appl. Phys. Lett.* **51** 118; Lur W, Cheng J Y, Chu C H, Wang M H, Lee T C, Wann Y J, Chao W Y and Chen L J 1989 *Nuclear Instruments and Methods in Physics Research B* **39** 297.

[3] Italyantsev G and Kuznetsov A Yu 1993 *Appl. Surf. Sci.* **73** 203.

[4] Krause-Rehberg R and Leipner H S 1998 *Positron Annihilation in Semiconductors*(Berlin: Springer Verlag); Hautojarvi P (ed.) 1979 *Positrons in Solids*(Berlin: Springer Verlag).

[5] Schultz P J and Lynn K G 1988 *Rev. Mod. Phys.* **60** 701.

[6] Amarendra G, Viswanathan B, Venugopal Rao G, Parimala J and Purniah B 1997 *Curr. Sci.* **73** 401.

[7] Abhaya S, Amarendra G, Reddy G L N, Rajaraman R, Venugopal Rao G and Narayanan K L 2003 *J. Phy.: Condens. Matter* **15** L713; Abhaya S, Amarendra G, Venugopal Rao G, Rajaraman R, Panigrahi B K, Sastry V S 2007 *Materials Science and Engineering B* **142** 62.

[8] Van Veen A, Schut H, de Vries J, Hakvoort R A and Ijpma M R 1990 *AIP Conf. Proc.* **218** 171.

[9] Abhaya S, Amarendra G, Padma Gopalan, Reddy G L N and Saroja S 2004 *J. Phys. D: Appl. Phys.* **37** 3140; Abhaya S, *Positron annihilation studies of metal silicides* (Univ. Of Madras, India, 2007)

[10] Knights A P, Ponjee M W G, Simpson P J, Zinke- Allmang M and Carlow G R 1997 *Semicond. Sci. Technol.* **12** 173.

[11] Abhaya S, Amarendra G, Panigrahi B K and Nair K G M 2006 *J. Appl. Phys.* **99** 033512.

[12] Ziegler J F, Biersack J P, Littmark U 2003 *The Stopping and Range of ions in Solids* (New York: Pergamon Press).

[13] Amarendra G, Rajaraman R, Venugopal Rao G, Nair K G M, Viswanathan B, Suzuki R, Ohdaira T, Mikado T 2001 *Phys. Rev. B* **63** 224112; Amarendra G 2002 *Defects Diffus. Semicond.* **200-202** 189.

[14] Abhaya S, Amarendra G, Kalavathi S, Panigrahi B K, Saroja S, Nair K G M, Sastry V S and Sundar C S 2008 *Eur. Phys. J. Appl. Phys.* **41** 107.

Journal of Physics: Conference Series **262** (2011) 012007 doi:10.1088/1742-6596/262/1/012007

Positron transport in molecular gases in crossed electric and magnetic fields

A Banković[1], S Dujko[1,2,3], RD White[3], G Malović[1], SJ Buckman[4] and ZLj Petrović[1]

[1]Institute of Physics, University of Belgrade, Pregrevica 118, POB 68, Zemun, Serbia

[2]Centre for Mathematics and Informatics (CWI), P. O. Box 94079, 1090 GB Amsterdam, The Netherlands

[3]ARC Centre for Antimatter-Matter Studies, School of Engineering and Physical Sciences, James Cook University, Townsville 4810, QLD, Australia

[4]ARC Centre for Antimatter-Matter Studies, Research School of Physics and Engineering, The Australian National University, Canberra, ACT 0200, Australia

ana.bankovic@gmail.com

Abstract. Transport properties of positron swarms drifting and diffusing in neutral gases under the influence of crossed electric and magnetic fields are investigated using a multi-term theory for solving the Boltzmann equation and Monte Carlo simulation technique. In the presence of magnetic fields the number of transport properties is increased compared to the situation when the positron swarm is acted on solely by an electric field. Since the longitudinal and transverse components of the drift velocity show different sensitivities with respect to the strength of the magnetic field, it is found that the negative differential conductivity effect in a crossed field configuration can be controlled through the variation of the magnetic field strengths. Various diffusion tensor elements also exhibit different sensitivities with respect to the magnetic field and also with respect to the positronium (Ps) formation process.

1. Introduction

In the last few decades there has been growing interest for studying positrons both theoretically and experimentally [1]. One of the major motivational factors for this was the revolutionary improvement of buffer-gas, Penning Malmberg traps for positrons made by Surko and co-workers [2,3]. The improved positron traps have provided high-flux, low-energy positron beams and thus the new era of positron physics was born. Perhaps the first response came from the atomic physicists with the provision of accurate cross sections for positron interactions with different gases [4-7], which has opened the possibility to model the transport processes of positrons in neutral gases and soft-condensed matter [8-10, 11], with the principal idea of improving the positron based technologies such as those employed in positron emission tomography. In this paper we investigate the transport properties of positron swarms in neutral molecular gases in a crossed electric and magnetic field configuration. We investigate the positron transport in crossed electric and magnetic fields as possible applications may use combined fields to achieve a better control of properties of the ensemble [12]. It is well known from the recent comprehensive electron studies that the transport properties of electrons can be significantly affected by the magnetic field, particularly under conditions when the cyclotron frequency dominated the collision frequency [13]. Using these arguments as a background, we have been motivated to check the basic features of positron transport in electric and magnetic fields.

The transport properties of positron swarms moving in neutral gases have been studied using a multi-term theory for solving the Boltzmann equation [14, 15] and a Monte Carlo simulation technique [16]. Both techniques have been developed for studying the electron swarms in varying configurations of electric and magnetic fields under hydrodynamic and non-hydrodynamic conditions. Numerical codes have passed all well-established benchmarks with high accuracy levels. In the framework of recent positrons studies, however, special attention has been paid to the correct treatment of positronium (Ps) formation. Ps formation is a non-conservative process unique to positrons which often has a cross section larger than the cross sections for elastic and various inelastic collisions. The kinetic effects induced by the non-conservative nature of collisions manifest themselves as a difference between the flux and bulk transport properties [17]. It should be emphasized that the duality of transport coefficients has been systematically ignored in the majority of previous work in the plasma modeling community, and in related electron studies. The differences between the flux and bulk transport properties observed in positron transport are often more pronounced than those for electrons, particularly for those gases with a large Ps formation cross-section. One of the most important kinetic phenomena observed both in electron and positron transport is negative differential conductivity (NDC). This phenomenon can be defined as a decrease in drift velocity of the swarm with increasing driving electric field. Both the nature and the manifestation of this phenomenon are different for electrons and positrons. In case of electrons [18-20], NDC is present in both the flux and bulk velocity components while for positrons it has been observed only in the bulk drift velocity. The differences between flux and bulk drift velocities can be more than two orders of magnitude [8-10], while for electrons they are generally 15-20% [13, 14]. Also, NDC is much more pronounced for positrons than for electrons and it originates purely from the non-conservative nature of Ps formation. The interesting situation occurs when the inelastic channels start to open in the vicinity of the threshold of Ps formation. Sometimes, for example in nitrogen [9], the competition between these channels leads to the disappearance of NDC, even though the cross section for Ps formation is as high as it is for the gases with very pronounced NDC. In what follows these issues are illustrated and discussed for positrons in molecular hydrogen.

2. Transport in a crossed field configuration

In a crossed electric and magnetic field configuration the drift velocity has two components: the longitudinal w_E and transverse $w_{E\times B}$. These two drift velocity components behave in a completely different manner and, as a consequence, the NDC effect can be controlled by the magnitude of magnetic field. The diffusion tensor has five non-zero elements, which all exhibit different sensitivities with respect to magnetic field and also with respect to Ps formation process. The magnetic cooling effect has also been observed – the mean energy of the swarm is a decreasing function of the magnetic field for all electric field strengths considered in this work.

2.1. Magnetic cooling effect

The magnetic cooling effect is a well known phenomenon observed in electron transport in real [13, 21, 22] and model gases [23]. This phenomenon results from an inability of the electric field to efficiently pump energy into the system [23] because the electrons change their direction of motion due to the magnetic field. In Figure 1 we investigate this phenomenon for positrons in molecular hydrogen. We observe that the mean energy is a monotonically decreasing function of B/N for a fixed value of E/N. Since our calculations have been performed for zero gas temperature, we observe that the mean energy is much less than the thermal energy at 293 K in the limit of low E/N. The profiles of the mean energy reflect the energy dependence of the cross sections and are shifted towards higher values of E/N and lower mean energies for an increasing B/N. Similar behavior has been observed for positrons in molecular nitrogen and water vapour. This effect can also be investigated using the effective field concept (eq. 23 of [24]).

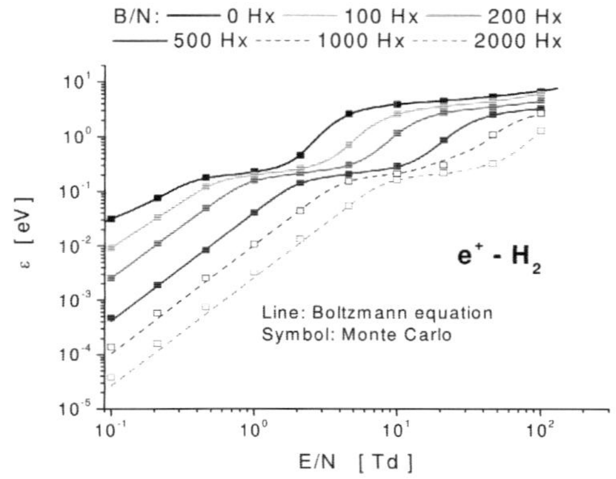

Figure 1: Variation of the mean energy of positrons in molecular hydrogen with E/N for various B/N.

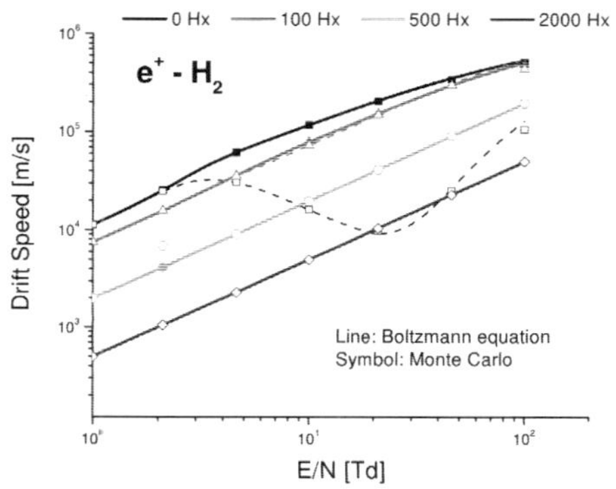

Figure 2: Variation of the drift speed of positrons in molecular hydrogen with E/N for various B/N. The flux values are represented by full lines and symbols, while the bulk values are given by dashed lines and open symbols.

2.2. *Is there NDC in* ***E***×***B*** *fields?*

As emphasized previously, when both the electric and magnetic fields are present, the drift velocity has two components: the longitudinal drift velocity component describes the drift along the electric field direction while the transverse component describes the drift along the **E**×**B** direction. Therefore, the drift can now be represented in terms of a drift speed and its direction is at some angle to the electric field (magnetic deflection angle). An interesting question arises: how does the magnetic field affect the drift speed under conditions when the cyclotron frequency is much less, of the same order, and much greater than the collision frequency? Figure 2 displays the flux and bulk components of the drift speed as a function of E/N for different values of B/N in these different regimes. We see that NDC effect has disappeared from the profiles for non-zero magnetic fields. For B/N of 100 Hx, a slight difference between the flux and bulk profiles still can be observed, but for the higher values of magnetic fields even this difference is absent. In order to understand the behavior of the drift speed it is essentially important to examine the longitudinal and transverse components of the drift velocity separately, as they exhibit completely different behavior in the presence of a magnetic field. In the profiles of $w_{\mathbf{E}\times\mathbf{B}}$ component there is a slight difference between the flux and bulk values but only for low magnetic fields, and this difference is in favor of bulk component. It is interesting to note that in the electron transport the difference between the flux and bulk components of $w_{\mathbf{E}\times\mathbf{B}}$ has never been observed. On the other hand, the profiles of the longitudinal component are lowered and shifted to the right compared to magnetic field free case. So, one could ask why has the NDC disappeared from the profiles of the drift speed in the presence of magnetic field? In the energy region where NDC exists for magnetic field free case, for the non-zero magnetic field profiles, the behavior of the drift speed is dominated by the component along the **E**×**B** direction. This is clearly evident from the profiles of the magnetic deflection angle shown in Figure 3. For B/N of 100 Hx, the flux magnetic deflection angle stays around 45° for the whole range of E/N considered, while the bulk value goes to 90° in the energy region where the Ps formation channel is open. This is the unique situation where due to the nature of collisions the flux and bulk drift velocities differ not only in magnitude, but also in direction.

3. Conclusion

In this work we have calculated various transport properties of positrons in neutral gases in crossed electric and magnetic fields using a multi-term theory for solving the Boltzmann equation and Monte Carlo simulation technique. The results obtained with these two different techniques are in very good agreement. We hope that present results and the multitude of interesting kinetic phenomena, induced by the synergism of magnetic field and Ps-induced non-conservative collisions, can be a good motivation for a further development of positron swarm experiments.

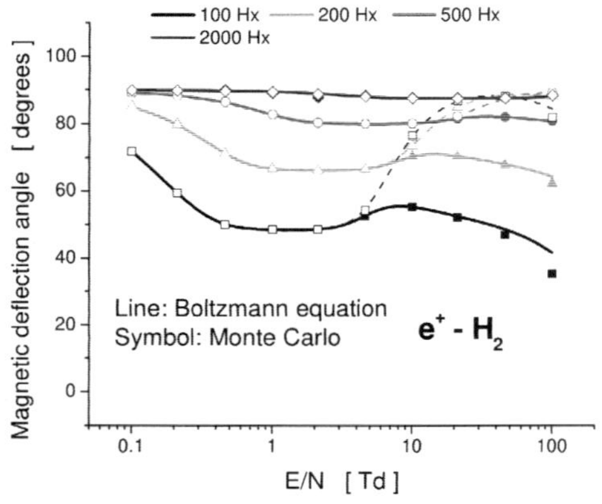

Figure 3: Variation of the magnetic deflection angle with E/N for various B/N. The flux values are represented by full lines and symbols while the bulk values are given by dashed lines and open symbols.

Acknowledgements

This work was funded by the MNTRS project 141025, COE Nonequilibrium processes in plasmas and environmental science, ARC Centre for Antimatter-Matter Studies and the International Science Linkages Program. Authors are grateful to J. Marler, for collaboration and advice and to C. Surko and M. Charlton for many useful comments.

References

[1] Charlton M and Humberston J 2000 *Positron Physics* (New York: Cambridge University Press)
[2] Surko CM, Passner A, Leventhal M, Wysoki FJ, 1988 *Phys. Rev. Lett.* **61** 1831
[3] Murphy TJ, Surko CM, 1992, *Phys. Rev. A* **46** 5696
[4] Sullivan JP, Gilbert SJ, Marler JP, Greaves RG, Buckman SJ, Surko CM, 2002 *Phys. Rev. A* **66** 042708
[5] Marler JP, Sullivan JP, Surko CM, 2005 *Phys. Rev. A* **71** 022701
[6] Sullivan JP, Makochekanwa C, Jones A, Caradonna P, Buckman S J, 2008 *J. Phys. B: At. Mol. Opt. Phys.* **41** 081001
[7] Makochekanwa C, Banković A, Tattersall W, Jones A, Caradonna P, Slaughter DS, Nixon K, Brunger MJ, Petrovic ZLj, Sullivan JP, Buckman SJ, 2009 *New J. Phys.* **11** 103036
[8] Šuvakov M, Petrović ZLj, Marler JP, Buckman SJ, Robson RE, Malović G, 2008 *New J. Phys.* **10** 053034
[9] Banković A, Marler JP, Šuvakov M, Malović G, Petrović ZLj, 2008 *NIMB* **266** 462
[10] Marler JP, Petrović ZLj, Banković A, Dujko S, Šuvakov M, Malović G and Buckman SJ, 2009 *Physics of Plasmas* **16** 057101
[11] White RD, Robson RE 2009 *Phys. Rev. Lett.* **102** 230602
[12] B. Schmidt, 1994 *Phys. Scr.* **T53** 30

Journal of Physics: Conference Series **262** (2011) 012007 doi:10.1088/1742-6596/262/1/012007

[13] Dujko S, Raspopović ZM, Petrović ZLj, 2005 *J. Phys. D : Appl. Phys.* **38** 2952
[14] S. Dujko, Ph. D. Thesis, James Cook University, Townsville, Australia, 2009
[15] Ness K. F, 1993 *Phys. Rev. E* **47** 323
[16] Šuvakov M, Ristivojević Z, Petrović ZLj, Dujko S, Raspopović ZM, Dyatko NA, Napartovich AP, 2005 *IEEE Trans. Plasma Sci.* **33** 532
[17] Robson RE, 1991, *Aust. J. Phys.* **44** 685
[18] Petrović ZLj, Crompton RW, Haddad GN, 1984 *Aust. J. Phys.* **37** 23
[19] Robson RE, 1986 *J. Chem. Phys.* **85** 4486
[20] Vrhovac SB, Petrović ZLj, 1996 *Phys. Rev. E* **53** 4012
[21] White RD, Robson RE, Ness KF and Makabe T, 2005 *J. Phys D:Appl. Phys.* **38** 997
[22] Dujko S, White RD, Ness KF, Petrović ZLj, Robson RE, 2006 *J. Phys.D: Appl. Phys.* **39** 4788
[23] Ness KF, 1994 *J. Phys. D: Appl. Phys.* **27** 1848
[24] Li B, White RD, Robson RE, Ness KF, 2001 *Ann. Phys.* **292** 179

Journal of Physics: Conference Series **262** (2011) 012008 doi:10.1088/1742-6596/262/1/012008

Fluid analysis of positrons in soft matter

G J Boyle[1], R D White[1], R E Robson[1,2]

[1] ARC Centre for Antimatter-Matter Studies, School of Engineering and the Physical Science, James Cook University, Townsville 4811, QLD, Australia

[2] ARC Centre for Antimatter-Matter Studies, Research School of Physical Sciences and Engineering, Australian National University, 2600 Canberra, Australia

E-mail: gregory.boyle@jcu.edu.au

Abstract.

Improvements to Positron Emission Tomography (PET) technology depend to a large extent on obtaining a better understanding of positron transport processes in soft condensed matter. The most economically computational, physically tenable method of carrying out such an investigation is the fluid approach, in which the Boltzmann kinetic equation is replaced by a hierarchy of low order, velocity moment equations, in which collision terms are approximated by extending 'momentum transfer theory' to allow for coherent scattering arising from the structure of the medium. We show how these fluid equations furnish structure-modifed versions of well known semi-empirical relations linking transport coefficients, and give numerical calculations for simple models.

1. Introduction

Positron Emission Tomography (PET) has become a well established and powerful, non-invasive tool for measuring biological processes in humans and animals [1]. However, most current models do not systematically address the inherent structure of the medium. This is critical as the biological materials that constitute the background media for PET procedures consist of soft condensed matter involving structure. The program at James Cook University aims to address this issue, and, although the present work involving the calculation of universal transport coefficients is not directly applicable to PET, it does represent a first step toward a complete transport model valid in structured media.

The approach to including structure in the framework of modern kinetic theory was proposed by Cohen and Lekner in 1967 [2], using a modified differential cross-section given by Van Hove [3]. Such microscopic information alone is insufficient for applications to the macroscopic world. The Boltzmann equation represents the fundamental equation from which macroscopic transport properties of interest, such as drift velocity and mean energy, can be determined [4]. However, the integro-differential nature of Boltzmann's equation makes finding a general solution for the velocity distribution function a formidable task.

The approach undertaken in this paper involves approximate 'moment' representations of the Boltzmann equation, otherwise known as the 'fluid description' [5]. The set of moment or balance equations can be found by multiplying the Boltzmann equation by some property of the swarm velocity, $\Psi(\boldsymbol{v})$, and integrating over all velocities. Different forms chosen for $\Psi(\boldsymbol{v})$ yield the balance equations for the different moments. Momentum transfer theory (MTT) [4] can be used to facilitate analytic solutions and simplify computations for the resulting coupled non-linear differential equations.

In Section 2 the fluid treatment of the new kinetic equation accounting for the structure of the medium are undertaken using MTT. For a special case, the generalized Wannier energy relation [4] and a criterion for the phenomena of structure-induced negative differential conductivity (NDC) [6] are derived analytically. In Section 3 a numerical scheme is used to solve the balance equations in soft-condensed matter. The Percus-Yevick model [7] is used to simulate structure, and investigate the effect of volume fractions on the spatially homogeneous transport coefficients.

2. Theory

2.1. Boltzmann equation and the material structure

The fundamental equation from which macroscopic transport properties can be determined is the Boltzmann equation,

$$\left(\frac{\partial}{\partial t} + \boldsymbol{v} \cdot \nabla + \frac{\boldsymbol{F}}{m} \cdot \frac{\partial}{\partial \boldsymbol{v}}\right) f = -J(f), \tag{1}$$

where f is the phase-space distribution function, $\boldsymbol{v}$ is the velocity and m is mass, all relating to the swarm particle, and $\boldsymbol{F}$ refers to the external forces applied to the system. The right hand side is the eponymous Boltzmann collision integral and describes the effect of collisions on the velocity distribution function f at a fixed position and velocity. The collision integral accounts for the different processes, and generally the structure of the medium is only included via the elastic component, which is modified to account for coherent scattering. Robson and White [8] give derivations for the first and second order differential cross-sections valid in soft condensed media, and found an expression for the collisional rate of change involving the dynamic structure factor $S(\boldsymbol{K}, \Omega)$, where $\boldsymbol{K}$ and Ω are the momentum and energy exchanges during a collision. In the 'single-scatterer approximation' $S(\boldsymbol{K}, \Omega)$, which is the Fourier transform of the Van Hove [3] space-time pair correlation function, contains all the necessary information about the excitations of the system [2]. The structure factor is purely determined by the medium, and can be found in neutron or x-ray scattering experiments. The integral quantity,

$$S(\boldsymbol{K}) \equiv \int_{-\infty}^{\infty} d\Omega \; S(\boldsymbol{K}, \Omega) \tag{2}$$

represents the static structure factor, which is evaluated at $K = 2k\sin(\chi/2)$ [8], i.e. the momentum transfer in an elastic collision in which the positron is scattered through an angle χ. For the isotropic scattering considered in this work, the momentum transfer cross section can be written as $\tilde{\sigma}_m(v) = \sigma_m(v)s(v)$, where $\sigma_m(v)$ is the momentum transfer cross section for single particle scattering, and $s(v)$ is the angle-integrated structure factor given by

$$s(v) = \frac{1}{2}\int_{-1}^{1} S\left(\frac{2mv}{\hbar}\sin\left(\frac{\chi}{2}\right)\right)(1-\cos\chi)d(\cos\chi). \tag{3}$$

which also plays an important role in this discussion.

2.2. Fluid treatment

A full and general solution to (1) is formidable. A fluid equation treatment can yield approximate absolute results and relationships between physically measurable properties. Here the problem of solving the Boltzmann equation for f in phase space is replaced by a low order set of approximate equations for the moments of f [5, 9]. The set of moment equations is found by multiplying the modified Boltzmann equation (1) by an arbitrary property of swarm particle velocity, $\Psi(\boldsymbol{v})$,

Journal of Physics: Conference Series **262** (2011) 012008 doi:10.1088/1742-6596/262/1/012008

and integrating over all velocities. Momentum transfer theory provides a way of evaluating the collision terms [4] with good qualitative and quantitative results. Essentially, the mathematical form of expressions derived for the constant collision frequency model are assumed to carry over to the general, energy-dependent collision frequency case. The momentum and energy balance equations can then be found [10]. The homogeneous balance equations represent coupled first order differential equations, which can only be solved analytically in some special cases.

2.3. Generalized Wannier energy relation and structure-induced NDC

The simple case of a steady, spatially uniform swarm undergoing elastic collisions and subject to an electric field only is now considered. All time and space derivatives vanish, and the momentum and energy balance take the form,

$$e\boldsymbol{E} = \mu\tilde{\nu}_m(\epsilon)\boldsymbol{W}, \tag{4}$$

$$\epsilon = \frac{3}{2}k_b T_0 + \frac{1}{2}m_0 W^2 \frac{\tilde{\nu}_m}{\nu_m}. \tag{5}$$

where e is the charge, $\boldsymbol{E}$ is the electric field, μ is the reduced mass, m_0 is the neutral mass, $\boldsymbol{W}$ is the flux drift velocity, ϵ is the mean energy, T_0 is the neutral temperature, and ν_m and $\tilde{\nu}_m$ are the elastic momentum collision frequencies, the latter for coherent scattering. Equation (5) is a generalization of the well known Wannier energy relation in gas transport theory derived by Wannier in 1953 [4]. If the interparticle spacing is large compared to the de Boglie wavelength of the positron then the angle-integrated structure factor $s(\epsilon) \to 1$, and subsequently $\tilde{\nu}_m \to \nu_m$ reducing to the dilute gas case expressions.

Another important phenomena that is predicted by transport theory is negative differential conductivity (NDC). NDC is characterized by a decrease in the drift velocity $\boldsymbol{W}$ despite an increase in the magnitude of applied electric field $\boldsymbol{E}$, i.e. $\frac{d\boldsymbol{W}}{dE} < 0$. It is well known in gas transport theory that, for some energy profiles, the inclusion of inelastic or non-conservative scattering cross sections can cause regions of NDC [11, 12]. For a condensed media there is a new type of NDC which is purely a consequence of the medium structure [6]. For the cold gas approximation (T = 0) it can be shown that,

$$\frac{1}{m_0\boldsymbol{W}}\left(\frac{\tilde{\nu}_m}{\nu_m}\right)^{-1}\left[1 - \frac{d\ln(\frac{\tilde{\nu}_m}{\nu_m})}{d\ln\epsilon}\right]\frac{d\epsilon}{d\ln E} = \frac{d\boldsymbol{W}}{d\ln E}. \tag{6}$$

By considering the signs of the constituents of the left hand side and right hand side of (6) it is evident that for structure-induced NDC to occur, then the inequality

$$\frac{d\ln(\frac{\tilde{\nu}_m}{\nu_m})}{d\ln\epsilon} > 1, \tag{7}$$

must be satisfied.

3. Results and discussion

A numerical scheme was developed to solve the coupled non-linear equations (4) and (5).

3.1. Percus-Yevick

In order to investigate the effects of structured media on positron transport, a model for the static structure factor, S, of the medium is required. The model of Percus and Yevick (with the Verlet Weis correction) provides a simple description of the structure behaviour with particularly good agreement for hard spheres [7, 13]. The important factor is the volume fraction, Φ, which is a measure of how tightly packed the particulates of the media are. Figure 1a shows the static

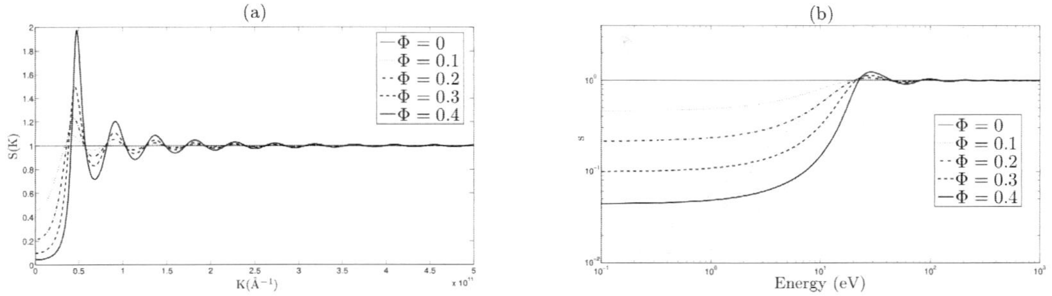

Figure 1: (a) Percus-Yevick structure factor for various volume fractions fractions Φ. (b) Angle-integrated structure factor for various volume fractions Φ.

structure value $S(K)$, for different values of Φ. In the limit as $K \to \infty$, $S \to 1$ and the dilute gas case is regained.

The relationship between the angle-integrated structure factor s and energy ϵ is shown in Figure 1b. It illustrates the oscillatory behaviour caused by structured media, and convergence to the dilute gas-phase-case for high energies. These features are echoed in the behaviour of the transport coefficients below.

3.2. *Maxwell Model*

The model considered here involves an elastic collision frequency which is energy independent i.e., $\sigma_m = 6\epsilon^{-\frac{1}{2}}$ Å^2, $m_0 = 4$ amu. and $T_0 = 0$ K. This is known as the Maxwell Model.

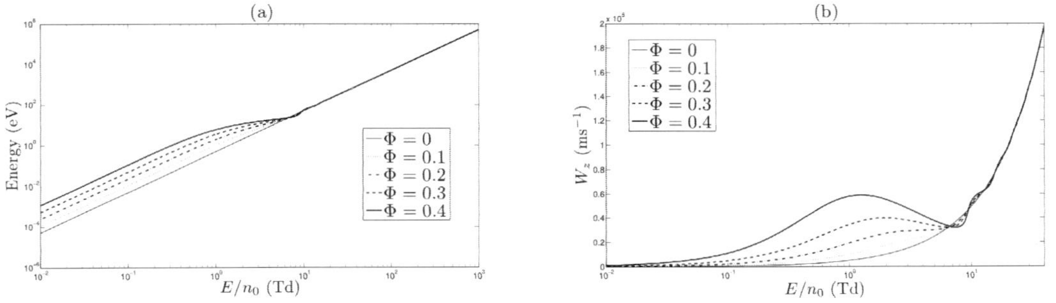

Figure 2: (a) Variation of mean energy with reduced electric field for various volume fractions Φ. (b) Variation of drift velocity with reduced electric field for various volume fractions Φ.

Figure 2a shows the variation of the mean energy with E/n_0 for different volume fractions. The profile for $\Phi = 0$ represents the dilute gas phase. $\tilde{\nu}_m$ is independent of energy and (4) gives $\boldsymbol{W} \propto \boldsymbol{E}$ as shown. When structure is included in the medium, the energy and drift equations become coupled and non-linear which give rise to large deviations from the dilute gas phase case behaviour. At low field strengths, the mean energy increases with the volume fraction, but as E/n_0 increases each of the profiles converge on the structureless line. Physically, as the energy of the incoming positron increases, its de Broglie wavelength decreases until it is effectively sampling a single medium particle during a collision and the structure effects are limited.

The variation in drift velocity with E/n_0 for various volume fractions is shown in Figure 2b. The larger volume fractions experience larger oscillations away from the dilute gas phase case, and again converge for high field strengths. The profiles in Figure 2b exhibit NDC for volume fractions above some critical value. A condition for the occurrence of this NDC was given in (7). For the Maxwell model, $\tilde{\nu}_m/\nu_m$ is effectively the angle-integrated structure factor s, and Figure 1b shows the log plot of s superimposed with straight lines of slope 1. There are energies for the $\Phi = 0.3$ and $\Phi = 0.4$ profiles for which the slope of $\ln s / \ln \varepsilon$ exceeds 1, but for lower Φ there are not. This coincides with the occurrence of NDC in Figure 2b for only the $\Phi = 0.3$ and $\Phi = 0.4$ profiles. It can be deduced from (4) that during NDC, a small increase E is accompanied by a rapid increase in $\tilde{\nu}_m$ resulting in an overall decrease in the drift velocity.

4. Conclusion

A fluid equation treatment of swarms in structured media was undertaken. Semi-analytic relations for drift velocity and mean energy were found from moments of the Boltzmann equation using MTT. By considering a simple steady-state situation, analytic expressions for the drift velocity and Wannier energy relation in structured media were derived. A condition for the occurrence of the new phenomena of structure-induced NDC was presented. Using a basic numerical code, the relationships between mean energy and drift velocity were investigated using the Percus-Yevick model to simulate structure. Excellent agreement was found between the numerical results and theory, and structure-induced NDC was clearly demonstrated.

Acknowledgments

The authors would like to thank the support of the Australian Research Council and the Centre for Antimatter-Matter studies.

References

[1] Cherry S, Sorensen J and Phelps M 2003 *Physics in nuclear medicine* (Philadelphia: Saunders)
[2] Cohen M and Lekner J 1967 *Phys. Rev.* **158** 305
[3] Hove L 1954 *Phys. Rev.* **95** 249
[4] Robson R 2006 *Introductory Transport Theory for Charge Particles in Gases* (Singapore: World Scientific)
[5] Viehland L and Mason E 1975 *Ann. Phys.* **91** 499
[6] White R and Robson R 2009 *Phys. Rev. Lett.* **102** 230602
[7] Verlet L and Weis J 1971 *Phys. Rev. A* **5** 939
[8] Robson R and White R *Transport Theory for Charged Particles in a Structured Medium* (In Preparation)
[9] Robson R 1986 *J. Chem. Phys.* **85** 4486
[10] Boyle G 2009 *Positron Transport in Soft Condensed Matter* (Hons. thesis: James Cook University)
[11] Robson R 1984 *Aust. J. Phys.* **37** 35
[12] Crompton R, Petrovic Z and Haddad G 1984 *Aust. J. Phys.* **37** 23
[13] Megen W and Pusey P 1991 *Phys. Rev. A* **43** 5429

Journal of Physics: Conference Series **262** (2011) 012009 doi:10.1088/1742-6596/262/1/012009

Monte carlo simulation of positron induced secondary electrons in thin carbon foils

L H Cai, B Yang, C C Ling, C D Beling and S Fung*

Department of Physics, The University of Hong Kong, Pokfulam Road, Hong Kong S.A.R., PR China

E-mail address: sfung@hkucc.hku.hk

Abstract. Emission of secondary electrons induced by the passage of low energy positrons through thin carbon foils was studied by the Monte Carlo method. The positron and electron elastic cross sections were calculated by partial wave analysis. The inelastic positron-valence-electron was described by the energy loss function obtained from dielectric theory. The positron-core-electron interaction was modelled by the Gryzinski's excitation function. Positron transport inside the carbon foil was simulated in detail. Secondary electrons created by positrons and high energy secondary electrons through inelastic interactions were tracked through the foil. The positron transmission coefficient and secondary electron yielded in forward and backward geometry are calculated and dependences on positron energy and carbon foil thickness are discussed.

1. Introduction

Monte Carlo methods have been extensively used to simulate the interaction between electrons with solids for several decades [1,2] and it has been applied to study the interactions between positron and aluminium [3, 4], copper [3] and gold [4]. The start signal is triggered by the secondary electrons generated by positrons passing through a thin carbon foil [5]. Although secondary electrons from thin carbon foil have been used in various types of time of flight mass spectrometers to provide a start signal, most of these experiments are induced by ions [6, 7]. Few results on positron induced secondary electron emission from carbon foil have been published by now.

In present work, we first describe the elastic and inelastic cross sections that are used for the Monte Carlo simulation. We calculated the secondary electron yield (SEY) and positron transmission coefficient of varied foil thickness. The energy distribution and angular distribution of forward positrons and secondary electrons are calculated.

2. Method

The Mott cross-section was adopted in the present simulation, for which the scattering induced phase shift of the electron spin was obtained by solving the Dirac equation [8]. Considering the correlation-polarization potentials at low energies and the exchange potential for electrons, the differential elastic cross sections of positron and electron are calculated [9] (see figure 1). Because the potentials for electron and positron are significantly different, the total elastic cross sections varied much at low energy (especially lower than 2 keV), as showed in figure 2. With increasing energy, the difference in elastic cross sections between electron and positron became smaller.

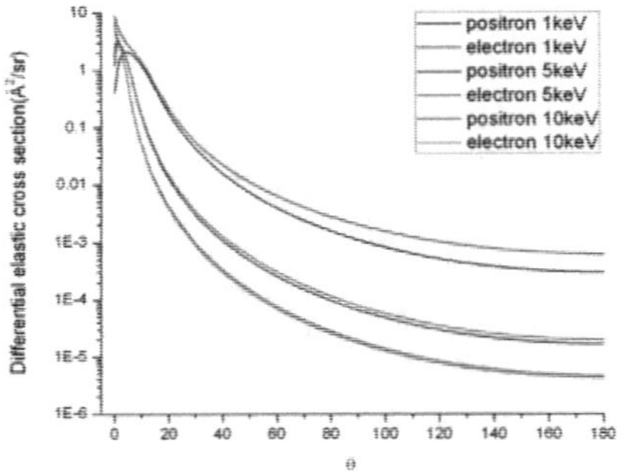

Figure 1. Differential elastic cross section of electron and positron in carbon.

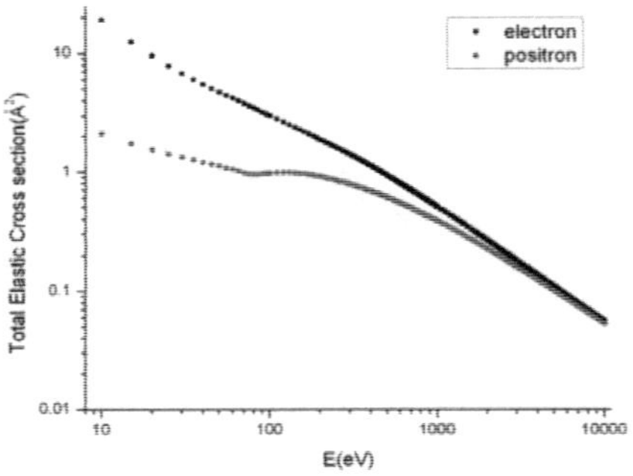

Figure 2. Total elastic cross section of various energies electron and positron in carbon sample.

For low energies, the conduction band excitations via plasmons excitation and electron hole pair creation are the dominant energy loss processes. Dielectric function model are extensively used to describe this process [10]. In this model, Energy Loss Function (ELF) $\mathrm{Im}\left\{\frac{-1}{\varepsilon(q,\omega)}\right\}$, determines the probability function of transferring energy $\hbar\omega$ and moment $\hbar q$ to the target sample. The Inelastic Mean Free Path (IMFP) can be calculated from integrating over particular region of (q,ω) plane, i.e.

$$\tau(E,\omega)=\frac{d\lambda^{-1}}{d\omega}=\frac{1}{\pi a_0 E}\int_{q^-}^{q^+}\frac{dq}{q}\mathrm{Im}\left\{\frac{-1}{\varepsilon(q,\omega)}\right\} \qquad (1)$$

$$\lambda_{\mathrm{in}}^{-1}=\int \tau(E,\omega)d\omega \qquad (2)$$

where $\hbar q^{\pm}=\sqrt{2m}(\sqrt{E}\pm\sqrt{E-\hbar\omega})$ is originated from the energy-momentum conservation.

The mean energy loss per unit length or stopping power, is given by [10]: $S=\int \omega\tau(E,\omega)d\omega$.

Ashley [10] proposed a simple momentum-dependent ELF, in which the influence of single particle interaction was considered for large momentum transfer. For small momentum transfer, interaction with plasmons was considered. The plasmons interaction was obtained from the optical data which was available in Ashley et. al. [11].

Considering the inner shell excitation, the total inelastic cross section for intermediate and high energy positrons was obtained from the Gryzinski's binary encounter theory [12]. The binding energies used in this paper were cited from reference [13], which took the atomic shielding effect into consideration.

Here, the dielectric function model was used to describe the inelastic collisions for projectile's energy is less than 2keV. The Gryzinski's excitation function is used for the remaining high energy range. Monte Carlo modeling of positron is similar to electron transport through the solid [1, 2]. Each positron and electron excitation is tracked step by step until it leaves the foil or its energy is less than 20 eV. The energy and the ejection angle of the positrons and the secondary electrons emitted from the surface are recorded.

3. Results and discussion

Positron induced secondary electron emission from thin carbon foils was investigated with different foil thicknesses between 10nm and 40 nm. Most of the positrons would penetrate through carbon foils. The secondary electron yield (SEY) is defined as the number of secondary electrons over the number of positrons. As shown in figure 3, the SEY varied with the foil thickness. It was observed that the thinnest foil (10 nm) had the lowest value. However, the SEY of the 30 nm thickness sample had the highest SEY if the positron incident energy was not larger than 5 keV. The 40 nm foil had the highest SEY when the positron incident energy was larger than 8 keV.

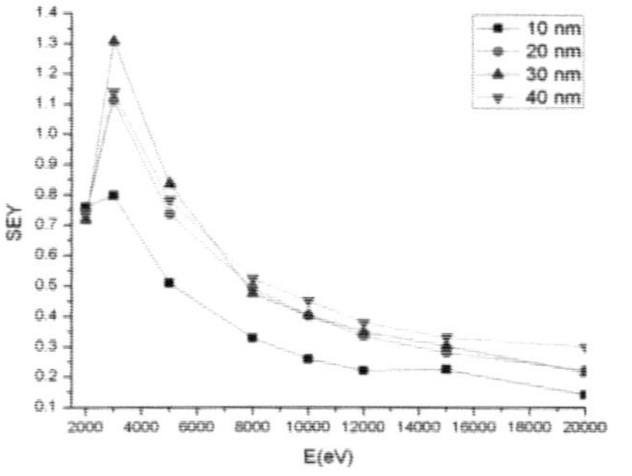

Figure 3. The thickness dependent secondary electrons yield of various energies positrons

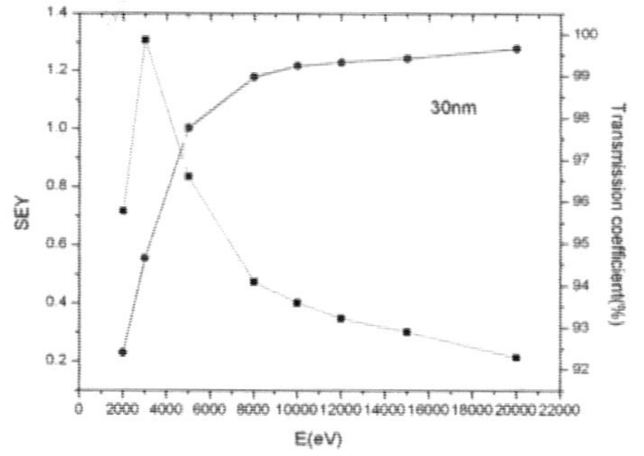

Figure 4. Positrons induced secondary electron yield and the transmission coefficient in carbon foil (30 nm thick)

Taking the 30 nm thickness case as an example, the transmission coefficient increased with the positron energy. It is worth stating that this value should be larger than experimental data, because back scattering from the surface was not considered in the simulation. The low energy positrons would have significant energy drop after passing through the foil while the high energy positrons would have little change (figure 5). Most of the positrons ejected from the foil at an angle of around 9 degrees (figure 6). The angular distribution became more dispersive as the incident positron energy decreased.

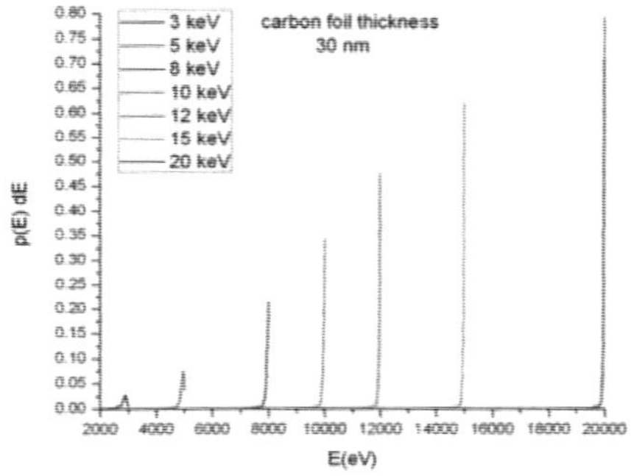

Figure 5. Normalized energy distribution of positrons passing through the thin carbon foil (30 nm thick).

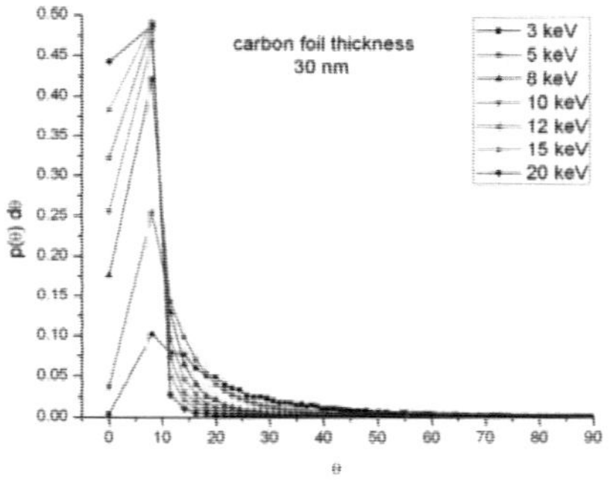

Figure 6. Normalized angular distributions of positrons passing through the thin carbon foil.

The energy of the forward emitted secondary electrons was usually lower than 60 eV. This result coincided with the findings of secondary electron spectra [14]. The energy of secondary electrons induced by different incident energies did not vary too much (figure 7). This was possibly because most of the secondary electrons emitted were confined in a thin region close to the sample surface [15].

If the incident positron is at 5 keV, the positron energy would decrease with the carbon foil thickness because of the inelastic interaction (figure 8). The foil thickness did not affect the energy distribution of the secondary electrons. Most of the secondary electrons had energy around 5 eV, as shown in figure 9. Similar to positron, the secondary electron angular distribution became more dispersive with foil thickness. Considering the secondary electrons yield and the positron energy and angular distribution, the 20 nm carbon foil seems to be the optimum choice.

4. Summary

Secondary electron induced by positrons was simulated by the Monte Carlo program. High secondary electron yields suggesting that it is promising to use the secondary electrons from carbon foils to start

Journal of Physics: Conference Series **262** (2011) 012009 doi:10.1088/1742-6596/262/1/012009

the lifetime counting. The energy of positrons decreased after penetrating through the carbon foil. The positrons' ejecting angle would deflect a little after passing through the foil. The secondary electron energy distribution was independent of the positron incident energies and the thickness of carbon foil.

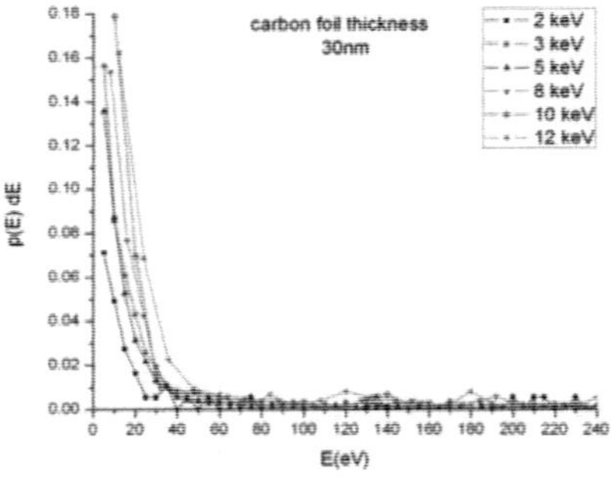

Figure 7. Normalized secondary electrons energy distribution.

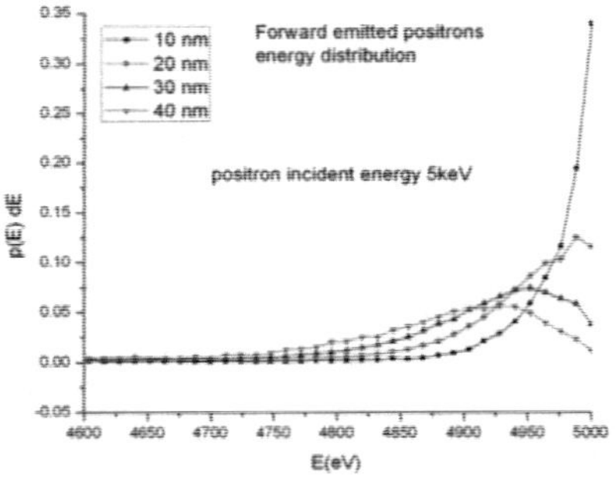

Figure 8. Normalized energy distribution of positrons passing through various thicknesses of carbon foils.

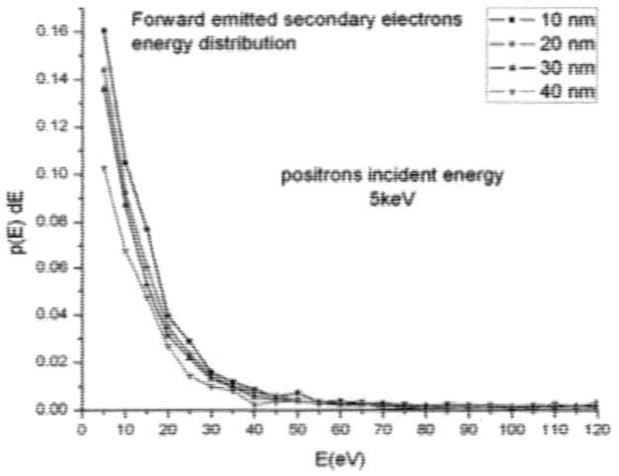

Figure 9. Normalized energy distribution of secondary electrons induced by 5 keV positrons.

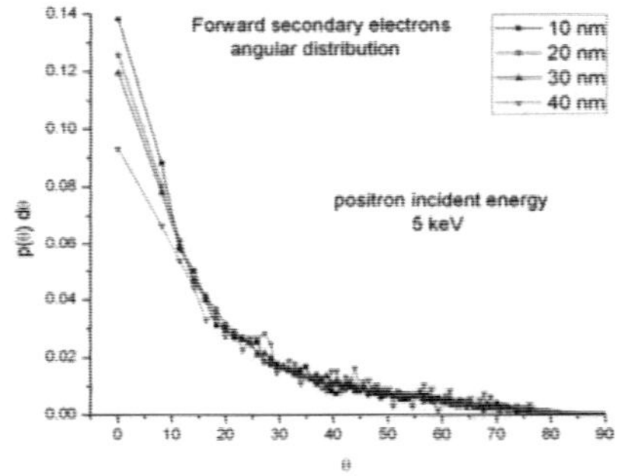

Figure 10. Normalized angular distribution of secondary electrons induced by 5 keV positrons.

References

[1] Shimizu R and Ding Z J 1992 *Rep. Prog. Phys.* **55** 487
[2] Dapor M 2003 *Electron-beam interactions with solids* (Berlin: Springer)
[3] Valkealahti S and Nieminen R M, 1983 *Appl. Phys.A* **32** 95
[4] Bentabet A and Fenineche N 2009 *J. Phys.: Condens. Matter* **21** 095403
[5] Chen D, Zhang J D, Cheng C C, Beling C D and Fung S 2008 *Appl. Surf. Sci.* **255** 122
[6] Ogawa H, Ohata T,Sakamoto N and Kaneko T 2007 *Phys. Rev. A* **76** 532
[7] Ritzau S M and Baragiola R A 1998 *Phys. Rev. B* **58** 2529
[8] Mott N F 1929 *Proc. R. Soc. Lond. A* **124** 425
[9] Salvat F, Jablonski A and Powell CJ 2005 *Comput. Phys. Commun.* **165** 157
[10] Ashley J C 1990 *J. Electron. Spectrosc. Relat. Phenom.* **50** 323
[11] Ashley J C 1979 *Thin Solid Films* **60** 361
[12] Gryziński M 1965 *Phys. Rev.* **138** A305
[13] Thomas D 1997 *Binding energies of electrons in atoms from H (Z=1) to Lw (Z=103)* Available from: http://www.chembio.uoguelph.ca/educmat/atomdata/bindener/elecbind.htm.
[14] Joy D C, Prasad M S and Meyer H M 2004 *Journal of Microscopy* **215** 77
[15] Knights A P and Coleman P G 1995 *Appl. Surf. Sci.* **85** 43

Journal of Physics: Conference Series **262** (2011) 012010 doi:10.1088/1742-6596/262/1/012010

DWBA calculation of positron impact ionization of argon

R I Campeanu[1], M Alam[1]
[1] Department of Physics and Astronomy, York University, Toronto, Canada M3J 1P3

E-mail: campeanu@yorku.ca

Abstract
The ionization of the 3p and 3s orbitals of argon by 200 eV incident positrons is studied with the DWBA approximation. Our TDCS results for the 3p orbital ionization are found to be in good agreement with recent CDW-EIS data.

1. Introduction

Triple differential cross sections for electron and positron impact single ionization of argon remains a formidable theoretical and computational challenge.

Positron experiments measuring TDCS are very new. In 1998 the University College London group measured TDCS in the forward direction for H_2 [1] and more recently the Missouri University of Science and Technology group obtained the full variation of TDCS with the ejected electron angles for argon [2]. The accuracy of the positron impact TDCS experiments is much smaller than in some of the recent electron impact TDCS experiments [3].

Theoretical work for positron impact TDCS in H_2 employed the 3C model [4-6]. This work managed to confirm the presence of the experimentally observed Electron Capture in the Continuum (ECC) peak. Theory and experiment disagreed in the position of the EEC peak.

For argon it is expected the need for theoretical models which contain the distortion and the screening of the positron and ejected electron wave functions. In the electron impact case the following theoretical models were employed: the distorted-wave Born approximation (DWBA) [7], the extension of DWBA and 3C models to 3DWBA [8], the Continuum Distorted Wave - Eikonal Initial State (CDW-EIS) model [9] and the distorted wave Born plus R matrix (RM2) model [10]. In spite of the high level of sophistication in some of these models the current status of agreement between theory and experiment in the electron impact case is not satisfactory [9].

Journal of Physics: Conference Series **262** (2011) 012010 doi:10.1088/1742-6596/262/1/012010

Very recently the CDW-EIS model was applied to the calculation of TDCS for positron impact ionization of the 3p orbital of argon [2]. The theoretical variation of TDCS with the ejected electron angles was compared with the experimental data. In this paper we study the same phenomenon employing a DWBA model.

2. Theory

The DWBA approximation is described very well in a number of papers, such as the one by Kheifets et al [11]. For relatively slow projectiles the distorting and screening effects on the continuum electron and positron states are very important and most of our work was dedicated to accounting for these effects.

For the ejected electron continuum states we included the attractive Hartree-Fock static potential of Ar+ and a Furness-McCarthy (FM) exchange potential. In this work we assumed that the exchange is triplet for all bound electrons. For the scattered positron continuum states there will be no exchange with the bound electrons, and the Ar+ static potential changes its sign relative to the electron case. For the incident positron continuum states we included the Hartree-Fock static potential of Ar, plus the target polarization potential.

We found that the convergence of the partial waves contributions was slow. In the orbital angular momentum expansions we needed to include 10 ejected electron continuum states and 50 incident positron continuum states.

In this work we considered the post-collision interaction (PCI) between the scattered positron and the ejected electron through a Gamow-type factor [12]. This factor lowers the TDCS in the electron impact case and raises them in the positron impact case.

3. Results and discussion

Figure 1 presents our results for positron impact ionization of the 3p and 3s orbitals of argon. We compare them with the recent CDW-EIS data of [2] for the 3p orbital ionization. To match the CDW-EIS results our DWBA data were divided by 1.447.

Figure 1 shows that the agreement between two theoretical models in the 3p orbital case. Unlike in the electron impact case, for positron impact ionization of argon the variation of TDCS with the ejected electron angles is very smooth showing a single binary peak at around 50 degrees and a recoil peak at around 220 degrees. The DWBA peaks are at slightly smaller angles than the CDW-EIS peaks. Also it is interesting that the 3p and 3s ionization TDCS have the peaks at the same angles.

A comparison with the experimental results can be made only after the theoretical results are convoluted over the experimental parameters. This work was done for the CDW-EIS 3p ionization data and it demonstrated that the convolution shifts the binary peak to 67 degrees and the recoil peak to 247 degrees, in quite good agreement with the experimental data.

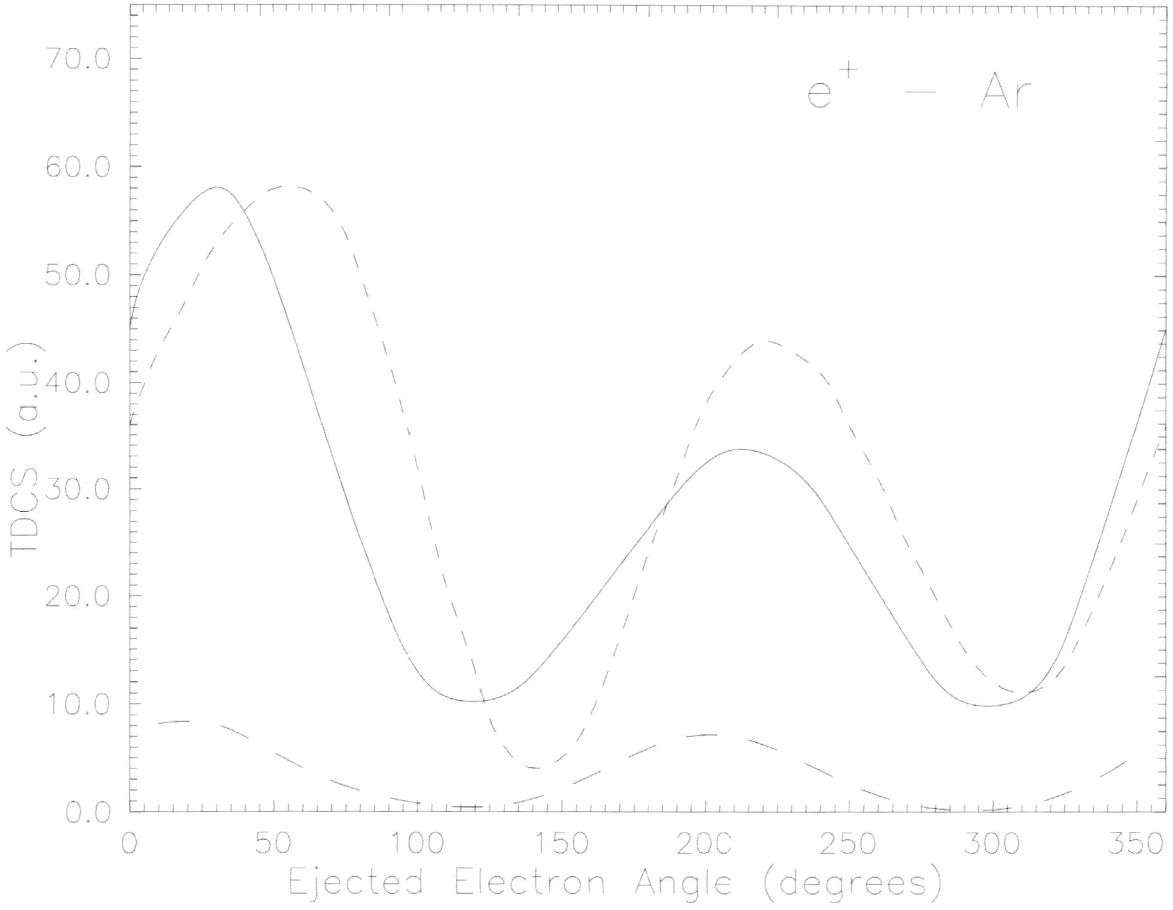

Figure 1. TDCS in atomic units for coplanar 200 eV positron impact ionization of the 3p and 3s shells of argon as a function of the ejected electron angles. The positron was scattered at 3 degrees and the ejected electron had 5.4 eV. For the 3p ionization the solid line corresponds to our DWBA calculation and the dashed line to the CDW-EIS calculations [2]. The long dashed curve corresponds to our DWBA data for the 3s ionization. Our DWBA data were divided by a factor of 1.447 to match the CDW-EIS results.

4. Conclusions

Our work shows that the DWBA model produces the same variation of TDCS with the ejected electron angle as the CDW-EIS model for the positron impact ionization of 3p orbital of argon. Our DWBA model included a complete account of the screening and distortion effects in the representation of the positron and ejected electron continuum states and an inclusion of the PCI effect.

Acknowledgements

This research was supported by a grant from the Natural Sciences and Engineering Council of Canada (NSERC). We want to thank R.Dubois, S.Otranto, R.P.McEachran and S.Ward for preliminary data and useful discussions.

References

[1] A Kover, G Laricchia, *Phys.Rev.Lett.* **80**(1998) 5309
[2] O G. de Lucio, S Otranto, R E Olson, R D Dubois, *Phys.Rev.Lett.* **104** (2010) 163201
[3] M Haynes, B Lohmann, *Phys.Rev. A* **64** (2001) 044701
[4] J Berakdar, *Phys.Rev.Lett.* **81**(1998) 1393
[5] J Fiol, V D Rodriguez, R Barrachina, *J.Phys.B* **34** (2001) 933
[6] A Benedek and R I Campeanu, *J.Phys.B* **40** (2007) 1589
[7] D A Biava, H P Saha, E Engel, R M Dreizler, R P McEachran, M A Haynes, B Lohmann, C T Whelan, D H Madison, *J.Phys.B* **35** (2002) 293
[8] A Prideaux, D H Madison, *Phys.Rev. A* **67** (2003) 052710
[9] S Otranto, *Phys.Rev. A* **79** (2009) 012705
[10] A Prideaux, D H Madison, K Bartschat, *Phys.Rev. A* **72** (2005) 032702
[11] A S Kheifets, A Naja, E M Staicu Casagrande, A Lahmam-Bennani, *J.Phys.B* **41** (2008) 145201
[12] C T Whelan, R J Allan, J Rasch, H R J Walters, X Zhang, J Roder, K Jung, H Ehrhardt, *Phys.Rev .A* **50** (1994) 4394

Journal of Physics: Conference Series **262** (2011) 012011 doi:10.1088/1742-6596/262/1/012011

Status report on the setup for the decay rate measurement of the negative positronium ion

Hubert Ceeh[1], Stefan Gärtner[2], Christoph Hugenschmidt[13], Klaus Schreckenbach[1], Dirk Schwalm[4], Peter Thirolf[2]

[1] Physik Department E21, Technische Universität München, D-85747 Garching
[2] Fakultät für Physik, Ludwig Maximillians Universität, D-85747 Garching
[3] Forschungsneutronenquelle Heinz Maier-Leibnitz FRM II, D-85747 Garching
[4] Max-Planck-Institut für Kernphysik, D-69029 Heidelberg

E-mail: `hubert.ceeh@frm2.tum.de`

Abstract. The negative positronium ion Ps^- is a bound system consisting of two electrons and a positron. Its three constituents are point-like and stable leptonic particles with equivalent mass, which are only subjected to the electroweak and the gravitational force. Hence Ps^- represents an ideal object to study the quantum mechanics of three-body systems.
We present a status report on a new a measurement setup, which was used to perform several high precision lifetime measurements at the NEPOMUC facility. The combined results of systematic test measurements are discussed and an outlook to future measurements as well as preliminary results of the most recent beam time will be given.

1. Introduction

The Ps^- apparatus, which was used in the present study was designed by Frank Fleischer in Heidelberg to produce Ps^- and to measure its decay rate [1]. It was successfully applied to measure the Ps^- decay rate in 2006 with unprecedented accuracy [2]. Since a low intensity positron source was used the measurement was limited by statistics. Therefore, the setup was slightly modified and transferred to the NEPOMUC facility in Munich in order to increase statistics. However, first experiments at NEPOMUC showed systematic deviations of the measured value for the decay rate compared to previous measurements. After the installation of the remoderator, which decreased the primary beam energy from ≈ 1000 eV to about 20 eV, further systematic examinations revealed an unexpected dependence of the measured decay rate on the energy of the incident positron beam. The technical details of the advanced Ps^- setup, that allowed us to eliminate this dependence are presented.

2. Experimental Setup

The beam-foil technique first presented by A. Mills was applied to produce Ps^- [3]. The principle of this measurement can be described as follows: Ps^- are produced and drifted with a certain energy over a variable decay gap. The intensity I of the surviving Ps^- ions is measured as a function of the gap width d: $I \propto e^{-\mu \cdot d}$. Hence we get the decay constant μ, which can be converted into the decay rate Γ as described below. The high-intensity positron beam provided

Journal of Physics: Conference Series **262** (2011) 012011 doi:10.1088/1742-6596/262/1/012011

by the NEPOMUC facility is directed onto a DLC foil[1] with an incident positron energy of 750 − 1200 eV. The positron subsequently picks up two electrons to create Ps^-. After the first electron is captured the emerging positronium can acquire an additional electron by inelastic scattering with a carbon atom. This is possible if the impact energy is insufficient to ionize the carbon atom but suffices to lift one of the outer shell electrons into the bound state of Ps^- (for a more detailed discussion see [1] or [4]). Therefore, the starting energy E_0 of the Ps^- ions is limited by the binding energy for the second electron, which is calculated to $E_b = 0.326675$ eV [5]. This could be confirmed by previous experiments where E_0 was found to be < 0.3 eV [1].

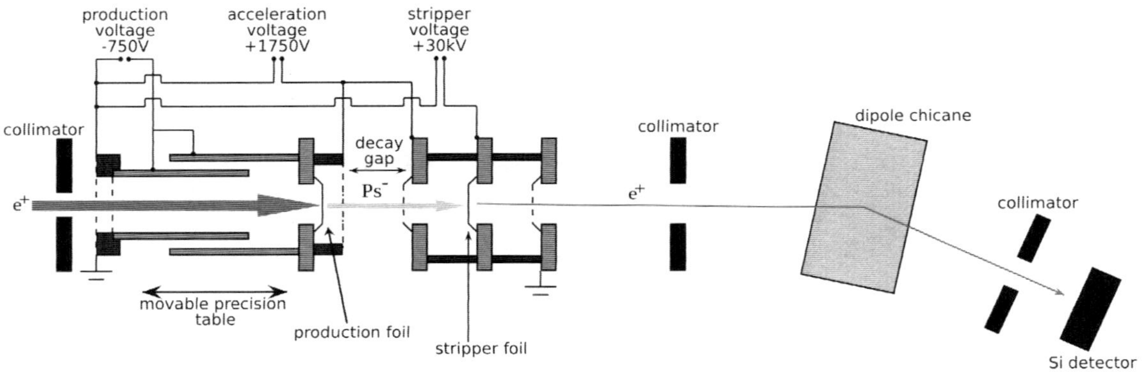

Figure 1. Schematics of the Ps^- setup. From left to right: Production and acceleration device, field-free decay gap and tandem-like stripping and detection unit.

The produced Ps^- ions are accelerated by an electric field between the production foil and a fine-meshed grid to their final energy of $(U_{accel} - U_{prod})e = 2500$ eV. Passing the acceleration grid the Ps^- ions enter the field-free decay volume which is terminated by a second fine-meshed grid. This field free decay volume represents the major difference to previous experiments, where the Ps^- acceleration took place over the whole decay gap. The variable distance between these two grids defines the length of the decay gap which can be adjusted by a high precision linear positioning stage.

The surviving Ps^- ions are accelerated towards a second DLC foil, to which a voltage of $U_{strip} = +30$ kV is applied. When the Ps^- ions impinge on the foil, the two electrons are stripped off and the remaining positron is again accelerated towards a grounded grid. At the exit of this tandem-accelerator-like setup the positrons have an energy of 40 keV and are magnetically guided towards a silicon particle detector. In order to increase the signal-to-noise ratio a chicane is mounted in front of the detector, which can only be passed by positrons with the right energy of 40 keV. Secondary electrons and ions are deflected and cannot reach the detector. This allowed us to increase the signal-to-noise ratio by a factor of 2 compared to the setup without the dipole. Consequently, a 40 keV positron signal in the detector serves as a fingerprint for a surviving Ps^- ion. The count rate of the surviving Ps^- ions is measured for a set of different gap distances in order to determine the decay constant μ.

The Ps^- vacuum decay rate Γ can hence be calculated from the measured decay constant μ:

$$\Gamma = \mu\beta\gamma c \tag{1}$$

[1] Diamond Like Carbon

Journal of Physics: Conference Series **262** (2011) 012011 doi:10.1088/1742-6596/262/1/012011

with

$$\beta = \sqrt{1 - \frac{1}{\left(1 + \frac{(U_{\mathrm{accel}} - U_{\mathrm{prod}})\mathrm{e} + E_0}{3 m_{\mathrm{e}} c^2}\right)^2}}\,, \tag{2}$$

the Lorentz factor $\gamma = (1 - \beta^2)^{-\frac{1}{2}}$, e the electron charge and m_{e} the electron mass.

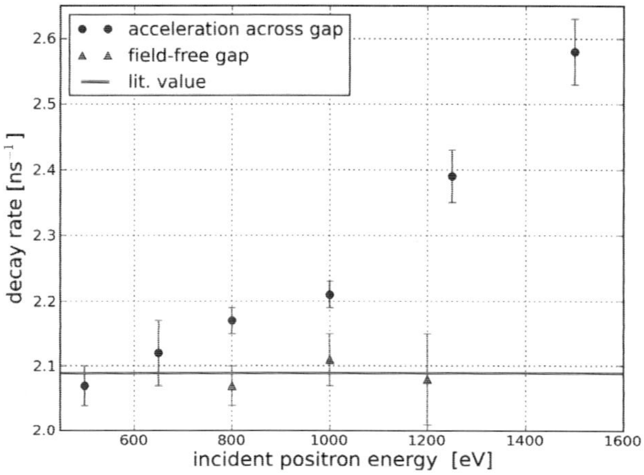

Figure 2. Dependence of the measured decay rate on the incident positron energy E (energy due to production voltage plus primary beam energy): values obtained with field-free decay gap are indicated by △, values obtained with Ps$^-$ acceleration over the decay gap are indicated by ○ .

3. Results

First measurements revealed an increase of the decay rate with the incident positron energy (see Fig.2). Using the setup with the field-free decay gap we were able to eliminate this dependence, as can be seen in Fig. 2. The results of a previous measurement of the Ps$^-$ decay rate Γ performed by Fleischer in Heidelberg could be confirmed within the error bounds. In two days of beam time at the NEPOMUC facility we were able to measure the decay rate with an accuracy of 1.1% [8] to:

$$\Gamma = 2.083(23)\,\mathrm{ns}^{-1}$$

This value is in good agreement with recent theoretical calculations for the decay rate of $\Gamma_{\mathrm{Th}} = 2.087963(12)\,\mathrm{ns}^{-1}$ [7] as well as with the value obtained in Heidelberg $\Gamma_{\mathrm{Exp}} = 2.089(15)\,\mathrm{ns}^{-1}$[2]. It was shown that the new setup ensures a constant detection probability for the surviving Ps$^-$ ions as well as a lower sensitivity to external perturbations such as stray magnetic fields, since all acceleration stages now operate at fixed positions in space and therefore with constant electric fields. This allows us also to resolve the problem with dependence of the measured value for the decay rate on the incident positron energy (see Fig. 2).

4. Discussion

The dependence of the measured decay rate on the incident positron beam energy is attributed to positrons which are transmitted through the production foil. The transmission probability itself, which depends on the positron energy, thickness and qulaity of the foil, is assumed to be constant. However, transmitted positrons are reflected in the acceleration[2] field and can

[2] This field is called "acceleration" field because it is used to accelerate the Ps$^-$ ions. For positrons it acts as a retarding field.

be reflected back onto the production foil with a certain probability, where they contribute to the production of Ps^-. Therefore, the probability that a transmitted positron produces Ps^- is dependent on the reflection efficiency, which varies with the strength of the acceleration field.

While the transmission probability is constant throughout the measurement, the reflection efficiency and therefore the number of transmitted positrons, which are reflected back onto the foil, varies with the gap width. Hence the Ps^- production rate can change with the gap width, if the transmitted positrons have a significant contribution to the overall Ps^- production rate. Theoretical calculations and simulations [8] suggest that this effect has no significant contribution for incident positron energies below 800 eV and for short gap widths, which was both the case in previous measurements [4], [2]. However for higher energies the contribution of transmitted positrons to overall Ps^- production rate is clearly not neglectable.

However, this effect can be completely overcome by using a field-free decay gap, which allows to keep the acceleration field constant as shown above. As can be seen in Fig. 2 the measured decay rate is independent of the incident positron energy when the setup with the field free decay gap is used. Hence an error due to a varying Ps^- production rate with the gap distance is eliminated by the new setup.

5. Outlook

The preliminary evaluation of a recently completed beam time of about 10 days shows that with the increased statistics we are able to increase the accuracy for the decay rate measurement by a factor of 2-3 compared to previous measurements:

$$\Gamma = 2.0869(67)\,\mathrm{ns}^{-1}$$

This measurement was performed using the setup with the field-free decay gap, which we presented here. A complete discussion of the measurement and the data and error treatment is to be published [9].

Furthermore, an experiment for the Ps^- photo detachment and the production of a mono energetic ortho-positronium beam is already in preparation and will soon be operational.

6. Acknowledgement

We would like to thank our colleagues from the FRM II positron group and the LMU for their encouragement during the exhausting beam times, as well as for the many interesting and fruitful discussions. We also would like to thank the DFG for funding this project (HA1101/13-1).

References

[1] Fleischer F, 2005, PhD Thesis, Max-Planck-Institut für Kernphysik Heidelberg
[2] Fleischer F et al., 2006, *Phys. Rev. Lett.* **96**, 063401
[3] Mills A P, 1981, *Phys. Rev. Lett.* **46**, 717
[4] Mills A P, 1983, *Phys. Rev. Lett.* **50**, 671
[5] Drake G W and Grigorescu M, 2005 *J. Phys. B* **38**, 3377
[6] Mills A P et al., 1990, *NASA Conference Publication*, **3058**, p. 213
[7] Puchalski M, Czarnecki A, Karshenhboim S, 2007, *Phys. Rev. Lett.* **99**, 203401
[8] Ceeh H, Diploma Thesis, 2009, *Universitätsbibliothek TU München*
[9] *to be published in P.R.A*

Development of high intensity source of thermal positrons APosS (Argonne Positron Source)

Sergey Chemerisov and Charles D. Jonah

Chemical Science and Engineering Division, Argonne National Laboratory, Argonne, IL, 60439 USA

chemerisov@anl.gov

Abstract. We present an update on the positron-facility development at Argonne National Laboratory. We will discuss advantages of using low-energy electron accelerator, present our latest results on slow positron production simulations, and plans for further development of the facility. We have installed a new converter/moderator assembly that is appropriate for our electron energy that allows increasing the yield about an order of magnitude. We have simulated the relative yields of thermalized positrons as a function of incident positron energy on the moderator. We use these data to calculate positron yields that we compare with our experimental data as well as with available literature data. We will discuss the new design of the next generation positron front end utilization of reflection moderator geometry. We also will discuss planned accelerator upgrades and their impact on APosS.

1. Introduction

Positron spectroscopies have huge unrealized potential in many fields of science, including chemistry, physics, materials science, surface science, biological sciences and nanoscience [1-7]. Advances in the use of positrons have been strongly inhibited by the low intensity of the available sources of thermalized positrons. Higher intensity positron beams are required to utilize new techniques such as positron microprobe, positron holography, to carry out gravitation experiments with antimatter and to do chemistry with antimatter [1]. The development of traps to confine positrons at high densities [2, 8], which have the potential to be portable and thus decouple positron production from positron applications also requires an intense source of positrons to allow accumulation of the high number of positrons in a reasonable amount of time. APosS utilizes a 20 MeV linac accelerator for the pair production of positrons from the x-rays formed by the interaction of high-energy electrons with matter. A conventional W moderator is used for positron thermalization. At APosS we have already demonstrated 3×10^7 e^+/sec slow positron flux [10] and are in the process of an upgrade that we expect to give us 3×10^9 e^+/sec.

The upgrade consists of three parts. First, we directly couple the positron target to the linac, second, we will increase the moderation efficiency of the fast positrons using reflection geometry and monocrystalline tungsten moderator and third, we will decelerate the positrons for better moderation efficiency. We are also working on an energy and power upgrade of the accelerator that will contribute to increased intensity of the APosS.

2. Present configuration of the positron beam line and proposed changes

Figure 1 shows the present configuration and the new configuration (yellow lines) of APosS. Presently, we are using positron beam line at table 2. A new positron front end will be installed at table 1. A transport line will connect the new converter/moderator chamber to the existing transport system at the table 2. By eliminating windows between the accelerator and the positron converter, we will be able to deliver maximum power (presently 16 kW) to the positron target without overheating the beam windows.

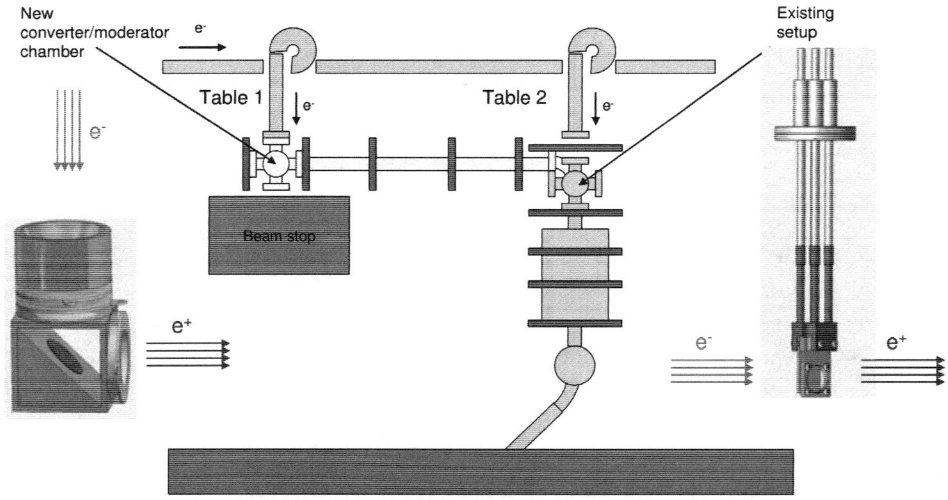

Figure 1. Present and proposed positron beam lines. On the right is the existing positron line. On the left is the new line design to utilize reflection mode of the moderation. An expansion of the new/existing converter/moderator assembly is shown on the sides

2.1. *Increase of moderator efficiency.*

The measured moderation efficiency for a tungsten moderator varies from 10^{-5} to $2x10^{-3}$ [11-13]. The value depends on geometry (transmission or reflection), crystalline structure (polycrystalline or monocrystalline), form (foil, mesh, fins etc) and thermal treatment. While no systematic study of the moderator efficiencies exists, the efficiencies reported by different authors can be summarized as follows: The efficiency of monocrystalline moderators is higher than polycrystalline moderators and reflection geometry is more favorable then transmission. Our recent calculations are consistent with the advantage of reflection geometry [15]. One can expect an increase of moderation efficiency by more than a factor of five if reflection geometry is used rather than transmission [12, 13]. With reflection geometry, it is possible to use a monocrystalline moderator, while a thin foil is needed to produce moderated positrons in transmission and a thin foil is difficult to obtain in monocrystalline form. Also, one can use higher power beams as the moderator in reflection geometry can be cooled.

2.2. *Deceleration of the fast positrons for better moderation efficiency.*

In [14, 15] we have calculated the relative moderator efficiency of a foil moderator for different energies of a positron impinged on the moderator. For these calculations, we use EGSnrc Monte Carlo code. It was assumed that the moderation efficiency would be proportional to the number of positrons whose energies were reduced below 2 keV closer to the surface than 1 μm. The geometry used for calculation shown on figure 2a. (The results are not dependent on the exact geometry chosen, as the relative distribution of the positrons in this system is independent of energy for positron energies greater than 20 keV.) Using this proxy we have calculated both relative reflection moderation efficiency and transmission moderation efficiency, and results are presented on figure 2b. We have compared those efficiencies with ones calculated if the positron energy spectrum is numerically shifted by 100 keV (to simulate electrostatic deceleration of the positrons). The 100 KeV value was chosen because one can use a short, double-gap, electrostatic decelerator and 100 kV power sources are commercially available.

Figure 2b shows the relative moderation efficiency calculated for original and shifted energy spectrum of positrons. One can expect 2.5 times increase in moderation efficiency for moderator in reflection mode due to deceleration of the positrons by 100 keV and four times for moderator in transmission mode.

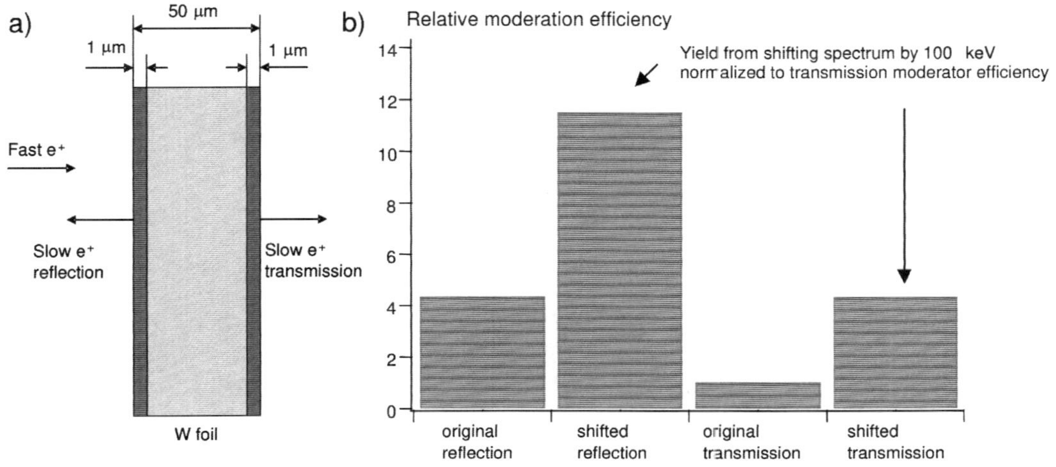

Figure 2. a) Geometry of the moderator used for EGSnrc calculations. The thickness of the W foil is 50 μm. 1μm layer at the surface was used to calculate relative moderation efficiency for reflection and transmission mode of moderation. b) Histogram of moderation efficiency relative to the original energy spectrum of positrons and moderator in transition mode.

2.3. Impact of the accelerator upgrade.

Traditionally, only high-energy electron accelerators were considered useful for efficient positron production, but recently, even low-energy 10-MeV accelerators have been considered for this purpose [17]. Where is the sweet spot for positron production using electron accelerator? To tackle this question we have used the approach similar to one we used for relative efficiency calculations. Results of the calculations were discussed in [18], and are presented on figure 3a.

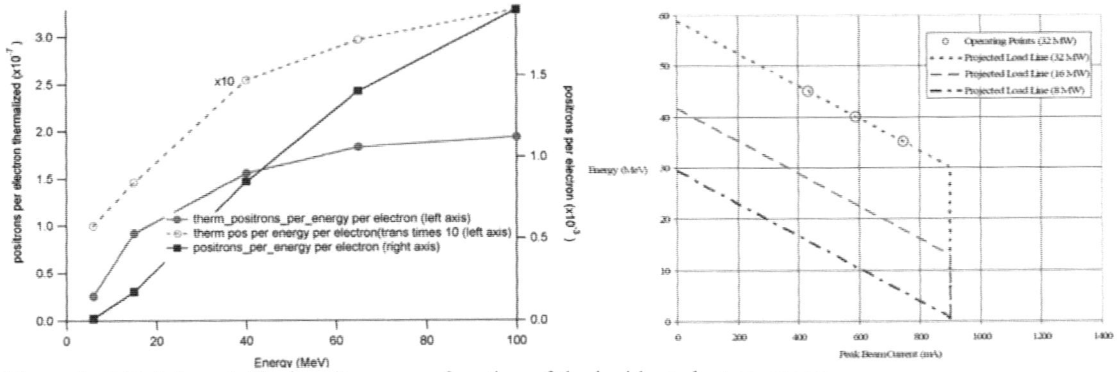

Figure 3. a) Relative yield of positrons as a function of the incident electron energy.
b) Projected accelerator upgrade parameters.

Because the efficiency of the generation process depends on the number of positrons formed for a given amount of electrical energy impinging on the target, we display our results as positrons per electron per energy. As we see, the yield of total positrons increases virtually continuously (closed

Journal of Physics: Conference Series **262** (2011) 012012 doi:10.1088/1742-6596/262/1/012012

squares) while the number of thermalized positrons appears to approach saturation at about 60 MeV both for reflected moderation (filled circles) or transmitted moderation (open circles).

We are in process of upgrading the accelerator-beam energy (projected electron beam energy presented on figure 3b. As one can see we will be at the sweet spot for efficient slow positron production ~40 MeV. The upgrade will also increase the maximum beam power on the target to 30kW. So improvements in accelerator will allow us to increase positron yield by a factor of 4 without any significant modifications to the positron setup.

3. Summary

We have updated the present status of the APosS and outlined planned upgrades of the facility. We have presented the results of the relative-moderator-efficiency calculations for moderators in transmission and reflection mode and have shown the increase of the positron yield from deceleration of the high-energy positrons.

Planned accelerator upgrades will allow us to increase APosS intensity. We are planning to achieve intensity of 10^9 e^+/s thermalized positron, with 2 cm beam diameter in 5 μs pulses with 300 Hz repetition rate.

The development of the high intensity positron beam line at APosS will allow one to conduct experiments on densities of antimatter that were not considered as even possible. Techniques like positron microprobe, positron holography and Positron Auger Spectroscopy will benefit from availability of intense positron flux. The highest density traps under design ($N \sim 10^{15}$) [9] can be filled in one day with the upgraded APosS.

Acknowledgements

We gratefully acknowledge discussions, assistance and help from Palakkal Asoka-Kumar, Ken Edwards, Richard Howell, Al Wagner and Lawrence Livermore National Laboratory. Work performed under the auspices of the Office of Science, Division of Chemical Science, US-DOE under contract number DE-AC02-06CH11357 and Air Force Research Labs.

References

[1] Jean P, Knodlseder J, Lonjou V, Allain M, Roques J P, Skinner G K, Teegarden B J and Vedrenne G 2003 *Astron. Astrophys.* **407** L55
[2] Surko C M and Greaves R G 2004 *Phys. Plasmas* **11** 2333
[3] Amoretti M *et al* 2002 *Nature* **419** 456
[4] Babrielse G *et al* 2002 *Phys. Rev. Lett.* **89** 213401
[5] Schultz P J and Lynn K G 1988 *Rev. Mod. Phys.* **60** 701
[6] Jean Y C, Mallon P E, SchraderD M (Eds.) 2003 *Principles and Applications of Positron and Positronium Chemistry* (Singapore, World Scientific Publishing Co) 235.
[7] Lee K H *et al* 1994 *Phys. Rev. Lett.*, **72** 1866-69
[8] Danielson J R, Weber T R and Surko C M 2006 *Phys. Plasmas* **13** 123502
[9] Surko C M and Greaves R G 2003 *Radiation Physics and Chemistry* **68** 419
[10] Chemerisov S *et al.* 2006 *Physica Status Solidi* (C) **4** 3957-3960
[11] Saito F *et al* 2002 *Appl. Surf. Science* **184** 13-15
[12] Vehanen A *et al.* 1983 *Appl. Phys.* A. **32** 2572
[13] Lynn K *et al* 1985 *Appl. Phys. Lett.* **47** 239-240
[14] Long J, Chemerisov S, Gai W, Jonah C D, Liu W and Wang H 2007 *IEEE Particle Accelerator Conference* 2165
[15] Chemerisov S, Jonah C D, Long J, and Gai W 2007, *Proceedings of AccApp'07* (LaGrange Park, American Nuclear Society) 326
[16] Kawrakow L 2000 *Med. Phys.* **27** 485-498
[17] Perez P and Rosowsky A 2004 *Nuclear Instr. and Meth. in Physical Research* A 532, 523-532
[18] Chemerisov S and Jonah C D 2009 *Material Science Forum* **607** 243-247

Low energy positron annihilation study of composite reverse osmosis membranes

Z Chen, K Ito, H Yanagishita, N Oshima, R Suzuki and Y Kobayashi

National Institute of Advanced Industrial Science and Technology (AIST),
Tsukuba, Ibaraki 305-8565, Japan

Email: zhezchen@gmail.com

Abstract. Positron annihilation with a slow positron beam was applied to the characterization of composite reverse osmosis membranes. The results, obtained at different positron incident energies, indicated that the membranes are asymmetric with respect to the pore structure, consisting of a thin top layer with little porosity and an underlying thick porous layer. A relationship between the longest positron lifetime near the membrane surface and the salt rejection rate was discussed in terms of the free-volume hole size for the thin top layer.

1. Introduction

Reverse osmosis (RO) membranes have been applied to the separation of various substances for water purification and chemical waste treatment. In a separation process using a RO membrane, pressurized water containing dissolved salts contacts the feed side of the membrane, and water is taken out with depletion of salt concentration on the other side. Separation of salts with RO membranes is considered to relate to molecular-level holes in the membranes, thus deeper knowledge with respect to the hole structure allows prediction of better membrane performance. Positron annihilation (PA) is known as a powerful tool for examining the hole structure of functional materials at a molecular level. [1-3]

In vacuum, *ortho*-positronium (*o*-Ps), a hydrogen-like triplet positron-electron pair, annihilates into 3γ rays with an intrinsic lifetime of 142 ns. If *o*-Ps is localized in a molecular-size hole, it annihilates with a lifetime much shorter than 142 ns through a 2γ pick-off process upon a collision with an electron at the hole wall. The *o*-Ps lifetime is documented to be well correlated with the hole dimension [4]. In addition, confinement of *o*-Ps to a smaller hole (especially sub-nm size) reduces more the probability of 3γ annihilation, a complementary process to 2γ annihilation. This enables one to analyze the layer structure of the composite membranes based on the 3γ annihilation probability observed with a variation of the positron incident energy [5-6].

In this work, PA with a variable-energy positron beam was applied to the characterization of the hole structure for the composite RO membranes, consisting of a thin top layer and an underlying thick substrate. We report results on the hole structure of RO membranes with different chemical structures of the top layers, obtained by using positron annihilation techniques, and discuss the positron results, in comparison with the salt rejection performances.

2. Experimental

Two kinds of commercially-available membranes (NTR729HF and LF10) were supplied by NITTO DENKO Corp., Japan. Both of them were obtained as composite membranes of polyvinyl alcohol and polyamide fabricated on polysulfone substrates through interfacial polymerization.

Positron annihilation γ-ray spectra were measured with a ^{22}Na source based, magnetically guided positron beam[7]. The fraction of 3γ decay annihilation was determined from the relative fraction of 3γ annihilation with energies lower than the 511-keV photo peak in the positron annihilation spectra observed at an incident energy between 0 and 10 keV. An intense pulsed-positron beam generated with an electron linear accelerator at AIST, was utilized for measuring lifetimes of positrons with an incident energy of 1.0 keV [8]. The recorded positron lifetime data were analyzed assuming three exponential components to deduce the average lifetime and relative intensity of *o*-Ps.

3. Results and Discussion

3.1. Positron annihilation γ-ray spectroscopy

Figure 1 shows the fraction of 3γ annihilation, $f_{3\gamma}$, as a function of positron incident energy for the RO membranes. For LF10, $f_{3\gamma}$ decreases with increasing incident energy from 0.6 keV to 1.0 keV. In the range of the incident energies from 1.0 keV to 2.5 keV, $f_{3\gamma}$ goes to a minimum value, meaning most *o*-Ps undergo 2γ annihilation at these incident energies. With further increasing the incident energy from 2.5 keV to approximately 5.0 keV, $f_{3\gamma}$ rises significantly, and then above 5.0 keV, LF10 membrane exhibits a flat stage in $f_{3\gamma}$ at these higher incident energies. As for NTR729HF, the lowest $f_{3\gamma}$ is observed in the energy range from 0.6 keV to 1.0 keV. With continued increase of the incident energy, $f_{3\gamma}$ increases dramatically. Fairly large $f_{3\gamma}$ are observed at the incident energies higher than 3.5 keV.

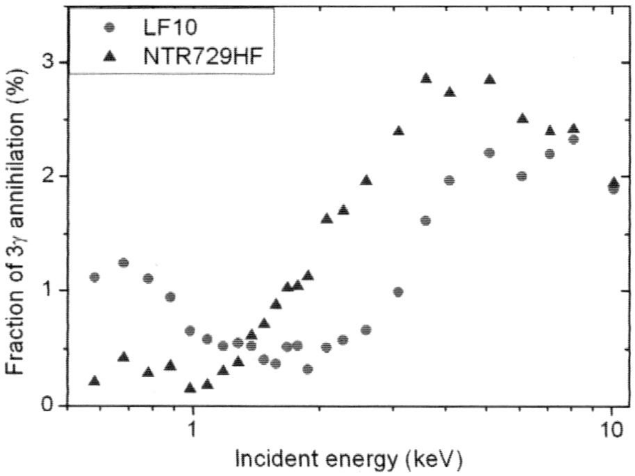

Figure 1. The fraction of 3γ annihilation $f_{3\gamma}$ as a function of incident energy for reverse osmosis membranes with different composition.

The slight increase of $f_{3\gamma}$ with reducing the incident energies below 1.0 keV for LF10 indicates some *o*-Ps formed near the surface, may escape out from the membrane, and self-annihilate in vacuum. With the increase of the incident energy, less *o*-Ps can escape out from the membrane, which results in the reduction of $f_{3\gamma}$. For both membranes, the lowest $f_{3\gamma}$, observed around an incident energy of 1.0 keV, indicates that most positrons undergo 2γ annihilation at corresponding depth in the membranes. Since 3γ annihilation occurs only when *o*-Ps exists in vacuum or in relatively large pores, the lowest $f_{3\gamma}$ indicates that most positrons annihilate in a pore-free layer at these incident energies and the higher $f_{3\gamma}$ indicates some positrons annihilate in a porous one. Thus the obtained results signify that most positrons annihilate in the dense top layer at the energies range from 1.0 keV to 2.5 keV and 0.6

keV to 1.0 keV for LF10 and NTR729HF, respectively, and both RO membranes have porous substrates, which leads to the higher 3γ annihilation at the incident energy higher than 3.0 keV.

3.2. Positron annihilation lifetime measurements and membrane performance

As seen in Figure 1, NTR729HF and LF10 membranes exhibit the lowest fraction of 3γ decay at the incident energies around 1.0 keV. This means in this energy range most positrons annihilate in the dense top layer, responsible for the separation. Therefore, we collected the positron lifetime data at the incident energy of 1.0 keV to investigate the free volume holes in the top layer for the RO membranes as well as to compare the lifetime data with the separation performance.

Figure 2 displays the raw positron lifetime data for the RO membranes acquired at an incident energy of 1.0 keV. A spectrum for Kapton which has no long-lived lifetime component is also shown for comparison. The difference in the slopes around 5 ns indicates that two RO membranes have free-volume holes with different sizes. Weak satellite peaks appear on the tail of the lifetime data, which may be due to the bunching of some unchopped slow positrons. In our measurements using a positron pulsing system, positrons from the accelerator are chopped with 20 MHz and further bunched into narrow-width pulses with a frequency higher than 20 MHz. During the data accumulation, a small number of unchopped positrons may be bunched continuously, resulting in the background with a time structure. To take into account this effect at analysis, lifetime data for Kapton were used as a background reference. The recorded data were confirmed to be well resolved into three lifetime components with a reasonable variance of fit.

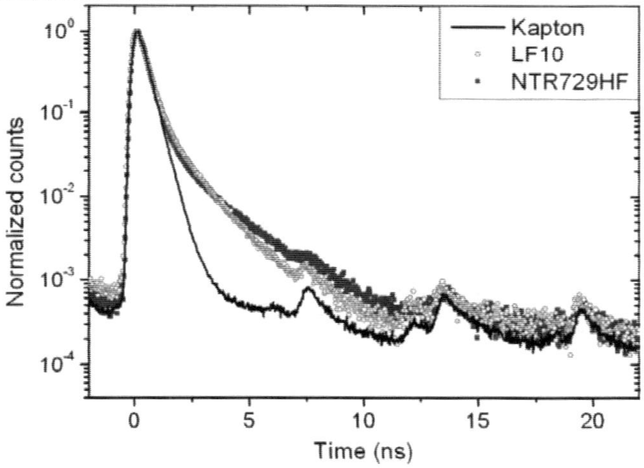

Figure 2. The positron lifetime data recorded with an incident energy of 1.0 keV for reverse osmosis membranes with different composition and Kapton reference.

Table 1. Average lifetime τ_3 and relative intensity I_3 of *o*-Ps, free-volume hole size, and NaCl rejection for the composite RO membranes

	τ_3 (ns)	I_3 (%)	Free volume hole size (nm)	NaCl rejection (%)
LF10	1.32	23.6	0.21	99.5
NTR729HF	2.12	14.9	0.30	75.0

From the obtained *o*-Ps lifetimes, the free-volume hole sizes of the top layers were calculated based on the Tao-Eldrup model [9,10], and are presented in Table 1 with the average lifetime τ_3 and relative intensity I_3 of *o*-Ps. The free-volume hole sizes in the top layer of NTR729HF and LF10 are 0.30 nm and 0.21 nm, respectively, and the hole of the former membrane is found about 1.5 times larger than the latter.

The most important indicator of the separation membrane performance is the rejection of specified salts. The desalination by RO membrane is normally explained in terms of Donnan and/or size exclusion effects. If the holes involved in the separation process are far smaller than a target salt, the transportation of the salt ions would be effectively restrained. In Table 1, the values of NaCl rejection for both membranes are also listed, which shows the separation performance of LF10 is much higher than NTR729HF. This result is consistent with the quantified hole sizes in the top layer of the two membranes, that is, LF10 with a smaller free-volume prevents salt ions more than NTR729HF.

4. Summary

The low-energy positron annihilation techniques have been applied to the characterization of the composite RO membranes of LF10 and NTR729HF. The obtained results suggested that both the membranes consist of a dense surface top layer on a porous substrate and an important role in the separation performance is played by the free volume sizes in the top layer. A combined use of the low-energy positron annihilation lifetime and γ-ray techniques was demonstrated to be useful for examining the hole structure of reverse osmosis membranes.

Acknowledgment

Prof. Y.Nagashima and his students of Tokyo University of Science are appreciated for the preparation of the positron moderator. Dr. Brian O'Rourke of AIST is appreciated for the suggestion to the manuscript. This work was supported by NEDO.

References

[1] Nagel C, Gunther-Schade K, Fritsch D, Strunskus T and Faupel F 2002 Free volume and transport properties in highly selective polymer membranes *Macromolecules* 35 2071-2077

[2] Wang CL, Weber MH, Lynn KG and Rodbell KP 2002 Nanometer-scale pores in low-k dielectric films probed by positron annihilation lifetime spectroscopy *Appl Phys Lett* 81 4413-4415

[3] He CQ, Muramatsu M, Ohdaira T, Oshima N, Kinomura A, Suzuki R and Kobayashi Y 2007 Tunable pores in mesoporous silica films studied using a pulsed slow positron beam *Radiat Phys Chem* 76 204-208.

[4] Eldrup M, Lightbody D and Sherwood JN 1981 The temperature-dependence of positron lifetimes in solid pivalic acid *Chem Phys* 63 51-58

[5] Jean YC, Hung WS, Lo CH, Chen H, Liu G, Chakka L, Cheng ML, Nanda D, Tung KL, Huang SH, Lee KR, Lai JY, Sun YM, Hu CC and Yu CC 2008 Applications of positron annihilation spectroscopy to polymeric membranes *Desalination* 234 89-98

[6] Cano-Odena A, Vandezande P, Hendrix K, Zaman R, Mostafa K, Egger W, Sperr P, De Baerdemaeker J and Vankelecom IFJ 2009 Probing the molecular level of polyimide-based solvent resistant nanofiltration membranes with positron annihilation spectroscopy *J Phys Chem B* 113 10170-10176

[7] Kobayashi Y, Zheng W, Chang TB, Hirata K, Suzuki R, Ohdaira T and Ito K 2002 Nanoporous structure of sputter-deposited silicon oxide films characterizaed by positronium annihilation spectroscopy *J Appl Phys* 91 1704-1706

[8] Suzuki R, Kobayashi Y, Mikado T, Ohgaki H, Chiwaki M, Yamazaki T and Tomimasu T 1991 Slow positron pulsing system for variable energy positron lifetime spectroscopy *Jpn J Appl Phys Pt B Lett* 30 L532-L534

[9] Tao SJ 1972 Positronium annihilation in molecular substances *J Chem Phys* 56 5499-5510

[10] Eldrup M, Vehanen A, Schultz PJ, Lynn KG 1985 Positronium formation and diffusion in crystalline and amorphous ice using a variable-energy positron beam *Phys Rev B* 32 7048-7064

Journal of Physics: Conference Series **262** (2011) 012014 doi:10.1088/1742-6596/262/1/012014

Digital setup for Doppler broadening spectroscopy

J Cizek, M Vlcek and I Prochazka

Charles University in Prague, Faculty of Mathematics and Physics,
V Holesovickach 2, CZ-18000 Prague 8, Czech Republic

E-mail: jakub.cizek@mff.cuni.cz

Abstract. New digital spectrometer for measurement of the Doppler shift of annihilation photons was developed and tested in this work. Digital spectrometer uses a fast 12-bit digitizer for direct sampling of signals from HPGe detectors. Analysis of sampled waveforms is performed off-line in software. Performance of the new digital setup was compared with its traditional analogue counterpart. Superior energy resolution was achieved in the digital setup. Moreover, the digital setup allows for a better control of the shape of detector signals. This enables to eliminate undesired signals damaged by pile-up effects or by ballistic deficit.

1. Introduction

Doppler broadening (DB) of the annihilation peak is very important observable frequently used in defect studies with slow positron beams. The Doppler shift in energy of annihilation photons can be precisely measured using HPGe detectors. A scheme of a traditional analogue DB spectrometer is shown in Fig. 1A. Signals from a charge sensitive preamplifier coupled with HPGe detector are amplified and shaped by a spectroscopy amplifier. Shaping into a pseudo-Gaussian form is performed by passing detector signals through active high-pass CR and low-pass RC frequency filters. Finally amplitudes of shaped pulses are converted to numbers by an analogue-to-digital converter.

Novel digital setup for DB spectroscopy developed in this work is shown in Fig. 1B. The analogue modules used in the traditional setup were replaced by a fast 12-bit digitizer which directly samples detector signals in the *real time*. Sampled waveforms are stored in computer and analyzed *off-line* in software. New digital setup has several advantages compared to the traditional analogue configuration: (i) sampled detector signals do not suffer from random noise added by analogue modules in the spectroscopic track, (ii) all detector signals are accessible for the analysis, i.e. the amount of recorded information is much higher than in the analogue setup, (iii) data analysis can be performed repeatedly in order to find the optimal strategy how to derive the required physical information, (iv) time consuming adjustment of analogue nuclear instrument modules is not necessary anymore. Performance of the digital setup was compared with performance of its analogue counterpart.

2. Experimental details

A $^{22}Na_2CO_3$ with activity of 1 MBq deposited on a 2 μm thick mylar foil was used as a positron source. A well annealed pure Al (99.99999%) was used as a reference sample. Both digital and analogue setup were tested using a HPGe detector Canberra GC 3519 with relative efficiency of 35 %.

In the traditional analogue configuration shown in Fig. 1A detector pulses are shaped in a spectroscopy amplifier Canberra 2020 and their amplitudes are converted to numbers by an analogue-to-digital converter Canberra 8713.

In the digital setup shown in Fig. 1B detector pulses are directly sampled by a 12-bit digitizer Acqiris DC440 (Aglient Technologies U1066A) with sampling frequency of 16.7 MHz (sampling period 60 ns). The digitizer is externally triggered by slow positive logic SCA signals from a constant fraction discriminator Ortec 473A fed by detector signals appropriately shaped in a delay line amplifier Ortec 460. Although external triggering is in principle not necessary, it has an advantage that position of pulse maximum occurs always at similar time. Moreover, uninteresting low-energy signals can be suppressed by adjusting the lower discrimination level of the constant fraction discriminator.

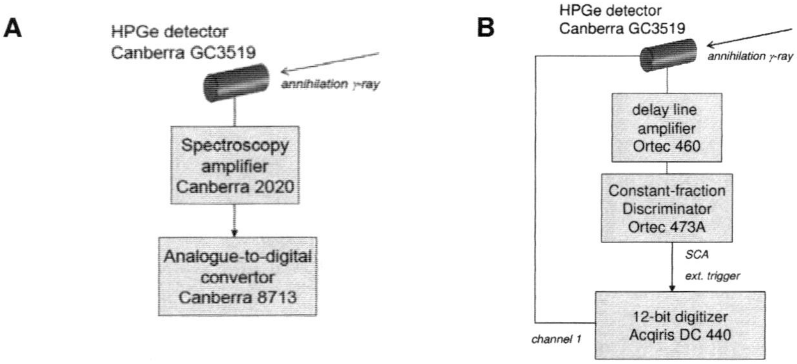

Figure 1. Scheme of (A) traditional analogue configuration for DB spectroscopy and (B) new digital setup.

3. Results and discussion

Examples of sampled detector pulses are shown in Fig. 2. Each sampled waveform consists of 2500 12-bit points. Fig. 2A shows the simplest case of a 'clean' pulse without any pile-up. More complex waveform consisting of a pile-up event, i.e. two overlapping pulses, is shown in Fig. 2B. Each waveform was fitted by a model function

$$f(t) = f_{main}(t - t_0) + f_{pile-up}(t - t_1) + f_{prec}(t) + bcg\ , \tag{1}$$

which consists of a main pulse $f_{main}(t)$, an overlapping random pile-up pulse $f_{pile-up}(t)$, an exponentially decaying base-line $f_{prec}(t)$ produced by some preceding pulse, and a constant base-line bcg. While $f_{main}(t)$ is always present, $f_{pile-up}(t)$ and $f_{prec}(t)$ are present only in complex waveforms, e.g. the waveform shown in Fig. 2B. The main pulse

$$f_{main}(t) = \left[\frac{1}{\beta_2\sqrt{2\pi}}\exp\left(-\frac{t^2}{2\beta_2^2}\right)\right] * \left[\beta_0 H_s(t - t_0)\exp(-\beta_1(t - t_0))\right] \tag{2}$$

is approximated by an exponential decay function with origin at time t_0 and decay rate β_1. Amplitude of the main pulse β_0 is the most important parameter because it is directly proportional to the energy of detected photon. The exponential decay function is multiplied by the Heaviside step function H_s and the product is subsequently convoluted (the symbol * denotes convolution) with a Gaussian with standard deviation β_2. This accounts for a finite energy resolution of HPGe detector. The pile-up pulse is modeled by a similar function as the main pulse

$$f_{pile-up}(t) = \left[\frac{1}{\beta_2\sqrt{2\pi}}\exp\left(-\frac{t^2}{2\beta_2^2}\right)\right] * \left[\beta_3 H_s(t - t_1)\exp(-\beta_1(t - t_1))\right]. \tag{3}$$

There are two additional fitting parameters related to the pile-up pulse: amplitude of the overlapping pulse β_3 and its position t_1. The tail of some preceding pulse is modeled by exponential decay function

$$f_{prec}(t) = \beta_4\exp(-\beta_1 t), \tag{4}$$

where the fitting parameter β_4 represents amplitude of the preceding pulse. Fitting of waveforms by the model function (1) is performed by the Newton-Raphson method. Examples of fit of a simple and a complex waveform, respectively, are shown in Fig 2A and 2B.

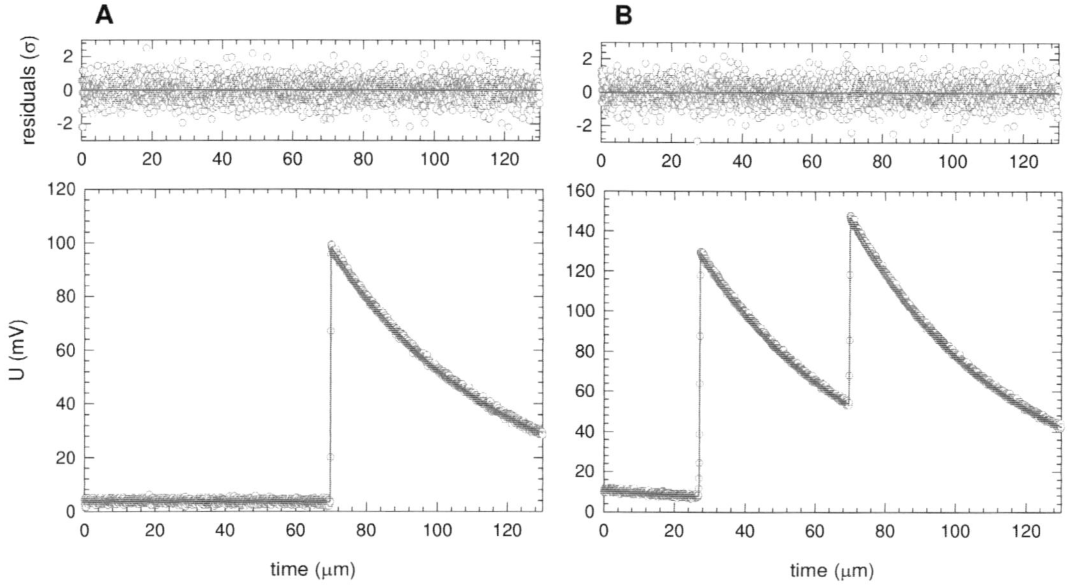

Figure 2. An example of sampled waveforms: (A) simple 'clean' pulse, (B) complex waveform consisting of a main pulse, an overlapping pile-up and an exponential tail from preceding pulse. Solid lines show a fit of waveforms by the model function (1). Residuals are shown in the upper panels.

The operations performed during analysis of each waveform can be summarized as follows:

1. read the waveform
2. fit the waveform by the model function (1)
3. perform χ^2 test to check if the waveform is well described by the model function (1). The waveform is accepted only if the calculated χ^2 value is smaller than a pre-selected value corresponding to the significance level of 0.01. Hence, distorted pulses damaged, e.g. due to ballistic deficit, do not pass the χ^2 test and are rejected at this point.
4. If the waveform was accepted, add the main pulse amplitude β_0 into histogram.

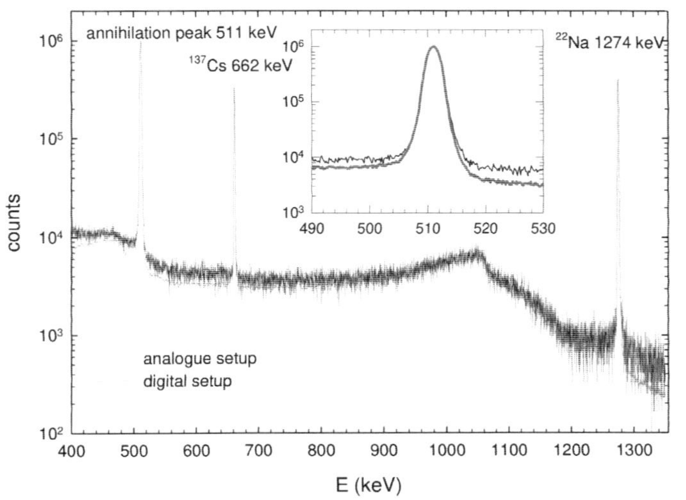

Figure 3. Comparison of energy spectrum measured using the traditional analogue configuration and the spectrum obtained with the new digital setup. The inset shows a detail of the annihilation peak.

Journal of Physics: Conference Series **262** (2011) 012014 doi:10.1088/1742-6596/262/1/012014

Accumulated histogram of main pulse amplitudes β_0 is calibrated using known energies of the annihilation γ-ray (511 keV) and the ^{22}Na starting photon (1274 keV). In order to improve the precision of energy calibration we used also γ-ray (662 keV) produced by auxiliary ^{137}Cs radioisotope. The calibrated energy spectrum obtained using this procedure is plotted in Fig. 3 together with the spectrum measured using the traditional analogue configuration. The inset in Fig. 3 shows a detail of the annihilation peak. One can clearly see in Fig. 3 that the digital setup suppresses background and improves the peak-to-background ratio compared to the traditional analogue configuration. This is due to a better control over the shape of sampled pulses which enables to reject events distorted by ballistic deficit or by pile-up effects.

Fig. 4 shows energy resolution (FWHM of γ-line) achieved using the analogue and the digital setup at various energies. It is well know that the width of a γ-line measured by a HPGe detector is directly proportional to the square root of energy deposited in detector by detected photon [1]. Indeed, points in Fig. 4 plotted versus the square root of energy fall on straight lines. Obviously γ-lines measured using the digital setup are narrower than corresponding lines measured by the traditional analogue configuration. The extrapolated energy resolution at 511 keV is (1.40 ± 0.05) keV and (1.21 ± 0.03) keV for the analogue and the digital setup, respectively. Hence, employment of digital technique leads to an improvement of energy resolution. This testifies that fitting of the whole detector pulse by a proper model function is the most precise way how to determine the energy of detected photons.

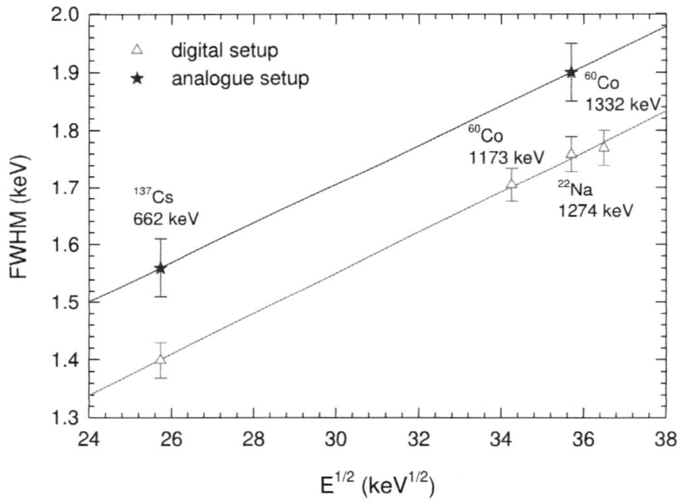

Figure 4. Energy resolution obtained using the traditional analogue configuration and the new digital setup. FWHM of γ-lines produced by ^{137}Cs, ^{60}Co and ^{22}Na radioisotopes are plotted as a function of the square root of energy of detected photons.

4. Conclusions

New digital setup for Doppler broadening spectroscopy has been developed in this work. Detector signals are directly sampled using a fast 12-bit digitizer. Sampled waveforms are analyzed in software. Performance of the new digital setup was compared with its traditional analogue counterpart. It was demonstrated that the digital Doppler broadening spectrometer enables to achieve superior energy resolution and improved peak-to-background ratio.

5. References

[1] Leo B W R 1994 *Techniques for Nuclear and Particle Physics Experiments* (Heidelberg: Springer)

Acknowledgments

This work was supported by the Academy of Science of the Czech Republic (project KAN300100801) and the Ministry of Schools, Youths and Sports of the Czech Republic (project MSM 0021620834).

Journal of Physics: Conference Series **262** (2011) 012015 doi:10.1088/1742-6596/262/1/012015

Ice and Atoms: experiments with laboratory-based positron beams

P G Coleman

Department of Physics, University of Bath, Bath BA2 7AY, UK

E-mail: p.g.coleman@bath.ac.uk

Abstract. This short review presents results of new positron and positronium (Ps) experiments in condensed matter and atomic physics, as an illustration of the satisfying variety of scientific endeavours involving positron beams which can be pursued with relatively simple apparatus in a university laboratory environment. The first of these two studies – on ice films - is an example of how positrons and Ps can provide new insights into an important system which has been widely interrogated by other techniques. The second is an example of how simple positron beam systems can still provide interesting information – here on a current interesting fundamental problem in positron atomic physics.

1. Introduction

Over the past forty years experimental positron beam investigations have developed from beginnings in atomic collision physics, in which for many years beam intensities of less than one positron per second were common [1], to a huge variety of studies in many fields using beams of up to 10^8 positrons per second – sometimes in large reactor or LINAC-based facilities [2-3]. Notwithstanding this impressive development, there still remain many interesting experiments which can be performed with what are today relatively standard laboratory-based positron beams. In this review two examples of such experiments are described, one with a simple beam system using a particle detector, and one with a standard medium-intensity magnetic-transport beam [4]. These are presented in the hope that science will continue to benefit from numerous small-scale endeavours as well as the ambitious, large-scale projects planned for the future.

2. Positron beam studies of ice films

While other techniques have provided insights into surface changes and pore properties of ice [5,6], VEPAS constitutes a sensitive method for studying the depth profile and evolution with temperature of both atomic and pore structure [7]. The annihilation of positrons in ice (either in the bulk or trapped in small open volume defects) is characterised by the lineshape parameter S, whereas the formation of Ps, which can be trapped in larger (meso-) pores or find its way to the surface via a pathway of interconnected pores, is characterised by the parameter F, a measure of the probability that Ps survives to decay into 3γ as ortho-Ps. The larger or more interconnected the pores are, the larger F will be.

2.1. Experimental method: ice films

By installing a simple copper cold finger attached to the head of a closed-cycle He refrigerator unit in the sample chamber of the magnetic-transport positron beam at the University of Bath [8], amorphous

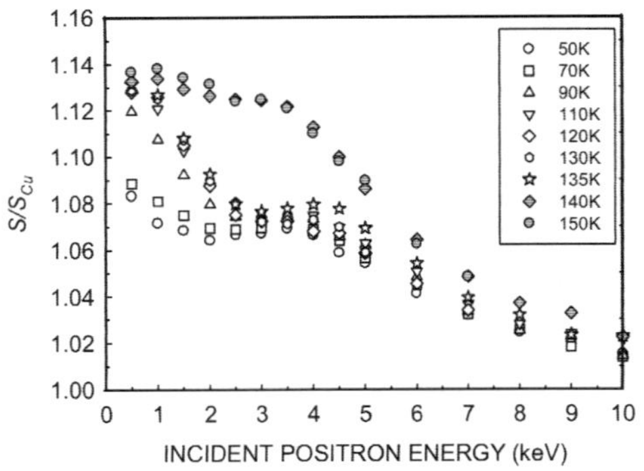

Figure 1. $S(E)/S(\text{Cu})$ for a 700nm thick amorphous ice film at 50-150K.

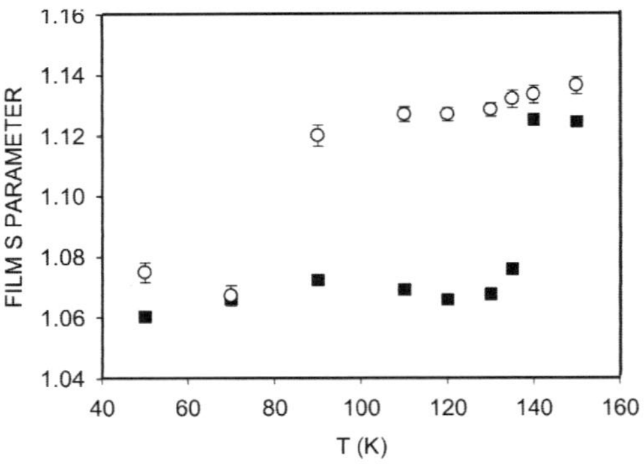

Figure 2. Fitted S for the 'bulk' film (■) and near-surface layer (o). Changes are irreversible.

solid water (ASW) films could be grown under controlled conditions by leaking vapour into the sample chamber via a needle valve from a side chamber containing pure water initially subjected to several freeze–pump–thaw cycles.

The parameter S was deduced from the 511 keV photopeak and the 3γ:2γ parameter R from the ratio of annihilation events in the 'valley' region between about 475-505 keV and in the photopeak.

2.2. *Positron results: phase changes*

S for an ASW film are shown in Fig. 1 as a function of sample temperature. By fitting the data assuming two layers (from 0 - 80 and 80 - 800nm) characteristic S values could be extracted [9]; these are plotted in Fig. 2. These data provide clear evidence of the sensitivity of S to the amorphous-crystalline phase transition at just below 140K. Unexpectedly, however, they also suggest that in the top 80nm region crystallisation occurs between 70-130K. This is corroborated by the S-W plot of Fig. 3; the large circles represent annihilation in Cu (Cu), ASW (A), the surface (S) and crystalline ice (C). Data points for positron energies below 2keV tend towards the point C at temperatures above ~70K.

2.3. *Positronium results: pore structure*

The parameter F, proportional to the probability that the implanted positrons will be annihilated as ortho-Ps, was calculated from the measured ratios R. F for the film featured in Figs. 1-3 is shown in Fig. 4. The decrease of F with T is interpreted as indicating a reduction in pore size and interconnectivity until, at ~120K, almost no Ps survives to decay into three gammas- indicative of pore collapse. Over the same temperature range the effective ortho-Ps diffusion length L_{Ps} falls from ~ 120 to 12 nm, consistent with a decrease in pore interconnectivity. Above 135K F increases irreversibly when the film undergoes crystallization, associated with an increase in the Ps diffusion length.

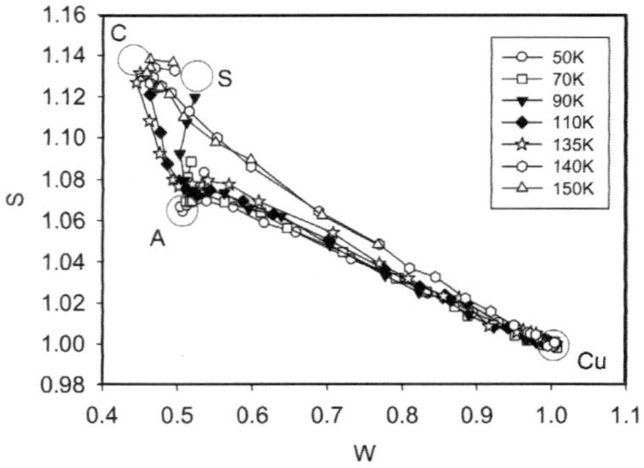

Figure 3. S-W plots at 50-150K.

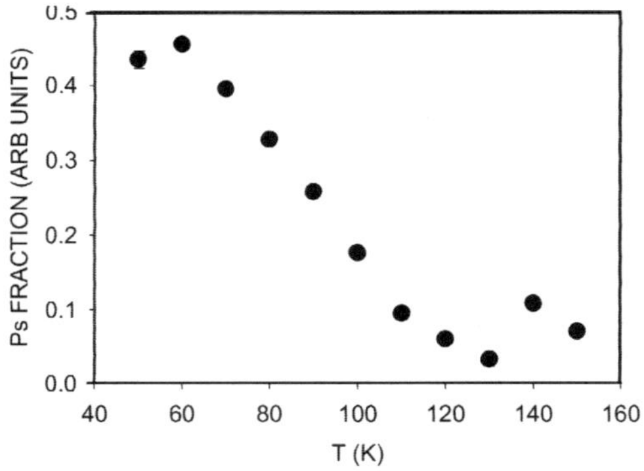

Figure 4. Fitted Ps fractions $F(T)$.

3. Elastic positron scattering above the Ps formation threshold

On the other side of the laboratory, a simple solenoid-guided positron beam system was constructed for final year undergraduate projects from hardware gathering dust on storage shelves. The simplicity of the system lent itself to straightforward positron-atom scattering measurements. However, this area of research is mature [10] and measurements have to be selected which are both possible with older technology but which are still able to add something meaningful to the field.

A candidate for such an experiment is the behaviour with positron energy of the elastic scattering cross section (Q_{el}) for positron-atom collisions in the energy range just above the Ps formation threshold. The early paper of Campeanu et al [11] proposed the existence of a cusp-like feature in Q_{el} at the Ps threshold energy in He. In an experiment in which both Q_{tot} (the total cross section) and Q_{Ps} were measured using the same apparatus, however, no significant cusp was observed [12]. A series of papers from UCL [13] concluded that there should be coupling between Q_{el} and Q_{Ps} in the noble gases near threshold, with cusps becoming more prominent with increasing Z.

3.1. Experimental method: elastic scattering

The source/moderator produced a 4mm-diameter beam of ~300 positrons s^{-1}. The energy spread of the beam was reduced to ~0.4eV FWHM by applying 2V above the moderator potential to a double 92%-transmission mesh held immediately in front of the moderator (Fig.5). This cut-off mesh also serves as an efficient reflector for back-scattered positrons. The positrons then pass through a 70mm-long gas cell, under the influence of a 5mT magnetic field. They pass through a cylindrical retarding field analyzer in the evacuated volume between gas cell and CEM detector; the total path length is approximately 400mm.

An example of the measured CEM count rate with and without as in the cell is shown in Fig. 6. The signal attenuation A_0 at V_{RFA} = 0V is due only to Ps formation events, and A_{tot} at V_{RFA} =7V is due to both elastic scattering and Ps formation). Therefore, in the limit of a thin gas cell, $Q_{el} = Q_{tot}$ (1-A_0/A_{tot}). This procedure was followed for He, Ne, Kr, Ar and Xe.

3.2. Results: elastic scattering

The results for He and Kr are shown in Fig. 7. In He and Ne there is a small but discernable cusp-like feature, the first in reasonable agreement with that seen by Caradonna et al [14]. In Ar, Kr and Xe Q_{el} rises significantly after the Ps threshold, levelling or falling slightly after a few eV [15]. This is not the expected cusp-like feature, in which the maximum is at the threshold energy.

Thus, once again, we have obtained an intriguing and unexpected result. One possible explanation is that there exists a virtual Ps state which enhances the branching into the final elastic channel. Another possibility is that the results of Fig. 8 are actually cusp-like features superimposed on a generally increasing cross section – but this requires that the energy calibration be in error by an unlikely 1eV or more. Recently Buckman et al [16] have reported measurements which suggest that cusps exist in all five gases; it will be interesting to probe why the results differ, but both groups agree

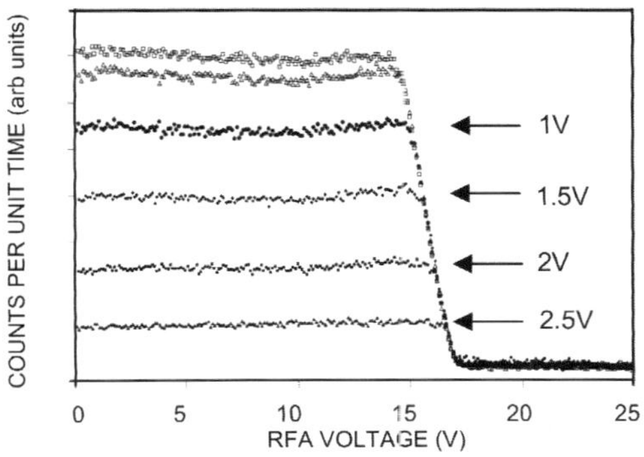

Figure 5. CEM count rate vs RFA voltage for cut-off potentials 1 - 2.5V.

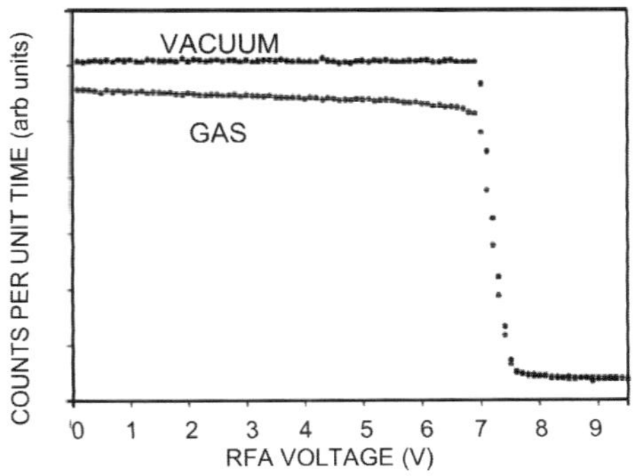

Figure 6. CEM count rate vs RFA voltage in vacuum and gas.

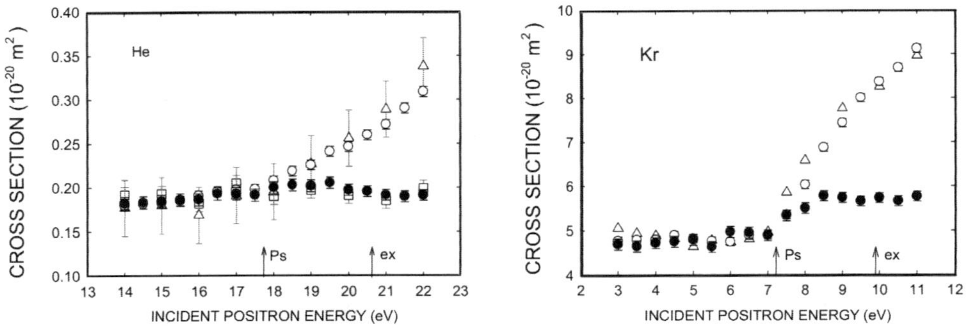

Figure 7. •: Q_{el} for positron scattering. O: Q_{tot}. Δ: Q_{tot} from earlier measurements. □ (He): Q_{el} from ref [23]. Threshold energies for atomic excitation and ionisation are shown.

that there is an interesting energy dependence of Q_{el} in the region of the Ps formation threshold.

4. Conclusions

The purpose of this short review is to underline the potential of simple positron beam measurements to have a significant research impact. The results of the ice film measurements are likely to be of considerable interest to physical chemists, astrophysicists, cryobiologists and modellers. In the case of atomic collisions, there are clearly still interesting current questions which can be answered – or at least illuminated - by using relatively simple positron beam systems. May laboratory-based positron beam systems long continue to play a useful and important role in developing our understanding of scientific and technological problems of the day.

Acknowledgements
The author is grateful to many people who have contributed to the research described in this paper, including Liz Lowry, Nicola Cheesman, Peter Jay, Prof. Yichu Wu, Alexis Kallis and Jing Jiang.

References

[1] Canter KF, Coleman PG, Griffith TC and Heyland GR 1972 *J Phys B* **5**, L167.
[2] Hugenschmidt C, Schreckenbach K, Stadlbauer M and Strasser B 2006 *Appl. Surf. Sci.* **252**, 3098.
[3] Krause-Rehberg R, Anwand W, Brauer G, Butterling M, Cowan T, Hartmann A, Jungmann M, Krille A, Schwengner R and Wagner A 2009 *Phys. Stat. Solidi* C **6** 2451.
[4] Coleman PG, 2003 *Principles and Applications of Positron and Positronium Chemistry* eds YC Jean, PE Mallon and DM Schrader (World Scientific: Singapore) p 37.
[5] Backus EHG, Grecea ML, Kleyn AW and Bonn M 2004 *Phys. Rev. Lett.* **92**, 236101.
[6] Raut U, Famá M, Teolis BD and Baragiola RA 2007 *J. Chem. Phys.* **127,** 204713.
[7] Weber MH and Lynn KG, 2003 *Principles and Applications of Positron and Positronium Chemistry*, Eds. YC Jean, PE Mallon and DM Schrader (World Scientific: Singapore) p.167.
[8] Chilton NB and Coleman PG 1995 *Meas. Sci. Technol.* **6** 53.
[9] van Veen A, Schut H, De Vris J, Hakvoort R A and Ijpma M R 1990 *AIP Conf. Series* **218** 171.
[10] Gribakin GF, Young JA and Surko CM 2010 *Rev. Mod. Phys.* accepted for publication.
[11] Campeanu RI, Fromme D, Kruse G, McEachran RP, Parcell LA, Raith W, Sinapius G and Stauffer AD 1986 *J. Phys. B: At. Mol. Phys.* **20,** 3357.
[12] Coleman PG, Johnston KA, Cox AM, Goodyear A and Charlton M 1992 *J. Phys. B: At. Mol.Opt. Phys.* **25,** L585.
[13] Meyerhof WE and Laricchia G 1997 *J. Phys. B: At. Mol. Opt. Phys.* **30,** 2221.
[14] Caradonna P, Jones A, Makochekanwa C, Slaughter DS, Sullivan JP, Buckman SJ, Bray I and Fursa DV 2009 *Phys. Rev. A* **80,** 032710
[15] Coleman PG, Cheesman N and Lowry ER 2009 *Phys. Rev. Lett.* **102**, 173201.
[16] Buckman SJ 2010 Private Communication.

Journal of Physics: Conference Series **262** (2011) 012016 doi:10.1088/1742-6596/262/1/012016

Development of a spin-polarised positron beam

PG Coleman and Alexis Kallis

Department of Physics, University of Bath, Bath, BA2 7AY, UK

E-mail: a.kallis@bath.ac.uk

Abstract. The development stages of a system for variable-energy spin-polarised positron beam spectroscopy are described. Methods for achieving maximum practical beam polarisation include a Na-22 source configuration with a low-*Z* backing to reduce positron backscattering, and positioning a 250μm beryllium foil in front of the source to absorb the slower and decelerate the faster beta positrons, the latter suffering significantly less depolarisation during moderation than the former. To switch the direction of spin polarisation of electrons in the target sample the direction of the positron guiding field was reversed, and strong rare earth magnets were placed behind the sample. Systematic problems associated with low beam intensity and electronic drifts are discussed. Results for single-crystal iron compare well with those expected from theory, but suggest a beam/target polarisation of ~ 5% of that expected. Prospects for future beam development and application are outlined.

1. Introduction

Variable-energy spin-polarised positron beam spectroscopy (VESPAS) has the potential for application in areas of current scientific and industrial interest such as in depth profiling nano-magnetic structures or of dilute magnetic species in a semiconductor layer developed for spintronics. The principle of the technique is that positrons are $\sim 10^2$ times more likely to annihilate electrons with opposite spin than with parallel spin [1]. Thus, if the incident positron beam is spin-polarised and the spin orientation of the polarisable electrons in a sample is reversed from the parallel to the anti-parallel direction, then the change in the shape of the two-photon annihilation gamma energy line centred at 511keV is essentially due only to the annihilation events involving the spin-polarised electrons. The magnitude of the lineshape change is reduced by partial beam/target polarization and by the non-negligible probability of three-photon annihilation events (originally discussed by Berko [2]). In a series of papers thirty years ago the Michigan group showed that beta positron polarisation survives the moderation process [3], and there have been programmes to develop polarised positron beams in the intervening years [4]. Recently Cassidy *et al* [5] found that their solid-neon-moderated beam had a polarisation of 22%.

A development programme aimed at maximizing both the beam and target polarisation, both of which are described in the following sections, has led to an understanding of the many problems attached to the implementation of VESPAS as a routine tool. While systematic problems such as those associated with electronic instabilities can be overcome, the fundamental problems of low signal rates and the consequences of the need for high magnetic fields at the target remain a challenge. Trial measurements have been made on a variety of samples, including iron in different states (high purity, annealed, single crystal), mu metal and solid oxygen (alpha phase) [6]. Experimental results for iron are compared with calculations.

2. Polarising the Positron Beam

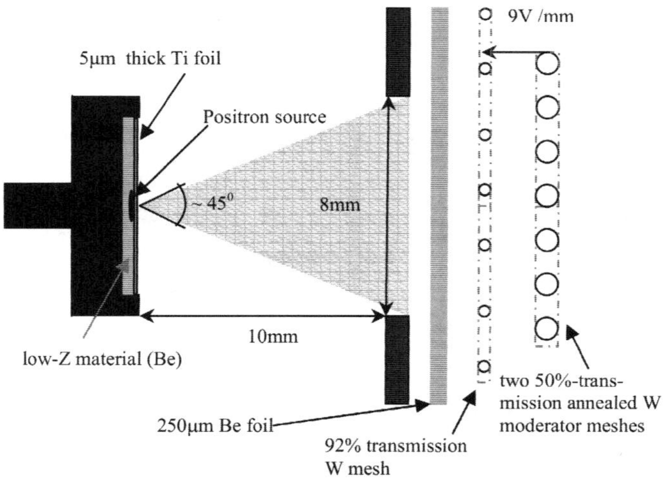

Figure 1: Sketch of source-moderator setup to maximise positron beam polarisation.

In order to obtain significant positron beam polarisation a number of modifications have been made to the standard source-moderator geometry [7], represented in Fig. 1. It is more probable that beta positrons emitted from the ^{22}Na source with higher energies retain their polarisation throughout their lifetime. Unfortunately it is the *lower*-energy beta positrons which are preferentially moderated to form the beam. To combat this, a thin Be foil was positioned in front of the source to absorb the slower beta positrons and slow down the faster, more highly polarised positrons. The foil thickness - 250μm – was chosen with reference to the results of ref. [3]. As beta positrons are spin-polarised in the emission direction the ^{22}Na source capsule was modified; the source is deposited on a 0.5mm-thick Be backing plate to minimise the backscattering of positrons (a Ta plate is usually used to maximize backscattering and thus positron flux) [3]. Positrons are polarised in their initial direction of emission.

The source was positioned at ~10mm from the moderator (normally at ~1mm) to decrease the angle of acceptance to 45°, which is well within the range suggested by Major [8]. As a result of installing the Be absorber and repositioning the source the positron beam intensity fell to below 5% of its original value, but the beam polarisation is expected to have increased to more than 50% [3].

3. Polarising the Sample

Various methods of orienting target electron spin directions have been investigated. Neodymium Grade N42 magnets [9] have been used behind or around the sample target to create a localised field in the direction required; magnetic lensing (Fig. 2) and an in-vacuum electromagnet were also trialled. The positron magnetic guiding field (100G) was also reversed. This rendered the operation of the curved-plate **ExB** filter [7] ineffective, a problem overcome by disconnecting the plates and using external permanent magnets to effect the required beam deflection.

Iron and mu-metal samples were expected to be 100% spin-polarised when placed immediately in front of a Nd N42 magnet, while only partially polarised by the other methods attempted. In the case of iron, the ferromagnetism is associated with the polarization of the 3d electrons. Antiferromagnetic α-phase solid oxygen, chosen because it is an insulator and because the polarisable electrons are in the outermost shell, could only be magnetised by the guiding field for practical reasons associated with rotating magnets and regrowing the sample; it was therefore expected that, while potentially interesting as a positron target, this sample would give a very small response.

Positrons of incident energy 30keV were used for these first measurements, to probe the well-known bulk magnetism of the samples and avoid any near-surface effects.

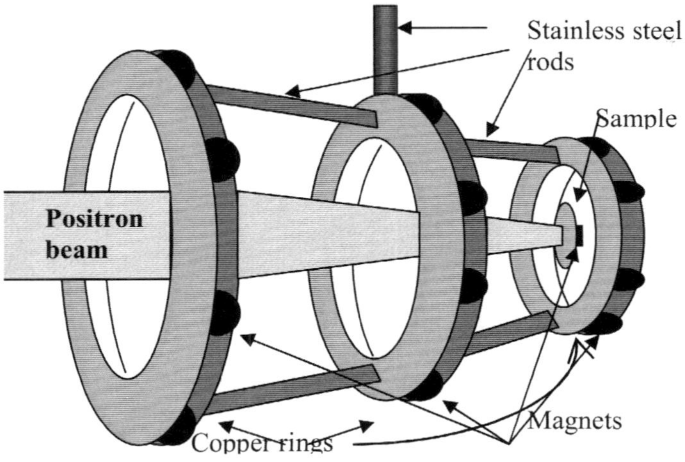

Figure 3: Experimental setup for polarising a sample using magnetic focusing.

The annihilation gamma photon count changed by up to 10% when the field direction and magnet orientation were reversed, mostly due to losses in the ExB plates. Careful subtraction of the background, which is not negligible in these low-signal-rate measurements, was therefore required to eliminate any consequent systematic effect on the results.

Measurements on Al and Si – of dimensions similar to the magnetic targets studied - were taken in order to investigate the stability of the system. False positive results were obtained when the detector resolution (and thus measured lineshape) was affected by variation in electronic noise, the use of very small samples and/or the presence of high magnetic field gradients in the vicinity of the sample (when positrons could proceed to, or be deflected into, annihilation sites other than the intended target).

4. Results and Discussion

Samples studied included pure Fe (99.999%) – both as-supplied and annealed – single-crystal Fe, μ-metal, and solid oxygen, in order to test the effectiveness of the system with samples of varying electronic structure.

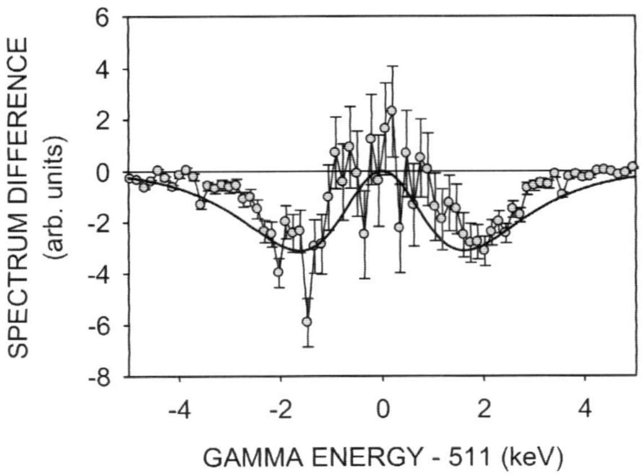

Figure 3: Normalised spectrum difference for a single crystal Fe in front of a Nd magnet. Solid line: theoretical result using the program MIKA-Doppler [10].

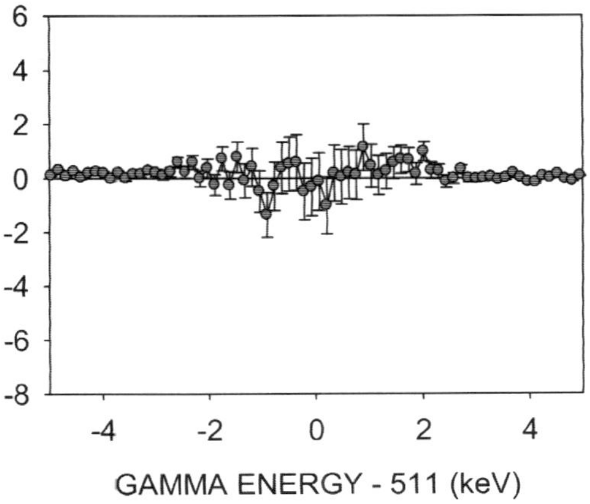

Figure 4: Normalised spectrum difference for α-phase solid oxygen (same scale as for Fig. 3).

Only single-crystal Fe held in front of a Nd magnet (field ~ 0.6T) gave a signal of the type expected when annihilation line spectra recorded with the external field in the two directions were background-subtracted, normalised and subtracted (Fig. 3, in which data points have been added in groups of 10). The solid line in Fig. 3 is the scaled difference calculated using the MIKA-Doppler package [10] – essentially the contribution to the annihilation line of 3d electrons – after convolution with the experimental detector resolution of 1.4 keV. While the shape of the difference is in reasonable agreement with the observation, the scaling factor of 0.025 suggests that the combined effective polarisation of the beam and target is about 5% of that expected.

No significant results were observed for any other samples, as exemplified by null result shown for α-phase solid oxygen (Fig. 4). The sample was grown in situ and was thought to be crystalline, with plenty of vacancies and lattice imperfections but essentially no free electrons; therefore polarisable electrons should make a significant contribution to the annihilation spectra. However, α-phase solid oxygen is anti-ferromagnetic [11] with a very small magnetic susceptibility, and it is concluded that the sample magnetisation was not sufficient for observation by the spin polarised positrons in the current system. For the pure Fe, pure annealed Fe and μ-metal samples, the failure to see an effect may be explained by the dominance of annihilation from vacancies and other trapping sites.

5. Conclusion

The results shown in Fig. 3 demonstrate that a spin polarised positron beam can be generated using a modified 'standard' magnetic-transport positron beam system, and that it can probe spin-polarised samples in magnetic targets. However, the very small response – 5% of that expected – suggests that either (a) the beam polarisation is only a few %, or (b) the magnetic field in the vicinity of the target is not strong enough. Having followed the recipe of Rich *et al* [3] it is felt unlikely that (a) is the case; a polarisation closer to 50-60% is expected. The recent observation of Cassidy et al. [5] suggests that a rare gas moderator, or any other moderator which does not rely on thermalisation and work function emission, may have a smaller depolarisation effect on the primary beta positrons. Increasing the magnetic field at the sample is problematic with the current system, and would require a complete redesign of the target chamber to accommodate a strong electromagnetic or, perhaps, a superconducting magnet. The transition between the 100G guiding field and a ~ 1T target field would, however, have to be carefully controlled. An appropriate choice of moderator, an intense positron source, and a strong (reversible) magnetic field would seem essential if VESPAS is to succeed as a valuable probe of magnetic films of current technological importance.

Acknowledgements

The authors are grateful to Dr Stephen Dugdale for performing the MIKA calculations, to Dr Daniel Wolverson for useful discussions, and to Harry Bone for invaluable technical support. This work was supported by EPSRC (UK) under grant no. EP/F029829/1.

References

[1] West RN 1974 *Positron Studies of Condensed Matter* (Taylor & Francis) p 2.
[2] Berko S 1967 *Phys. Rev.* **127**, 2166.
[3] Rich A, Van House J, Gidley DW, Conti RS and Zitzewitz PW 1987 *Appl. Phys.*A **43**, 275.
[4] Kumita T *et al*, 1997 *Appl. Surf. Sci.* **116**, 1.
[5] Cassidy DB, Meligne VE and Mills, AP Jr. 2010 *Phys. Rev. Lett.* **104**, 173401.
[6] Collins MF 1966 *Proc. Phys. Soc.* **89,** 415.
[7] Chilton NB and Coleman PG 1995 *Meas. Sci. Technol.* **6** 53.
[8] Major J 2000 in *Positron Beams and their Applications*, ed. PG Coleman (World Scientific) p 259.
[9] Neodymium grade N42 magnets from e-Magnets UK (http://e-magnetsuk.com).
[10] Torsti T et al 2006 *Phys. Stat. Solidi* B **243**, 1016.
[11] Stephens PW and Majkrzak CF 1986 *Phys. Rev. B* **33,** 1.

Study of helium evolution in nanocrystalline titanium films by slow positron beam analysis

A H Deng, Y L Zhou, L R Zhang, and Q Hou

Department of Physics, Sichuan University, Chengdu 610064, P. R. China

E-mail: ahdeng@scu.edu.cn

Abstract. In this paper, the evolution and depth profiles of vacancy-type defects were investigated by slow positron beam (SPB) analysis. Films of helium-containing titanium were deposited on Si substrates by direct current magnetron sputtering under the helium/argon (He/Ar) gases ambiance. Different helium concentrations were introduced into the sample films by altering the He/Ar inflow ratios (Q_{He}/Q_{Ar}). It was found that the defect-characteristic parameter S from positron-electron annihilation peaks rise with the increase of the ratio of Q_{He}/Q_{Ar}. Homogenous He-related defect profiles and their evolution deep into around ~400nm were found by SPB measurements, together with the analysis of Rutherford backscattering (RBS) and elastic recoil detection (ERD).

1. Introduction

Helium can be produced in tritium storage materials and structural materials of fission reactors, accelerators and future controlled fusion reactors. Its insolubility and high mobility of the helium in the materials are crucial to study its formation and evolution in materials. Helium atoms can not be naturally dissolved in the materials, but are easily trapped by dislocation, vacancies and voids. With the increment of helium concentration, helium bubbles in the materials will be formed by migrating, diffusing and coalescence of helium atoms. Helium may change microstructure of the materials, and even cause serious degradation of some macroscopic mechanical properties of the materials, such as swelling and helium-induced brittleness[1-3]. Therefore, studies of the behavior of helium in materials are critically important in material applications. However, helium behavior studies in materials are quite complicated, many problems, such as the evolution of helium-related defects and mechanism of helium brittleness, are still under investigation.

There are three traditional methods to artificially introduce helium into materials: tritium ageing [4], neutron irradiation [5] and helium ion implantation [6]. In recent years, a magnetron sputtering technique [7-9] which works in a mixture of helium (He) and argon (Ar) gas ambient has been developed. This method can introduce helium into the growing metal films. It has many advantages, such as controllable helium concentration and a uniform He distribution in depth.

Many experiments have been used to study the behavior of helium in titanium film. As a sensitive method for studying vacancy-like defects, positron annihilation technique has been effectively

Journal of Physics: Conference Series **262** (2011) 012017 doi:10.1088/1742-6596/262/1/012017

employed to probe microscopic characteristics of defects, such as concentration, size, charge states and chemical surroundings [10, 11] of the defects. Studies on helium behavior in metal films by SPB technique [8, 9] have been carried out in recent years. In this paper, He-Ti films with different helium concentrations were prepared by the magnetron sputtering technique. The methods of RBS and ERD analysis were used to determine the thickness and helium depth profile in the films, respectively. The SPB analysis was adopted to study the evolution of helium-related defect in the titanium films.

2. Experiment

The Ti films containing helium were deposited on the silicon (100) pieces by the magnetron sputtering coating equipment [8, 9]. A thin Pd layer (~10nm) was deposited onto the surface of Ti film in order to avoid surface contamination and He-Ti layer oxidization. With different ratios of helium and argon influx rates (Q_{He}/Q_{Ar}), various He-content samples were deposited on the silicon (Si) pieces for further analysis experiments. The ERD and RBS measurements were carried out at the NEC 9SDH-2 tandem accelerator in Fudan University in Shanghai. The thickness of He-Ti films measured by RBS was about 470 nm for eight different He-Ti samples. The ERD results show that the helium content increases with the increment of the influx ratio of Q_{He}/Q_{Ar} and the distribution of helium is homogenous in the different depth of the Ti film. The SPB experiments were carried out in the institute of high energy physics in Beijing. The total counts of 2.0×10^5 were collected in the 511keV annihilation line at each incident energy. The value of the S-parameter was defined to be central region of the γ spectrum from 510.56 to 511.44keV divided by the total region of the γ spectrum.

3. Results

For the samples with different inflow rates of Q_{He}/Q_{Ar} (a for 0 (pure Ti); b for 1.0; c for 2.0; d for 4.0; e for 8.0; f for 12.1; g for 18.0 and h for 25.0), the S parameter as a function of positron incident energy is shown in Figure 1. The S parameter measurement results were found to be depending on the flow rate Q_{He}/Q_{Ar}. The corresponding eight solid lines in figure 1 are the fitting results of VEPFIT [12], respectively. The electron momentum information at different depth is represented by the S values. The corresponding positron mean implantation depth of different energy approximately is noted in the top of x-axis in figure 1.

When the energies of positron beam are very low (E≤~1keV), positrons are mostly probing the extremely thin Pd surface protection layer. Since the protection-layers for all the eight samples are nearly the same, no obvious difference is observed among these low energy S-E curves (E≤1keV), where a higher surface S value is around 0.504 with $S_{Pd}\approx0.494$ characterizing the Pd layer. As the increasing of the positron beam energy, positrons will penetrate into the He-Ti layer and annihilate there, giving the structure-related information of the region by the corresponding specific S values. For those positron beam energies ranging from around 2keV to 10keV, all S-E curves are quite smooth and flat (except h), with S value distribution being uniform in depth. These results indicate that the defect profiles of the helium-related in these Ti films are fairly uniform. When the energies of the incident positrons are above 10keV, the S-E curves for the eight samples tend to merge. This implies that positrons are implanted dominantly into the Si substrate, giving the structural information of the Si substrate.

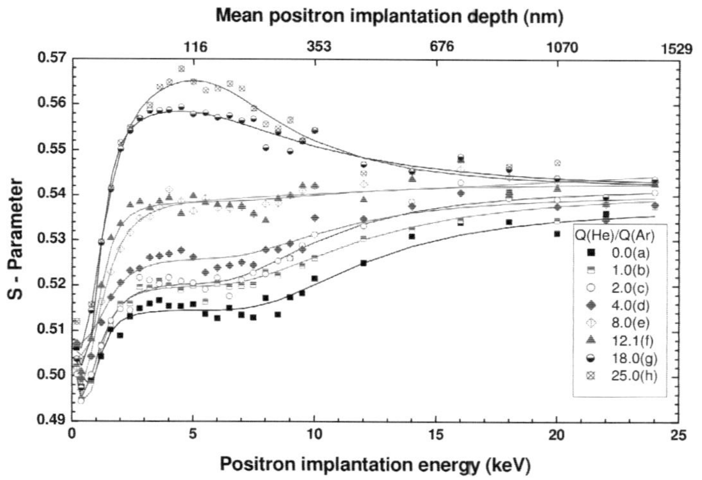

FIG. 1. The S-parameter versus the positron implantation energy for Ti films with various helium concentrations. The solid lines are the fitting results of VEPFIT.

4. Discussions

For He-Ti layer (~2keV-~10keV) grown on substrates with different flow rates of Q_{He}/Q_{Ar}, an first increasing, following saturation and further increment tendency of $S_{He\text{-}Ti}$ values have been observed from figure 1. When the flow rate of Q_{He}/Q_{Ar} is increased from 1 to 4, a first increment tendency of S values was observed compared with pure Ti film. This is because higher helium content is favoring for the penetration of more He ions into the Ti film, where the Ti vacancies will be filled by those He. After all the Ti vacancies were effectively filled by the He, further increasing of He into the Ti-film will make the excessive He aggregate. Eventually, He-Ti-vacancy complex and He bubbles will co-exist, with the bubble concentration getting higher. This induces the slow increment of S-values.

From Figure 1, the increasing of S values became obvious, when the flow rate of Q_{He}/Q_{Ar} was raised from 4 to 8 and 12.1, due to the annihilation of more positrons with the He-bubbles formed. With the increase of the flow rate of Q_{He}/Q_{Ar}, the size of He bubbles grows gradually and the bubble density will also increase, which gives rise to the increase of $S_{he\text{-}Ti}$ value. These results indicate that applying the flow rate of Q_{He}/Q_{Ar} is more favorable for migration and aggregation of helium in growing Ti films.

Moreover, those helium atoms in the Ti lattice that localize in a specific region may be affected further by the increase of the flow rate of Q_{He}/Q_{Ar} from 12.1 to 18 and 25. Larger bubbles may form through the migration and the coalescence of small bubbles. Some bubbles may grow bigger while the concentration of helium bubble will also increase upon the further helium content increment. Therefore, the magnitude of S increases continuously.

Further studies will be carried out on the annealing and temperature scanning behavior of the He-Ti films. In order to understand the evolution of helium bubble and the phenomena of helium brittleness further, the other relevant techniques, such as slow positron beam lifetime measurements, the TEM and thermal desorption spectrometry, are needed to determine the size and internal pressure of He bubbles and their progressive evolution. In the meantime, in order to understand the physical mechanism of helium evolution further, the study of helium behavior in material by the use of Monte Carlo simulation [13, 14] is also in progress.

5. Conclusions

In this paper, helium-charged Ti films with different helium concentration were deposited by the He-Ar magnetron co-sputtering method. ERD result demonstrates the uniform distribution of helium in the bulk. The results of SPB reveal that the evolution of defects is related to helium content in He-Ti films, and it was found that the S parameter representing defect profiles of the helium-related in the Ti films increase with the influx rate Q_{He}/Q_{Ar}. S-E Data fitting results based on the three-layer model are in good agreement with the experimental data. Our investigations have demonstrated that SPB measurements are useful adjunct to conventional techniques in studies of the behavior of helium introduced into materials through magnetron sputter deposition in a He/Ar environment.

Acknowledgements

The author gratefully acknowledges the support of K.C.Wong Education Foundation, Hong Kong. This work was also supported by the National Scientific Foundation of China under Grant Nos. 10775102 and 10775101.

References

[1] Ullmaier H 1983 *Radiat. Eff.* **78**, 1
[2] Trinkaus H, Singh B N 2003 *J. Nucl. Mate.* **323**, 229
[3] Cowgill D F 2005 Fusion Sci. Technol. **48**, 539
[4] Weave H T and Camp W J 1975 Phys. Rev. B **12**, 3054
[5] Rau R C and Ladd R L 1969 J. Appl. Phys. **40**, 2899
[6] Rajainmäki H, Linderoth S S, Hansen H E, Nieminen R M and Bentzon M D 1988 Phys. Rev. B **38**, 1087
[7] Pang H C, Luo S Z, Long X G, An Z, Liu N, Duan Y M, Wu C X, Yang B F, Wang P L, Zheng S X 2006 Chin.Phys.Lett. **23**, 12
[8] Liu C Z, Shi L Q, Zhou Z Y, Hao X P, Wang B Y, Liu S, Wang L B 2007 J. Phys. D: Appl. Phys **40**, 2150
[9] Liu C Z, Zhou Z Y, Shi L Q, Wang B Y, Hao X P, Zhao G Q 2007 Chin.Phys.Lett. **24**, 8
[10] Shultz P J and Lynn K G 1988 Rev. Mod. Phys. **60**, 701
[11] Puska M J and Nieminen R M 1994 Rev. Mod., Phys. **66**, 814
[12] Veen A van, Schut H, Vries J de, Hakvoort R A and Ljpma M R 1990 *Positron beams for Solids and Surfaces* edited by Schultz P J, Massoumi G R, and Simpson P J, AIP conf. Proc. **218**, 171
[13] Zhou Y L, Hou Q, Wang J, Deng A H 2009 Chin. Phys. Lett. **26**, 9
[14] Hou Q, Zhou Y L, Wang J, Deng A H 2010 J. Appl. Phys. **107**, 084901

Positron and positronium studies of silica aerogel

CJ Edwardson, MDW Grogan, TA Birks, PG Coleman

Department of Physics, University of Bath, Bath BA2 7AY, UK

E-mail: C.J.Edwardson@bath.ac.uk

Abstract. Variable-energy positron annihilation spectroscopy (VEPAS) has been used to probe the structure of gold nanoparticles suspended in a silica aerogel matrix by measuring the annihilation parameters S and W, and the ortho-positronium (o-Ps) fraction parameter F, as a function of incident positron energy. Correction methods have been developed to improve the sensitivity of F by removing background contributions to the number of recorded o-Ps annihilation events. Charging effects have also been observed and investigated.

1. Introduction

Aqueous colloids (suspensions) of gold nanoparticles in water have attracted attention because of applications in photonics technologies such as ultrafast switching, surface-enhanced Raman scattering, and plasmonic sensing [1]. Silica aerogel is a highly porous form of silica glass made of a rigid, open network of tiny intersecting silica strands about 5 nm thick. The average size of the pores between the strands is 40 nm, and over 90% of the total volume is air. The refractive index of aerogel is ~ 1.05 and the optical loss due to Rayleigh scattering can be as low as 0.1 dB/cm at a wavelength of 900 nm [2].

Gold nanoparticle-implanted aerogels can be incorporated into waveguides using techniques developed by Bath's Centre for Photonics and Photonics Materials (CPPM) [3,4]; the resulting devices are stable with time, as the nanoparticles are 'locked in' by the silica network.

Past positron studies such as those by Hyodo and colleagues have looked at silica aerogel surface and bulk positronium (Ps) interactions using positron age–momentum correlation [5] as well as angular correlation of annihilation radiation, positron lifetime and Doppler broadening [6]. Other studies have used silica aerogel as a Ps formation medium [7]. Initially, the study described here was aimed at using VEPAS to investigate the gold nanoparticles within the aerogel medium, but it was found that the positron response was dominated by the structure of the aerogel itself.

2. Experimental details

2.1. Aerogel formation

In previous reports, doped aerogels have suffered from nanoparticle aggregation or other limitations. The CPPM group made sol using aqueous gold colloid, resulting in a purple-looking wet gel and aerogel, Fig. 1 (middle). Using TEM it was confirmed that the purple colour is correlated with nanoparticle aggregation due to the presence of methanol. A two-step procedure was developed to stop this in wet gels of the same composition. Using the modified wet gel process prevents aggregation while retaining control over gel time, allowing nanoparticles of diameter ~ 40 nm to be embedded at various concentrations in aerogel, Fig. 1 (sample (3), bottom) and Fig. 2. The samples used during this experiment were those shown in Fig. 1, with ~30% gold in (2) and (3). The samples were all initially cleaved in half down their lengths to reveal as clean a surface as possible for the measurements.

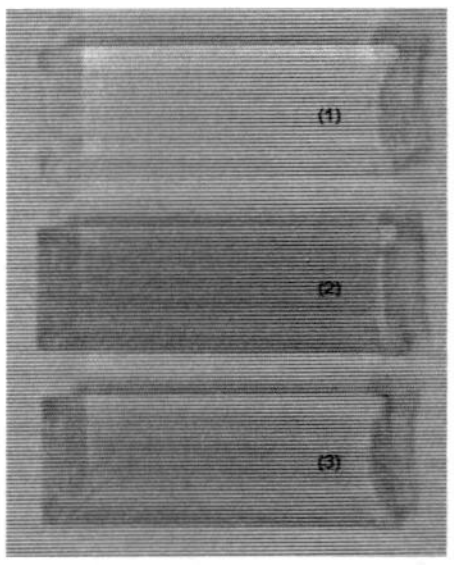

Figure 1. Aerogel samples: (1) (top) with no nanoparticles: (2) (middle) with clustered nanoparticles: (3) (bottom) with ~ 40nm diameter spherical nanoparticles.

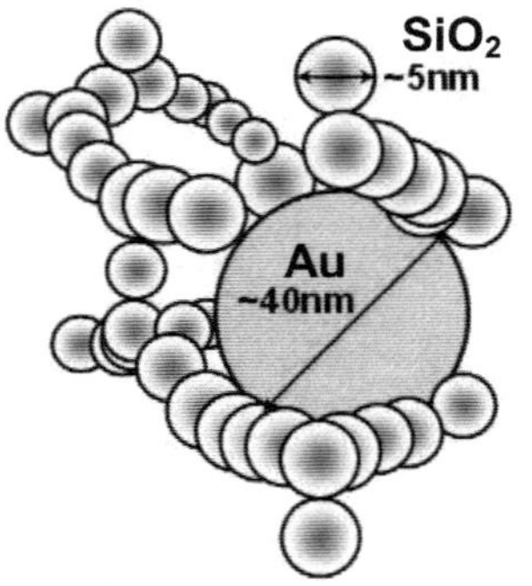

Figure 2. Aerogel: a rigid open network of tiny intersecting silica strands with air-filled pores between them (on average about 40 nm across). Over 90% of the total volume is air.

2.2. Slow positron beam

All the measurements described herein were performed on the magnetic-transport positron beam at the University of Bath [8]. The energy spectra of annihilation gamma photons from the sample were recorded by a Ge detector, yielding the Doppler-broadened linewidth parameters S and W [9] and the parameter $P:V$ – the ratio of spectrum counts in the 511keV annihilation photopeak to those in the valley region between ~ 400-500 keV - for each incident positron energy E from 0.5 – 30 keV (corresponding to mean depths of 0.13 – 92 μm). Beam energy and position, together with all aspects of data collection, are computer-controlled.

3. Data analysis

3.1. Normalisation method for S and W

Doppler broadening parameters can be affected by drifts in the Ge detector resolution and electronic instabilities. The resultant changes in both the S and W parameters can be larger than the real changes within and between data sets, rendering comparisons meaningless. In order to minimize the effect of such drifts the sample is positioned on the holder above a reference sample – here the aerogel sample without nanoparticles - in the vertical plane with a ~3 mm gap between. The two orthogonal trim coils, normally used to position the beam at different energies, then deflect the ~8 mm diameter 30 keV beam by ~11 mm from the base of the sample of interest to the top of the reference sample, and spectra are collected (and S and W evaluated) alternately for the two bulk samples. The average ratio between the parameters at 30 keV is then used to evaluate normalised S and W relative to the reference sample. Figs. 3 and 4 show the difference between the results before and after this procedure.

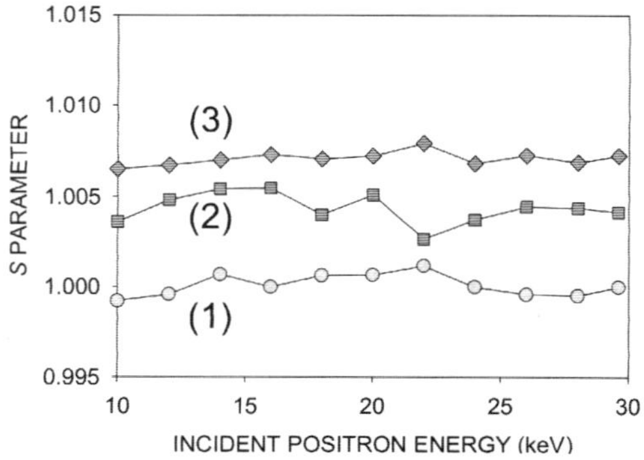

Figure 3. S parameters for samples (1)-(3) normalised to average bulk S value for sample (1).

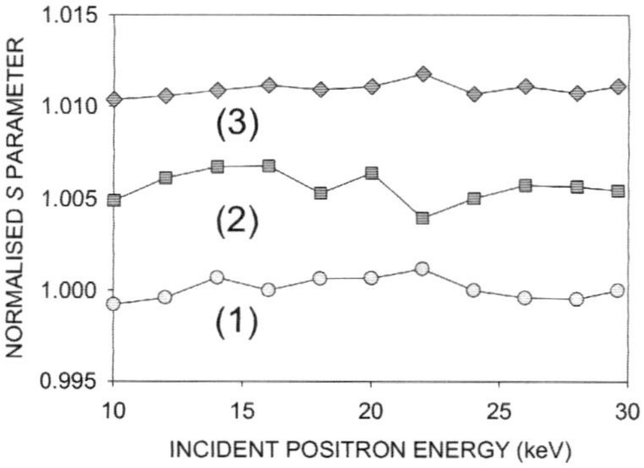

Figure 4. S-Parameter data normalised using the alternating sample method.

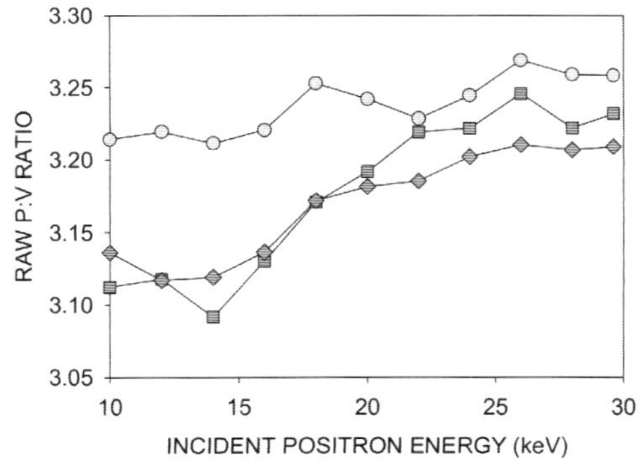

Figure 5. Raw bulk measurements for P:V ratio for all samples.

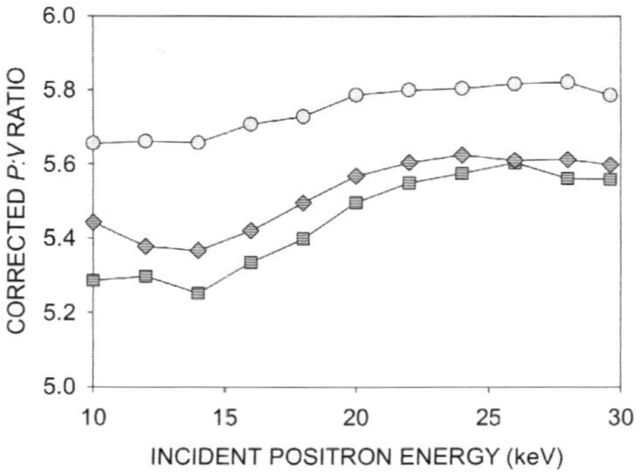

Figure 6. P:V ratios after correction method was applied

3.2. Correction method for P:V ratio

The *P:V* ratio is particularly sensitive to the probability that ortho-positronium (o-Ps) is formed in the aerogel and survives to decay into three gamma photons. (Higher *P:V* means less o-Ps.) o-Ps decay photons contribute to the valley counts *V*.

V, however, also contains a background contribution associated with the peak *P* (due to incomplete charge collection in the Ge detector) as well as from other background sources. To correct for this a background spectrum (collected with the slow positron beam switched off) was measured and background counts V_B subtracted from a Si spectrum (with no Ps) to find the contribution V_P to *V* from incomplete charge collection as a fraction of *P*. Using this information, V_B and V_P were computed and subtracted from each measured *V*, which was then used to calculate the 'true' P:V ratio.

Fig. 5 shows the raw *P:V* ratios for the three samples. The corrected ratios in Fig. 6 show a fourfold increase in sensitivity to o-Ps annihilation between samples 1 and 2. The raw *P:V* for Cz Si had an average *P:V* = 8.4 whereas when the corrected ratio was 18.9.

4. Charging Effects

Charging effects in sample (1) were investigated by consecutively repeating one *S*(*E*) run (E = 0.5 to 30 keV) ten times (Fig. 7). Each run took 12.5 h. The lower *S* parameters at $E < 1.5$ keV are associated with epithermal positron annihilation. Over time the charging effect increased, with the formation of a saturated state in the 1.5–4.5 keV region, shown also in the *S-W* plot in Fig. 8. This effect causes the *P:V* ratio to increase (and o-Ps formation to decrease) dramatically with time in the 1.5–4.5 keV region. The subsurface field appears to be drifting positrons toward the surface at epithermal energies, with a consequently lower Ps formation probability.

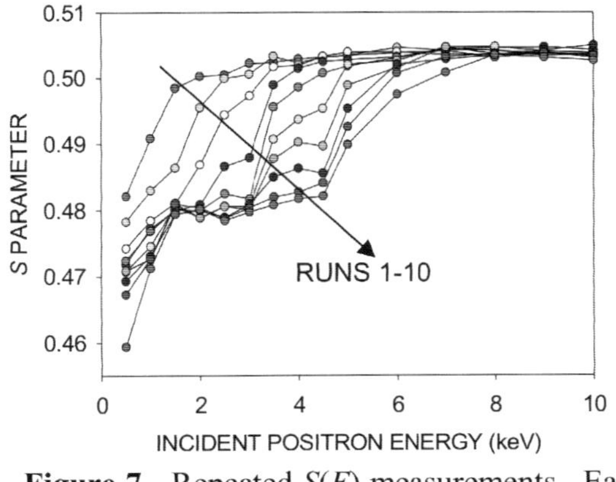

Figure 7. Repeated *S*(*E*) measurements. Each run lasted 12.5 h.

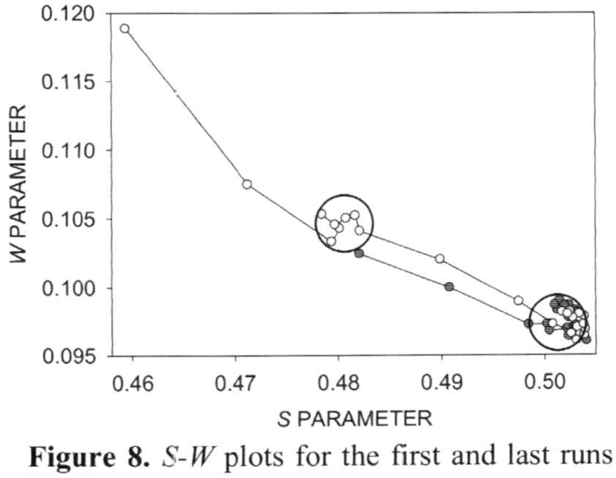

Figure 8. *S-W* plots for the first and last runs of Fig.7.

Journal of Physics: Conference Series **262** (2011) 012018 doi:10.1088/1742-6596/262/1/012018

5. Final Results

Whereas there is a real difference between the normalised *S*-Parameters for the three cleaved aerogel samples (Fig. 4), there is a much smaller difference between corrected *P:V* ratios the two gold samples (Fig.6), but there is a difference between the gold and plain silica aerogel sample.

To test whether the gold nanoparticles could be a catalyst for increased o-Ps formation, more samples were obtained with various gold nanoparticle concentrations between 0 and 29% (Fig. 9). The first set of results used the standard normalisation method described above. In the second set measurements were made separately at the centre of the samples with no normalisation. The third set is identical to the first except that the centres of each sample are measured rather than the ends. The fourth set comprises two repeats of the first measurement of the 14% Au sample. The fifth set is a data point from normalisation run of the original cleaved sample (3). The point at 0% Au is for sample (1).

Some of the results show agreement but others have differences far outside statistical fluctuations, suggesting that positrons and Ps are very sensitive to differences in the structure of the aerogel itself.

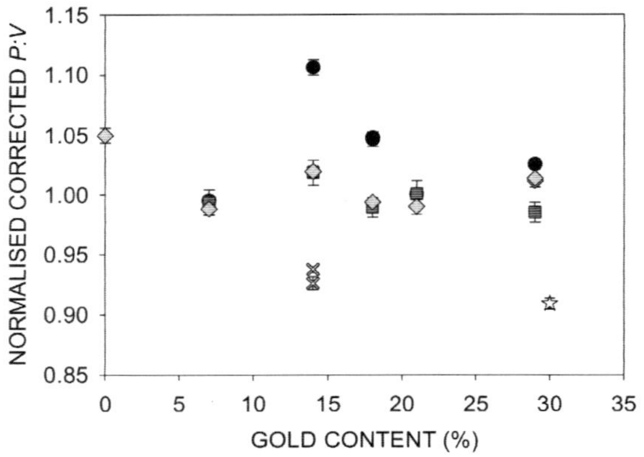

Figure 9. Normalised P:V ratios for different concentrations of gold nanoparticles and for different locations on the samples. ● – sample ends: □ - individual runs: ◊ - sample centres: ✕ – repeats on sample ends: ★ - cleaved sample.

6. Conclusions

Doped silica aerogel was investigated initially to look at the positron response to the gold nanoparticles, the *P:V* ratio being the parameter of choice. Techniques were developed to improve both the sensitivity and reliability of the *P:V* ratio and *S*-Parameter. The findings suggest, however, that while *P:V* was sensitive to the presence of the nanoparticles, there is little or no sensitivity to aggregation. Further investigations with different concentrations of gold nanoparticles revealed that the differences and similarities we originally found appear to be dependent upon the on the face and lateral part of sample probed by the positrons. We have to conclude that the structure of the silica aerogel itself may dominate VEPAS results, and further work may lead to information on aerogel structure of importance to its novel applications in photonics and other fields.

References

[1] Souza RF, Alencar MARC, da Silva EC, Meneghetti MR and Hickmann JM 2008 *Appl. Phys. Lett.* **92,** 201902.
[2] Grogan MDW, Rollings MD, Xiao LM, Wadsworth WJ, England R, Maier SA and Birks TA 2010 *Quantum Electronics and Laser Science*, OSA Technical Digest (CD), JThE21.
[3] Xiao L, Grogan MDW, Leon-Saval SG et al. 2009 *Opt. Lett.* **34,** 2724.
[4] Grogan MDW, Leon-Saval SG, Williams R, England R and Birks TA 2009 *CLEO*, OSA Technical Digest (CD), CMW3.
[5] Suzuki N, Oonishi T, Hyodo T et al 2002 *Appl. Phys. A* **74,** 791.
[6] Saito H and Hyodo T 1999 *Phys. Rev. B* **60**, 11070.
[7] Nagashima Y, Morinaka Y, Kurihara T et al 1998 *Phys. Rev. B* **58** 12676.
[8] Chilton NB and Coleman PG 1995 *Meas. Sci. Technol.* **6,** 53.
[9] van Veen A, Schut H and Mijnarends PE 2000 *Positron Beams and their Applications* ed P G Coleman (World Scientific: Singapore) p 191.

Journal of Physics: Conference Series **262** (2011) 012019 doi:10.1088/1742-6596/262/1/012019

Theoretical aspects of studies of oxide and semiconductor surfaces using low energy positrons

N G Fazleev[1,2], **W B Maddox**[1] **and A H Weiss**[1]
[1] Physics Department, University of Texas at Arlington, Arlington, TX 76019, USA
[2] Physics Department, Kazan Federal University, Kazan 420008, Russian Federation

E-mail: fazleev@uta.edu

Abstract. This paper presents the results of a theoretical study of positron surface and bulk states and annihilation characteristics of surface trapped positrons at the oxidized Cu(100) single crystal and at both As- and Ga-rich reconstructed GaAs(100) surfaces. The variations in atomic structure and chemical composition of the topmost layers of the surfaces associated with oxidation and reconstructions and the charge redistribution at the surfaces are found to affect localization and spatial extent of the positron surface-state wave functions. The computed positron binding energy, work function, and annihilation characteristics reveal their sensitivity to charge transfer effects, atomic structure and chemical composition of the topmost layers of the surfaces. Theoretical positron annihilation probabilities with relevant core electrons computed for the oxidized Cu(100) surface and the As- and Ga-rich reconstructed GaAs(100) surfaces are compared with experimental ones estimated from the positron annihilation induced Auger peak intensities measured from these surfaces.

1. Introduction

Oxide and semiconductor surfaces play a fundamental role in modern science and technology. Unlike conventional Auger techniques that average over several atomic layers, positron annihilation induced Auger electron spectroscopy (PAES) has a very high degree of top layer selectivity due to the fact that the positrons are trapped just outside the surface prior to annihilation, hence is ideal to probe the near surface region [1]. Recently this technique has been applied to study the oxidized Cu(100) single crystal to gain a better understanding of the corrosive and catalytic processes associated with thin film metal oxides and the GaAs(100) surface that exhibits a variety of surface reconstructions, roughly classified as As-rich and Ga-rich phases [2]. The experimental annihilation probability, p_{nl}, of surface trapped positrons with Cu *3p* core electrons estimated from the measured intensity of the positron annihilation induced Cu $M_{2,3}VV$ Auger peak shows a large increase as the oxidized Cu(100) single crystal is subjected to a series of isochronal anneals in vacuum up to annealing temperature 300 ºC, which then decreases monotonically as the annealing temperature is increased to ~550 ºC [3]. The PAES spectrum from GaAs(100) displays six As and three Ga Auger peaks corresponding to $M_{4,5}VV$, M_2M_4V, $M_{2,3}M_{45}M_{45}$ Auger transitions for As and $M_{2,3}M_{4,5}M_{4,5}$ Auger transitions for Ga. Experimental positron annihilation probabilities p_{nl} with As *3p*, *3d*, and Ga *3p* core electrons estimated from the Auger peak intensities are equal to 0.51%, 0.92%, and 0.33%, respectively [4]. The experimental positron work function, Φ_p, is equal to -0.6±0.1 eV [4]. Because of the variety of reconstructions of the GaAs(100) surface, positron annihilation probabilities are expected to be affected by atomic structure and elemental composition of the topmost layers of the surface [5].

This paper presents the results of theoretical studies of positron surface states and annihilation characteristics of surface trapped positrons with relevant core electrons at the oxidized Cu(100) surface and at both As-rich and Ga-rich reconstructed GaAs(100) surfaces. Such studies are indispensable for clarifying localization and spatial extent of the positron surface-state wave function at oxide and semiconductor surfaces, and the surface atomic structure and chemical composition dependence of probabilities of annihilation of surface-trapped positrons with relevant core-level electrons, the information needed to explain the experimental PAES data.

2. Theory

The positron potential $V(\mathbf{r})$ at a semiconductor surface contains an electrostatic Hartree potential $V_H(\mathbf{r})$ and a correlation part $V_{corr}(\mathbf{r})$. $V_H(\mathbf{r})$ is constructed as a superposition of atomic Coulomb potentials. Atomic calculations were performed self-consistently within the local-spin-density approximation [6] using the exchange-correlation functional from Ref. [7]. We followed the method of Weinert and Watson [8] to account for the effects of charge redistribution at the surface and placed the atoms in a "compensating" potential well extending from the atom center out to one Wigner-Seitz radius and then linearly ramping to a value of 0.00 Ry at twice the Wigner-Seitz radius and beyond. The atomic wave functions then provided the overlapping electron densities and corresponding atomic Coulomb potentials via Poisson's equation. $V_{corr}(\mathbf{r})$ is described in the bulk using the local density approximation (LDA) [5, 9]. Outside the surface $V_{corr}(\mathbf{r})$ is described by the image-type potential that is constructed to have the same corrugations as the total electron density [5, 10]. Positron surface states were obtained by solving Schrödinger's equation numerically with the boundary condition that the positron wave function vanishes sufficiently far into the bulk or the vacuum. Positron annihilation rates, $\lambda_{n,l}$, with specific core electrons, described by quantum numbers n and l, and the total annihilation rate, λ, (and corresponding positron surface state lifetime, τ,) are calculated taking electron-positron correlation effects explicitly into account using the LDA [5, 9]. The probability of annihilation of surface trapped positrons, $p_{n,l}$, with electrons in the core level n,l is obtained dividing $\lambda_{n,l}$ by λ : $p_{n,l} = \lambda_{n,l} / \lambda$.

3. Results and discussion

Calculations of positron surface states and annihilation probabilities of surface trapped positrons with relevant core electrons, $p_{n,l}$, were performed for the oxygen adsorption at surface hollow sites of the unreconstructed Cu(100) surface. We considered 1/8, 1/4, 1/2 and one monolayer (ML) oxygen coverages. Calculations were also performed for the missing row (MR) reconstructed Cu(100) surface with 0.5 ML on-surface and 0.5 ML sub-surface oxygen coverages. In the latter case the adsorbed O atoms occupy surface hollow and subsurface octahedral sites. All surfaces were fully relaxed in calculations. The computed binding energy, E_b, for a positron trapped at the oxidized Cu(100) surface, the positron surface-state lifetime, τ, and the annihilation probabilities of the surface trapped positron with relevant core electrons, $p_{n,l}$, are given in Table 1.

Table 1. Theoretical positron binding energy, E_b, positron surface-state lifetime, τ, annihilation probabilities of the surface trapped positron with relevant core electrons, $p_{n,l}$, computed for the oxidized Cu(100) surface.

System	E_b (eV) (theory)	τ (ps) surface state (theory)	$p_{n,l}$ (%) surface (theory) Cu 3s	Cu 2p	Cu 2s+2p	O 1s
Cu(100)	2.75	461	1.766	6.990	8.756	
Cu(100) + O (1/8 ML)	3.14	503	1.203	4.806	6.009	0.030
Cu(100) + O (1/4 ML)	3.12	495	1.043	4.172	5.215	0.067
Cu(100) + O (1/2 ML)	3.08	470	0.748	2.991	3.740	0.142
Cu(100) + O (1 ML)	2.94	448	0.398	1.595	1.993	0.227
Cu(100)+O (0.5 ML on-surf. +0.5 ML sub-surf.)	3.02	394	0.322	1.406	1.728	0.089

It follows from Table 1 that E_b, τ, and $p_{n,l}$ are sensitive to atomic structure and elemental composition of the topmost layers of the oxidized Cu(100) surface. The calculated $p_{n,l}$ with Cu *3s* and *3p* (O *1s*) core electrons increase (decrease) with the decrease of the oxygen coverage. These results are consistent with the observed increase (decrease) of the intensity of the Cu $M_{2,3}VV$ (O KLL) Auger signal as a result of annealing up to 300 °C, which causes O atoms to desorb or diffuse into the bulk. We interpret the decrease (increase) of the intensity of the Cu $M_{2,3}VV$ (O KLL) Auger signal observed after the annealing temperature exceeds 300 °C as being due to diffusion of oxygen atoms from the bulk back to the surface where it would produce a re-oxidation of the surface. Calculations confirm that the oxygen overlayer pushes the positron wave function from the top layer of Cu atoms away into the vacuum reducing (increasing) its overlap with Cu (O) atoms. Thus, re-oxidation leads to a large decrease (increase) in the positron annihilation probabilities with Cu *3s* and *3p* (O *1s*) core electrons, and hence to a large decrease (increase) in the intensity of Cu $M_{2,3}VV$ (O KLL) Auger signal in agreement with the experiment.

Calculations of positron surface states, positron work function and annihilation characteristics of the surface trapped positron with relevant core electrons were also performed for both Ga-rich and As-rich, ideally terminated, non-reconstructed GaAs(100) surfaces, and for the GaAs(100) surfaces with c(2×8), (2×4) and c(4×4) reconstructions. Theoretical positron binding energy, E_b, positron work function, Φ_p, and positron annihilation probabilities with relevant core electrons, $p_{n,l}$, computed for both As-rich and Ga-rich GaAs(100) surfaces are given in Tables 2 and 3, respectively.

Table 2. Theoretical positron binding energy, E_b, positron work function, Φ_p, and annihilation probabilities of the surface trapped positron with relevant core electrons, $p_{n,l}$, computed for the As-rich GaAs(100) surface.

Surface	E_b (eV) (theory)	Φ_p (eV) (theory)	Annihilation probability (%) Core level			
			As 3p	As 3d	Ga 3p	Ga 3d
Nonreconstructed GaAs(100)	2.63	-0.07	0.50	1.63	0.08	0.37
Reconstructed GaAs(100)-c(2×8)	2.54	-1.17	0.43	1.30	0.18	0.76
Reconstructed GaAs(100)-(2×4)	2.58	-0.95	0.36	1.15	0.16	0.71
Reconstructed GaAs(100)-c(4×4)	2.59	-0.68	0.66	2.06	0.01	0.04

It follows from the results presented in Tables that the computed values of E_b and Φ_p are sensitive to the surface atomic structure and chemical composition of the top layers of the reconstructed GaAs(100) surface. Calculations confirm that localization and spatial extent of the positron surface-state wave function are both strongly affected by charge transfer, surface reconstructions and elemental content of the topmost layers of the GaAs(100) surface (see Figures 1 and 2). Theoretical $p_{n,l}$ given in Tables 2 and 2 are also found to be sensitive to reconstruction and chemical composition of the topmost layers of the GaAs(100) surface.

Table 3. Theoretical positron binding energy, E_b, positron work function, Φ_p, and annihilation probability of surface trapped positrons with core level electrons, $p_{n,l}$, computed for the Ga-rich GaAs(100) surface.

Surface	E_b (eV) (theory)	Φ_p (eV) (theory)	Annihilation probability (%) Core level			
			As 3p	As 3d	Ga 3p	Ga 3d
Nonreconstructed GaAs(100)	1.76	-0.72	0.06	0.18	0.53	2.27
Reconstructed GaAs(100)-c(2×8)	1.77	-0.66	0.37	1.15	0.15	0.62
Reconstructed GaAs(100)-(2×4)	1.80	-0.17	0.30	1.02	0.14	0.61
Reconstructed GaAs(100)-c(4×4)	1.39	-0.74	0.02	0.07	0.50	2.22

It follows from Tables II and III that the experimental positron work function and annihilation probabilities of the surface trapped positron with As *3d*, *3p*, and Ga *3p* core electrons agree best with theoretical Φ_p and $p_{n,l}$ computed for the Ga-rich reconstructed GaAs(100)-c(2×8) surface: -0.66 eV, 0.39%, 1.18%, and 0.15%.

4. Conclusions

The results of PAES studies of the oxidized Cu(100) single crystal and the GaAs(100) surface have been analyzed by performing calculations of positron surface and bulk states and positron annihilation characteristics taking into account the charge redistribution at the surface and surface reconstructions. Localization and spatial extent of the positron surface-state wave function, the positron binding energy, work function and annihilation characteristics of the surface trapped positron with relevant core-level electrons have been found to be sensitive to charge transfer, atomic structure and chemical composition of the topmost layers of the surfaces. Possible explanations have been proposed for the observed behavior of the intensity of positron annihilation induced Auger peaks measured from the oxidized Cu(100) single crystal and the GaAs(100) surface. The obtained results indicate that PAES could serve as an important surface diagnostic tool capable of distinguishing different oxide and semiconductor surfaces and defining their state of reconstruction.

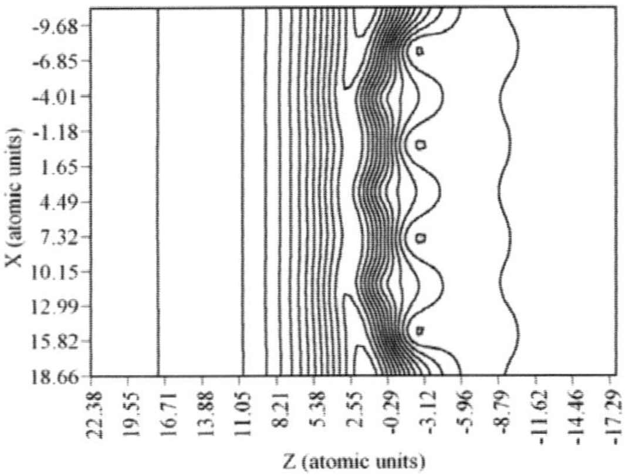

Figure 1. Contour plot in the X-Z plane for Y = 0 of the positron surface state wave function at the As-rich reconstructed GaAs(100)-(2×4) surface. Contours are separated by 0.0012 atomic units.

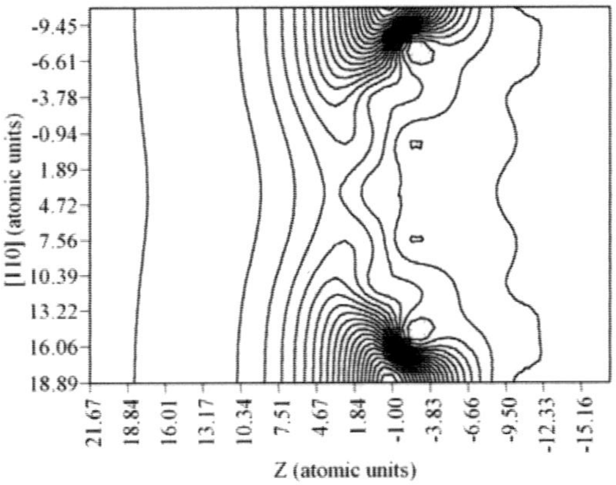

Figure 2. Contour plot in the X-Z plane for Y = 0 of the positron surface state wave function at the Ga-rich reconstructed GaAs(100)-(2×4) surface. Contours are separated by 0.0014 atomic units.

Acknowledgments

This work was supported in part by the National Science Foundation Grant #DMR-0907679, the Robert A. Welch Foundation Grant Y-1100, and Ministry of Education and Science of Russian Federation within the Program "Development of scientific potential of higher education" (Grants 2.1.1/2985 and 2.1.1/3199).

References

[1] Koymen A R, Lee K H, Yang G and Jensen K O 1993 *Phys. Rev. B* **48** 2020
[2] Mönch W 1995 *Semiconductor Surfaces and Interfaces*, 2nd edition, Berlin: Springer Verlag
[3] Nadesalingam M 2007 *PhD Dissertation*, University of Texas at Arlington
[4] Starnes S G 2001 *MS Thesis* University of Texas at Arlington
[5] Fazleev N G, Fry J L and Weiss A H 2004 *Phys. Rev. B* **70** 165309
[6] Gunnarsson O and Lundqvist B I 1976 *Phys. Rev. B* **13** 4274
[7] Ceperley D M and Adler B J 1980 *Phys. Rev. Lett.* **45** 566
[8] Weinert M and Watson R E 1984 *Phys. Rev. B* **29** 3001
[9] Boronski E and Nieminen R M 1986 *Phys. Rev. B* **34** 3820
[10] Nieminen R M and Puska M J 1983 *Phys. Rev. Lett.* **50** 281

Journal of Physics: Conference Series **262** (2011) 012020 doi:10.1088/1742-6596/262/1/012020

Homogeneous porous silica for positronium production in AEgIS

R Ferragut[1,2†], A Dupasquier[1], A Calloni[1], G Consolati[2,3], F Quasso[3], M P Petkov[4], S M Jones[4], A Galarneau[5] and F Di Renzo[5]

[1] LNESS and Dipartimento di Fisica, Politecnico di Milano, via Anzani 42, 22100 Como, Italy

[2] Istituto Nazionale di Fisica Nucleare, via Celoria 16, 20133 Milano, Italy

[3] Dipartimento di Fisica, Politecnico di Milano, piazza L. da Vinci 32, 20133 Milano, Italy

[4] Jet Propulsion Laboratory - NASA, California Institute of Technology, 4800 Oak Grove Drive, Pasadena, CA 91109, U.S.A.

[5] Institut Charles Gerhardt - MACS, ENSCM, 8 Rue Ecole Normale, 34296 Montpellier, France

E-mail: rafael.ferragut@polimi.it

Abstract. Positronium (*Ps*) formation measurements in homogeneous porous silica (Xerogel of 85 mg cm^{-3} and swollen MCM-41 of 390 mg cm^{-3}) were performed at different temperatures (8K-293K) by means of a variable energy positron beam equipped with a Ge detector. The results indicate that Xerogel and swollen MCM-41 samples have a high *Ps* production, which is independent on the temperature. An estimation of the ortho-*Ps* mean diffusion length was obtained by measuring samples capped with an Al film (~110 nm). An efficient formation of cooled *Ps* atoms is a requisite for efficient production of antihydrogen, with the aim of a direct measurement of the Earth gravitational acceleration of antimatter, a primary scientific goal of AEgIS (Antimatter Experiment: gravity, Interferometry, Spectroscopy, CERN). Porous materials with open pores at the surface are necessary to produce a high yield of *Ps* atoms as well as to cool *Ps* through collisions with the inner walls of the pores before emerging in the free space outside the target. The results indicate that Xerogel and swollen MCM-41 are good candidates for an efficient formation of cold *Ps* atoms within the framework of the AEgIS project.

1. Introduction

Some fundamental questions of modern physics relevant to unification of gravity with the other fundamental interactions, models involving vector and scalar gravitons, matter–antimatter symmetry can be enlightened via experiments with antimatter [1]. The AEgIS experiment has been approved at the CERN Antiproton Decelerator to directly measure the gravitational acceleration g of a beam of cold antihydrogen [2-4]. Gravitational interaction between matter and antimatter has never been tested experimentally.

† To whom any correspondence should be addressed.

An efficient formation of cooled positronium (*Ps*) atoms is a requisite for the production of antihydrogen ($\overline{H}$) in the AEgIS experiment. Porous materials are necessary to produce a high yield of *Ps* atoms as well as to cool *Ps* through collisions with the inner walls of the pores. The collisions between *Ps* and the internal surface of the pores involve a weak coupling to phonons or other surface modes. The velocity distribution of the *Ps* atoms coming out of the target should be the order of 10^4 m/s to allow *Ps* laser excitation to a Rydberg state (Ps^*) [5, 6] and for efficient $\overline{H}$ formation, which requires that the relative velocity of antiprotons and Ps^* must be not higher than the classical orbital velocity of the positron in the Rydberg *Ps* atom. The formation of cold Rydberg antihydrogen $\overline{H}^*$ will be possible by means of the charge exchange reaction between Ps^* and cold $\overline{p}$

$$Ps^* + \overline{p} \rightarrow \overline{H}^* + e^-, \tag{1}$$

with a residual electron (more details in Refs. 4 and 7).

The porous materials tested in the present work (Xerogel and swollen MCM-41) were characterized by means of a variable energy positron beam with a high purity germanium detector. The results indicate that Xerogel and swollen MSM-41 are appropriate candidates for efficient formation of cold *Ps* atoms for the AEgIS project.

2. Experimental procedure and methodology

Silica Xerogel samples were fabricated using a procedure developed for particle capture collectors of the NASA Stardust project [8]. The sample density was controlled by mixing different amounts of acetonitrile into the sol. The solution was then dried at room temperature, which allows the solvent to be removed while the network gel remains in a highly expanded state in the form of a three dimensional array of filaments. The density of the prepared Xerogel sample is 85 mg cm^{-3} with a porosity of about 96%. MCM-41 samples were prepared according to Refs. 9 and 10. The MCM-41 samples were synthesized at 115°C by using cetyltrimethylammonium bromide (CTAB, Aldrich), 1,3,5-trimethylbenzene (TMB, Aldrich), pyrogenic silica (Aerosil 200 Degussa), sodium hydroxide (Prolabo), and deionized water in molar ratios 1 SiO_2/0.26 NaOH/0.035 $NaAlO_2$/0.1 CTAB/20 H_2O/*x* TMB. The material was then filtered, washed with water, and dried at 80°C for 24 h. The solids were then calcined in air at 550°C for 8 h. The density of the prepared swollen MCM-41 samples is 390 mg cm^{-3} with a mesoporous and macroporous volume of 1.65 cm^3 g^{-1} and 0.5 cm^3 g^{-1}, respectively.

The measurements of *Ps* formation yield were performed by means of a monoenergetic positron beam using the well-known "3γ method" [11-13]. The positron implantation energy was variable from 0.9 keV up to 18 keV. The measurements were performed at different temperatures from 8 K up to room temperature (RT≡ 293 K) with a vacuum level between 10^{-8}-10^{-10} mbar. The samples were put in contact with the cold-finger of a cryostat. A high purity germanium (HPGe) detector was used to obtain the annihilation γ spectrum. From the annihilation spectrum it was possible to obtain the $R(E)$ parameter

$$R(E) = \frac{V}{P}, \tag{2}$$

where *P* and *V* are the integrated counts (after background subtraction) in the peak (511 ± 4.25 keV) and valley area (after the Compton edge, from 350 keV up to 500 keV) of the energy γ spectrum. When there is no pick-off annihilation, i.e. when the *Ps* is emitted in vacuum, the positronium fraction $f(E)$ can be obtained from $R(E)$ using the relationship:

$$f(E) = \left[1 + \frac{P_1(R_1 - R(E))}{P_0(R(E) - R_0)}\right]^{-1}, \tag{3}$$

where P_0 and R_0 are the parameters when the positronium fraction is nil ($f = 0$) and P_1 and R_1 are the parameters when 100% of the positrons becomes *Ps* ($f = 1$). The calibration for 0% and ~100% of *Ps* in vacuum was performed using a Ge single crystal (100) at 1000 K [12]. In all other cases we report the result obtained from Eq. (3) as "3γ fraction" $F_{3\gamma}$ ($f(E) = F_{3\gamma}$). We warn however that this nomenclature, which is adopted by other authors [14,15], is somewhat misleading, since the true fraction of positrons annihilated in 3γ is smaller by a factor ¾. Nevertheless $F_{3\gamma}$ is the useful parameter to assess the potential of a material to emit o-*Ps* that has lost kinetic energy by collision while meandering through open pores.

3. Results and discussion

Figure 1 shows the 3γ fraction $F_{3\gamma}$ as a function of the positron implantation energy in Xerogel and MCM-41 samples (full symbols) obtained at room temperature (RT). The *Ps* yield is very high at all implantation energies. Similar results were found in Aerogel with a density of 150 mg cm^{-3} [16]. The decrease of $F_{3\gamma}$ observed in Fig. 1 (full symbols) at low implantation energies is symptomatic of *Ps* escape in the vacuum. This is an instrumental effect related to a decreased efficiency of detection of ortho-*Ps*.

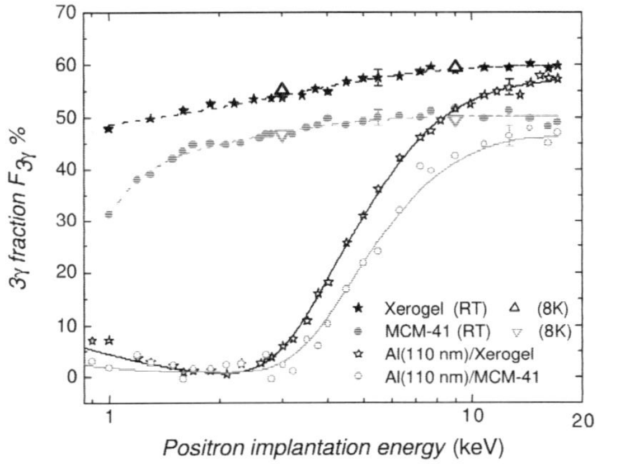

Figure 1. Positronium 3γ fraction $F_{3\gamma}$ as a function of the positron implantation energy in Xerogel and MCM-41 in homogeneous and capped samples. The dashed lines are only a visual guide. The continued lines are fits of the VEPFIT model. Error bars are shown for one point only in each evolution.

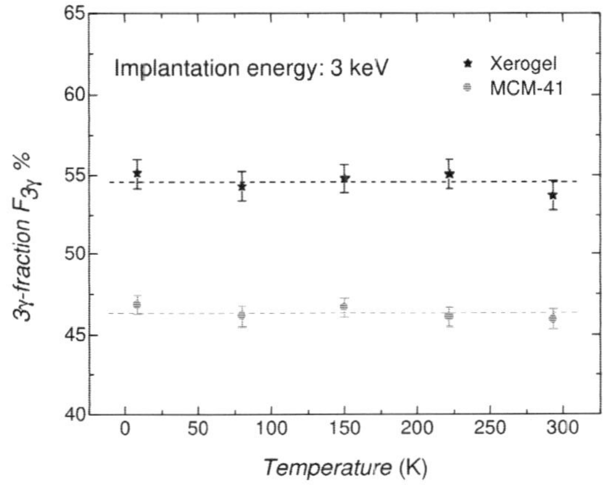

Figure 2. Positronium 3γ fraction $F_{3\gamma}$ as a function of the temperature for positron implanted inside the sample at 3 keV in Xerogel and swollen MCM-41 samples.

The porous samples were capped with an Al film of 110 nm to avoid the *Ps* escape (open symbols in Fig. 1). *Ps* formation at about 2 keV is almost zero in capped samples. In this case about 100% of positrons are implanted inside the Al film where *Ps* formation is not expected. The experimental data were fitted by VEPFIT [17] with a two-layer model (Al/porous material). VEPFIT takes into account the positron implantation distribution as a function of the implantation energy and the material density. The continuous line through the experimental data in Fig. 1 corresponds to the best-fit values of *Ps* diffusion lengths. The ortho-*Ps* diffusion lengths were estimated in (3.5 ± 0.3) μm and (0.56 ± 0.05) μm in Xerogel and MCM-41 samples, respectively. This model is based on the assumption that the diffusion coefficient does not depend of the *Ps* temperature during cooling. A full quantum mechanical treatment is required to properly describe the regime in which *Ps* is confined in a system whose dimensions are of the same order as the *Ps* de Broglie wavelength. However, these high diffusion lengths are consistent with a high ortho-*Ps* emission outside the material also when positrons are implanted at relatively high energies. *Ps* atoms are emitted from the pore walls with high kinetic energy (1-2 eV). Low *Ps* kinetic energies are desirable for AEgIS. The mean kinetic energy of ortho-*Ps* was estimated in Xerogel at room temperature. A value of 36 meV was obtained by means of Coincidence Doppler Broadening with and without the application of a static magnetic field (~0.3 T)

[18] (with a procedure similar to that used by Nagashima et al. [19]). This result indicates that ortho-*Ps* cooling in homogeneous porous silica is feasible. Measurements of the ortho-*Ps* emission energy in vacuum for these materials using *Ps* time of flight (*Ps*-TOF) are underway in homogeneous silica samples. *Ps*-TOF results were recently reported in Refs. 20 and 21 in silicon and silica ordered nano-channels films. Mills et al. [22] observed ortho-*Ps* thermalization also at low temperature in SiO_2 compressed powder (180 mg cm^{-3}). The results of Ref. 22 indicate: 2% at 4.2 K, 8% at 77 K and 12% at 300 K of thermal ortho-*Ps* with positron implantation energies of 19 keV, 10 keV and 7 keV, respectively. These high implantation energies require rather thick samples (several microns, for materials with low energies), which is a condition easily met with homogenous porous samples. However a compromise must be reached between efficient cooling, which requires deep positron implantation, and high ortho-*Ps* yield outside the target, which is favoured by implantation depths not large in comparison with the ortho-*Ps* diffusion length. Low-density silica homogeneous materials are promising target candidates under two aspects: high formation and long survival of ortho-*Ps* inside the pores.

The temperature dependence of the *Ps* yield was studied since the AEgIS experiment will be performed at very low temperature (100 mK). Figure 1 shows that at 3 and 9 keV the values of $F_{3\gamma}$ measured at RT and at 8 K (open triangles) in Xerogel and MCM-41 samples are independent on the temperature within the experimental errors. More detailed information of the *Ps* yield as a function of the temperature in both materials for positrons implanted in the sample at 3 keV is presented in Fig. 2. These results indicate that in homogeneous silica-based materials the high *Ps* production does not depend on temperature. A similar result was found in silica films [21].

References

[1] Hughes R J 1993 *Nucl. Phys.* A **558** 605
[2] AEgIS web site: http://aegis.web.cern.ch/aegis/
[3] Kellerbauer A et al (AEgIS collaboration) 2008 *Nucl. Instr. & Methods* B **266** 351
[4] Testera G et al (AEgIS collaboration) 2008 *Proc. of Cold Antimatter Plasmas and Application to Fundamental Physics Conference (Okinawa)* vol 1037 (AIP Conference Proceedings) p 5
[5] Giammarchi M G et al (AEgIS collaboration) 2009 *Hyperfine Interact.* **193**, 321
[6] Castelli F, Boscolo I, Cialdi S, Giammarchi M G and Comparat D 2008 *Phys. Rev.* A **78** 052512
[7] Bonomi G et al (AEgIS collaboration) 2009 *Hyperfine Interact.* **193**, 297
[8] McDonnell J A M et al 2000 *Adv. Space Res.* **25** 315
[9] Kresge C T, Leonowicz M E, Roth W J, Vartuli J C and Beck J S 1992 *Nature* **359** 710
[10] Beck J S, Vartuli J C, Roth W J, Leonowicz M E, Kresge C T, Schmitt K D, Chu C T W, Olson D H, Sheppard E W, McCullen S B, Higgins J B and Schlenker J L 1992 *J. Am. Chem. Soc.* **114** 10834
[11] Marder S, Hughes V, Wu C S and Bennet W 1956 *Phys. Rev.* **103** 1258
[12] Mills A P Jr 1978 *Phys. Rev. Lett.* **41** 1828
[13] Lynn K G and Welch D O 1980 *Phys. Rev.* B **22** 99
[14] Liszkay L et al 2008 *Appl. Phys. Lett.* **92** 063114
[15] Petkov M P, Weber M H, Lynn K G and Rodbell K P 2001 *Appl. Phys. Lett.* **79** 3884
[16] R. Ferragut et al 2010 *J. Phys.: Conf. Ser.* **225**, 012007
[17] van Veen A, Schut H, Clement M, de Nijs J M M, Kruseman A and Ijpma R 1995 *Appl. Surf. Sci.* **85** 215
[18] Ferragut R et al (to be published)
[19] Nagashima Y et al 1995 *Phys. Rev.* A **52** 258
[20] Mariazzi S, Bettotti P and Brusa R S 2010 *Phys. Rev. Lett.* **104** 243401
[21] Crivelli P et al 2010 *Phys. Rev.* A **81** 052703
[22] Mills A P Jr, Shaw E D, Chichester R J and Zuckerman D M 1989 *Phys. Rev.* B **40** 2045

Journal of Physics: Conference Series **262** (2011) 012021 doi:10.1088/1742-6596/262/1/012021

Positron storage in micro-traps with long aspect ratio: results of computer simulations*

Paola Folegati[1,2], Jia Xu[2], Marc H Weber[2] and Kelvin G Lynn[2]

[1]Polo Regionale di Como, Politecnico di Milano, Via Anzani 42, 22100 Como, Italy

[2]Washington State University, Pullman, WA 99164-2711

paola.folegati@wsu.edu, kgl@wsu.edu

Abstract. A feasibility study of a non–conventional Penning-Malmberg trap is presented. The study is based on simulations performed by the WARP code. [1,2] Comparisons with previous simulations done with the Charged Particle Optics (CPO) programs [3] are presented under homogeneous magnetic field conditions; however, real B fields do not have a significant effect.. The analysis has been carried out on micro-traps of radii ranging from 10 to 100 micrometers and length of 126 mm. Different configurations of the applied electric fields have been considered and an axial uniform magnetic field of 7 T was used in most of the simulations. For the final device, the microtraps should be combined into a multiple parallel tube array. Results from the simulations with the WARP and CPO codes indicate that decreasing the microtrap radius allows an increase in the total number of positrons per macrotrap. This is due to the increased number of micro-tubes when reducing the beam radius.

1. Introduction

During the last years, positron physics is becoming increasingly more important from the technological point of view for applications in materials analysis [4]. Many of these applications require the use of a significant number of positrons and sometimes require the possibility of tuning their energy to test different depths inside a material (defect profiling). Slow positron beams have been constructed to meet these requirements and have been used for many years [5]. The possibility of storing large numbers of positrons may have other applications: (1) allow the study of Bose condensation of positronium atoms [6]: (2) provide a means for studying electron–positron plasmas in parameter regimes in astrophysics [7] (3) allow the creation of antihydrogen [8]; and (4) production of intense pulses of positronium atoms [9]. Storing positrons in "portable" devices could provide an attractive alternative to radioactive positron sources for these and other uses[10,15]. Various techniques have been considered for storing large numbers of positrons including the use of electron–positron plasmas [10] and confinement in specially designed Penning–Malmberg traps [11]. The advanced technology that can be used for the accumulation of positrons in Penning–Malmberg traps makes it one of the most attractive candidates for the accumulation of a large numbers of positrons [12]. The unavoidable limitations imposed on the number of particles that can be stored in such devices has led to the proposal of a new type of trap with long aspect ratio and portability [14,15,16,17,18].

2. Objectives

Charged plasmas can reach thermodynamic equilibrium and be confined in Penning-Malmberg type trap by static magnetic and electric fields for very long times – for an infinite time in principle. Losses arise because of trap imperfections associated with the trap components [13]. Our goal is to evaluate the maximum number of positrons that can be stored in a trap with large length-to-diameter aspect ratio and micrometer diameter (Fig. 1). A large number N_{tubes} of these microtraps, each with conducting walls, are than stacked in parallel to fill a macrotrap of volume V_{macro} inside a 7 Tesla superconducting magnet. The diameter is shrunk to micrometers, while keeping the length fixed. N_{tubes} is increased to fill V_{macro} (of 50 mm diameter and 126 mm length) with a 75% fill factor. Gain is achieved when the total number of positrons $N = N_t \times N_{tubes}$ exceeds the number of positrons that can be stored in a single trap of the size of V_{macro}. N_t is the number of positrons in a single microtrap. In addition, the conducting walls of the microtraps generate image potentials that shield positrons in one tube from those in all other tubes. This lowers the required electric potential at the end caps, bringing portable traps closer to reality.

The schematic diagram for our simulations is shown in Fig. 1. It is a cylindrical microtrap L_t=126 mm length and variable radii R_t from 5 to 50 microns (containing a positron plasma source L_S=70 mm in length, centered in the trap cylinder, and radius $R_S=R_t/\sqrt{3}$ to keep constant a scale factor of 3 between the cross sectional areas). Static electric and magnetic fields are always used. A uniform magnetic field, parallel to the trap axis, is used for the radial confinement and an electric field for the azimuthal confinement. The model device has a central grounded cylinder, three thin (1mm length) electrodes at increasing potential (2.5, 5 and 7.5 V respectively) on each side and two end-caps at 10 V. The trap simulated in CPO code does not have the grids on the end caps, but the result is the same when added.

This objective is approached within standard physics equations or codes however careful attention is paid to the aspect ratio and trap dimensions. This means limitations on the number of positrons that can be stored in the trap come essentially from two factors:

1) the Brillouin limit, which is an upper limit on the density that can be stored in a given magnetic field :

$$n_B = \frac{B^2}{8\mu_0 mc^2} \qquad (1)$$

where B is the magnetic field in the trap, m is the mass of the plasma particles, and

2) the space-charge limit,

$$V_S \cong \left(1.4 \times 10^{-7}\right)\frac{N_t}{L}\left[1 + 2\ell n\frac{R_w}{R_p}\right] \qquad (2)$$

where R_p and R_t are the trap radius and source radius, respectively, N_t is the total number of particles in the microtrap, L is the source length, V_s represents the maximum potential generated by the charge itself, and if L is in meters the result is in volts.

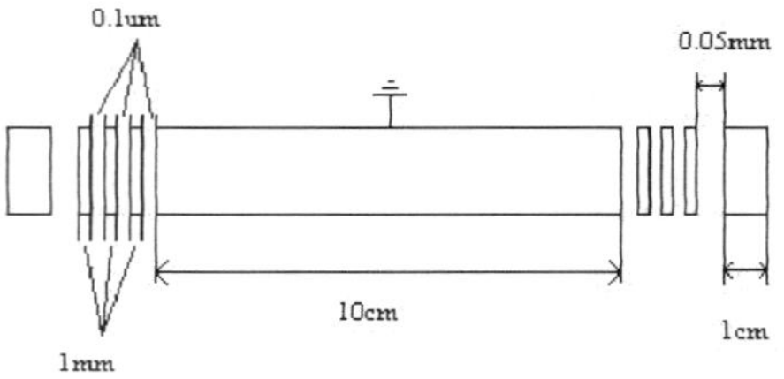

Fig.1. Schematic diagram of the trap used in WARP simulations

3. Computational tools

Two software tools were used and compared to model the positrons stored in microtraps. Charged Particle Optics (CPO) builds in single charge particle optics and adds space charge effects. Warp is designed to include space charge of high density beams and can was used for plasmas. Here, the code is used at lower density and very narrow trap dimensions. Results from each are compared.

Journal of Physics: Conference Series **262** (2011) 012021 doi:10.1088/1742-6596/262/1/012021

3.1 CPO
The CPO program package uses Boundary Element Method or surface charge method to simulate the motions of charged particles in the electromagnetic fields. It has proven to agree with the experimental results based on the simulations of low-energy ion dynamics. To evaluate the effects of space charge in the tube, the space charge tube method [4] is utilized in the ray tracing. After several test runs, a reasonable number of rays (normally 36) have been found to reach a stable result for the simulations. The code uses several iterations to evaluate the space charge effect by applying the space charge effect of the current iteration upon the particles for the next iteration. A number of rays can be traced representing the motions of charged particles in the beam. The mechanics of space charge tube method is when the program processes the space charge effect of one ray, a small tube surrounding the trajectories of each ray and the space charge of the ray will be deposited uniformly in the tube. The space charge deposited in the tube will be used for next iteration and the radius of the tube can be specified by the user.

In our case, the magnetic attraction between the rays is excluded by the CPO program. The formation of positronium (Ps) on background impurities in the vacuum system is not accounted in our simulations as the vacuum is planned to be better than 10^{-11} torr and the positron energy below the threshold for Ps formation.

In CPO all rays start at the vertical middle plane of the microtrap with randomized initial angles with respect to the horizontal axis of the tube to spread the rays across the length of the trap to simulate a plasma. Each ray simulation finishes a complete oscillation – an iteration – ending at the middle plane. A shell program[1] based on CPO program was written using C++ language to feed the final current and averaged energy of all rays from the current iteration back into CPO for the next iteration. In every iteration all rays are uniformly distributed within a circle with the radius with the circle radius $1/\sqrt{3}$ of the radius of the tube. The energy of the rays for the next iteration is based on the averaged final energy for the current iteration. The radius of the beam for the next iteration is set as 0.7* the maximum radius of rays of the current iteration. Then, the number of trapped particles is the flight time of rays staying the tube multiplying the current that the rays carry for all the rays.

3.2 WARP
The WARP code in its latest three-dimensional version, WARP3d, is a multidimensional particle-in-cell originally developed to simulate creation and propagation of high-current, space-charge dominated beams and also allows the study of plasmas. It has been used in many applications and was used for comparing the electron trapping experiments[20]. Geometry of the trap, materials constituting the trap walls and fields can be set by the user. Different boundary conditions can be applied for the solution of Poisson's equation (Dirichlet, Neumann). The WARP code allows for the use of multipile fields, time-varying fields and periodic structures necessary to simulate particle accelerators (linacs or colliders). Not only including the space-charge effects, the WARP code used Langevin collision operator to treat the Coulomb collision during the simulation[21,22].

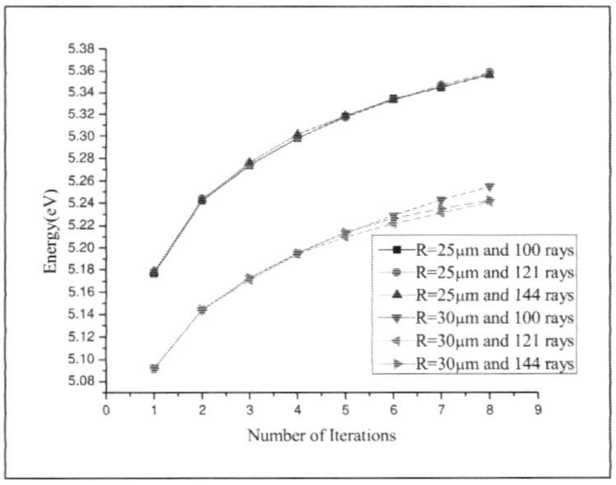

Fig.2 The energy increase as iterations proceeds with the particles trapped in a microtube with a radius of 50μm. After 1.2μs (8 iterations), the energy reaches about 5.3eV which is much smaller than 10V of end cap potentials. Initial conditions for the beam shown in this figure are 25μm and 30μm and denotes the initial radius of the beam.

4. Results

4.1CPO simulations

4.1.1 Energy increase of positrons: In Fig.2, a typical example shows the averaged energy of particles increases as a function

---
[1] A loop program was constructed by setting up the initial conditions of each ray in one iteration based on the final conditions of each ray in previous iterations, but the CPO program does not read in the initial conditions of all rays in each iteration, and the ray tracing time will take very unpractical long times.

of time in a 50μm microtrap. In Fig.2, the maximum averaged energy of the beam reaches up to 5.3eV, much lower than the end cap potentials 10V. In all cases of CPO code, regardless of the radius of the microtrap and the initial conditions of the beam, the averaged energy of the beam never exceeds about 6.4eV after about 0.003ms (20 iterations). The end cap potentials can trap the positrons with 6.4eV as well since they are much higher than the final energies of positrons. In all CPO results, no positrons are lost nearby the end caps of the tube, which confirms the increase of the energy is not a major issue for the loss of the positrons within the current time duration. The results also show the positrons always hit the electrode radically. This evidence is also found in WARP code results.

4.1.2 The number of trapped particles as a function of time: The number of particles that can be trapped in a microtrap as a function of time is shown in Fig.3. There are three phases for the behaviors of the beam: (1) At first the radius of the beam expands due to the space charge effect but does not reach the radius of the tube. Thus, there is no particle loss in this phase. (2) Then the radius of the beam reaches up to the radius of the tube, the beam starts to significantly lose positrons based on the slope of the curve shown in Fig. 2. (3) Finally, the beam reaches an equilibrium state and there is no further positron loss. In all the cases from CPO and WARP[2] codes, the beam or plasma arrives at equilibrium status.

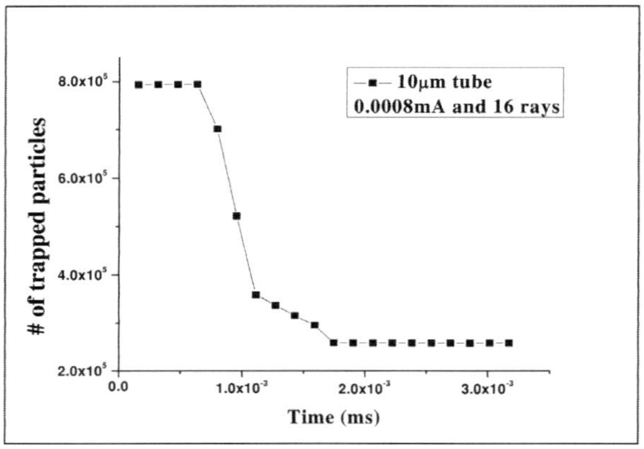

Figure 3 The number of trapped particles as a function of time. 16 rays carrying 0.8μA are traced in a tube with 10μm radius.

As the radius of one microtrap is decreased, the total number of positrons that can be trapped in one macrotrap increases as shown in Table 1. The CPO program has difficulty doing the ray tracing when the size of the tube is narrowed down to around 10μm and smaller. The number of trapped positrons in a 10μm microtrap in Table 1 shows preliminary results. More simulation work will be done.

4.2 WARP simulations

The WARP code [1,2,18] requires the definition of the number of physical particles (positrons in this case) per "simulation particle" which is named "positron weight". For this weight, a reasonable compromise has to be chosen at 100 positrons per simulation particle to limit computation time. The fraction of positrons lost in trap wall collisions was evaluated for different "weights" from 10 to 100, and the results suggest that a positron weight equal to 100 can be a reasonable choice. In all cases of WARP code, the positrons with 0.025eV of initial energy start from the source of cylindrical in the microtrap with random initial angle with respect to magnetic field ranging from 0 degree to 180 degree, while in CPO code all the positrons with 5eV start from a plane with the initial angle from 0 to 90 degree with respect to the magnetic field.

Simulations at fixed density and changing the trap size have been made to compare with CPO results, and they resulted in good agreement. Increasing the density and decreasing the trap radius up to values that are more difficult to be reached with the CPO code were simulated. The choice of parameters was based on the assumption that the behavior of a trap should be the same as positron containment for a constant value of the ratio between electrostatic and thermal energy. The electrostatic energy is indeed a space charge effect affecting the radial containment and increases as the square of the trap radius. This is the reason for going to smaller radii which allows an increase the density. The results of the simulations

[2] The Coulomb scattering time is estimated to be around 2ps for $2.0*10^{17}m^{-3}$ of plasma density and 0.025eV of positrons as used in WARP simulations, and the simulation time is larger than 100ns much longer than Coulomb scattering time. With the increased plasma density, the Coulomb scattering time will be reduced further. Thus the positron loss from Coulomb scattering is not a major factor for positron loss.

are also shown in Table 1. In the case of WARP the densities have been chosen keeping fixed the ratio between electrostatic and thermal energy.

Table 1. The number of trapped positrons with radii of microtraps. The macrotrap has a 2.5cm radius,70 mm source length and about126 mm trap length. A fill factor of 75% is accounted for in evaluating the number of microtraps contained in a macrotrap, and therefore the total number of positrons that can be stored. The first four rows refer to CPO results, the last four to WARP results.

	R=10μm		R=30μm		R=50μm	
Trap type	CPO	Warp	CPO	Warp	CPO	Warp
Density n (m^{-3})	4.8×10^{16}	5×10^{18}	4.5×10^{16}	5×10^{17}	2.3×10^{16}	2×10^{17}
Loss % from initial total number of positrons	16.7% from 1.8×10^{6}	0% from 3.7×10^{7}	14.3% from 1.4×10^{7}	0% from 3.3×10^{7}	0% from 1.9×10^{7}	0% from 3.7×10^{7}
Number of positrons per Microtrap N_t	1.5×10^{6}	3.7×10^{7}	1.2×10^{7}	3.3×10^{7}	1.9×10^{7}	3.7×10^{7}
microtraps in a macrotrap N_{tubes}	4.7×10^{6}	4.7×10^{6}	5×10^{5}	5.2×10^{5}	1.8×10^{5}	1.8×10^{5}
Total number of positrons in macrotrap	7.5×10^{12}	1.7×10^{14}	6×10^{12}	1.7×10^{13}	3.4×10^{12}	6.6×10^{12}

5. Conclusions

From these preliminary modelling results, we can confirm the advantage of smaller trap radii for increasing the total number of trapped positrons. The discrepancies between CPO and WARP are less than an order of magnitude but the overall conclusions remain the same that with decreased size of one microtrap more positrons can be trapped in one macrotrap. In the future, we will work on the simulations of positrons in a tube with a radius even smaller than 10μm and with longer simulation time.

6. Acknowledgements

The authors acknowledge very productive discussions and help from David Grote for the WARP simulations and Frank Read for the CPO simulations. *Funding for this research provided by the Office of Naval Research Award No. N00014-10-1-0543, Dr. Harold S. Coombe

7. References

[1] D P Grote et al, Proc. of the 1995 International Symposium on Heavy Ion fusion, Fusion Engineering and Design, **32-33** (1996) 193-200
[2] D P Grote, A Friedman, I Huber, W M Fawley, J L Vay, Nucl. Instrum. Methods Phys. Res. **A 415**, 428 (1998)
[3] CPO programs, free versions available at http://www.electronoptics.com/
[4] F H Read, A Chalupka and N J Bowring, "The charge-tube method for space-charge simulations." The International Journal for computation and Mathematics in Electrical and Electronic Enginnering 18, no. 4 (1999): 548-555.
[5] P J Schultz. and K G Lynn, Rev. Mod. Phys. **60** (1988) 701.
[6] Paul Coleman, Positron Beams and its applications, ed. World Scientific Books, 2004
[7] P M Platzman, A P Mills Jr, Phys. Rev. B **49**,454–458, 1994.
[8] R G Greaves, C M Surko, 2002. In: Anderegg, F., et al. (Ed.), Non-Neutral Plasma Physics, Vol. IV. American Institute of Physics, Melville, NY, pp. 10–23.
[9] M Amoretti, C Amster, G Bonomi, et al., 2002. Nature **419**, 456.
[10] C M Surko, M Leventhal, W S Crane, et al., 1986. Rev. Sci. Instrum. **57**, 1862–1867.
[11] R G Greaves, S J Gilbert and C M Surko, Appl. Surf. Sci. **194** (2002) 56.

Journal of Physics: Conference Series **262** (2011) 012021 doi:10.1088/1742-6596/262/1/012021

[12] R G Greaves, C M Surko, 1997. Phys. Plasmas **4**, 1528–1543.
[13] C F Driscoll, J K Malmberg, Phys. Rev. Lett. **50** (1983) 167.
[14] R G Greaves, S J Gilbert, C M Surko, Appl. Surf. Sci. **194** (2002) 56.
[15] J Xu, M H Weber and K G Lynn Proc. of the Int. Enrico Fermi School 2009,"Physics with many Positrons", in press
[16] C M Surko and R G Greaves, Radiation Physics and Chemistry,**68**,419 (2003)
[17] J R Danielson, T R Weber and C M Surko, Phys. Plasma, **13**, 123502(1996)
[18] R.G. Greaves, C.M. Surko, 2002. In: Anderegg, F., et al. (Ed.), Non-Neutral Plasma Physics, Vol. IV. American Institute of Physics, Melville, NY, pp. 10–23.
[19] A Friedman and S P Auerbach. Numerically induced stochasticity. Journal of Computations Physics, **93**:**171**, 1991
[20] K. Gomberoff, J S Wurtele, A Friedman, D P Grote and J.L. Vay J. Comput. Phys. **225** 1736-1752(2007)
[21] W M Manheimer, M Lampe and G Joyce J. Comput. Phys. **138 563**(1997)
[22] B I Cohen L Divol A B Langdon and E A Williams Phys. Plasma, **13**, 022705(2006)

Journal of Physics: Conference Series **262** (2011) 012022 doi:10.1088/1742-6596/262/1/012022

Clay particles - potential of positron annihilation lifetime spectroscopy (PALS) for studying interlayer spacing

N Fong[1], P Guagliardo[2], J Williams[2], A Musumeci[3], D Martin[3] and SV Smith[1]

[1] Centre for Antimatter-Matter Studies, Australian Nuclear Science and Technology Organisation (ANSTO), Menai, NSW 2234, Australia

[2] Centre for Antimatter-Matter Studies, School of Physics, University of Western Australia, WA 6009, Australia

[3] ARC Centre of Excellence for Functional Nanomaterials, Australian Institute for Bioengineering and Nanotechnology, The University of Queensland, QLD, Australia

Email: nfg@ansto.gov.au

Abstract. Characterisation of clays is generally achieved by traditional methods, such as X-ray diffraction (XRD) and transmission electron microscopy (TEM). However, clays are often difficult to characterise due to lack of long-range order, thus these tools are not always reliable. Because interlayer spacing in clays can be adjusted to house molecules, there is growing interest to use these materials for drug delivery. Positron annihilation lifetime spectroscopy (PALS) was examined as an alternative tool to characterise a series of well-known clays. XRD of two layered double hydroxides; MgAl-LDH and MgGd-LDH, natural hectorite, fluoromica and laponite, and their PALS spectra were compared. XRD data was used to calculate the interlayer d- spacing in these materials and results show a decrease in interlayer spacing as the heavy metal ions are substituted for those of large ionic radii. Similar results were obtained for PALS data. This preliminary study suggests PALS has potential as a routine tool for characterising clay particles. Further work will examine the sensitivity and reliability of PALS to percent of metal doping and hydration in clay microstructure.

1. Introduction

Natural and synthetic clay nanoparticles are currently used in a range of applications such as cosmetics and pharmaceuticals [1]. In addition to having properties such as high aspect ratios, natural abundance and/or ease of manufacture, clay nanoparticles may be chemically manipulated to afford properties tailored for specific applications. One significant application that has sparked interest in nanoparticles in the medical field is in drug delivery, and more specifically for gene therapy. The active agent is expected to fit between the layers of the clay particle and be released by exchange or breakdown of the clay nanoparticles *in vivo*.

The nanoscale dimensions of these particles may be utilized in gene therapy to enable penetration of nucleic acids or drugs through the cell wall and into the cells' cytoplasm or nucleus to prevent and/or treat disease [3]. The transport and release of the drug into the cell is highly dependent on the interaction between the biomolecule and nanoparticle, as well as their interaction with the physiological environment [4]. It is therefore of key importance to understand the physical and chemical properties of the nanoparticle carriers, and the effect of different ion dopings on their structure.

A range of inorganic nanoparticles have been used for DNA delivery [5]. Among these, LDHs have recently attracted much attention for use as drug/gene carriers. They may be represented by the formula $[M_{1-x}^{2+}Mx^{3+}(OH)_2](Ax \cdot yH_2O)$, where M^{2+} and M^{3+} are divalent and trivalent metal ions, and A- can be any anion [6]. They possess favorable properties such as low cytotoxicity, excellent biocompatibility [7, 8], and a high surface area for drug loadings [9].

Other significant clay particles include silicates belonging to the smectite family, including fluoromica and hectorite. These nanoparticles have a 2:1 layer structure consisting of two tetrahedral sheets sandwiching an octahedral sheet through sharing of apical oxygens [10].

Interlayer spacings and morphology of the clay particles are usually determined using XRD and TEM, however small platelet size and lack of long-range order can interfere with full analysis [11]. As PALS is unlikely to be affected by these properties it is of interest to test the feasibility of PALS as an alternative tool for characterizing clays.

2. Experimental procedure

Two types of LDH nanoparticles were synthesized to assess the effect of different ions within their structure on particle interlayer spacing. The structural characteristics of these nanoparticles as well as other commonly studied clay nanoparticles were assessed using XRD and PALS.

MgAl-LDH and MgGd-LDH were synthesized using a co-precipitation method similar to that reported by Xu *et al.* [12]. Gadolinium was used in the synthesis of MgGd-LDH since Gd (2.54 Å) has a larger atomic radius than Al (1.82 Å), which was expected to affect the structural properties of the LDH. In brief, MgAl-LDH was prepared by mixing 7.5 mL of a mixed metal salt solution containing $MgCl_2$ (0.4 M) and $AlCl_3$ (0.2 M) and adding it to 30 mL base solution containing Na_2CO_3 (0.03 M) and NaOH (O.3 M), under vigorous stirring. Following 10 mins of stirring at room temperature, an LDH slurry was obtained by washing and centrifuging at 4500 rpm for 5 mins at 25°C three times. The LDH slurry was re-dispersed in 26.5 mL Milli-Q water and hydrothermally treated. More specifically, the dispersion was transferred into a Teflon lined stainless steel autoclave and placed into a preheated oven at 100°C for 4 hours, resulting in a stable, homogenous suspension of LDH nanoparticles. MgGd-LDH was made in a similar manner to that described above, except the $AlCl_3$ was replaced by $Gd(NO_3)_3$. Following hydrothermal treatment, the LDHs were allowed to cool for at least 3 hours and freeze-dried for characterization with XRD and PALS.

Natural hectorite $(Na_{0.4}Mg_{2.7}Li_{0.3}Si_4O_{10}(OH)_2)$ laponite $(Na^{+}_{0.7}[(Si_8Mg_{5.5}Li_{0.3})O_{20}(OH)_4]^{-}_{0.7})$ and fluoromica $(K[Mg_{2.44}Mn_{0.24}][Si_{3.82}Mn_{0.18}]O_{10}F_2)$ (ME-100, CBC Japan) were used as received.

PALS measurements were carried out using a fast-fast spectrometer with detectors comprising of a BC418 plastic scintillator coupled to a Burle 8850 photomultiplier tube. The positron source consisted of ~30 μCi of $^{22}NaCl$ sealed in 7μm thick Kapton foil (grade HN from Goodfellow) and was sandwiched between two identical samples. The time resolution of the system is about 200 ps as determined from analysis of a spectrum of high purity, annealed nickel. The spectra comprise of at least 2 million counts and have been analysed using PAScual version 1.3.0 [13].

XRD studies were performed using a Philips PW1830 Bragg–Brentano X-ray diffraction system with Cu Kα radiation used at $\lambda = 1.5418$ Å, goniometer = 173 mm and tension at 40 kV and 40 mA. Measurements were taken between 4° and 40° at a rate of 0.008 °/s and a step size of 0.04°. Results were expressed as average nanometre distances between silicate layers, calculated from Bragg's law, $\lambda = 2d\sin\theta$, where λ=wavelength of the incident X-ray beam, 2θ=diffraction angle and d= interlayer distance.

3. Results and Discussion

Two types of LDH clay particles (MgAl-LDH and MgGd-LDH) were synthesized in a similar manner with Al^{3+} substituted by the large cation Gd^{3+} for the latter clay. Natural hectorite, laponite and fluoromica were purchased and used without further modification. Though similar in structure the laponite has a higher ratio of Si^{4+} (1.46 Å) and Mg^{2+} (1.72 Å) atoms compared with hectorite, and fluoromica has similar structure but contains different types and size cations, K^{+} (2.77 Å) and Mn^{2+} (1.79 Å) ions in its structure providing different interlayer spacing.

Lifetime spectra for the LDH and clay particles are shown in Figure 1. Significant differences between the spectra suggest that PALS is sensitive to the structural differences in the samples. The spectra could be deconvoluted into either three or four discrete lifetime components. The longer components, τ_3 and τ_4, which exceed 1 ns are more interesting for the present study as they arise from o-Ps pick-off annihilation in nano-pores.

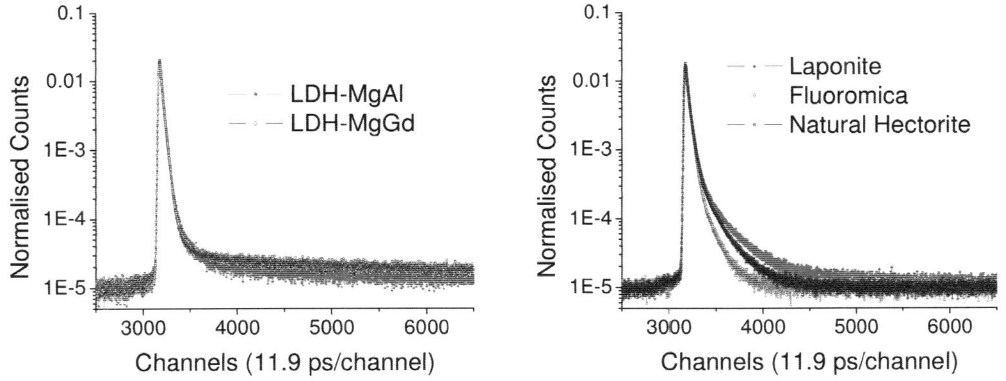

Figure 1. Positron lifetime spectra for LDH (Left), and smectite clay particles (Right) studied.

The o-Ps lifetimes, intensities and their pore sizes for spherical and square channel geometries using the RTE model [14] are shown in Table 1, along with their XRD interlayer spacings. The pore sizes may provide an indication of the size of free volume holes present in the interlayer space. For the LDH materials XRD *d*-spacings of 7.7 and 3.8 Å were observed for MgAl-LDH, and these decreased by a factor of ~1.6 in MgGd-LDH, when the larger Gd ion was substituted into the layered structure. PALS showed that both LDH particles could be deconvoluted into four components. The lifetimes and the corresponding intensities are smaller for LDH synthesised with Gd compared to Al. This could be due to the larger spatial extent of Gd compared to Al. This is in qualitative agreement with the XRD results.

τ_3 equates to sub-nanometre pores, with diameters in the range of 6.1 and 5 Å, for MgAl-LDH and MgGd-LDH, respectively. The τ_4 values are significantly larger and are apparent in the long tail of the spectra, shown in Figure 1, which extends to the end of the channel range. Converting τ_4 to a pore size gives dimensions significantly larger than the interlayer spacing derived from XRD, for any geometry assumed. If a square channel geometry is assumed then this gives rise to 25 Å and 18 Å pores for MgAl-LDH and MgGd-LDH, respectively.

Significant nano-porosity was also observed in the smectite clay particles. Fluoromica and hectorite could be deconvoluted into three components. If a square channel geometry is assumed, then τ_3 equates to channel widths of 3.9 Å and 5.1 Å, for fluoromica and hectorite respectively. The intensity of this component (I_3) is 8.8% for fluoromica and 12.2% for hectorite. This suggests that both the size and concentration of free volume holes in the interlayer spacing is larger for hectorite. This is consistent with LDH findings indicating substitution of larger metal ions into the clay structure decreased pore size. However, PALS results for fluoromica were not consistent with XRD findings, which showed a larger *d*-spacing in fluoromica than hectorite. Differences between XRD and PALS may be attributed the level of hydration of samples for both types of analyses.

Four components could be extracted for laponite, indicating the presence of large pores not observed in the other smectite clays. It was noted that the τ_3 pore sizes for laponite and hectorite are not significantly different. This can be attributed to their similar chemical composition. In contrast the presence of large pores in laponite indicates an effect of the higher large metal ion concentrations compared with hectorite. These values also complement XRD findings, where laponite has the largest *d*-spacing among the clays studied. The o-Ps intensity for laponite is also larger than any of the other particles studied, $I_3+I_4 = 15.4\%$, suggesting the highest concentration of pores.

Table 1-. o-Ps lifetimes and intensities for the clay particles studied, pore dimensions calculated using the RTE model for spheres and square channels and XRD basal spacing

	PALS									**XRD**	
					Pore diam.		***Channel width***			***d-spacing* (Å)***	
Sample	**τ_3# (ns)**	**I_3# (%)**	**τ_4# (ns)**	**I_4# (%)**	**d_3 (Å)**	**d_4 (Å)**	**r_3 (Å)**	**r_4 (Å)**	**2θ**	**001**	**002**
LDH-MgAl	1.82 (0.05)	3.1 (0.2)	40 (1)	4 (1)	6.1	33	4.4	25	11.4 23.3	7.7	3.8
LDH-MgGd	1.35 (0.06)	2.3 (0.4)	24 (1)	3 (1)	5	21	3.5	18	18.7 29.7	4.8	3
Natural Hectorite $Na_{0.4}Mg_{2.7}Li_{0.3}Si_4O_{10}(OH)_2$	2.24 (0.01)	12.2 (0.1)	-	-	6.9	-	5.1	-	7.7 19.5	11.4	4.5
Laponite $Na^+_{0.7}[(Si_8Mg_{5.5}Li_{0.3})O_{20}(OH)_4]^-_{0.7}$	2.50 (0.04)	13.5 (0.1)	18.4 (0.4)	1.9 (0.1)	7.3	16.6	5.5	13.6	7.0 19.6	12.7	4.5
Fluoromica $K[Mg_{2.44}Mn_{0.24}][Si_{3.82}Mn_{0.18}]O_{10}F_2$	1.55 (0.01)	8.8 (0.1)	-	-	5.5	-	3.9	-	7.2 9.3	12.3	9.5

*Distance between layers. d_{001} and d_{002} refer to the first two basal reflections. # Mean (SD)

The identification of such structural properties of nanoparticles is a key step to understanding their behavior as drug delivery vehicles. Large and numerous pores may facilitate high concentrations of drug loading and sustained release compared with particles with smaller voids. Further investigation into the effect of different ion and/or drug dopings on nanoparticle structure is warranted.

4. Conclusions

The interlayer spacing of clay nanoparticles was assessed using PALS and conventional XRD methods, and their results were compared. PALS showed promise as a useful alternative tool for analysing the microstructure of clay nanoparticles, demonstrating sensitivity in detecting differences in pore sizes in response to changes in metal ion dopings within the clay structures.

References

[1] Cerretero MI, Pozo M 2009 *Appl Clay Sci,* **46** 23
[2] Auffan M, et al. 2009 *Nature Nanotech.* **4** 634
[3] Xu, ZP, Zeng QH, Lu GQ, Yu AB 2006 *Chem. Eng. Sci.* **61** 1027
[4] Sokolova V and Epple M 2008, *Angew. Chem. Int. Ed.* **47** 1382
[5] Davide P et al. 2004 *Angew. Chem. Int. Ed.* **43** 5242
Choy J-H, Kwak S-Y, Jeong Y-J, Park J-S 2000 *Angew. Chem. Int. Ed.* **112** 4207
[6] Kovanda F, Kolousek D, Cílová Z, Hulínský V 2005 *Applied Clay Science* **28** 101
[7] Xu ZP, Walker TL, Liu K-L, Cooper HM, Lu GQM, Bartlett PF 2007 *Int. J Nanomedicine* **2** 163
[8] Kriven WM, Kwak S-Y, Wallig MA, Choy J-H 2004 *MRS Bulletin* **29** 33
[9] Choy J-H, Jung J-S, Oh J-M, Park M, Jeong J, Kang Y-K, Han O-J 2004 *Biomaterials* **25** 3059
[10] Kloprogge JT, Komarneni S, Amonette JE 1999 *Clays Clay Miner.* **47** 529
[11] Doeff MM, Reed JS 1998 *Solid State Ionics* **113-115** 109
[12] Tyner KM, Schiffman SR, Giannelis EP, 2004. *J.Control. Release* **95** 501
[13] Pascual-Izarra C et al 2009 *Nucl. Instr. Meth. Phys. Res. A* **603** 456
[14] Dull T L *et al* 2001 *J. Phys. Chem. B* **105** 4657

Journal of Physics: Conference Series **262** (2011) 012023 doi:10.1088/1742-6596/262/1/012023

Characterisation of metakaolin-based geopolymers using beam-based and conventional PALS

P Guagliardo[1], J Roberts[2], E R Vance[3], R Weed[2], A D Sergeant[1], A Howie[1], P Wilkie[1], M Went[2], J Sullivan[2], J Williams[1], S Samarin[1] and S Buckman[2]
[1]Centre for Antimatter-Matter Studies, School of Physics, University of Western Australia, Crawley, WA, 6009, Australia
[2]Centre for Antimatter-Matter Studies, Research School of Physical Sciences, Australian National University, Canberra, ACT, 2600, Australia
[3]Institute for Materials Engineering, ANSTO, Menai, NSW, 2234, Australia

E-mail: guaglp01@student.uwa.edu.au

Abstract. The nano-porosity of metakaolin-based geopolymers and the effect of heat-treatment on porosity have been studied with conventional and beam-based positron annihilation lifetime spectroscopy (PALS). Conventional PALS found significant nano-porosity in the geopolymers, as indicated by the presence in the PALS spectrum of two long lifetime components, $\tau_3 = 1.58$ ns and $\tau_4 = 47$ ns, associated with pore diameters of approximately 0.5 and 3 nm respectively. The lifetime of the shorter component was found to decrease monotonically with successive heat treatments of 300°C and 600°C. Beam-based PALS, conducted at 5 keV, also indicated two long lifetime components, $\tau_3 = 4.84$ ns and $\tau_4 = 54.6$ ns. These are significantly longer than those observed by conventional PALS and the monotonic decrease of τ_3 with successive heat treatments was not observed. As the beam-based PALS probed only the near-surface region, with an average implantation depth of about 350 nm, these results suggest that the near-surface structure may vary significantly from that of the bulk. This could be an inherent property of the samples or an artefact caused by surface effects or sample outgassing.

1. Introduction

Geopolymers are a class of inorganic polymers that are based on aluminosilicates. They are a potential replacement for traditional building materials such as Portland cement because of their superior fire and acid resistance and their manufacture requires a much lower calcining temperature and produces only 10-20% of the CO_2 per unit mass compared to ordinary cement. Other potential applications include, for example, immobilisers for radioactive and toxic waste, as refractory and fire resistant products, and composite materials for aircraft and automobile interiors [1].

Geopolymers are usually produced by adding a reactive aluminosilicate precursor, such as fly ash or metakaolin, to a highly alkaline silicate solution in order to facilitate the break-up of the primary aluminosilicate framework, leading to polymerisation and solidification. Then curing at 25-90°C in a humid atmosphere for hours or days completes the process. If the Si/Al molar ratio is ≥ 1.5 the resulting structure does not display long-range atomic order; it is amorphous with respect to X-ray

[1] To whom any correspondence should be addressed.

Journal of Physics: Conference Series **262** (2011) 012023 doi:10.1088/1742-6596/262/1/012023

powder diffraction. The amorphous regions are comprised of SiO_4 and AlO_4 tetrahedra distributed randomly along cross-linked polymeric chains creating a complicated network of pores [1-3].

The mechanical properties of cementitious materials are heavily influenced by porosity, since the volume and size distribution of pores control both their strength and durability [4, 5]. A detailed characterisation of the pore structure and the factors that affect it are thus crucial to advancing the design of these materials; however, classical porosimetry methods such as mercury intrusion and gas adsorption cannot always be relied upon to extract information on pores smaller than a few nanometres [3-5]. Positron annihilation lifetime spectroscopy (PALS) possesses several advantages over these conventional techniques. It is sensitive to both open and closed porosity, and pores in the size range of 0.1-10 nm typically can be resolved. In this paper the porosity of metakaolin-based geopolymers and the effect of heat treatment on porosity has been investigated by both beam-based and conventional PALS.

2. Experimental procedure

The geopolymers studied have Si/Al and Na/Al molar ratios of 2.0 and 1.0 respectively. Metakaolin is produced by heating kaolinite at 750°C for 15 hours in air, which renders the kaolinite X-ray amorphous. The batch compositions consisted of 33.7% metakaolin, 63.0% sodium silicate D solution and 3.3% deionized water. The sodium silicate solution was added to a mixture containing the various additives, and after agitating for 5 minutes, the composite solution was poured into a polycarbonate cylindrical jar and covered with a screw cap. The samples were shaken for 5 minutes on a vibrating table to de-air. They were cured at ambient for 24 hours and then at 40°C for a further 24 hours. To ascertain the effect of heat treatment on porosity, two of the samples were heated in air at either 300°C or 600°C for 3 hours.

Conventional PALS measurements were obtained using a fast-fast spectrometer with detectors comprising of a BC418 plastic scintillators coupled to a Burle 8850 photomultiplier tube. In these measurements about 30 μCi of $^{22}NaCl$ was deposited on 7μm Kapton foil (grade HN from Goodfellow) and sandwiched between two identical geopolymer samples. The time resolution of the system is about 200 ps as determined from analysis of a spectrum of high purity, annealed nickel. The spectra comprise of at least 2 million counts and have been analysed using PAScual version 1.3.0 [6].

Beam-based PALS was performed at the Australian Positron Beamline [7], which is based on the Surko buffer-gas trap system. This technique has been implemented to produce sub-nanosecond positron pulses, enabling positron lifetime studies of materials to be performed. For gamma detection a BaF_2 scintillator and an XP2020Q photomultiplier are employed. Timing information is obtained using a Comtech P7889 Fast Timing Card and the timing resolution of the system was approximately 800 ps. Analysis of lifetime spectra has been carried out with custom-written Matlab software CAMSfit [7].

3. Results and discussion

Table 1 shows results obtained with conventional and beam-based PALS for the three geopolymer samples studied, one without any heat treatment, and the others after being heated to either 300°C or 600°C. With conventional PALS, four lifetime components could be extracted from the spectra for each sample. The two shortest components are associated with p-Ps and free positron annihilation in the bulk of the material, as well as annihilations in the source material and supporting foil. One of the longer lifetime components (τ_3 = 1.58 ns) extracted from the spectrum of the untreated geopolymer is close to the approximate 1.7 ns o-Ps lifetime of free water [8]. Similar lifetimes have also been observed in cement and the intensity of this spectral component has been used to monitor changes in hydration over time [4, 5].

Heat treatments of 300°C and 600°C led to a monotonic decrease in τ_3 and after heating to 300°C the intensity of this component (I_3) decreased by a factor of about 1.5. The decrease in the lifetime and intensity after heat treatment may be due to the removal of free water from pores, because heating to greater than 150°C is sufficient for this purpose [9]. The decrease in the lifetime after the 600°C heat treatment could be due to the early stages of a transition to a dense glass ceramic. However, the intensity (I_3) is approximately the same as that of the 300°C sample. Using the RTE model [10] these longer lifetimes correspond to pore diameters in the range of 0.5 nm.

The behaviour of τ_3 is in broad agreement with the results of Vance et al [11], where conventional PALS was performed on metakaolin geopolymers subjected to heat treatments over a larger temperature range. However, Vance et al did not observe the fourth lifetime component, τ_4, of about 50 ns observed in these samples, corresponding to pore sizes in the range of 3 nm. The intensity of this component varied from 7 to 9%, however, it did not show a dependence on heat treatment, as identical results were obtained for the untreated geopolymers and the 600°C sample. The absence of strong temperature dependence could suggest that this component is associated with o-Ps annihilation in empty pores, that is, pores that do not contain water.

There are differences in the lifetimes and intensities obtained with conventional and beam-based PALS. Like conventional PALS, beam-based PALS produces two long lifetime components; however, there are significant differences in the lifetimes measured using the two techniques. τ_3 for beam-based PALS is larger by a factor of 2.8-4.4, and the monotonic decrease in lifetime with heating seen with conventional PALS is not observed. The trend in the intensity of this component is similar however to conventional PALS and I_3 decreases by a factor of about 2.3 after the 300°C heat treatment. Using the RTE model, this lifetime can be equated to pores with diameter 0.9-1.2 nm. The τ_4 values from beam-based PALS are in closer agreement to those from conventional PALS; however, they still differ by about 14%. As with the data from conventional PALS, this component does not display a strong dependence on heat treatment and would correspond to pores of diameter 2.8-3.4 nm.

In conventional PALS the energy spectrum of ^{22}Na positrons has its peak at 0.545 MeV - hence positrons penetrate deep into the sample and the signal is averaged over a large volume (thus in this case positrons are implanted within about 1mm of the geopolymers' surface). In contrast, beam-based PALS provides the means to study near-surface layers and to obtain information as a function of depth. In this study the measurements were conducted at an incident energy of 5 keV; for which 99.9% of positrons thermalise within 1μm of the surface and the mean implantation depth is about 350 nm [12]. The differences observed could suggest that the porosity in the near surface region differs from that of the bulk material. However, these differences may also be the result of outgassing, or surface effects associated with beam-based PALS. Outgassing of the samples could selectively modify the near surface region, giving rise to significant differences between the two methods. Beam-based PALS was performed at approximately 10^{-7} Torr and outgassing of the samples was observed; conventional PALS measurements were then carried out in air. PALS studies of zeolites and mesoporous silicates performed at different vacuum levels reveal that very long lifetime components emerge when measurements are performed under vacuum conditions [14]. This is largely attributed to the quenching of ortho-Positronium by air; such dramatic differences are not seen in the data presented here. Only a moderate increase in τ_3 is seen in the beam-based data and the τ_4 values are comparable. Surface effects associated with Positronium escaping from the surface and annihilating outside of the viewing angle of the detector may distort beam-based PALS data. To reduce the contribution from surface effects, and to facilitate a better comparison between beam-based and conventional PALS, future studies will be performed over a wider range of energies (0.5-20 keV).

4. Conclusion

Studies of metakaolin-based geopolymers with conventional and beam-based PALS reveal two long lifetime components, indicating that the geopolymers contain significant nano-porosity. τ_3 obtained with conventional PALS is close to that of free water and a monotonic decrease in the lifetime and intensity of this spectral component is observed upon heating to 300°C and then 600°C, which may signal the removal of free water. τ_3 for beam-based PALS conducted at 5 keV is larger by a factor of 2.8-4.4 and the decrease in lifetime with heating is not observed. This could suggest that the porosity in the near-surface region is greater than the average porosity in the underlying bulk. However, we cannot eliminate sample outgassing or surface effects as the underlying cause of these differences.

Journal of Physics: Conference Series **262** (2011) 012023 doi:10.1088/1742-6596/262/1/012023

Table 1. Lifetimes and intensities obtained with conventional and beam-based PALS for the three geopolymers studied. Standard deviations are given in the brackets.

Geopolymer treatment	Conventional PALS				Beam based PALS at 5 keV			
	τ_3 (ns)	I_3 (%)	τ_4 (ns)	I_4 (%)	τ_3 (ns)	I_3 (%)	τ_4 (ns)	I_4 (%)
No heat	1.58 (0.02)	6.2 (0.1)	47 (2)	7.2 (0.2)	4.84	3.1	54.6	6.0
300°C	1.42 (0.03)	4.1 (0.3)	50 (2)	9.2 (0.6)	6.22	1.3	58.5	6.4
600°C	1.34 (0.03)	4.3 (0.2)	47 (2)	7.4 (0.2)	3.70	0.9	41.4	4.1

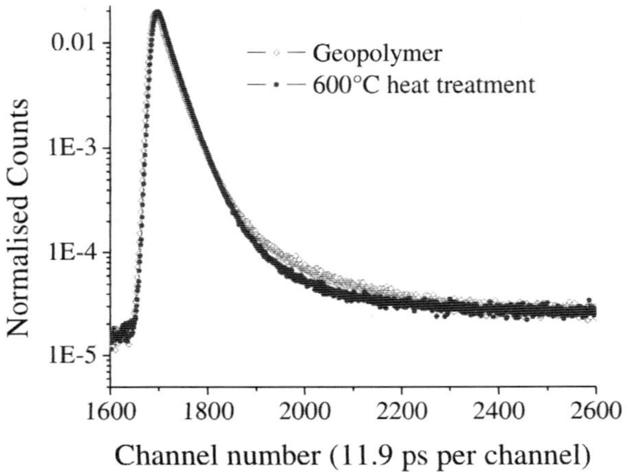

Figure 1. Lifetime spectra obtained with conventional PALS of an untreated geopolymer and a geopolymer heated at 600°C for 3 hours

5. References

[1] Davidovits J 1991 *J. Thermal. Anal.* **37** 1633
[2] Barbosa V F, MacKenzie K J, and Thaumaturgo C 2000 *Int. J. Inorganic Mater.* **2** 309
[3] Lloyd R R *et al* J 2009 *Microporous and MesoporousMaterials* **126** 32
[4] Consolati G, Quassi F 2003 *Rad. Phys. Chem.* **68** 519
[5] Consolati G, Dotelli G and Quasso F 1999 *J. Appl. Phys.* **86** 4225
[6] Pascual-Izarra C *et al* 2009 *Nucl. Instr. Meth. Phys. Res. A* **603** 456
[7] Sullivan J P *et al* 2010 *Meas. Sci. Technol.* **21** 085702
[8] Patro A P, Sen P 1972 J.Phys C; *Solid State Phys.* **5** 3273
[9] Perera D S, Vance E R *et al* 2005 *Advances in Ceramic Matrix Composites X* (Westerville, OH: American Ceramic Society) p 87
[10] Dull T L *et al* 2001 *J. Phys. Chem. B* **105** 4657
[11] Vance E R *et al* 2008 *J. Am. Ceram. Soc.* **91** 664
[12] Koybayashi Y *et al* 2008 *Appl. Surf. Sci.* **255** 174
[13] Parrott L J 1992 *Cement and Concrete Research* **22** 1077
[14] He Y J *et al* T 2001 *J. Phys.: Condens. Matter* **13** 2467

Journal of Physics: Conference Series **262** (2011) 012024 doi:10.1088/1742-6596/262/1/012024

Operation and testing of the PULSTAR reactor intense slow positron beam and PALS spectrometers

Ayman I Hawari[1], David W Gidley[2], Jeremy Moxom[1], Alfred G Hathaway[1] and Saurabh Mukherjee[1]

[1]Nuclear Reactor Program, Department of Nuclear Engineering, North Carolina State University, P.O. Box 7909, Raleigh, NC 27695, USA

[2]Physics Department, University of Michigan, 450 Church Street, Ann Arbor, MI 48109, USA

E-mail: ayman.hawari@ncsu.edu

Abstract. An intense slow positron beam has been established at the 1-MW PULSTAR nuclear reactor. The beam is operational generating mono-energetic positrons with an energy of 1-keV. The maximum measured intensity slightly exceeds 10^9 e+/s. The beam is operated routinely with an intensity of approximately 5×10^8 e+/s. The positrons are generated through gamma-ray pair production interactions in two back-to-back banks of tungsten converter/moderators. The gamma-rays are produced in the PULSTAR core and by thermal neutron capture in a cadmium shroud that surrounds the tungsten. The primary utilization of the PULSTAR positron beam is the characterization of nanoscale structure in materials. Consequently, the beam has been equipped with two state-of-the-art PALS spectrometers. The first spectrometer is dedicated to measurements in materials such as metals and semiconductors. This spectrometer is based on pulsing and bunching of the primary beam and is currently operating with a timing resolution of approximately 390 picoseconds. The second spectrometer is dedicated to measurements in materials where positronium formation is promoted. The timing resolution of this spectrometer is designed to be ~ 0.5 nanosecond with an on-sample spot size of 1-2 mm. For both spectrometers, the energy of the positrons can be varied to allow depth profiling with on-sample intensity exceeding 10^6 e+/s.

1. Introduction

The use of positrons and positronium has proven to be a powerful nondestructive technique to characterize the properties of materials at the nanometer scale. The utilization of low energy beams of positrons wherein the positron's implantation depth can be controlled vastly expands the possible applications by enabling the study of thin films, surfaces, interfaces, and depth-dependent phenomena [1,2]. In particular, beam-based positron spectroscopies have demonstrated great utility in characterizing nanoporous thin film materials and nanocomposites [3]. Introducing and controlling nanometer-sized pores is a rapidly developing area in nanotechnology with wide-ranging implications from H storage and CO_2 catalysis to next-generation interlayer dielectrics in microchips [4]. Depth-profiling will be essential to understanding the role of free volume in: (1) controlling transport properties in polymer electrolytic membranes in fuel cells and thin polyamide control layers in reverse osmosis membranes; (2) correlating with electrical properties in organic semiconductors; and (3) determining thermal and mechanical properties of thin polymer films and nanocomposites. Despite this promising and demonstrated potential, low positron beam rates and/or poor accessibility to beams have limited widespread adoption of positron techniques. To address such issues a new high rate

Journal of Physics: Conference Series **262** (2011) 012024 doi:10.1088/1742-6596/262/1/012024

positron beam attached to two different positron/positronium lifetime spectrometers has been constructed at the PULSTAR nuclear reactor on the campus of North Carolina State University.

2. The Intense Slow Positron Beam

The positron beam-line is located in beam tube number 6 of the PULSTAR reactor. This beam is currently operational, generating quasi mono-energetic positrons with a mean energy of around 1 keV and a spread of nearly 30 eV. To date the maximum measured intensity slightly exceeds 10^9 e^+/s and the beam is operated routinely with an intensity of approximately 5×10^8 e^+/s. The positrons are generated through gamma-ray pair production interactions in two back-to-back banks of tungsten converter/moderators. Each tungsten bank is made of interlocking tungsten strips (with 1-cm pitch) and is 22-cm in diameter and 2.5 cm thick. Each individual bank underwent four hours of vacuum annealing at 2200 K. The moderators are placed within an aluminum vacuum chamber that is located at the end of a magnetic solenoid tube and inserted into the 30-cm x 30-cm square beam tube. Gamma-rays produced by the core interact in the moderators producing positrons via pair production reactions. In addition, 0.25 mm thick sheets of cadmium, which shroud the tungsten moderator portion of the vacuum chamber, are used to harden and intensify the gamma-ray spectrum by generating gamma-rays through thermal neutron (n,γ) capture reactions. Positrons extracted from the moderators are focused and accelerated by a six element electrostatic lens system to an energy of 1 keV. Subsequently, the positrons are injected into a 60 Gauss (0.006 T) magnetic solenoid for transport outside the reactor biological shield and through two magnetic beam switches allowing the beam to be directed into either of two positron annihilation lifetime spectrometers (PALS) spectrometers. Figure 1 shows the intense positron beam facility (including the spectrometers) and the measured intensity of the generated positrons. Figure 2 shows the image and profile of the beam exiting the beam-line and into the spectrometers. The FWHM measured in a field of 80 G is 2.3 cm.

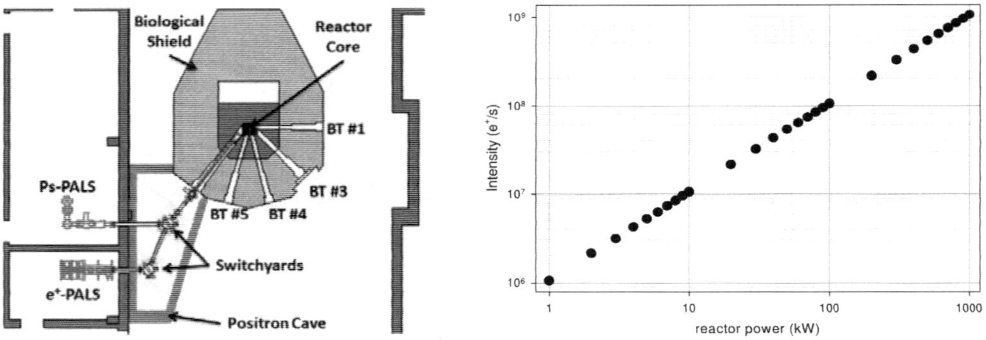

Figure 1. The layout of the slow positron beam facility including the two PALS systems (left). The positron count rate as a function of reactor power (right).

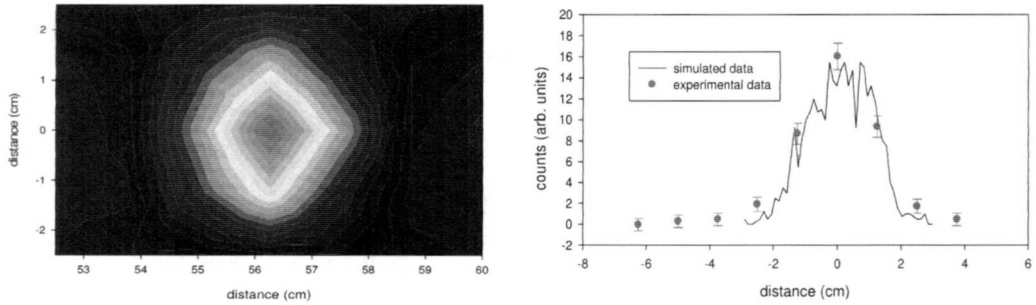

Figure 2. An image of the beam at the end of the beam-line (left). The beam profile of the beam as compared to simulations (right).

3. The lifetime spectrometers

The first of the two spectrometers is dedicated to measuring positron lifetimes on the order observed in materials such as metals and semiconductors (i.e., lifetimes in the range of 300 picoseconds or less). This instrument utilizes time varying electric fields to generate narrow bunches (< 300ps) of positrons with a 50 MHz rep. rate and the beam is magnetically guided to the sample, which can be biased to negative potentials to vary the implantation energy. Positrons entering the spectrometer are first implanted into a 1 cm diameter, 100 nm thick single crystal tungsten remoderator foil to which a 50 MHz ramp voltage is applied. This acts as a pre-buncher for the main bunching stage. The velocity modulation is optimized by changing a DC offset placed on the moderator ensuring that the focal point of the pre-buncher is at the entrance of the main buncher. This is a double gap device comprising a 9 cm long tube with a 50 MHz, 280 V sine wave, applied to achieve time focussing at the sample. The length of the bunching electrode was calculated so that positrons entering the buncher exit in phase with the applied sine wave, the amplitude of which is chosen to result in optimum time focusing at the sample.

At present, the main buncher has been tested with the remoderation stage withdrawn, i.e., using positrons from the primary beam. Since these positrons have a wide range of transverse energies a retarding potential was applied to a repeller tube prior to the buncher to reduce the longitudinal energy spread at the buncher and improve the time resolution. A lifetime spectrum for Si has been collected using this system and is shown in Figure 3 along with a picture of the spectrometer as it currently stands. The spectrum was fitted with the PALSFIT [5] software giving a resolution estimate of 390 ps.

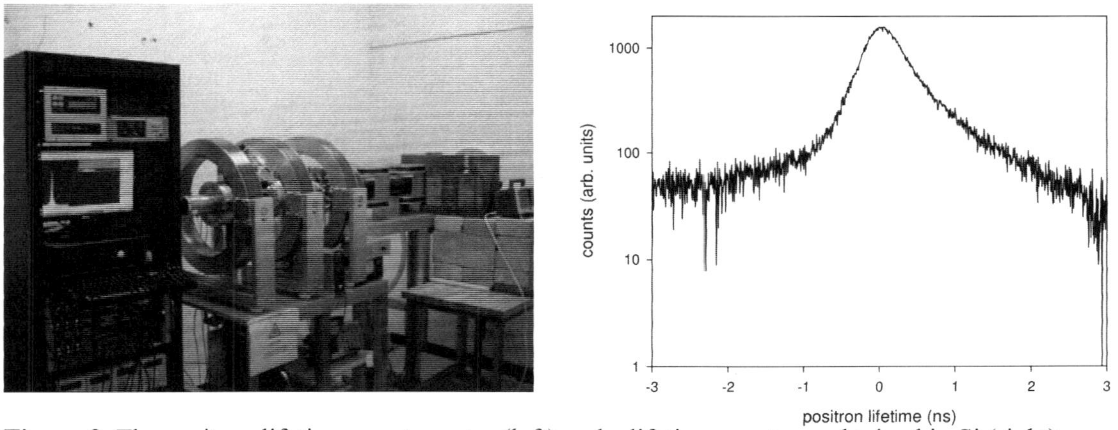

Figure 3. The positron lifetime spectrometer (left) and a lifetime spectrum obtained in Si (right).

The second spectrometer is dedicated to the measurement of lifetimes in materials where positronium formation is promoted. In this case, the main positron beam is accelerated to 8 keV after it leaves the magnetic field and is electrostatically focused to a diameter of 20 mm on a 100 nm thick tungsten transmission foil remoderator (the small reduction in beam diameter from 30 mm to 20 mm is due to the large (500 eV) spread in the longitudinal beam energy in the 0.006 T guiding field). After acceleration to 5 keV the remoderated beam is focused to a 1-2 mm spot on a second transmission foil remoderator and after further focusing the final beam on target can be ~1 mm diameter at even the lowest beam implantation energies. The predicted on-target positron rate is 5×10^6 /s, which is still sufficient for the spectrometer to be dead-time limited in the individual pulse-counting mode. Random background rates further limit the overall coincidence data rate to $\sim10^5$ events/s. A typical lifetime spectrum will thus be acquired in less than one minute.

The start signal for the lifetime measurement using this spectrometer is obtained from a channel electron multiplier array (CEMA) detecting secondary electrons, emitted when positrons impact the target. The positron beam passes through a co-axial hole in the CEMA and strikes the sample target

Journal of Physics: Conference Series **262** (2011) 012024 doi:10.1088/1742-6596/262/1/012024

about 1 cm behind it. A subsequent annihilation gamma-ray is detected in a bank of ten fast-plastic scintillation detectors and a time digitizer measures and records the time interval for each event. Timing resolution is limited to approximately 0.5 ns due to the energy and angular distribution of the reemitted secondary electrons but this is acceptable for measurement of positronium lifetimes in polymers and insulators. Figure 4 below shows this spectrometers current setup.

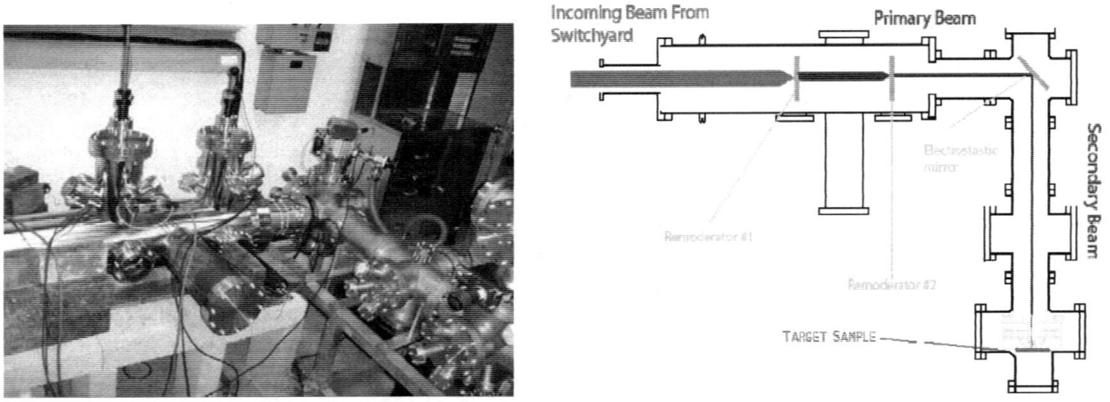

Figure 4. The Ps-PALS spectrometer (left) and a schematic showing the main components (right).

4. Conclusions

An intense slow positron beam has been established at the PULSTAR nuclear reactor of North Carolina State University. This beam operates with a routine intensity of around 5×10^8 e^+/s, a diameter of 2.3 cm FWHM and an energy of 1-keV with an overall spread of approximately 30 eV. Two positron/positronium lifetime spectrometers are currently undergoing testing. The first spectrometer has demonstrated a time resolution of 390 ps and further improvements are anticipated. Doppler broadening techniques can also be performed with each spectrometer. Once completed, this facility will provide a positron beam with energies up to 30 keV and a millimeter spot size for applications in nanophase characterization of materials.

References

[1] Schultz P J and Lynn K G 1988 *Rev. Mod. Phys.* **60** 701.

[2] Coleman P 2000 *Positron Beams: and Their Applications* (Sigapore: World Scientific).

[3] Gidley D, Peng P and Vallery R 2006 *Annu. Rev. Mater. Res.* **36** 49.

[4] Gidley D W, Peng H G, Vallery R S, Soles C L, Lee H J, Vogt B D, Lin E K, Wu W L, and Baklanov M R 2007 Porosity of Low Dielectric Constant Materials *Dielectric Films for Advanced Microelectronics* ed M Baklanov, M Green and K Maex (London: John Wiley & Sons) chapter 3 pp 85-136.

[5] Plsen J V, Kirkegaard P, Pedersen N J and Eldrup M 2007 *Phys. Status Solidi* C **4** 4004.

Journal of Physics: Conference Series **262** (2011) 012025 doi:10.1088/1742-6596/262/1/012025

A study of defects in electron- and ion-irradiated ZrCuAl bulk glassy alloy using positron annihilation techniques

F Hori[1], N Onodera[1], Y Fukumoto[1], A Ishii[1], A Iwase[1], A Kawasuso[2], A Yabuuchi[2], M Maekawa[2] and Y Yokoyama[3]

[1]Department of Materials Science, Osaka Prefecture University, 1-1, Gakuen-cho, Naka-ku, Sakai, Osaka 599-8531, Japan,

[2]Advanced Science Research Center, Japan Atomic Energy Agency, 1233, Watanuki-cho, Takasaki, Gunma, 370-1292, Japan

[3]Institute for Materials Research, Tohoku University, 2-1-1, Katahira, Aoba-ku, Sendai 980-8577, Japan

E-mail: horif@mtr.osakafu-u.ac.jp

Abstract. Free volume changes in $Zr_{50}Cu_{40}Al_{10}$ bulk glassy alloys irradiated by 200 and 2.5 MeV Xe ions, 180 keV He ions, and 2 MeV electrons were investigated at room temperature using positron annihilation lifetime and Doppler broadening techniques. In addition, a slow positron beam was used to probe the change in free volume in the 180 keV He ion-irradiated sample. X-ray diffraction revealed that no crystallization took place in any of the irradiated samples. The Doppler broadening spectra from the annihilated gamma rays remained essentially constant in all ion-irradiation cases; however, an extremely minor change of positron mean lifetime was detected in each case. For electron- and He ion-irradiated samples the positron lifetime increased, and the opposite was seen in heavy-ion irradiated samples. The Doppler broadening S parameter increased with He-ion radiation dose, and the depth profile correlated well to the damage profile.

1. Introduction

Bulk glassy alloys show promise for various applications because they possess superior mechanical properties such as hardness, strength, corrosion resistance and micro-formability [1-3]. We propose that the properties of bulk glassy alloys can be further improved via irradiation treatments. To date, thin-film metallic glasses treated with electron- and ion radiation have been extensively studied [4-8]. In these thin films, both electron [4] and ion irradiation [7-8] was found to crystallize the metallic glass. However, it has become apparent that the local structure and properties in thin films are different to bulk glassy alloys, and furthermore, it is known that glassy alloys contain internal stresses. We suspect that the larger sample size has a direct influence on the effects of radiation on bulk glassy alloys, and have previously reported on the effects of electron-and swift heavy-ion radiation on the

Journal of Physics: Conference Series **262** (2011) 012025 doi:10.1088/1742-6596/262/1/012025

free volume and mechanical properties of ZrCuAl bulk glassy alloys [9]. In the present work we investigate the influence of radiation species on the radiation effects and free volume changes during low-energy He-ion radiation of ZrCuAl bulk glasses using positron annihilation techniques.

2. Experimental procedure

$Zr_{50}Cu_{40}Al_{10}$ bulk glassy alloy samples were fabricated by the tilt casting method in an arc furnace and cut into 0.6-mm-thick disks of 8 mm in diameter following the procedure given in [9]. The samples were irradiated at room temperature with 200 and 2.5 MeV Xe ions, 2 MeV electrons, or 180 keV He ions using a tandem-type accelerators at the Japanese Atomic Energy Research Agency (JAEA) Takasaki, the Wakasa-Wan Energy Research Center (WERC), and JAEA Tokai Japan, respectively. The total doses of radiation used are listed in Table 1. The irradiation fluence of He ions used was 1.0 $\times 10^{15}$, 5.0 $\times 10^{15}$ and 1.0 $\times 10^{16}$ ions/cm^2. Positron annihilation lifetime measurement was performed for the electron-irradiated and 200 MeV Xe ion-irradiated samples using a classical sandwich configuration (CSC) of the positron source between two identical samples. Furthermore, coincidence Doppler broadening (CDB) measurements were performed for electron, He and Xe ion-irradiated samples using the CSC method. Positron lifetime and Doppler broadening measurements for the 180 keV He and the 2.5 MeV Xe ion-irradiated samples were carried out in vacuum using slow positron beam apparatus with an energy range of 1 to 30 keV, which corresponds to mean implantation depth of incident positrons from 0.05 to 1.3 mm in this ZrCuAl alloy. The statistical error of the fits was less than 1 ps, however, the overall accuracy was 2 ps due to the reproducibility of the experiments. The region of $|p_L| < 3.6 \times 10^{-3}$ mc in the Doppler broadening spectrum was selected as the window for the S parameter, where p_L is an electron-momentum, c the speed of light and m the electron-mass at rest. X-ray diffraction (XRD) was performed using a Rigaku Ultima IV to determine the crystallinity of the samples.

3. Results and discussion

No crystalline peaks were observed in any of the XRD patterns from the bulk glassy alloys, whether obtained either before or after irradiation. This is different to behavior reported for thin-film metallic glasses where nanocrystals are formed after both light- and heavy-ion irradiation [4, 8]. This difference must therefore be due to the difference in sample size.

All the positron lifetime spectra were analyzed based on the one lifetime component shown in Table 1. The positron lifetime increases only after electron and 180 keV He irradiation. This indicates that the radiation has increased the open volume size (free volume). On the other hand, the positron lifetime was unchanged by irradiation with heavy Xe-ions

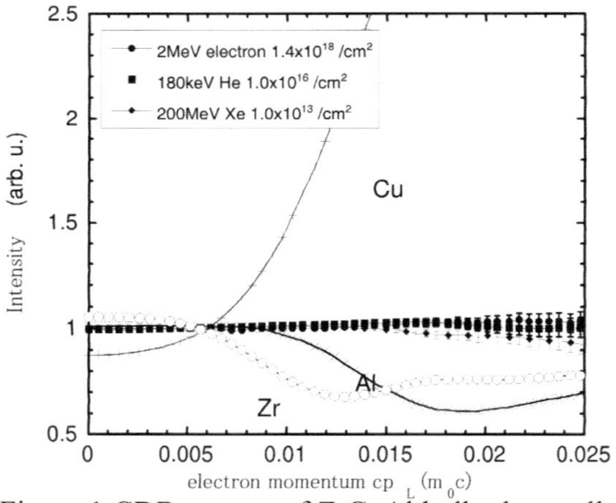

Figure 1 CDB spectra of ZrCuAl bulk glassy alloy irradiated with electron, He and 200MeV Xe ions, expressed as a ratio of CDB intensity of irradiated specimens to that of the as-prepared specimen.

Table 1 Positron lifetime of bulk ZrCuAl glassy alloy before and after irradiation.

	total dose (/cm^2)	incident positron energy	positron lifetime difference $\Delta\tau$ (ps)
before irradiation		15 keV, CSC	[191]
180 keV He	1 x 10^{16}	15 keV	+23
2.5 MeV Xe	1 x 10^{14}	15 keV	-3
200 MeV Xe	1 x 10^{13}	CSC	-3
2 MeV electron	1 x 10^{18}	CSC	+6

CSC: classical sandwich configuration method using RI source at either energy of 2.5 or 200 MeV. In the case of the 200 MeV Xe ion-irradiated sample measured by the CSC method, the change in positron lifetime of 3 ps is underestimated because the maximum damage depth of ~14 μm is shallower than implantation depth of positrons emitted from the ^{22}Na radioactive source. We suggest that this difference in behavior can be attributed to the difference in displacements per radiation particle (dpp), because the dpp estimated for 2 MeV electrons and 180 keV He ions is approximately three or more orders of magnitude less than that for heavy Xe ions. Under these radiation sources, an increasing positron lifetime indicates the introduction of vacancy-type defects, and decreasing positron lifetime reveals the disappearance and shrinking of free volume due to structural relaxations and/or by crystallization. Similar changes due to structural relaxation and crystallization caused by heat treatment have been reported elsewhere [10-12]. In our experiments, however, the XRD results show that crystallization does not take place under any irradiation investigated, therefore, the decrease of positron lifetime with Xe-ion irradiation can only be due to structural relaxations, with the energy provided not through heat treatment but via the particle energy. It follows then that the relaxation processes caused by irradiation are not necessarily the same as those from the movement of thermally activated atoms. No significant changes were observed in the CDB spectra for any irradiation, as shown Figure 1. In the case of He- and Xe-ion irradiation using the CSC method, not all positrons annihilate within the damaged layer, hence the change in CDB ratio is small. It has been reported that the CDB ratio changes during crystallization but does not change during structural relaxation [12, 13]. Therefore, we can consider these irradiation treatments are not especially effective for local atomic reordering around the free volume.

Figures 2(a), (b), and (c) show the Doppler broadening S parameter as a function of incident positron energy and the damage profile from 180 keV He-ion irradiation on

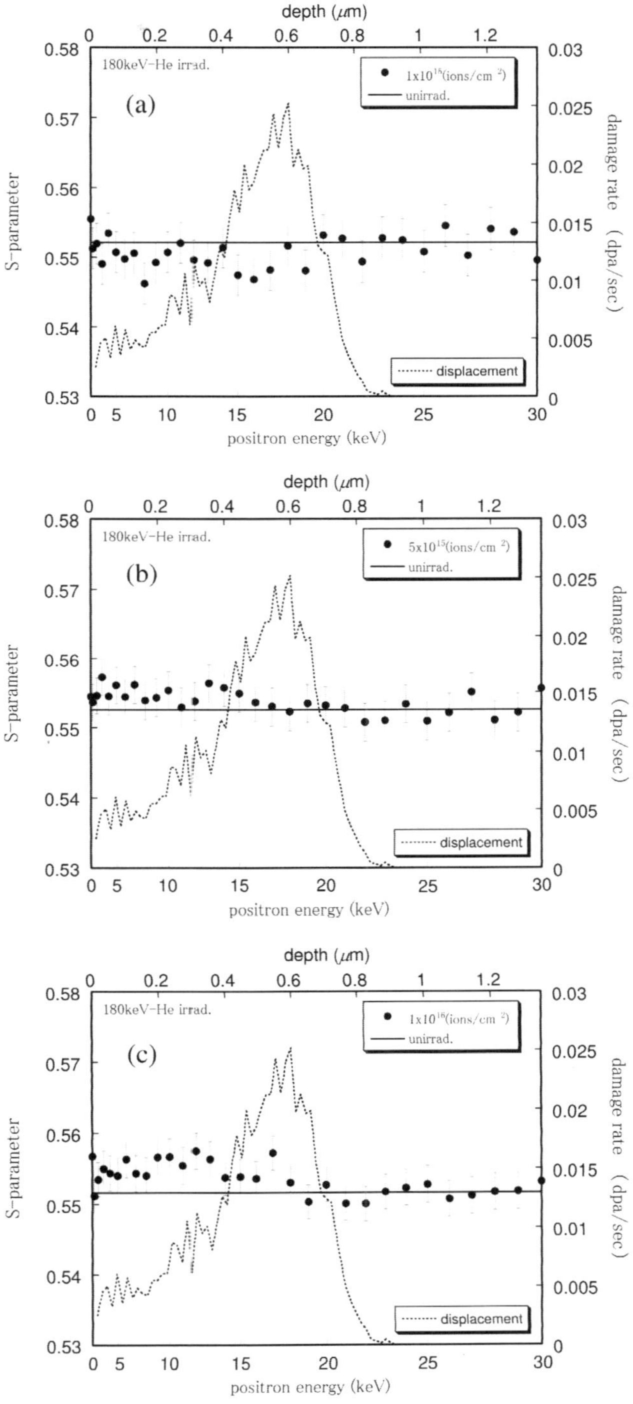

Figure 2 Positron annihilation Doppler S parameter in ZrCuAl bulk glassy alloy as a function of depth after 180 keV He-ion irradiation. Also shown is the irradiation damage profile. Total dose: (a) 1x10^{15} ions/cm^2, (b) 5x10^{15}, and (c) 1x10^{16} ions/cm^2.

the ZrCuAl bulk glassy alloy. The S parameter bulk value is defined as the mean value in the non-irradiated area between 22 and 30 keV in each sample, and the damage profile was calculated using SRIM2006 code [14]. The maximum damage depth for this radiation was estimated to be approximately 0.8μm. In Figure 2(a), the S parameter does not increase despite the high irradiation fluence of of 1×10^{15} ions/cm^2. However, at higher doses the S parameter is slightly increased in the damaged layer from the surface to 0.6 μm, as shown in Figures 2(b) and (c). This result reveals that the He-ion radiation is effective at introducing vacancy-type atomic defects producing excess free volume. According to the positron lifetime results given in Table 1, the average size of the open volume in the bulk glassy alloy before, and then after He-ion irradiation, corresponds roughly to single or di-vacancy, or larger than metallic tri-vacancy, respectively [15]. Considering the fact that in this case the irradiated material is not crystalline but amorphous, a conclusive determination of the number of vacant sites included in the open volume and its structure cannot be made. Consequently, all that can be concluded is that the open volume, which is not necessarily the same intrinsic free volume in the as-prepared glassy alloy, increases with 180 keV He-ion irradiation.

4. Summary

Free volume changes in a $Zr_{50}Cu_{40}Al_{10}$ bulk glassy alloy after electron and ion irradiation were studied by employing positron annihilation spectroscopy. Different trends in the free volume change were observed for the various radiation species. Specifically, the free volume increased only under electron and 180 keV He-ion radiation. The increase in free volume within the damaged region was measured during 180 keV He-ion irradiation using the slow positron beam technique. We found that no crystallization in the bulk glassy alloy took place under any radiation source or by changing irradiation parameters such as energy, flux, total dose or irradiation temperature. We also found that the radiation effect in ZrCuAl bulk glassy alloy is dependent on the displacement per radiation particle.

Acknowledgement

Authors would like to express cordial thanks to technical staffs for electron and ion irradiation at Japan Atomic Energy Agency. This work was also performed under the inter-university cooperative research program of the Advanced Research Center of Metallic Glasses, Institute for Materials Research, Tohoku University.

References

[1] Inoue A 2000 *Acta. Mater.* **48** 279
[2] Löffler J F 2003 *Intermetallics* **11** 529
[3] Johnson W L 2002 *JOM.* **54** 40
[4] Nagase T, Hosokawa T, Takizawa K and Umakoshi Y 2009 *Intermetallics* **17** 657
[5] Petrusenko Y, Bakai A, Borysenko V, Astakhov A and Baarankov D 2009 *Intermetallics* **17** 246
[6] Sugita K, Matsumoto M, Mizuno M, Araki H and Shirai Y 2008 *J. Phys. Conf. Ser.* **106** 1
[7] Dunlop A, Jaskierowicz G and Rizza G 2003 *Phys. Rev. Lett.* **90** 1
[8] Carter J, Fu E G, Bassiri G, Dvorak B M, Theodore N D, Xie G, Lucca D A, Martin M, Hollander M, Zhang X and Shao L 2009 *Nucl. Instrum. Method B* **267** 1518
[9] Yokoyama Y, Fukaura K and Inoue A 2002 *Intermetallics* **10** 1113
[10] Ishii A, Hori F, Iwase A, Fukumoto Y, Yokoyama Y and Konno T J 2008 *Mater. Trans.***49** 1975
[11] Hori F, Yano T, Yokoyama Y, Akeno Y and Konno T J 2007 *J. Alloys Comp.* **434-435** 207
[12] Ishii A, Iwase A, Fukumoto Y, Yokoyama Y, Konno T J and Hori F 2009 *J. Alloys Comp.* **480** 462
[13] Yano T, Yorikado Y, Akeno Y, Hori F, Yokoyama Y, Iwase A, Inoue A and Konno T J 2005 *Mater. Trans.* **46** No.12, 2886
[14] http://www.srim.org/
[15] Puska M J and Nieminen R M 1983 *J. Phys. F: Met. Phys.* **13** 333

12th International Workshop on Slow Positron Beam Techniques (SLOPOS12) IOP Publishing
Journal of Physics: Conference Series **262** (2011) 012026 doi:10.1088/1742-6596/262/1/012026

KEK-IMSS Slow Positron Facility

T Hyodo[1], K Wada, A Yagishita, T Kosuge, Y Saito, T Kurihara, T Kikuchi, A Shirakawa, T Sanami, M Ikeda, S Ohsawa, K Kakihara and T Shidara
High Energy Accelerator Research Organization (KEK)
1-1 Oho, Tsukuba, Ibaraki, 305-0801, Japan

E-mail: toshio.hyodo@kek.jp[1]

Abstract. The Slow Positron Facility at the Institute of Material Structure Science (IMSS) of High Energy Accelerator Research Organization (KEK) is a user dedicated facility with an energy tunable (0.1 - 35 keV) slow positron beam produced by a dedicated 55MeV linac. The present beam line branches have been used for the positronium time-of-flight (Ps-TOF) measurements, the transmission positron microscope (TPM) and the photo-detachment of Ps negative ions (Ps^-). During the year 2010, a reflection high-energy positron diffraction (RHEPD) measurement station is going to be installed. The slow positron generator (converter/ moderator) system will be modified to get a higher slow positron intensity, and a new user-friendly beam line power-supply control and vacuum monitoring system is being developed. Another plan for this year is the transfer of a ^{22}Na-based slow positron beam from RIKEN. This machine will be used for the continuous slow positron beam applications and for the orientation training of those who are interested in beginning researches with a slow positron beam.

1. Introduction

The slow positron beam at High Energy Accelerator Organization (KEK) started in 1991 sharing the 2.5 GeV electron linac beam with high energy TRISTAN experiments and the synchrotron radiation source for the storage ring of the Photon Factory[1-3]. This beam was used for the positronium time-of–flight (Ps-TOF) measurements, where the linac was operated in a short pulse mode (10ns pulse width and 50 pulses/s at a power of 2 kW. In this experiment the energy distributions of Ps emitted from a SiO_2, KI, and MgO single crystals were measured [4]. Overall time resolution of the TOF machine was about 20ns, short enough to measure *ortho*-positronium time-of-flight whose intrinsic lifetime in vacuum is 142ns. The results showed that Ps is emitted from the SiO_2 surface by two mechanisms: emission of Ps formed in the bulk and Ps formation on the surface, with energies 1 and 3 eV, respectively.

Then, in 1997, as a result of the reformation of KEK, Institute of Materials Structure Science was established in KEK, and the Photon Factory became a part of it, and the Slow Positron Facility was formally affiliated to the Photon Factory [5]. At the same time, in conjunction with a new high energy physics experiments called KEKB project, an upgrade of the linac (the KEKB J-linac) aiming at increasing the energy to 8.5 GeV started. As a consequence, the location of the slow positron generator (converter/moderator assembly) was moved to the point where the energy of the accelerated electrons was 1.5 GeV. The linac was operated at a power of 0.75 kW. In addition, because the machine time of the KEKB J-linac available for the slow positron experiments was expected to be

Journal of Physics: Conference Series **262** (2011) 012026 doi:10.1088/1742-6596/262/1/012026

very short, a linac of 55 MeV and 600 W dedicated to the Sow Positron Facility was constructed. It was placed side by side with the KEKB beam line to share the slow positron generator. Unfortunately, however, it suffered from a trouble in cooling water system of the slow positron generator before the dedicated linac became in full operation. It was then realized that the maintenance of the generator system was very difficult as long as it was located near the busy KEKB beam line. Thus in 2001 it was decided to move the dedicated linac to an end of the slow positron experiment hall *i.e.*, the present location [6]. It operates in a short pulse (width 22ns, $2x10^5$ e^+/s) and a long pulse (width 1μs, $6x10^6$ e^+/s) modes.

2. Present status

The present layout of the branches of the beam line, and the experimental stations are schematically shown in Figure 1. The positronium time-of–flight (Ps-TOF) measurement apparatus was moved to the present experiment hall after a small modification. It has been used for the investigation of the structure of porous materials in collaboration with Mills' group[7, 8].

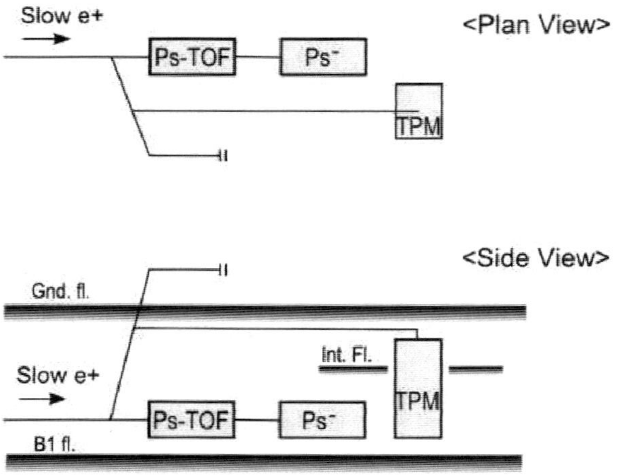

Figure 1. Schematic diagram of the present layout of the stations connected to the slow positron beam line. Ps TOF: positronium time-of-flight. TPM: transmission positron microscope. Ps^-: photo-detachment of positronium negative ion.

In 2005 development of a transmission positron microscope (TPM) by Fujinami group was started. In this machine, the magnetically guided slow positron beam is focused on a transmission-type Ni remoderator for a brightness enhancement, behind which the beam is made free from the magnetic field and guided electrostatically [9]. It has a switch system which makes it possible to observe exactly the same part of the specimen by both the transmission electron microscope (TEM) and TPM . Magnified real space images (up to x10,000) and diffraction patterns of a Au foil observed by the electrons and the positrons have been compared [10].

Another research currently running is photo-detachment of positronium negative ions (Ps^-). Nagashima developed a method of creating a high fraction (1%) of Ps^- by bombarding positrons on an alkali metal coated W [11]. It was accomplished with a ^{22}Na based slow positron beam at his laboratory in Tokyo University of Science. The Ps^- is accelerated by an electric field and blue-shifted annihilation γ-rays from the moving Ps^- give a clear evidence for their existence [12]. The photo-detachment is performed by using the pulsed positron beam (12 ns FWHM, 50 pulses/s) at KEK and an intense photon beam from a Q-switched ND:YAG laser (1 J/pulse at 1064 nm, 12 ns FWHM, 25 pps) synchronized to the positron bunch [13]. Observed reduction of the Doppler shifted Ps^- signal indicates that the Ps^- ions are converted into neutral Ps, of which 3/4 is *ortho*-Ps which does not annihilate into 2□. It is expected that the photo-detachment process will be useful for the production of an energy tunable Ps beam that can be use in UHV environment.

3. Plans for near future

During the year 2010, the converter/moderator at the end of the dedicated linac is going to be renewed, expecting to get better yield of the slow positrons. At present, the high energy electron bombardment on a water cooled Ta target of 4mm thick produces bremssstrahlung photons and the following pair creation. The positrons are then moderated by seven strips of W foils of thickness 25 μm placed perpendicular to the electron beam. The slow positrons are extracted in the direction perpendicular to the incident electron beam. The new version uses 4mm thick Ta for a converter, whose electrostatic potential may be varied from 100 V to 35 kV as the present one. This determines the energy of the slow positrons incident on a grounded sample. The new W moderator foils will be tow-fold and each has a honeycomb structure as shown in Figure 2. Electrostatic bias will be applied between the Ta frame and the back W arrangement, between the back and front W arrangements and between the front W arrangement and the extraction mesh electrode. Each bias can be varied independently from 0 to 10 V.

In addition, the control system for the beam line solenoids and Helmholtz coils and the vacuum monitoring system are being revised to make them more user-friendly.

Also two new stations are going to be constructed; one will be connected to an existing branch of the linac based beam, and the other will stand alone. The planned layout is schematically shown in Figure 3 (It also includes an ACAR machine which is not going to be completed soon).

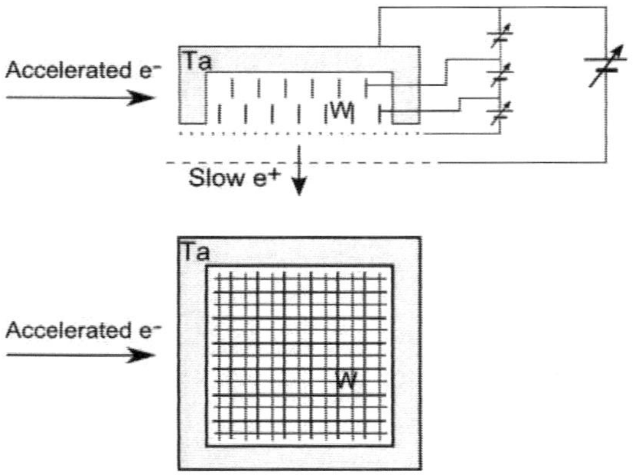

Figure 2. Schematic diagram of the new converter/moderator assembly.

The station to be connected is for the reflection high energy positron diffraction (RHEPD). The RHEPD is a positron version of the reflection high energy electron diffraction (RHEED), but has a quite different characteristic. RHEPD was brought into practical use by Kawasuso [16]. The major difference comes from the positivity of the crystal potential for the positron. Since the crystal potential of materials for the electrons is negative, the electrons incident on a surface at a small glancing angle will be sucked into the materials. It means that the actual incident angle on the crystal lattice plane inside is greater than the initial incident angle on the surface. This causes the electrons to interact with the deeper region under the surface. Also it is impossible to observe a diffraction of the lowest indices. On the contrary, positrons feel positive crystal potential such that they cannot go into the material if incident with a small glancing angle (total reflection) [17]. Thus it is very sensitive to the topmost layer of atoms of the surface and it is possible to observe the lowest index diffraction spots. Fukaya *et al.* have published several results with their RI-based slow positron beam [18]. If RHEPD works with the linac based intense slow positron beam, fruitful results are expected.

Journal of Physics: Conference Series **262** (2011) 012026 doi:10.1088/1742-6596/262/1/012026

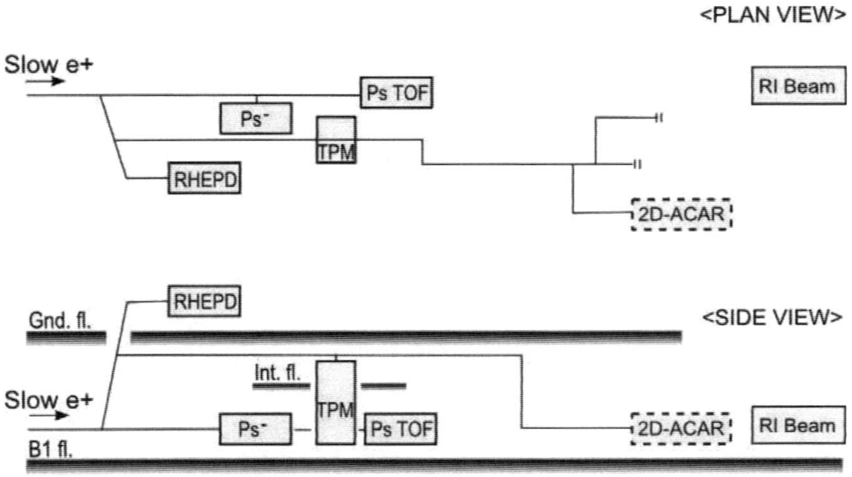

Figure 3. Schematic diagram of the layout of the stations connected to the slow positron beam line in near future. RHEPD: reflection high-energy positron diffraction. RI Beam: ^{22}Na-based slow positron beam. 2D-ACAR: two dimensional angular correlation of annihilation radiation.

The ^{22}Na-based slow positron beam is to be moved from RIKEN in Wako. It has been used, for example, to detect the positron-impact X-ray emission cross sections near threshold by Nagashima *et al.* [19]. It was observed that Cu-K emission shows good agreement with the theory, but not so much in the case of Ag-L emission. The beam was also used to analyse the ion-beam irradiated polymers by Saito *et al.*[20]. The slow positron depth profile, the AFM observation and the analysis with the nano-cutting machine SAICAS were combined to investigate poly-lactic acid irradiated by He^+ ions at 150 keV. The AFM showed that the surface of the irradiated part of the sample subsided with the amount increasing with the ion dose. The SAICAS showed that the ion irradiation depth also increased with the dose. It is not the case for metals, semiconductors, or ionic crystals, where the depth of the effect of ion irradiation is essentially independent of the dose. The SAICAS data and the positron energy dependence of the annihilation □□ray S-parameter was combined to obtain the density of the ion irradiated part of the sample. The result showed that the density was reduced to be lower than the half of the original value. This machine will be open for the public use after a new ^{22}Na source is obtained. It is expected to serve for the researches where the use of a continuous positron beam is essential, as well as for the users who want to use the slow positrons for the first time.

4. Summary

A brief history of the KEK-IMSS Slow Positron Facility, its present status and the plans for near future have been given. Recent results with the Ps time-of-flight, the transmission positron microscope and the photo-detachment of positronium negative ions are shown. In the near future, reflection high-energy positron diffraction (RHEPD) and a ^{22}Na-based slow positron beam will be installed.

References

[1] Asami A, Enomoto A, Kobayashi H, Kurihara T, Nakahara K and Shidara T 1992 *Mater. Sci. Forum* **105-110** 1833

[2] Shidara T, Enomoto A, Kamitani T, Kobayashi H, Kurihara T, Shirakawa A, Hirayama H, Kanazawa I, Asami A and Nakahara K 1992 *Mater. Sci. Forum* **175-178** 205

[3] Kurihara T, Shirakawa A, Enomoto A, Shidara T, Kobayashi H and Nakahara K 1995 *Appl. Surf. Sci.* **85** 178

[4] Nagashima Y, Morinaka Y, Kurihara T, Nagai Y, Hyodo T, Shidara T and Nakahara K 1998 *Phys. Rev.* B **58**, 12676-12679
[5] Kurihara T 2001 *Mater. Sci. Forum* **363-365** 433
[6] Kurihara T, Nagashima Y, Shidara T, Nakajima H, Ohsawa S, Ikeda M, Oogoe T, Kakihara K, Ogawa Y, Shirakawa A, Furukawa K, Sanami T and Enomoto A 2004 *Mater. Sci. Forum* **445-446** 486
[7] Tanaka H K M, Kurihara T and Mills A P Jr 2005 *Phys. Rev.* B **72**, 193408
[8] Tanaka H K M, Kurihara T and Mills A P Jr 2006 *J. Phys.: Condens. Matter* **18** 8581
[9] Fujinami M, Jinno S, Fukuzumi M, Kawaguchi T, Ogawa K and Akahane T 2008 *Anal. Sci.* **24** 73
[10] Fujinami M 2010 presented at SLOPOS-12, this volume
[11] Nagashima Y, Hakodate T, Miyamoto A and Michishio K 2008 *New J. Phys.* **10** 123029
[12] Mills A P Jr 1981 *Phys. Rev. Lett.* **46** 717
[13] Tachibana T, Michishio K, Terabe H, Wada K, Hyodo T, Kurihara T, Yagishita A and Nagashima Y 2010 *Nucl. Instr. and Meth. in Phys. Res.* A **621** 670
[14] Michishio K, Tachibana T, Terabe H, Wada K, hyodo T, Kuga T, Yagishita A, Igarashi A and Nagashima Y 2010 presented at SLOPOS-12, this volume
[15] Nagashima Y, Michishio K, Tachibana T and Terabe H 2010 presented at SLOPOS-12, this volume
[16] Kawasuso A and Okada S 1998 *Phys. Rev. Lett.* **81** 2695
[17] Fukaya Y, Kawasuso A and Ichimiya A 2009 *Phys. Rev.* B **79** 193310
[18] Fukaya Y, Hashimoto M, Kawasuso A and Ichimiya 2008 A *Surf. Sci.* **602** 2448
[19] Nagashima Y, Saito F, Itoh Y, Goto A and Hyodo T 2004 *Phys. Rev. Lett.* **92** 223201
[20] Saito F, Fujii Y, Nagashima Y, Yotoriyama T, Nakao A, Suzuki Y, Iwaki M, Nishiyama I and Hyodo T 2007 *phys. stat. sol.* (c) **4** 3718

Journal of Physics: Conference Series **262** (2011) 012027 doi:10.1088/1742-6596/262/1/012027

Characterizing free volumes and layer structures in polymeric membranes using slow positron annihilation spectroscopy

Y C Jean[1], Hongmin Chen[1], Sui Zhang[2], Hangzheng Chen[2], L James Lee[3], Somia Awad[1], James Huang[4], Cher Hon Lau[2], Huan Wang[2], Fuyun Li[2], and Tai-Shung Chung[2]

[1] *Department of Chemistry, University of Missouri–Kansas City, Kansas City, MO 64110*

[2] *Department of Chemical and Biomolecular Engineering, National University of Singapore, Singapore 117576*

[3] *Department of Chemical and Biomolecular Engineering, The Ohio State University, Columbus, OH 43210*

[4] *Yoshimura Industries Co. Ltd. P.O. Box 46-200, Taichung Industrial Park, Taichung, Taiwan*

Email: jeany@umkc.edu

Abstract.. Positron annihilation spectroscopy coupled with a newly built slow positron beam at National University of Singapore has been used to study the free volume, pore, and depth profile (0 - 10 μm) in cellulose acetate polymeric membrane at the bottom and top sides of membranes for ionic separation in water purification applications. The S and R parameters from Doppler broadening energy of annihilation radiation representing free volumes (0.1-1 nm size) and pores (>1 nm-□m) as a function of depth have been analyzed into multilayers, i.e. skin dense, transition, and porous layers, respectively. The top side of membrane has large free volumes and pores and the bottom side has a skin dense layer, which plays a key role in membrane performance. Positron annihilation lifetime results provide additional information about free-volume size and distribution at the atomic and molecular scale in polymeric membrane systems. Doppler broadening energy and lifetime spectroscopies coupled with a variable mono-energy slow positron beam are sensitive and novel techniques for characterization of polymeric membrane in separation applications.

1. Introduction

An asymmetric polymeric membrane system has a multilayered structure usually consisting of a thin surface or dense skin layer, an intermediate layer, and a highly porous substrate [1]. While polymeric materials and layer systems have been used a long time for separation, purification, and storage, the basic understanding of physical structures and membrane interactions is still primitive. There is limited understanding of the causes for performance in polymeric membrane systems, i.e. permeability and selectivity, at the molecular level. Positron annihilation spectroscopy (PAS) was developed as a useful tool to probe the nanoscopic properties of polymeric materials [2]. Most investigations in polymeric membranes and film systems have been reported mainly in the bulk and few in thin films

[1] To whom any correspondence should be addressed.

using positron annihilation lifetime and Doppler broadening energy techniques [3-12]. Using a variable mono-energy positron beam (from kT to several ten keV) [13] coupled with positron annihilation lifetime and momentum density measurements is capable of probing depth profiles from the surface, interfaces and to the bulk. In this paper, we present the results of both Doppler broadening energy spectroscopy (DBES) and positron annihilation lifetime (PAL) spectroscopy using a newly built variable mono-energy positron beam at National University of Singapore in one of important membrane systems, i.e. cellulose acetate (CA) on both sides of the membranes for water filtration applications [14-16]. In addition to the use of PAS technique, characterization of this CA membrane from FESEM, ATR-FTIR, AFM measurements has been made and reported elsewhere [17].

2. Experiments

2.1 Polymeric Membrane Preparation

We first prepared polymer solutions by dissolving cellulose acetate, 22.5 wt% CA-398-10 (supplied by Eastman Chemical Company) to a mixed solution of acetone (5 wt%) and NMP (72.5 wt%). Solutions were then cast on a glass plate with a 50 μm casting knife, followed by immediate immersion in a tap water bath at 25°C. After removal of residual solvents, membranes were ready for use or underwent heat treatment prior to use. The prepared CA membranes were removed from the glass plate, dried and two sides of membranes were measured in positron experiments, the top side without contact with the glass and the bottom side directly interacted with the glass plate. While both two sides have skin layers, the top layer has been known to be thicker (~0.9 μm) but with nm pores and the bottom layer contains a dense thin skin layer (~0.1 μm), which controls the separation performance. Figure 1 shows the FESEM cross sectional structure of the double-layer CA membrane assemble and our focus is on the bottom side of the membrane, which plays a key role for membrane performance. Detailed experimental procedures could be found elsewhere [17].

2.2. Positron Annihilation Spectroscopy and Slow Beam

A newly built slow positron beam at the Department of Chemical and Biomolecular Engineering, National University of Singapore (NUS) is shown in Figure 2. This beam was designed and built similar to that at the University of Missouri-Kansas City (UMKC) [18,19] and at Chung Yuan Christian University (CYCU) [20] in Taiwan. This new beam has been calibrated and its performance is as good as or better than that of existing radioisotope beams: slow positron converting efficiency = 6 x 10^{-4}, beam diameter < 5 mm using PC-control focus for positron energies 0-30 keV.

Two positron annihilation spectroscopies were used in this study: Doppler energy spectroscopy (DBES) and positron annihilation lifetime (PAL) spectroscopy. The S and W parameters from DBES represent the relative value of the free volume (Å to 1 nm) depth profile from o-Ps 2γ radiation (pick-off annihilation) in polymeric systems. The other parameter R, which is the 3γ to 2γ annihilation ratio and it gives information about the existence of large pores (nm to μm), where o-Ps undergoes 3γ annihilation. The S parameter in Si waffle was measured this new beam and the result is compared with those obtained at UMKC [18,19] and CYCU [20] beams. Figure 3 shows good agreements on data obtained among those three slow positron beams. The beam-PAL spectrometer uses the secondary electrons as the starting signal and the lifetime resolutions were found to be about 450 ps at the positron energy > 2 keV. The PAL data contain quantitative information on the free-volume and pore properties in polymeric systems. Analyzed o-Ps pick-off lifetimes (τ_3) from PATFIT fitting are on the order of 1-5 ns in polymeric materials, and were used to calculate the mean free-volume radius (Å to nm) from an established semi-empirical correlation equation [21] based on Tao's spherical-hole infinite potential quantum model [22]. The long o-Ps lifetime, τ_4 contains pore (>nm-μm) information. Detailed descriptions of slow positron beam and data analysis could be found elsewhere [2,18,19,21].

3. Results and Discussion

Figure 4 shows the results of S and R parameters vs. positron incident energy (which is also expressed as the mean depth according to an established equation [13]) for both the top and bottom sides of CA membrane. From the S data, we have the following observations: (1) S near the surface increases sharply as the positron energy increases; (2) S reaches a maximum and then decreases. The detailed variations contain multilayer structural and chemical composition information. We have employed a computer program VEPFIT [23] to fit the S vs. depth data. We found that the results from 3-layer model fits have good chi-squares (< 1.5) and give stable results and reasonable error bars as lines plotted in Figure 4. These three layers are assigned as: the dense skin layer, the transition layer between the dense skin and the porous layer, and the porous layer of CA membrane. The resolved 3-layer structure is also consistent with the findings of most asymmetric membranes [24,25]. The results of VEPFIT fitted layer thickness are shown in Table 1 for both top and bottom modes of CA membrane. It is found that the top side of the membrane has a much thicker skin layer (852 nm) than that of the bottom side (98 nm). This result is consistent with the FESEM result as shown in Figure 1.

The R parameter represents 3 γ annihilation radiation which is either from o-Ps localized in large pores (> nm- μm) of CA membrane or in the beam chamber. The R result vs depth has the following features: (1) Near the surface, R is large due to the backscattering and back diffusion of o-Ps to the vacuum; (2) it decreases as the positron enters the surface of CA and reaches a low value due to the pick-off 2γ annihilation in the dense skin layer; and (3) R increases as the positron enters the transition and then at a large value in the porous region of CA. We employed the VEPFIT program to fit R vs depth in 3-layer model and the result of the skin, transition, and porous layers is shown in Table 1. In general, we found the layer thickness and structure are consistent with that from S fit, i.e. bottom side of CA has thinner skin and transition layers than top side. However, the thickness (260 nm) of the dense skin layer of bottom side from R fit is larger than that of S fitted result (98 nm) while the transition layer from R fit (865 nm) is smaller than S fitted result (2210 nm). This difference is understandable from the intrinsic difference between R and S, which are from 3γ- and 2γ- o-Ps annihilation, respectively. And the layer structures from R and S are schematically shown in Figure 5, which reflect the pores (> nm) and free volumes (< nm) information, respectively.

We analyzed the obtained PAL spectra into four lifetime components: (τ_1= 0.125 ns, constrained), (τ_1 ~0.45 ns), (τ_3 ~ 2-3 ns), and (τ_4 ~ 12-40 ns). The o-Ps pick-off lifetime (τ_3) of CA polymer is in agreement with the values of common polymers as o-Ps annihilation lifetime and the corresponding free-volume radius ranges from 0.25 to 0.40 nm. The long o-Ps lifetime (τ_4) of CA membrane is also consistent with porous membranes on the order of > nm. The variations of τ_3 and τ_4 vs. depth are shown in Figure 6 and the corresponding intensities, I_3 and I_4 are shown in Figure 7. Those results are consistent with that expected from the multi-layer structures as resolved by S and R parameters above. The result of τ_3, τ_4, I_3 and I_4 at 1.0 and 1.5 keV of positron energy (30 nm, 58 nm mean depth) of the skin layer is shown in Table 2. This result indicates that heating (annealing) reduces the pores and possibly free volume content of CA membrane. We also analyzed the positron lifetime distributions using the MELT program and the τ_3 results (with corresponding free-volume radius distributions) are shown in Figure 8. The kinetic radius of water (1.3 Å) [26] is much less than free volumes of CA , however, the hydrated ions, such as Na^+, and Cl^- are between 1.5 Å and 3.5 Å [27], which are in the range of free-volume radius distribution of CA. The free-volume size distribution is valuable information for the understanding of ionic selectivity in membranes.

Conclusion

Positron annihilation spectroscopy coupled with a variable energy slow positron beam has been used to determine the layer structures and free-volume properties in an asymmetric CA membrane for forward osmosis (FO) membrane separations. We found a 3-layer structure existed in the bottom side of CA membrane prepared on the glass: a dense skin (96 nm), and a transitional layer from dense skin to porous (0.1-2 μm) and a porous CA layer. Free-volume sizes and distributions of asymmetric membranes are useful information to study the membrane performance, particularly for selectivity.

Journal of Physics: Conference Series **262** (2011) 012027 doi:10.1088/1742-6596/262/1/012027

Positron annihilation spectroscopy coupled with a slow positron beam is a useful method for quantitative analysis of free volume and pore size and distributions for FO water purification and may provide guidance for the design of multi-layer membranes for separation applications.

Acknowledgment

We wish to express sincere gratitude to the supports of National University of Singapore, NSF-Nanoscale Science and Engineering Center for Affordable Nanoengineering of Polymeric Biomedical Devices (NSEC-CANPBD) at the Ohio State University and YMT Industrial Inc.-Taiwan.

References

[1] Mulder M *Basic Principle of Membrane Technology* 2nd ed. 1996 (Kluwer Academic Publisher, The Netherlands)
[2] For example, see Jean Y C, Mallon P E, Schrader D M ed. 2003 *Principles and Applications of Positron and Positronium Chemistry* (World Sci. Pub., Singapore)
[3] Kobayashi Y, Kojima I, Hishita S, Suzuki T, Asari E, Kitajima M 1995 *Phys. Rev. B* **52** 823
[4] Jean Y C, Zhang R, Cao H, Yuan J P, Huang C M, Nielsen B, Asoka-Kumar P 1997, *Phys. Rev. B* **56** R8459
[5] Consolati G, Pegoraro M, Quasso F, Severini F 2001 *Polymer* **42** 1265
[6] Dlubek G, Redmann F, Krause-Rehberg R 2002 *J. Appl. Polym. Sci.* **84** 244
[7] Shantoravich V P, Kevnida I B,Yamplospkii Y P, Anlatiev A Y 2000 *Macromolecules* **33** 7453
[8] Gunther-Schade K, Fritsch D, Strunskus T, Faupel F 2000 *Macromolecules* **35** 207
[9] Dutta D, Bhattacharyya A, Ganguly B N 2003 *J. Membrane Science.* **224** 127
[10] Winberg P, Eldrup M, Pedersen N J, Van Es M A, Maurer F H J 2005 *Polymer* **46** 8239.
[11] Wang X Y, Willmore F T, Raharjo R D, Wang X, Freeman B D, Hill A J 2006 *J. Phys. Chem.* **B**, 110 16685
[12] Pujari P K, Sen D, Amarendra G, Abhaya S, Pandey A K, Dutta D, Mazumder S 2007 *Nucl. Instrum. Methods Phys. Res. Sect. B* **254** 278
[13] For examples, see Coleman P ed. *Positron beams and Their Applications* 2000 (World Sci. Pub., Singapore); Schultz P J, Lynn K G 1989 *Rev. Mod. Phys.* **60** 701
[14] Yang Q, Wang K Y, Chung T S, 2009, *Separ. Purif. Technol.* **69** 269
[15] Mi B X, Elimelech M 2010 *Environ. Sci. Technol.* **44** 2022
[16] Teoh M M, Wang K, Bonyad S, Chung T S 2010 *Innovation Magazine* **9**, 1
[17] Zhang S, Wang K Y, Chung T S, Chen H, Jean Y C, Amy G 2010 *J. Membrane Science* **36** 522
[18] Zhang R, Chen H, Cao H, He Y, Yuan J P, Sandreczki T C, Jean Y C, Suzuki R, Ohdaira T, Nelsen B 2000 *Radiat. Phys. Chem.* **58** 639
[19] Chen H, Zhang R, Li Y, Zhang J, Wu Y C, Sandreczki T C, Mallon P E, Suzuki R, Ohdaira T, Gu X., Nguyen T, Jean Y C 2004 *Materials Science Forum* **445-446** 274
[20] Hung W S, Lo C H, Cheng M L, Chen H, Liu G, Chakka L, Nanda P, Tung K L, Huang S H, Lee K R, Lai J Y, Sun Y M, Wu Y C, Zhang R W, Jean Y C 2008 *Applied Surface Science* **255** 210
[21] Jean Y C 1990 *Microchemical J.* **42** 72
[22] Tao S J 1972 *J. Chem. Phys.* **56** 5499
[23] Van Veen A, Schut H H, de Vries J, Hakvoort H A, Ijpma M R 1990 *AIP Conf. Proc.* **218**, 171 (AIP, New York)
[24] Chen H, Hung W S, Lo C H, Huang S H, Cheng M L, Liu G, Lee K R, Lai J Y, Sun Y M, Hu C C, Suzuki R, Ohdaira T, Oshima N, Jean Y C 2007 *Macromolecules* **40** 7542
[25] Hung, W S, De Guzman M, Huang S H, Lee K R, Jean Y C, Lai J Y 2010 *Macromolecules* **43** 6127
[26] Van der Bruggen B, Schaep J, Wilms C, Vandecasteele J 1999 *J. Membrane Science* **156** 29
[27] Su J C, Yang C, Teo J F, Chung T S 2010 *J. Membrane Science* **355** 36

Table 1. Results of S and R values and thicknesses (L) from multilayer analysis of CA membranes using VEPFIT.

Mode*	***S1***	***S2***	***S3***	***L1(nm)***	***L2***(nm)**
Top	0.4841±0.0003	0.4783±0.0016	0.4830±0.0012	852±530	2778±1792
Bottom	0.4824±0.0006	0.4793±0.0004	0.4813±0.0009	98±67	2210±1583

Mode*	***R1***	***R2***	***R3***	***L1(nm)***	***L2***(nm)**
Top	0.4010±0.0003	0.4600±0.0004	0.4200±0.0004	693±52	4884±269
Bottom	0.3994±0.0003	0.4210±0.0003	0.4310±0.0003	260±38	865±112

* Top mode refers to the characterization of the top side of the membrane facing to the air, and bottom mode refers to the characterization of the bottom side of the membrane facing the glass plate.
** L_1 and L_2 are the boundary lengths of the dense and transition sub-layers, respectively.

Table 2. PAL results from PATFIT analysis for the bottom side of CA membrane cast on a glass plate measured at 1.0 and 1.5 keV positron energies (mean depth 30 and 58 nm), respectively.

Energy (keV)	**Membrane**	**τ_3 (ns)**	**I_3 (%)**	**τ_4 (ns)**	**I_4 (%)**
1.0	**CA-90°C-bottom**	2.02±0.02	25.42±0.53	12.97±0.54	2.31±0.08
1.0	**CA-RT-bottom**	2.05±0.03	28.26±0.56	13.55±0.55	2.15±0.07
1.5	**CA-90°C-bottom**	2.14±0.02	25.97±0.34	13.31±0.83	1.28±0.07
1.5	**CA-RT-bottom**	2.17±0.02	27.64±0.35	13.88±0.84	1.14±0.06

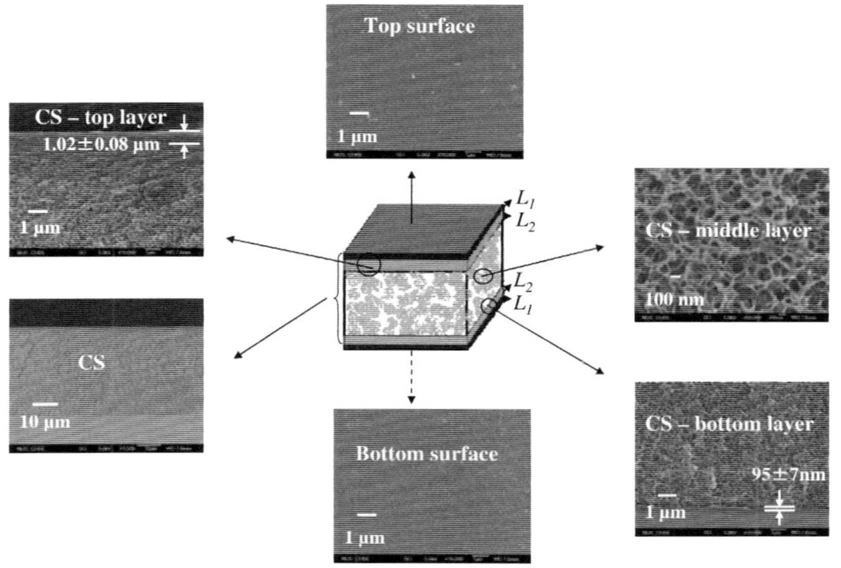

Figure 1. Schematic diagram and FESEM images of the CA membrane cast on glass plate.

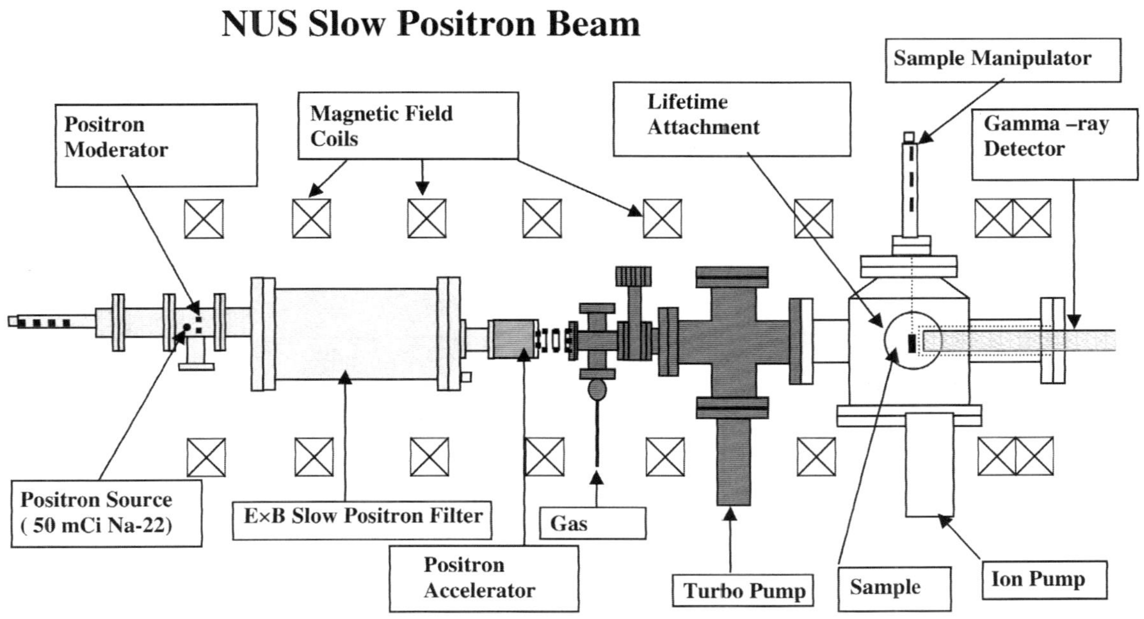

Figure 2. A schematic diagram of variable mono-energy (0-30 keV) slow positron beam at National University of Singapore. This beam includes Doppler broadening energy (DBES) and positron annihilation lifetime (PAL) spectrometers for the free-volume and pore depth profile determination.

Journal of Physics: Conference Series 262 (2011) 012027 doi:10.1088/1742-6596/262/1/012027

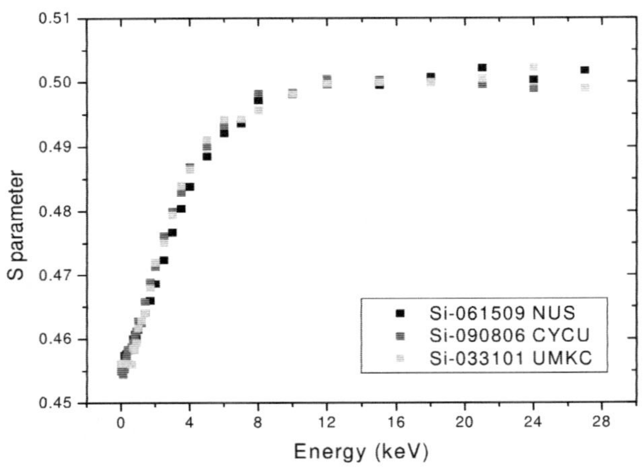

Figure 3. Comparison of S parameters obtained in the newly built slow positron beam from Si at National University-Singapore (NUS) with existing two beams at University of Missouri-Kansas City (UMKC) [18,19] and at Chung-Yuan Christian University (CYCU) in Taiwan [20].

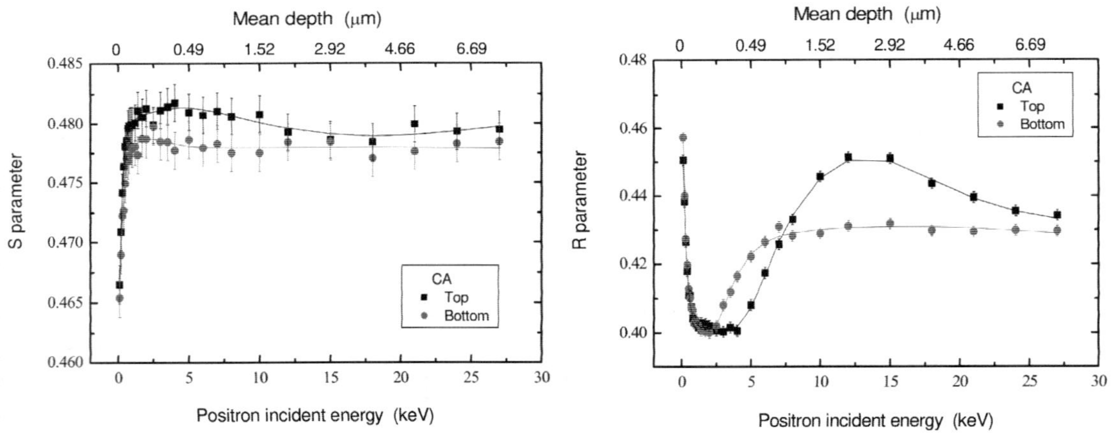

Figure 4 The S (left) and R parameter (3γ/2γ ration) for the bottom side of CA membrane.

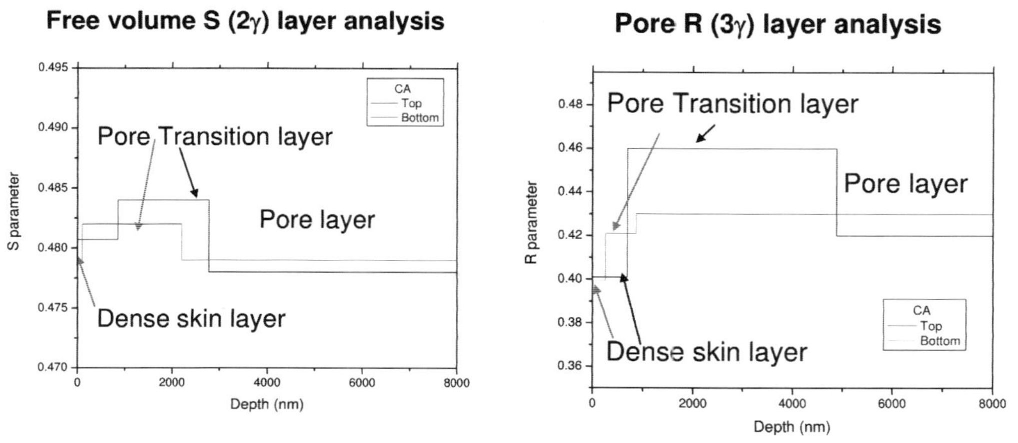

Figure 5. Three layers are resolved from S (left) and R (right) parameters data of Figure 4: skin, transition from skin to porous, and porous CA membrane as indicated vertical lines as boundaries from the VEPFIT analysis.

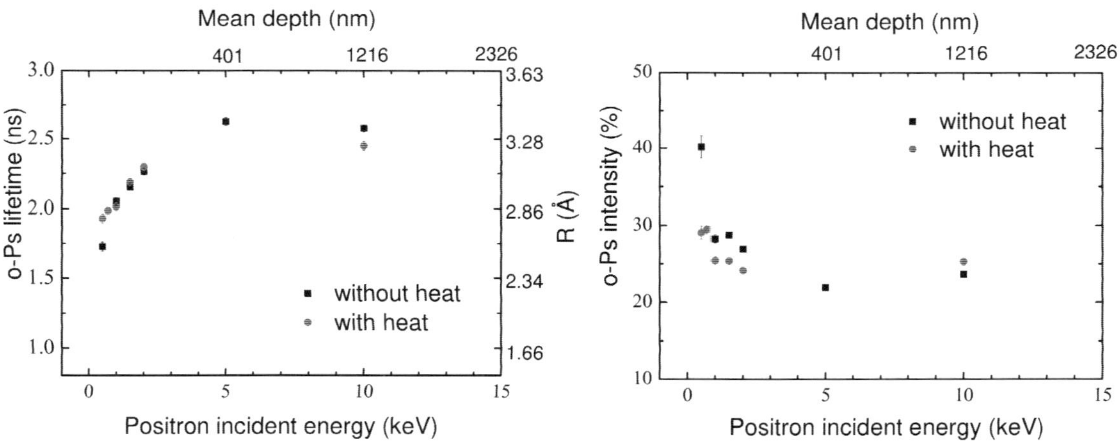

Figure 6. Pick-off o-Ps 2γ annihilation lifetime (τ_3) and intensity (I_3) vs. positron energy (depth) in the bottom mode of CA membrane cast on a glass plate with and without 15 min anneal at 90°C.

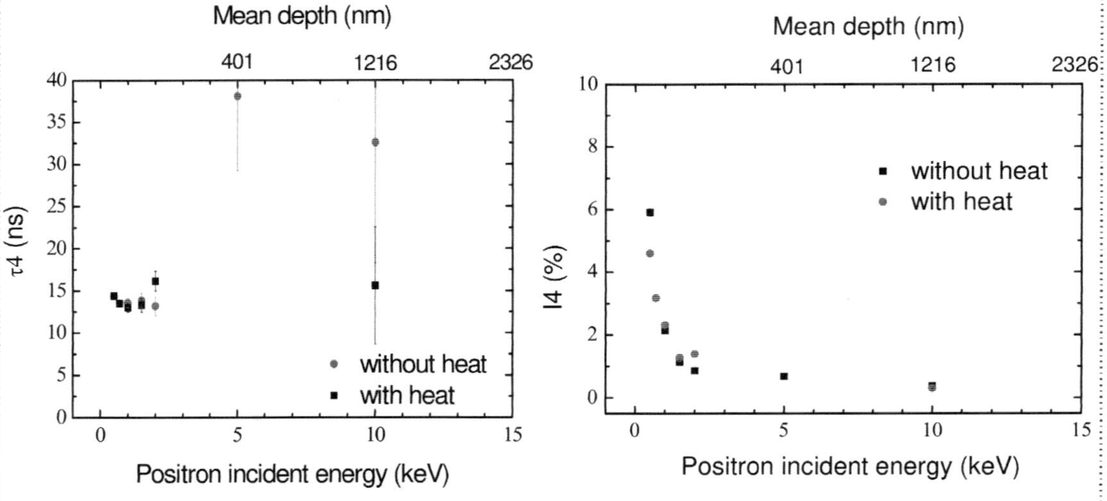

Figure 7. Long o-Ps 3γ annihilation lifetime (τ_4) and intensity (I_4) vs. positron energy (depth) in the bottom mode of CA membrane cast on a glass plate with and without 15 min anneal at 90°C.

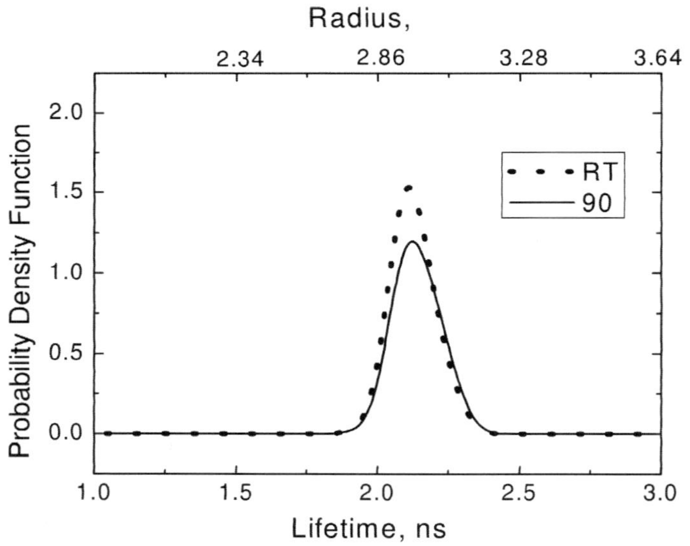

Figure 8. The o-Ps lifetime distributions (τ_3) or free volume radius of the bottom mode of CA membranes cast on a glass plate with and without anneal at 90°C for 15 min.

12th International Workshop on Slow Positron Beam Techniques (SLOPOS12) IOP Publishing
Journal of Physics: Conference Series **262** (2011) 012028 doi:10.1088/1742-6596/262/1/012028

Convergent close-coupling approach to positron and antiproton collisions with atoms

A S Kadyrov, I B Abdurakhmanov, R Utamuratov, A V Lugovskoy, D V Fursa, I Bray, and A T Stelbovics

ARC Centre for Antimatter-Matter Studies, Curtin University, GPO Box U1987, Perth, WA 6845, Australia

E-mail: a.kadyrov@curtin.edu.au

Abstract. Recent developments in application of the convergent close-coupling approach to antimatter-matter scattering are outlined. These include positron collisions with alkalis and helium, in the ground or metastable states, as well as extension of the method to heavy projectiles, such as antiprotons.

1. Introduction

Understanding of antimatter induced reactions with atoms (and molecules) remains a significant challenge, driven by the quest to advance fundamental knowledge of the underlying physical mechanisms, as well as the desire to control and to apply them. In this respect, ionization (or simply breakup) is particularly important. Theoretical treatment of antimatter-matter interactions may seem similar to those involving only matter. In the case of positron and electron scattering on atoms, the latter has the complexity of exchange arising from the indistinguishability of electrons, while the former has the complexity of rearrangement due to formation of positronium (Ps). Thus the positron-atom scattering system has two natural centres whereas the electron scattering system has only one. The two problems share similar complexity in the treatment of the target excitation and the three-body continuum.

The convergent close-coupling (CCC) method has been very successful in treating electron-atom scattering problems. The technique relies on a sufficiently large single centre expansion of the total wave function in order to correctly take into account all possible atomic excitation and ionisation processes. Convergence in transitions of interest is obtained by simply increasing the basis size. The same idea can be applied to positron scattering except now two-centre expansions are required. Such expansions incorporate the required atomic and Ps channels but make the calculations far more difficult and time-consuming.

Here we report on recent progress in applications of the CCC method to antimatter-matter interactions. In particular, we focus on developments in positron collisions with alkalis and metastable helium and extension of the CCC method to antiproton-atom scattering.

2. Positron scattering on metastable helium

For some targets, particularly excited ones, the Ps-formation channel is open starting from zero collision energy. Positron scattering on the 2^3S metastable state of helium is an example of such a system where the Ps-formation threshold is negative (-2.06 eV). This system has not been yet studied neither theoretically nor experimentally. Recently, we have reported two-centre CCC results for positron scattering on the ground state of helium [1] where a good agreement between the theory and experimental data has been seen. Here we apply this formalism to positron scattering on metastable helium. The motivation for this study is twofold. First, we would like to check the applicability of the two-centre CCC approach to a system with a negative Ps-formation threshold. Second, the recent experimental achievements on electron scattering from metastable helium [2], in the group that also has a positron beam, suggests that in a not too distant future there might be experimental data to compare with our calculations. Moreover, there are still unresolved differences between theory and experiment on electron scattering from metastable states of He [3, 4]. Alternative studies using positrons instead of electrons may therefore shed more light on the situation.

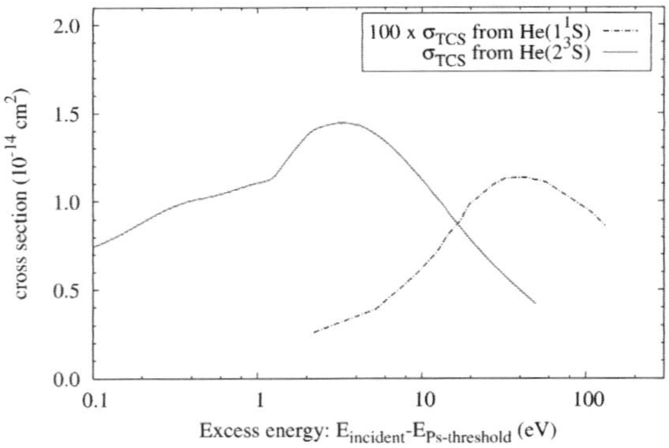

Figure 1. Comparison of the total cross sections for e^++He(1^1S) and e^++He(2^3S) scattering as functions of the excess energy above the Ps-formation threshold.

Comparison of the total cross sections for positron scattering from He(2^3S) and He(1^1S) is given in Fig. 1. Relative magnitudes of cross sections vary substantially depending on the processes. The total scattering cross section for He(2^3S) is larger than that for He(1^1S) by more than two orders of magnitude. Though not shown, for Ps formation this ratio turns out to be around 50, while for breakup of He(2^3S) it is around 10. The large relative ratio of cross section for He(2^3S) and He(1^1S) shows the importance of scattering from excited states in understanding positron scattering from helium [5].

3. Positronium formation in positron-alkali collisions

The two-center CCC method has been implemented for the case of positron scattering with alkali atoms. This is another example of the case with the negative Ps-formation threshold. To calculate cross sections for positron-alkali scattering we employ the frozen-core approximation. The target is considered as an hydrogen-like atom with one active electron interacting with an inert ion core. This approach is valid for impact energies less than 60 eV where the inner-shell electrons do not participate directly in a collision event. The core electrons provide screening of the nucleus and take part in exchange with the active electron. Also, the virtual core excitation is accounted for phenomenologically by introducing a small polarization potential. The core wavefunctions are obtained by using the Hartree-Fock method. These functions are required to calculate effective potentials for positron and active-electron interactions with the core. We use the equivalent local approximation to take into consideration the exchange part of the electron-ion interaction. Target wavefunctions are computed by diagonalizing the target atom Hamiltonian in the basis of Laguerre functions. This provide a good representation of low lying discrete atomic states and also a square-integrable representation of the target continuum. Finally, we formulate a set of momentum-space coupled Lippmann-Swinger equations for the T-matrix and solve them.

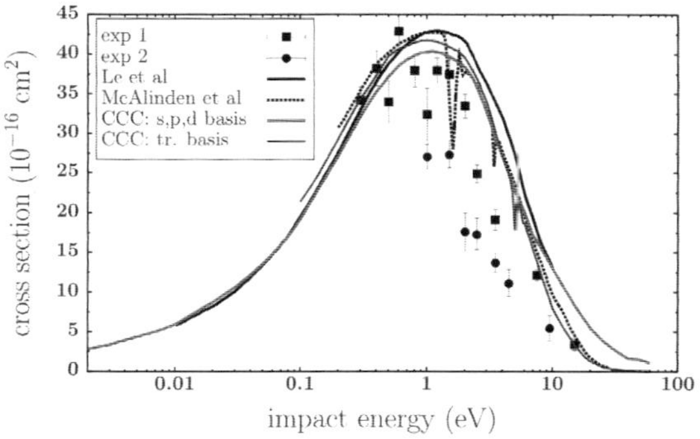

Figure 2. Positronium formation cross section for e^+-Li along with the experimental points [6] and theoretical calculations [7, 8].

The cross sections have been calculated for a broad range of energies of practical interest [9]. Convergence in the calculated cross sections has been acheived by increasing the basis sizes and orbital angular momentum of the included states for each of the centers. Figure 2 shows the positronium formation cross section. We compare our calculations with the experimental data [6] and other theoretical results [7, 8]. The red line was obtained with the full basis including ten s, nine p and eight d states for both target and Ps centres. To calculate the blue curve we used the truncated pseudostate basis with three eigenstates (1s, 2s and 2p) for positronium and twenty nine states (2s-9s, 2p-9p, 3d-9d and 4f-9f) for lithium. We see that all theoretical curves are in an overall qualitative agreement. As expected our truncated-basis calculations agree better with the results by McAlinden *et al* [7]. than our convergent full-basis cross sections. The differences can be attributed to the fact that we take into account the exchange part of the electron-electron interaction and also use close but different representation of the positronium and lithium states. The exchange interaction was also taken into account in the hyper-spherical close-coupling calculations [8]. The authors obtained the resonance-free energy dependence for the positronium formation cross section. We see that our and their results (black solid line) are very close to each other and differ slightly in the peak magnitude.

4. Antiproton-impact ionization of hydrogen

In antiproton-hydrogen collisions at the keV energy region the protonium formation channel is negligible [10]. This greatly simplifies the problem as the scattering system essentially has only one natural centre. Consequently, a standard one-centre expansion of the total wave function using sufficiently large orthogonal Laguerre basis is used. Such a basis allows to correctly account for all the atomic excitation channels as well as ionization. In the case of light projectiles the three-dimensional CCC integral equations are reduced to one dimensional ones using the partial-wave expansion. However, for heavy projectiles like antiproton the partial-wave method becomes impractical. Instead, we use the transformation into the impact-parameter space. This results in one-dimensional integral equations for the scattering amplitudes in the impact parameter representation [11]. Cross sections for the elastic scattering, excitation and ionization are calculated from the scattering amplitudes in impact-parameter representation.

Fig. 3 shows our results for the total ionization cross section (TICS) for antiproton-impact ionization of hydrogen in comparison with the experimental data [12] and other calculations [13, 14, 15, 16]. The experimental data are available only above 31.2 keV, however a controvercy seems to exist at lower energies where the close-coupling methods [13, 14, 15] predict the TICS several times larger than that predicted by the CDW-EIS [16] approach. As a test of our expansion basis, we also reproduce the exact first Born TICS obtained in the full wave treatment by Bates and Griffing [17]. An excellent agreement with the exact Born result indicates that the size of our pseudobasis is sufficiently high to reproduce the physics of the process. Our full results (solid black curve) are in an excellent agreement with the experiment

Journal of Physics: Conference Series **262** (2011) 012028 doi:10.1088/1742-6596/262/1/012028

throughout the energy range considered. They show maximum around 30 keV with rapid decline with decreasing energy, whereas the results of some other groups [13, 14, 15] predict much larger cross sections at low energies. Only the results of the high-energy CDW-EIS method show the same trend as our results. However, they are 50% smaller. New measurements of cross sections at low energies would clarify the situation at this region.

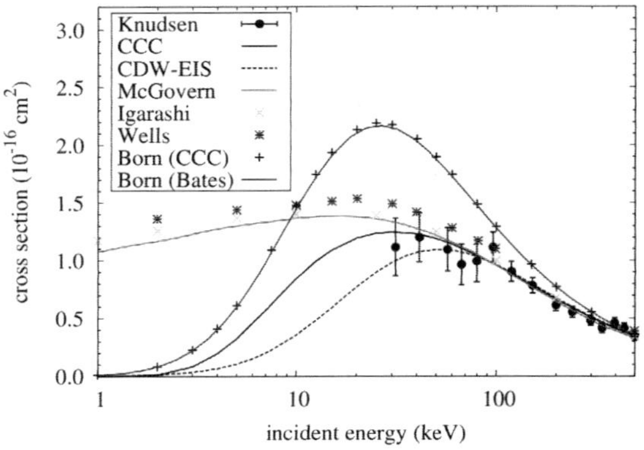

Figure 3. The total cross section for antiproton-impact ionization of hydrogen. The CCC calculations have been obtained using 1176 pseudostates with $n_{max} = 20$ and $l_{max} = 8$.

5. Conclusion

We have given a brief account of the latest convergent close-coupling calculations of antimatter-matter interactions. Positron scattering on metastable helium shows that presence of excited states in the experimental target gas chamber may significantly enchance the cross sections. The results for positron scattering on lithium are in an overall agreement with available theoretical calculations. Meanwhile, the cross section for hydrogen ionization by antiproton significantly differs from the others and reopens the discussion on its low-energy behaviour.

Acknowledgments

This work was supported by the Australian Research Council.

References

[1] Utamuratov R, Kadyrov A S, Fursa D V, Bray I and Stelbovics A T 2010 *J. Phys. B* **43** 125203
[2] Uhlmann L J, Dall R G, Truscott A G, Hoogerland M D and Buckman S J 2005 *Phys. Rev. Lett.* **94** 173201
[3] Lagus M E, Boffard J B, Anderson L W and Lin C C 1996 *Phys. Rev. A* **53** 1505–1518
[4] Piech G A, Chilton J E, Anderson L W and Lin C C 1998 *J. Phys. B* **31** 859–872
[5] Utamuratov R, Kadyrov A S, Fursa D V, Bray I and Stelbovics A T *submitted for publication*
[6] Surdutovich E, Johnson J M, Kauppila W E, Kwan C K and Stein T S 2002 *Phys. Rev. A* **65** 032713
[7] McAlinden M T, Kernoghan A A and Walters H R J 1997 *J. Phys. B* **30** 1543
[8] Le A T, Bromley M W J and Lin C D 2005 *Phys. Rev. A* **71** 032713
[9] Lugovskoy A V, Kadyrov A S, Bray I and Stelbovics A T *submitted for publication*
[10] Schultz D R, Krstic P S, Reinhold C O and Wells J C 1996 *Phys. Rev. Lett.* **76** 2882–2885
[11] Abdurakhmanov I B, Kadyrov A S, Bray I and Stelbovics A T *submitted for publication*
[12] Knudsen H, Mikkelsen U, Paludan K, Kirsebom K, Møller S P, Uggerhøj E, Slevin J, Charlton M and Morenzoni E 1995 *Phys. Rev. Lett.* **74** 4627–4630
[13] McGovern M, Assafrão D, Mohallem J R, Whelan C T and Walters H R J 2009 *Phys. Rev. A* **79** 042707
[14] Igarashi A, Nakazaki S and Ohsaki A 2000 *Phys. Rev. A* **61** 062712
[15] Wells J C, Schultz D R, Gavras P and Pindzola M S 1996 *Phys. Rev. A* **54** 593–604
[16] Crothers D S F and McCartney M 1992 *Comp. Phys. Comm.* **72** 288 – 294
[17] Bates D R and Griffing G W 1954 *Proc. Phys. Soc. (London)* **67** 663

Journal of Physics: Conference Series **262** (2011) 012029 doi:10.1088/1742-6596/262/1/012029

Defect profiles in ion-irradiated metal samples by slow positron beams in comparison with simulation profiles

A Kinomura[1], R Suzuki[1], T Ohdaira[1], N Oshima[1], K Ito[1], Y Kobayashi[1] and T Iwai[2]

[1]National Institute of Advanced Industrial Science and Technology (AIST), 1-1-1 Umezono, Tsukuba, Ibaraki 305-8568, Japan
[2]Nuclear Professional School, School of Engineering, The University of Tokyo, 2-22 Shirakata-Shirane, Tokai-mura, Ibaraki 319-1188, Japan

E-mail: a.kinomura@aist.go.jp

Abstract. Metal (Fe and Ni) samples ion-irradiated at 100 – 773 K were characterized by Doppler broadening measurements with slow positron beams to investigate depth profiles of irradiation-induced defects. Obtained results were compared with defect profiles calculated by Monte-Carlo simulation. Defects profiles obtained by positron measurements were always deeper than those by the simulation. The difference between measured and simulated profiles was observed even in the case of the Fe sample irradiated at 100 K where vacancies are immobile. The origin of such differences was discussed with respect to vacancy diffusion, depth calculation, surface contamination and ion channeling.

1. Introduction

Ion irradiation causes radiation damage that considerably affects properties of materials. Defects induced by ion irradiation diffuse and react with each other, leading to recombination or evolution to secondary defects. It is important to know the distribution and the motion of defects in ion-irradiation studies. Positron annihilation spectroscopy using energy-variable slow-positron beams has been extensively used as a depth-dependent characterization technique for defects (vacancies) in ion-irradiated layers [1,2]. Some of the previous studies on ion irradiation to Si [3,4] or steel [5] reported that defect profiles given by positron measurements are deeper than calculated defect profiles. It should be noted that implanted positrons diffuse and spread in samples before annihilation and thus defect profiles must be deduced by taking account of positron diffusion. Even after the diffusion calculations, obtained defect profiles were deeper than the calculated profiles in these studies. Moreover, defects were detected after irradiated layers were precisely etched off [3]. The origin of the deep defect profiles is not well understood but they were ascribed to diffusion of defects or channeling of ions [6]. In this study, we investigate pure metal samples (Fe and Ni) to compare defect profiles by positron measurements with defect profiles by numerical simulation.

2. Experimental procedure

Pure Fe (99.994%) and Ni (99.994%) were cut into a size of 1.5 cm × 1.5 cm and mechanically mirror-polished. To remove mechanical damage, the samples were annealed in quartz tube furnaces at 973 K for 0.5 h in flowing H_2 atmosphere (Fe) or at 1173 K for 1h in vacuum (Ni). Radiation damage was introduced by 150 keV Ar^+ irradiation at room temperature (RT), 573 K and 773 K or 200 keV C^+ irradiation at 100 K. These ion species were chosen owing to the limitation of available ions of the equipments. We believe that the chemical effects of Ar or C ions do not affect the results because the Ar (C) concentrations are sufficiently low at the depths of interest (e.g., >200 nm in the case of Ar). The ion beam was defocused and adjusted to the samples to achieve uniform and continuous irradiation. Incident angles of Ar^+ and C^+ beams were 0° and 30° from the normal to the sample surface, respectively. Doppler broadening of annihilation radiation was used as a slow-positron measurement technique at energies below 30 keV. Obtained S-E curves were analyzed by the VEPFIT code [7]. Depth profiles of vacancies were calculated by the TRIM (SRIM) code based on a Monte-Carlo method [8].

3. Results and discussion

Figure 1 shows the Doppler broadening S parameter as a function of positron energy for the Fe samples irradiated with 150 keV Ar^+ to a dose of 5×10^{15} cm^{-2} at 573 K and 773 K, together with S parameters for the unirradiated Fe sample. Solid lines are fits to S parameters of the three samples. At positron energies above 5 keV, S parameters of the irradiated samples were higher than those of the unirradiated samples owing to defects induced by ion irradiation. Low S parameters below 5 keV were presumably caused by oxygen atoms recoiled from oxide layers on the Fe samples.

The S-E curves were imported in the VEPFIT code and the depth dependence of S parameters was calculated as shown in figure 2. Firstly, we determined the S parameter and diffusion length of bulk from the unirradiated sample and fixed them in the analysis. Then, initial parameters for other layers (S parameter, diffusion length and layer thickness) were approximately determined. Starting from the initial parameters, a fitting process for each single layer was performed layer by layer from the surface layer to deep layers to obtain optimum fitting results.

A 3-layer model was applied to describe the sample structure for the irradiated samples in figure 1, where the top and middle layers correspond to the irradiation-induced defects and the 3rd layer corresponds to bulk Fe. The increase in S parameters (i.e., damage layers) extends to 400 or 450 nm. Similar experiments were performed for Fe at RT and Ni as summarized in table 1 [9]. The calculated depth was always deeper than the projected ranges and the simulated profiles.

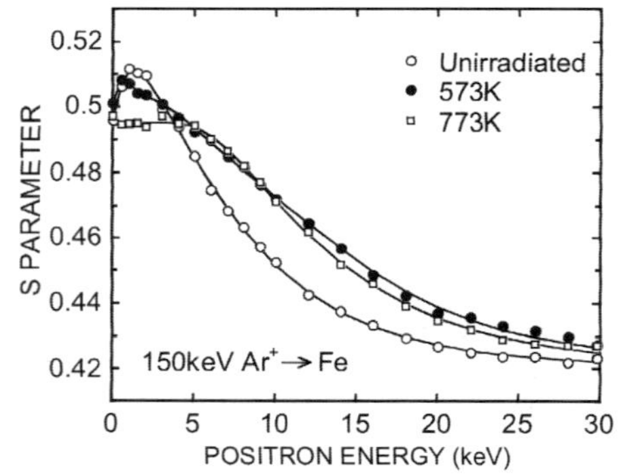

Figure 1. Doppler-broadening S parameter as a function of positron energy for Fe samples irradiated with 150 keV Ar^+ to a dose of 5×10^{15} cm^{-2} at 573 and 773 K.

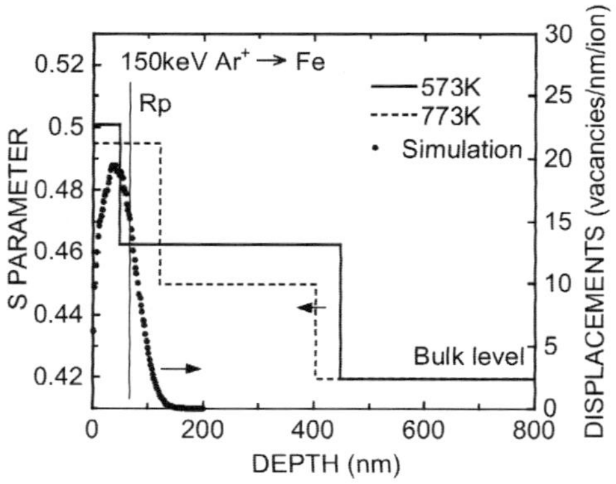

Figure 2. Depth dependence of the S parameter corresponding to the S-E curves in figure. 1. A simulated vacancy profile (closed circles) and a projected range (Rp) are shown for comparison.

The difference between the measured and simulated profiles could be explained as a result of vacancy diffusion. Vacancy diffusion is a temperature dependent process. Diffusion of monovacancies in pure Fe occurs at room temperature because of its relatively low activation energy

Journal of Physics: Conference Series **262** (2011) 012029 doi:10.1088/1742-6596/262/1/012029

(<0.6 eV). On the other hand, monovacancy diffusion in Ni is negligible at room temperature as calculated in our rate-equation analysis assuming an activation energy of 1.0 eV. However, there was no clear temperature dependence in both Fe and Ni cases in the temperature range of RT to 773 K.

Table 1. Calculated depth of damage layers in Fe and Ni after 150 keV Ar^+ irradiation. The projected ranges of Ar^+ were 68 and 61 nm in Fe and Ni, respectively.

Substrate	Fe			Ni		
Temperature (K)	RT	573	773	RT	573	773
Dose (cm^{-2})	1×10^{15}	5×10^{15}	5×10^{15}	1×10^{15}	5×10^{15}	5×10^{15}
Depth (nm)	5.2×10^2	4.5×10^2	4.0×10^2	6.0×10^2	4.7×10^2	5.5×10^2

To further investigate the temperature dependence, the Fe sample was irradiated at 100 K where Fe monovacancies are immobile [10]. By using an in-situ positron measurement system at the University of Tokyo [11], an Fe sample was irradiated with 200 keV C^+ to a dose of 5×10^{14} cm^{-2} at 100 K, followed by the Doppler broadening measurement at 100 K in the same vacuum chamber. Since the 150 keV Ar^+ beam was not available with the system, the 200 keV C^+ beam was used.

Figure 3 shows the result of the Doppler-broadening measurements for the Fe sample irradiated at 100 K. The S-E curve for the unirradiated condition (open circles) was obtained at 100 K from the same sample before the irradiation. The S parameters increased after the irradiation in all the energies except for the near-surface region below 1 keV. The S-E curves in figure 3 were analyzed and the depth dependence of S parameters was calculated as shown in figure 4. A 4-layer model was applied to describe the sample structure, where the 4th layer corresponds to bulk (undamaged) Fe in this case. The diffusion length and the S parameter of the bulk Fe at 100 K were obtained from the S-E curve of the sample before irradiation. The increase in the S parameters shows that the damage layer extends to 570 nm. The depth of the damage layer was, again, deeper than the projected range of Ar^+ (199 nm) in Fe and the simulated defect profile. These results suggest that the vacancy diffusion cannot account for the difference between the measured and simulated profiles.

Positron implantation profiles in this study were calculated by a Makhovian function [2] with the empirical parameters reported by Vehanen et al. [12]. It is known that the Makhovian parameters are slightly material- and energy-dependent [2]. However, we estimated that the errors originating from such dependences were less than 18% in the energy range of 5 – 30 keV. It is unlikely that the differences between the measured and simulated profile are attributed to the errors of the Makhovian function.

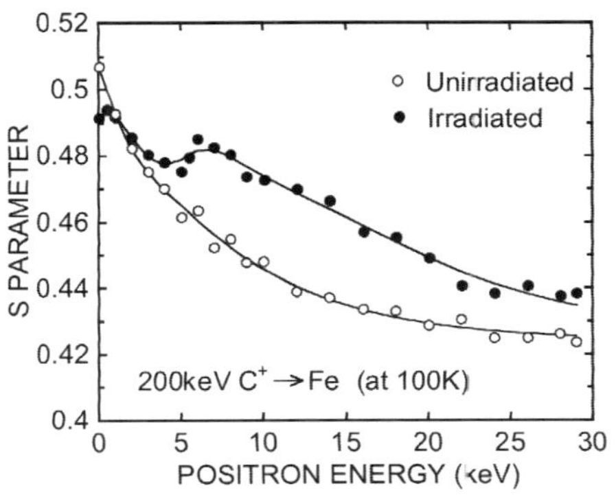

Figure 3. Doppler-broadening S parameter as a function of positron energy for an Fe sample irradiated with 200 keV C^+ to a dose of 5×10^{14} cm^{-2} at 100 K.

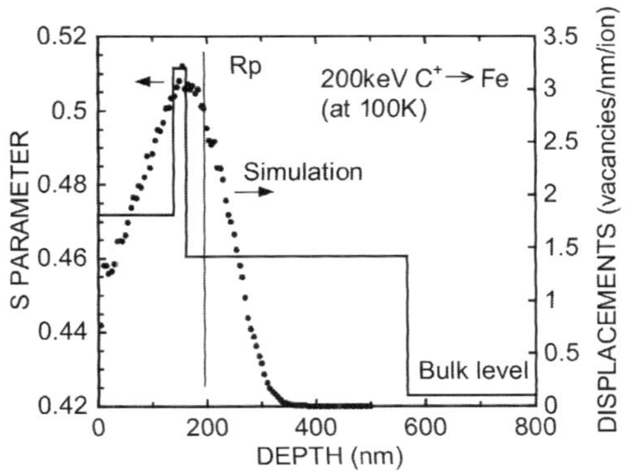

Figure 4. Depth dependence of the S parameter corresponding to the S-E curve in figure. 3. A simulated vacancy profile (closed circles) and a projected range (Rp) are shown for comparison.

It is necessary to take account of the influence of the recoil atoms from surface contaminations such as native oxide and hydrocarbon originating from vacuum systems. Maximum recoil energies of H, C and O atoms by 150 keV Ar^+ can be calculated as 14.6, 107 and 123 keV, respectively. The projected ranges of such recoils were calculated to be 137 nm (C in Fe) at most. This depth is close to the simulated defect profiles for Ar^+ and much shallower than the measured defect profiles. Thus the influence of the contamination atoms is negligible.

As the samples used in this study are in the form of polycrystalline with grain sizes of the order of 10 μm, it is necessary to consider channeling effects of ion beams. When incident ions are aligned to crystalline axes of each grain, ion ranges can be extended to several times the projected ranges in amorphous samples [13]. Indeed, the influence of channeling on Doppler broadening measurements was clearly demonstrated in the case of As^+ implantation in Si [14]. The observed difference between the measured and simulated defect profiles can be attributed to the channeling effect but further investigations are required to confirm it.

4. Summary

Defect profiles in Fe and Ni samples ion-irradiated at 100 K – 773 K were characterized by Doppler-broadening measurements and compared with defect profiles calculated by a Monte-Carlo simulation. Measured defect profiles were always deeper than simulated defect profiles. The difference was observed in the case of irradiation to Fe at 100 K where vacancies are immobile. The origin of such differences cannot be explained by vacancy diffusion, errors of positron implantation depths and recoils from the surface impurities. The effect of channeling to polycrystalline grains was suggested as a possible explanation.

Acknowledgements

A part of this study was financially supported by the Budget for Nuclear Research of the Ministry of Education, Culture, Sports, Science and Technology, based on the screening and counseling by the Atomic Energy Commission. We would like to thank H Schut (Delft Univ. of Technology) for allowing us to use the VEPFIT code. We also thank T. Yoshiie, Q. Xu (Kyoto Univ.), H. Abe and M. Hasegawa (Tohoku Univ.) for their helpful advice on this study.

References

[1] Asoka-Kumar P, Lynn K G and Welch D O 1994 J. Appl. Phys. **76** 4935
[2] Schultz P J and Lynn K G 1988 Rev. Mod. Phys. **60** 701
[3] Fujinami M and Chilton N B 1993 J. Appl. Phys. **73** 3242
[4] Uedono A, Wei L, Dosho C, Kondo H, Tanigawa S, Sugiura J and Ogasawara M 1991 Jpn. J. Appl. Phys. **30** 201
[5] Aruga T, Takamura S, Nakata K and Ito Y 1995 Appl. Surf. Sci. **85** 229
[6] Coleman P G and Knights A P 1999 Appl. Surf. Sci. **149** 82
[7] van Veen A, Schut H, Clement M, de Nijs J M M, Kruseman A and Ijpma M R 1995 Appl. Surf. Sci. **85** 216
[8] Ziegler J F, Biersack J P and Littmark U 1985 *The Stopping and Range of Ions in Solids* (New York: Pergamon Press)
[9] Kinomura A, Suzuki R, Ohdaira T, Oshima N, Ito K and Kobayashi Y, Surf. Coat. Technol. (accepted for publication).
[10] Doyama M 1997 Mater. Chem. Phys. **50** 106
[11] Iwai T, Ito Y and Koshimizu M 2004 J. Nucl. Mater. **329-333** 963
[12] Vehanen A, Saarinen K, Hautojärvi P and Huomo H 1987 Phys. Rev. B **35** 4606
[13] Davies J A, Eriksson L and Whitton J L 1968 Canadian J. Phys. **46** 573
[14] Hirata K and Kobayashi Y 1998 Nucl. Instrum. Methods in Phys. Res. B **136-138** 763

Journal of Physics: Conference Series **262** (2011) 012030 doi:10.1088/1742-6596/262/1/012030

Defect investigations of micron sized precipitates in Al alloys

B Klobes[1], B Korff[1], O Balarisi[1], P Eich[1], M Haaks[1], I Kohlbach[1], K Maier[1], R Sottong[1] and T E M Staab[2]

[1] Helmholtz-Institut für Strahlen- und Kernphysik, Nußallee 14-16, D-53115 Bonn, Germany
[2] Fraunhofer ISC, Neunerplatz 2, D-97082 Würzburg, Germany

E-mail: `klobes@hiskp.uni-bonn.de`

Abstract. A lot of light aluminium alloys achieve their favourable mechanical properties, especially their high strength, due to precipitation of alloying elements. This class of age hardenable Al alloys includes technologically important systems such as e.g. Al-Mg-Si or Al-Cu. During ageing different precipitates are formed according to a specific precipitation sequence, which is always directed onto the corresponding intermetallic equilibrium phase. Probing the defect state of individual precipitates requires high spatial resolution as well as high chemical sensitivity. Both can be achieved using the finely focused positron beam provided by the Bonn Positron Microprobe (BPM) [1] in combination with the High Momentum Analysis (HMA) [2]. Employing the BPM, structures in the micron range can be probed by means of the spectroscopy of the Doppler broadening of annihilation radiation (DBAR). On the basis of these prerequisites single precipitates of intermetallic phases in Al-Mg-Si and Al-Cu, i.e. Mg_2Si and Al_2Cu, were probed. A detailed interpretation of these measurements necessarily relies on theoretical calculations of the DBAR of possible annihilation sites. These were performed employing the DOPPLER program. However, previous to the DBAR calculation the structures, which partly contain vacancies, were relaxed using the ab-initio code SIESTA, i.e. the atomic positions in presence of a vacancy were recalculated.

1. Introduction

Aluminium based alloys combine light weight with high strength and are thus extensively used in all application areas, in which this combination is indispensable. Among the multiplicity of established Al alloys the class of age-hardenable alloys stands out, since their favourable mechanical properties are induced by the agglomeration and precipitation of alloying elements [3]. For instance, alloys based on the Al-Cu and Al-Mg-Si systems (the so-called AA2xxx and AA6xxx series, respectively) are frequently applied as structural materials in the aerospace or automotive industry. Usually, the hardening structures are formed after solution heat treatment and quenching yielding a supersaturated solid solution together with an excess concentration of quenched-in vacancies. In the course of subsequent ageing the supersaturated solid solution is decomposed via vacancy driven diffusion processes and agglomerates and precipitates are formed. The succession of structurally different precipitates depends on the exact alloy composition as well as on the applied heat treatment. Nevertheless, the precipitation sequence of each Al alloy system is generally governed by an intermetallic phase, onto which the evolution of precipitates is directed [4, 5]. Concerning Al-Cu and Al-Mg-Si the corresponding phases are θ (Al_2Cu)

and β (Mg_2Si), respectively. Although Positron Annihilation Spectroscopy (PAS) has been used successfully to investigate precipitation hardening in Al alloys for decades [6, 7, 8], the defect state of individual precipitates has not been explored yet. This type of investigation cannot be performed utilizing radioactive sources, since these probe bulk regions. Instead one has to employ a slow positron beam having a spatial resolution in the micron range, in order to assure positron annihilation inside single precipitates. These prerequisites are met by the Bonn Positron Microprobe (BPM), which is used to demonstrate how the defect state of individual, micron sized precipitates can be probed. For the interpretation of the experimental data theoretical calculations of the Doppler broadening of annihilation radiation (DBAR) were performed. Additionally, radioactive source based DBAR measurements support the uniqueness of the findings obtained by positron microscopy.

2. Experimental - Materials and Methods

Specimens were made from two laboratory alloys exhibiting the compositions Al-1.7at%Cu and Al-1.11at%Mg-0.77at%Si which were cast using high purity base materials. Each alloy was subjected to a heat treatment starting with a solution heat treatment at 768 K for the Al-Cu alloy and at 833 K for the Al-Mg-Si alloy, respectively. Subsequently, the Al-Cu alloy was slowly (0.5 K/min) cooled to room temperature (RT), whereas the Al-Mg-Si sample was aged at 623 K for 48 hours. Each of these steps was conducted in high vacuum at about $1 \cdot 10^{-6}$ mbar. Previous to positron microscopy, optical as well as scanning electron microscopy combined with energy dispersive X-ray (EDX) mapping were employed for surface investigations of the prepared samples and the identification of θ and β (cf. [9] for details concerning β and [10] for evidence concerning θ). The DBAR of individual precipitates was measured using the finely focused positron beam provided by the BPM. Its monoenergetic positron beam of 30 keV exhibits a mean penetration depth of about 3.4 microns in aluminium. Throughout this work the beam diameter was about 7.5 microns and the resolution in terms of FWHM of the HPGe detector used to measure the Doppler broadening was 1.25 keV at the 478 keV γ-line of ^{7}Be, which was used for electronic stabilization. Since the DBAR is recorded by a single Ge detector, the original data was processed employing the HMA, which reliably allows for accessing the high momentum region. Additional DBAR measurements based on the positron source ^{68}Ge were carried out using similar prepared samples of appropriate thickness. Especially in the high momentum region the interpretation of experimental DBAR data often necessitates standards for comparison, which are not accessible in the case of embedded precipitates. Therefore, theoretical calculations of the DBAR were performed by means of the programm DOPPLER [11] within general gradient approximation (GGA) [12] and using the enhancement factor given by [13]. Calculated DBAR curves were convolved with a Gaussian function of appropriate width in addition. Previous to the DBAR calculations the structures containing vacancies were relaxed, i.e. the atomic coordinates were recalculated, using the ab-initio code SIESTA [14] based on density functional theory. This procedure is described in detail for the β case in [15].

3. Results and Discussion

First of all, β and θ precipitates were identified optically. In particular, β precipitates could be discovered easily because of their *chinese script* morphology and their bluish appearance. Due to the applied heat treatment the surface of individual precipitates was often larger than the positron microbeam diameter. Since additional EDX maps obtained using 25 keV electrons showed no Al signal, it is reasonable to assume that positron annihilation took place exclusively inside β precipitates. In contrast, θ precipitates exhibited a needle-like shape and a width of 3 microns on average. Thus, using the BPM positrons are expected to annihilate in θ, in the alloy matrix and at the corresponding interface in addition. The DBAR ratio-plots of a single β precipitate and a corresponding matrix measurement using the positron microbeam of the BPM are displayed in figure 1a, in which also a conventional DBAR measurement of a similar sample is shown. Both the matrix measurement obtained by the BPM and the conventional DBAR measurement do not indicate any significant annihilation in vacancy-like defects or precipitates. Merely, due to the high momentum region a small influence of alloying atoms or incoherent phase boundaries can be assumed. In contrast to this, an individual β precipitate yields a significant

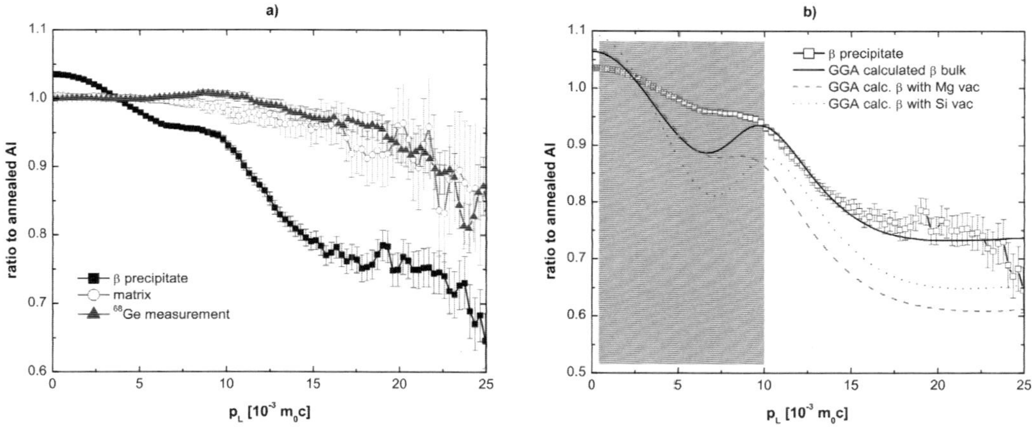

Figure 1. Ratio-plots of DBAR measurements and calculations concerning β in Al-Mg-Si. a) Directing the positron beam onto an individual precipitate yields a significant curve containing information not accessible using conventional radioactive source based DBAR. b) Comparison of theoretical calculations of possible annihilation sites in β with the experimental data. The area of systematic errors of the calculations is marked in grey.

ratio-plot. Since it exhibits an increase in the low momentum region compared to annealed Al one might be apt to attribute this feature to the existence of vacancies, which is a common approach concerning Al alloys. However, since the DBAR definitely originated from the inside of a β precipitate, theoretical calculations have to be used for a meaningful interpretation. Therefore, DBAR calculations of positron annihilation in bulk β and in vacancies on either sub-lattice of Mg_2Si were performed, which are shown in figure 1b. Though the employed approach suffers from systematic errors in the low momentum region, the high momentum part usually can be used for interpretation, since there mainly core electrons not taking part in the bonding process contribute to positron annihilation. On the basis of these calculations it is possible to exclude positron annihilation in vacancies of β. In the light of the excess Si above the stoichiometric composition of Mg_2Si this finding was not be expected. The analogue procedure yielded the ratio-plots displayed in figure 2a for the Al-Cu alloy. While the matrix measurement again reveals only little influence of alloying elements or other possible structures affecting positron annihilation, the agglomeration of Cu in θ is obvious due to the increase in the high momentum region. The corresponding DBAR calculations are shown in figure 2b. Though the shape of the calculated ratio-plots is similar to the θ measurement in the high momentum region, the amplitudes do not match. Since the positron microbeam covered more than just a single θ precipitate, it is most probable that annihilation events in the matrix or at the incoherent interface of θ dominate, which shift the ratio-plot towards the matrix measurement. Nevertheless, this is also a nice example of chemical sensitivity in the micron range achieved by positron annihilation.

4. Conclusions

The presented measurement and calculations show that the defect state of single, micron sized precipitates can be uniquely accessed using positron microscopy. Employing GGA calculations of

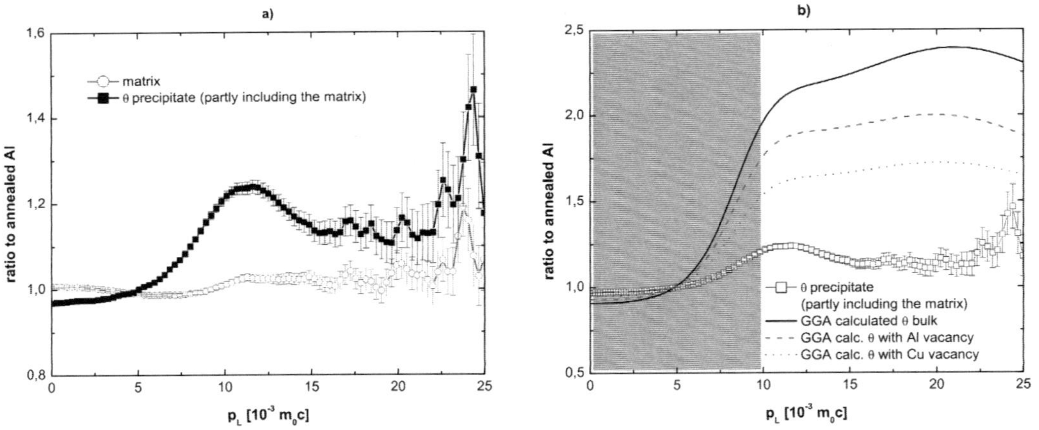

Figure 2. Ratio-plots of DBAR measurements and calculations concerning θ in Al-Cu. a) The agglomeration of Cu inside θ is obvious comparing the precipitate and the matrix measurement, though the precipitate measurement may be affected by matrix and interface contributions. b) Comparison of theoretical calculations of possible annihilation sites in θ with experiment.

the DBAR β precipitates could be shown to grow without incorporation of structural vacancies. Though the experimental results concerning θ precipitates in Al-Cu do not provide a basis for a similar conclusion, these measurements clearly show the possibility of achieving chemical sensitivity in the micron range using positrons.

Acknowledgments

Parts of this work have been financially supported by the German Research Foundation (STA 527/3-2). We would also like to thank the Institute of Applied Materials at the Helmholtz Zentrum Berlin für Materialien und Energie for producing the alloys.

References

[1] Greif H, Haaks M, Holzwarth U, Männig U, Tongbhoyai M, Wider T, Maier K, Bihr J and Huber B 1997 *Appl. Phys. Lett.* **71** 2115–2117
[2] Haaks M, Staab T and Maier K 2006 *Nucl. Instrum. Methods Phys. Res.* A **569** 829
[3] Hornbogen E 2001 *J. Light Met.* **1** 127–132
[4] Starink M and Yan J 2006 *Mater. Sci. Forum* **519-521** 251–258
[5] Ravi C and Wolverton C 2004 *Acta Mater.* **52** 4213–4227
[6] Dlubek G 1987 *Mater. Sci. Forum* **13-14** 11–32
[7] Dupasquier A, Kögel G and Somoza A 2004 *Acta Mater.* **52** 4707–4726
[8] Somoza A, Dupasquier A and Ferragut R 2009 *Phys. Status Solidi* C **6** 2295–2300
[9] Klobes B, Korff B, Balarisi O, Eich P, Haaks M, Maier K, Sottong R, Hühne S, Mader W and Staab T 2010 *Phys. Rev.* B **82** 054113
[10] Ringer S and Hono K 2000 *Mater. Charact.* **44** 101–131
[11] Alatalo M, Kaupinnen H, Saarinen K, Puska M, Mäkinen J, Hautajärvi P and Nieminen R 1995 *Phys. Rev.* B **51** 4176
[12] Alatalo M, Barbiellini B, Hakala M, Kaupinnen H, Korhonen T, Puska M, Saarinen K, Hautojärvi P and Nieminen R 1996 *Phys. Rev.* B **54** 2397
[13] Barbiellini B, Puska M, Torsti T and Nieminen R 1995 *Phys. Rev.* B **51** 7341
[14] Soler J, Artacho E, Gale J, Garcia A, Junquera J, Ordejón P and Sánchez-Portal D 2002 *J. Phys.: Condens. Matter* **14** 2745
[15] Staab T 2009 *Phys. Status Solidi* B **246** 1587

Journal of Physics: Conference Series **262** (2011) 012031 doi:10.1088/1742-6596/262/1/012031

Probing the formation of silicon nano-crystals (Si-ncs) using variable energy positron annihilation spectroscopy

A P Knights[1,5], J D B Bradley[1], O Hulko[1], D V Stevanovic[1], C J Edwards[2], A Kallis[2], P G Coleman[2], I F Crowe[3], M P Halsall[3], and R M Gwilliam[4]

[1]Department of Engineering Physics, McMaster University, Hamilton, Ontario, L8S4L7, Canada.

[2]Department of Physics, University of Bath, Bath, BA2 5A7, United Kingdom.

[3]School of Electrical and Electronic Engineering, The University of Manchester, Manchester, M13 9PL, United Kingdom.

[4]University of Surrey Ion Beam Centre, University of Surrey, Guildford, GU2 5XH, United Kingdom.

Abstract. We describe preliminary results from studies of the formation of silicon nano-crystals (Si-ncs) embedded in stoichiometric, thermally grown SiO_2 using Variable Energy Positron Annihilation Spectroscopy (VEPAS). We show that the VEPAS technique is able to monitor the introduction of structural damage. In SiO_2 through the high dose Si^+ ion implantation required to introduce excess silicon as a precursor to Si-nc formation. VEPAS is also able to characterize the rate of the removal of this damage with high temperature annealing, showing strong correlation with photoluminescence. Finally, VEPAS is shown to be able to selectively probe the interface between Si-ncs and the host oxide. Introduction of hydrogen at these interfaces suppresses the trapping of positrons at the interfaces.

1. Introduction

Low dimensional silicon continues to be of considerable interest for applications such as silicon based solid state lighting, non-volatile memories and dielectric engineering [1]. Specifically, silicon nano-crystals (Si-ncs) formed in the dielectric material SiO_2 or Si_3N_4 have been shown to possess a range of properties not usually associated with the silicon bulk. Such nano-crystals are formed via phase separation in silicon-rich dielectric. This precursor material may be fabricated using a number of standard processes such as plasma enhanced chemical vapour deposition [2] or sputtering, both suitable for high-volume, large area applications. The most controllable fabrication technique (and thus the preferred method for the methodical study of Si-ncs) is ion implantation [3]. In this case, excess silicon is introduced into stoichiometric, thermally grown SiO_2, with phase separation taking place during a subsequent high temperature (>1000 C) annealing step. In this current work, we show

---

[5] Author to Whom Correspondence should be addressed, email: aknight@mcmaster.ca

Journal of Physics: Conference Series **262** (2011) 012031 doi:10.1088/1742-6596/262/1/012031

the significant potential of Variable Energy Positron Annihilation Spectroscopy (VEPAS) for obtaining novel information on the formation and light emitting potential of Si-ncs formed via high dose Si^+ ion implantation of SiO_2. We demonstrate that VEPAS shows a strong relationship between the damage contained in the SiO_2 (resulting from the implantation process) and the formation mechanics of the Si-ncs. We confirm previous hypotheses that the interface plays a significant role in light emission. This is achieved through the observation of the VEPAS signal for Si-ncs with and without interface passivation.

2. Experimental

2.1.Sample preparation

All of the samples reported here were prepared via high dose silicon ($>1x10^{16}cm^{-2}$) Si^+ ion implantation into thermally grown SiO_2 thin films on a low-doped *p*-type silicon substrate; followed by a high temperature (>1000°C) anneal in N_2. A fraction of the films were subsequently annealed at 500°C for 10 minutes in forming gas (N_2:H_2 95:5). Details of specific sample preparation are provided in the text.

2.2. Sample Characterization

Confirmation of the presence of Si-ncs was obtained via Transmission Electron Microscopy (TEM) using a conventional CM-12 microscope operated at 120 kV. Cross-sectional specimens oriented along the {110} zone axis were prepared by mechanical polishing, followed by ion milling. Dark-field examinations were carried out with two beam diffraction condition ($\mathbf{g}$ = 220) relative to the Si substrate.

Photoluminescence (PL) measurements were performed using the 405nm line of an InGaN/GaN laser diode operating at 50mW. The spectra were collected using a CCD array. Variable Energy Positron Annihilation Spectroscopy (VEPAS) was performed using the University of Bath slow positron beam, details of which are described elsewhere [4]. The annihilation spectra were analysed in terms of the classic *S*-parameter for incident positron energies ranging from 0.1-30keV.

3. Results and Discussion

Figure 1 shows an example of a TEM image of a collection of Si-ncs, in this case for a sample prepared using an implantation dose of $8x10^{16}cm^{-2}$ and energy of 80keV. Following ion implantation the sample was annealed at a temperature of 1150°C for 50s.

Figure 1. Dark field TEM image of Si-ncs in SiO_2; (a) wide view of distribution; (b) enlarged view of Si-nc.

Phase separation leading to the formation of the Si-ncs is confirmed, with the Si-ncs showing as light regions (in the dark field image). In this case the mean diameter of the Si-ncs was approximately 3nm. Annealing at temperatures in the range 1000-1200°C and times from 1-100secs produced similar arrays of Si-ncs (as observed by TEM), albeit their concentration and mean diameter varied with each thermal treatment. The description of this variation will be the subject of a future publication.

Journal of Physics: Conference Series **262** (2011) 012031 doi:10.1088/1742-6596/262/1/012031

The structure of Si^+ ion implanted SiO_2/Si (SiO_2 film thickness = 500nm) as a function of annealing time was characterized using VEPAS for a sample set again prepared using an implantation energy and dose of 80keV and $8x10^{16}cm^{-2}$ respectively. The annealing was performed at 1100°C for times ranging from 1-100secs. The *S*-parameter (normalized to bulk silicon) versus incident positron energy is shown in figure 2. Data for annealing times between 1 and 100 seconds (exclusive) showed a slowly varying trend of reduction in *S*-parameter for the region between 1-7keV, and is not shown in order to maintain clarity.

The data for the unimplanted SiO_2/Si film is consistent with a film thickness of 500nm with strong positron tarpping at the SiO_2/Si interface. Following the relatively high dose Si^+ ion implantation the *S*-parameter corresponding to the end of range of the implanted ions increases (positron energy ~3.5keV), signifying the likely introduction of large open volumes, or at least a significant change in the film composition/structure. Somewhat remarkably, even for an annealing time of only 1sec, there is a significant drop in *S*-parameter at an energy (1-5keV) which would be consistent with the expected depth at which phase separation of excess Si would occur. This reduction in *S*-parameter continues to an annealing time of 100 secs, after which negligible evolution in the shape of the *S-E* data is observed (data not shown for clarity). Consistent with previous reports [5] we ascribe the 'dip' in the VEPAS data at ~2.5keV with annihilations which take place at the interface of the Si-ncs and the host SiO_2 (the Si-nc/SiO_2 interface closely resembles that of a thin silm SiO_2/Si interface).

Photoluminescence data shown in figure 3 was obtained from the same samples as those used to obtain the VEPAS data. For the as-implanted sample there is a measurable but small luminescence signal centred at 670nm which likely results from luminescent defects. After annealing for 1 sec the formation of Si-ncs is confirmed by the large increase in signal strength and the shift in emission wavelength to ~770nm. Further annealing for 100secs results in a small red-shift of emission wavelength (consistent with an increase in size of the average diameter of the Si-ncs) and an increase

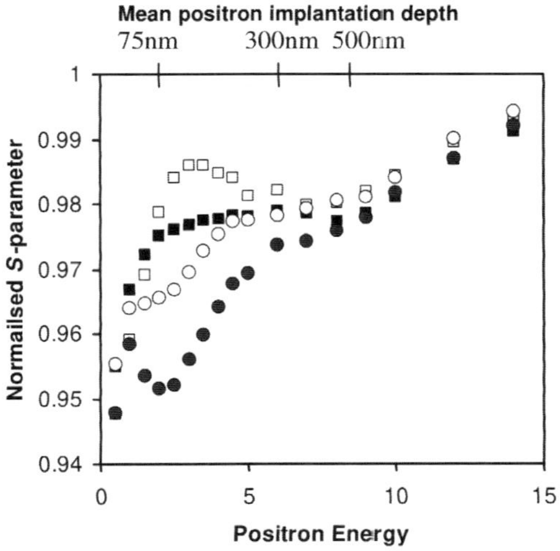

Figure 2. Normalized *S*-parameter versus positron energy (and mean positron implantation depth) for samples implanted with $8x10^{16}cm^{-2}$ at an ion energy of 80keV and annealed at 1100°C for 1 sec (open circles) and 100sec (closed circles). Data for the unimplanted film shown as closed squares, and for the as-implanted (unannealed) shown as open squares.

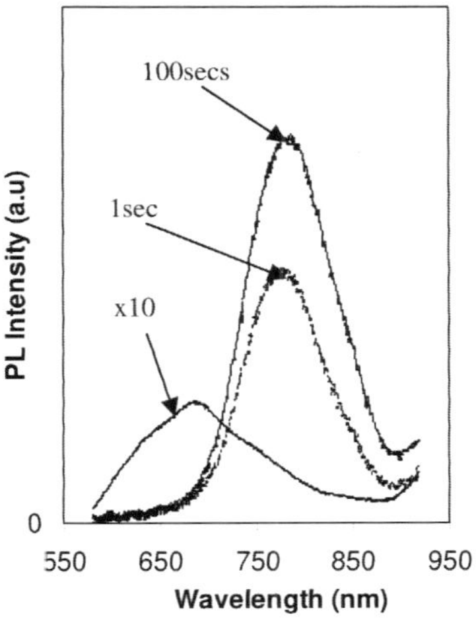

Figure 3. Photoluminescence data for samples implanted with $8x10^{16}cm^{-2}$ and annealed at 1100°C for 1 sec (dotted line) and 100sec (dashed line). Data for the as-implanted sample is shown as a solid line (data has been multiplied by 10 times to allow comparison).

in luminescence intensity. This increase occurs with concomitant reduction in structural damage of the SiO_2 shown by figure 2. This increase is likely due to the removal of non-radiative recombination centres. The VEPAS technique thus provides an important method through which the removal of luminescence suppression may be monitored.

Finally, we discuss the important role which VEPAS provides in relation to the characterization of the Si-nc/SiO_2 interface. In figure 4 we show data similar to that in figure 2. In this case the excess silicon has been introduced via a 40keV Si^+ ion implantation to a dose of $4x10^{16}cm^{-2}$. The thickness of the SiO_2 film was nominally 100nm. The implantation again induces structural damage to the oxide film indicated by an increased *S*-parameter in the energy range 1.5-3keV. Following annealing at 1100°C for 100secs in N_2 the formation of Si-ncs results in a 'dip' in the data, centred at ~2keV, associated with trapping of positrons at the Si-nc/oxide interface. Following a second anneal at 500°C for 600secs in forming gas (containing 5%H_2), the trapping of positrons at the interface is significantly suppressed, and hence the 'dip'is removed. This is consistent with the passivation of defects at the Si-nc surface, an effect which has been documented as inducing a large increase in the luminescence yield [6].

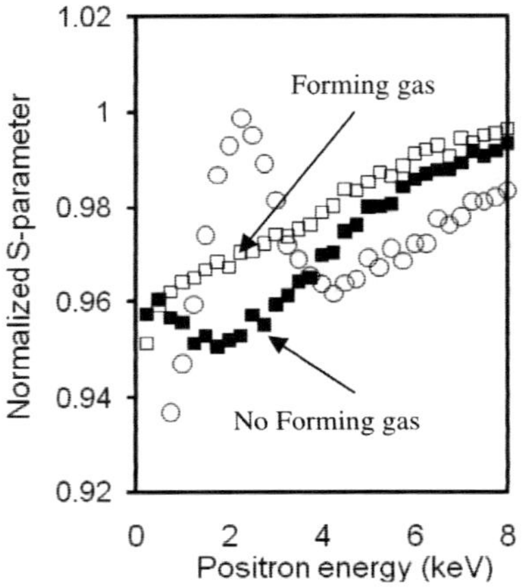

Figure 4. Normalized *S*-parameter versus positron energy for samples implanted with $4x10^{16}cm^{-2}$ at an ion energy of 40keV and annealed at 1100°C for 100 sec in N_2 (closed squares); 1100°C for 100sec N2, followed by forming gas anneal at 500oC for 600secs (open circles). Data for as-implanted (unannealed) is shown as open circles.

4. Conclusion

We have presented preliminary data on the use of VEPAS to probe the formation of Si-ncs in SiO_2. There are few techniques (if any) which can provide similar depth-resolved information on this technologically important system. VEPAS monitors the removal of luminescence suppressing defects following high dose ion implantation. Further, it is able to sensitively probe the interface of Si-ncs and the host SiO_2.

5. References

[1] Daldosso N, Pavesi L 2009 *LaserPhotonics Rev.* **3** 508.
[2] Milgram J N, Wojcik J, Mascher P, and Knights A P 2007 *Opt. Express* **15** 14679.
[3] Knights A P, Milgrarn J N, Wojcik J, Mascher P, Crowe I, Sherliker B, Halsall M P, Gwilliam R M 2009 *Phys. Status Soidi. A* **206** 969.
[4] Chilton N B and Coleman P G 1995 *Meas. Sci. Technol.* **6** 53.
[5] Pi X D, Coleman P G, Harding R, Davies G, Gwilliam R M 2005 *J. Appl. Phys.* **12** 8155.
[6] Comedi D, Zalloum O H Y, Wojcik J, and Mascher P 2006 *IEEE J. Sel. Top.Quantum Electron.* **12** 1561.

Journal of Physics: Conference Series **262** (2011) 012032 doi:10.1088/1742-6596/262/1/012032

Positron annihilation study of the disordered phase of hydrogen adsorbed on the Ni (111) surface

S Komagata[1], F Hori[1], A Iwase[1], H Suzuki[2], K Hirota[2], I Kanazawa[2], K Fukutani[3], K Nozawa[4] and F Komori[4]

[1]Department of Materials Science, Osaka Prefecture University, 1-1 Gakuen-cho, Naka-ku, Sakai, Osaka 599-8531, Japan

[2]Department of Physics, Tokyo Gakugei University,4-1-1, Koganei-shi, Tokyo 184-8501, Japan

[3]Institute of Industrial Science, University of Tokyo,4-6-1, Meguro-ku, Tokyo 153-8505, Japan

[4]Institute of solid state Physics, University of Tokyo,5-1-5, Kashiwa-shi, Chiba 277-8581, Japan

E-mail: komagata-s@mtr.osakafu-u.ac.jp

Abstract. The adsorption of hydrogen on transition-metal surface has been studied by both experimental and theoretical methods. Hydrogen adsorbed Ni (111) surface had measured by using slow positron beam apparatus. The positron annihilation γ -ray energy spectra emitted from hydrogen-adsorbed Ni (111) surfaces are analyzed. We have discussed the stability of the disorder phase of adsorbed hydrogens on Ni (111) surfaces. Positron annihilation results showed that the disordered phase is unstable until about 300K.

1. Introduction

The motivation to investigate hydrogen-surface interaction phenomena arises from various viewpoints. First, we have to mention the field of heterogonous catalysis in which a great part of the reactions involve hydrogen as a reactant or product molecule. On the hydrogen-chemisorbed Ni(111) surface, two temperature-programmed- desorption (TPD) peaks, β_1 and β_2, have been observed [1]. The structure, which corresponds to the β_2 state of the hydrogens on Ni (111) surface, is (2×2)-2H superstructure. On the other hand, the structure of the disordered state of adsorbed hydrogens of the β_1 state is not yet known.

Reemitted slow-positron spectroscopy [2] will provide important informations on adsorbed light atoms on surfaces. In this study, we have measured the changes in positron annihilation spectra from hydrogen-adsorbed Ni (111) surface due to different adsorbing conditions of hydrogens on Ni (111) surface. Oishi et al. [3] have observed that the number of reemitted slow positrons from hydrogen-saturatedly adsorbed Ni(111) surface is reduced much in comparison with that from the clean Ni(111) surface. Recently, we have measured reemitted slow positron yields from hydrogen adsorbed Ni (111) surface on different adsorbing conditions [4]. In this study, we have analyzed the positron annihilation spectra from the disorder state of adsorbed hydrogen, and discuss the stability of the disordered state. Especially, it is shown that the disorder state is unstable abut ~300K.

Journal of Physics: Conference Series **262** (2011) 012032 doi:10.1088/1742-6596/262/1/012032

2. Experiment

The slow positron beam is composed of two stages. The first stage positron-beam system is composed of a ^{22}Na positron source (~3mCi) and a W single crystal (100) foil of 1□m thickness. The W foil has been annealed in an Ultra High Vacuum (UHV) chamber at ~2300K, is attached 5mm apart from the isotope for moderating positrons. The slow positrons are extracted from the moderator using an accelerating grid of 15V, and are guided along the magnetic field through the solenoid tube of the first stage. The extracted slow positron beam is guided in vacuum along a magnetic field toward the measurement chamber. The field (60~80 Gauss) is supplied by solenoid coils directly wound around the vacuum transport tube, and is supplemented with Helmholtz coils at several positions where the chamber of the ^{22}Na source cannot be backed for making UHV. Thus, we set an Al block with a cylindrical hole for differential pumping, whose diameter is 10mm and length is 40mm, at the end of the first stage. The slow positrons are transpoted through the hole and are injected into the UHV experimental chamber. When the gate valve between the Al cylinder and the transport tube is opened, the pressure of the UHV chamber increases to 4.0×10^{-10}Torr from 2.0×10^{-10}Torr.

For the slow-positron spectroscopy, it is necessary to transfer the beam from the magnetic field of the transport system to a magnetic field-free region in the measurement chamber. To extract the beam into the magnetic field free space without changing the beam size, an electrostatic accelerator is used in the magnetic field. Without acceleration, the radius of the cyclotron motion becomes larger as the magnetic line of force expands at the end of the magnetic transport tube. The extracted positron beam is further transported to the measurement chamber by electrostatic Einzel lenses. A simulation of the trajectories of the positrons has been performed. Finally the positron beams with a diameter of a few mm and an energy of 2 keV are transported to the center of the measurement chamber for the spectroscopy. The intensity of the slow positron beam thus obtained is approximately 0.6×10^4 cps.

We installed the LEED and AGE optics in the experimental chamber. Positron annihilation spectra are measured with a high-purity germanium detector. The sample is suspended by W wires, and can be annealed by direct current through the wires. The highly pure hydrogen and argon gases can be introduced to the UHV chamber through a variable leak valve. The base pressure of the UHV chamber was 2.6×10^{-10} Torr. The specimen surface is cleaned by repeated argon ion bombardment. No contamination was checked by the Auger electron spectra. Then cleaned Ni (111) surface was annealed for 30 min at 1100K to eliminate defects in 2.0×10^{-9} Torr. After the sample cools to room temperature , a sharp (1×1) LEED pattern was observed. The sample was cooled by liquid nitrogen and then hydrogen gas at some conditions was introduced at 180K. When hydrogen gas 1.0×10^{-8} Torr is introduced in the chamber for 5min at 180K, typical (2×2)-2H LEED pattern was observed. The positron annihilation γ-ray energy spectra from hydrogen-saturated adsorbed Ni(111) are measured at 180K to 350K.

3. Results and Discussions

The order-disorder problem of a honeycomb lattice such as H/Ni(111) has been considered theoretically [5, 6]. Nagai et al [7] have discussed the order-disorder transition in H/Ni(111) on the basis of position-space renormalization-group (RG) calculations. They considered the following inter particle (hydrogen) interactions; (A) nearest-neighbor exclusion, second-neighbor repulsion, and third-neighbor attraction, which was previously proposed by Domany et al [8], (B) nearest-neighbor exclusion, second- and third-neighbor repulsions, and further-neighbor interactions up to the sixth-neighbor one. When the interaction parameters involver are suitably adjusted, both the interaction (A) and (B) lead to the phase diagrams in good agreement with the experiment by Christmann et al. [1] In both theoretical and experimental phase diagrams [1, 7], there in only one phase of disorder (liquid) in hydrogen-coverage region $\theta > 0.6$. Oishi et al, [3] have observed that the reemitted slow positron yield is reduced in the disordered phase on hydrogen-saturatedly adsorbed Ni(111). Origin of the reduction of reemitted slow positron yield in the disorder phase is not confirmed. It is expected theoretically that charge flows from the surface to wards the hydrogen addatom, causing a dipole moment and, in turn, an electrostatic

Journal of Physics: Conference Series **262** (2011) 012032 doi:10.1088/1742-6596/262/1/012032

repulsion between the hydrogen addatoms [9]. That is, in high coverage case, the repulsion between the adsorbed hydrogen atoms remarkably. In addition, the repulsion is strongest for those hydrogen atoms that are located in a large distance away from the surface, because the surface electrons cannot screen the repulsion interaction among there hydrogen atoms.

Recently we have observed that counts of the annihilation spectra around 511keV emitted from hydrogen-saturatedly adsorbed Ni (111) increase in comparison with those from hydrogen-lightly adsorbed Ni (111) [10]. This is coexistent with reduction of reemitted slow positrons from hydrogen saturatedly adsorbed Ni (111). It is seen that the value of s-parameter decreases remarkably on the hydrogen saturated adsorbed Ni (111) surface [11]. The interaction between highly dense adsorbed hydrogen atoms and reemitted slow positron might disturb strongly emission of thermalized slow positrons into vacuum, and as a result positron annihilation of two □-ray emitted from the surface increases. Furthermore, the collision between highly dense adsorbed hydrogen atoms and orth-positroniums might induce more trapping of ortho-positroniums on hydrogen-saturatedly adsorbed Ni (111) surface. Fig.1 shows counts of the positron annihilation spectrum around 511keV emitted from hydrogen-adsorbed Ni(111) surface at each temperature. Counts at 180K and 300K are attributed to the positron annihilation γ-ray spectra emitted from Ni(111) surface of the hydrogen exposure ~80L ($1L=1.0\times10^{-6}\times1s$). As shown Fig.2 the value of s-parameter increases. In this condition (saturatedly hydrogen-adsorbed surface), (1×1) LEED pattern was observed at 180K as shown in Figure 3 (a). This shows the disorder phase of adsorbed hydrogens. As shown in Fig.1 when temperature from 300K to 350K, counts of the positron annihilation γ-ray spectrum decrease remarkably. Also (1×1) LEED pattern was observed at 350K. Subsequently when this sample is kept on hydrogen exposure at 180K, (2×2)-2H LEED pattern appears as shown in Fig.3 (b). The results show that the disorder phase of adsorbed hydrogens is unstable above ~300K and the hydrogen gas phase might exist above ~300K. That is, if the phase above ~300K is hydrogen-gas state, the interaction between adsorbed hydrogens and reemitted slow-positrons decreases in comparison with one of hydrogen-saturated adsorbed state (the disorder phase). Thus reemitted slow positron yield increases and, as a result, the counts of the positron annihilation γ-ray spectrum emitted from the surface decreases at 350K, furthermore, the collision between adsorbed hydrogens and ortho-positroniums is reduced at 350K, and, as a result, more ortho-positroniums are emitted from the surface to vacuum. Thus the value of s-parameter increases, and the three-□ annihilation rate from the surface decrees, when temperature increases from ~300K to 350K. That relative contribution of para-positronium in the spectrum seems to increase on the hydrogen-lightly adsorbed Ni (111) surface, because escaping of the ortho-positronium from the hydrogen-lightly adsorbed Ni (111) surface increases in comparison with that from the hydrogen -saturatedly one [10]. The present results suggested strongly that there is a boundary line between the disorder phase and the gas phase in the phase diagram.

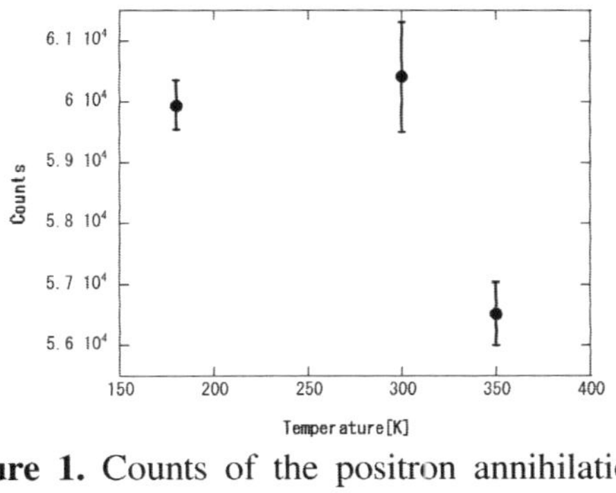

Figure 1. Counts of the positron annihilation spectrum around 511keV emitted from

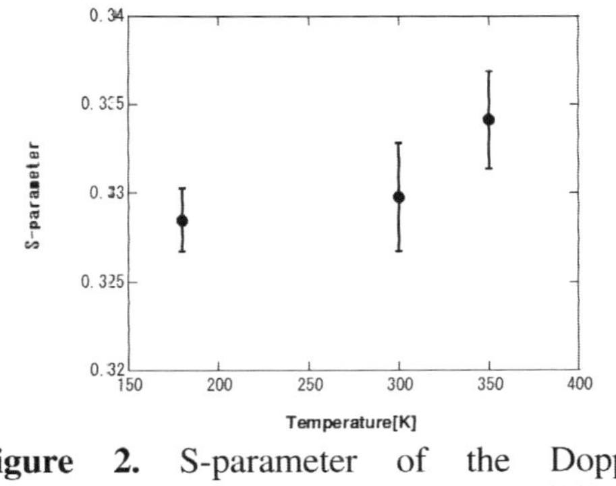

Figure 2. S-parameter of the Doppler broadening spectra of the positron annihilation

hydrogen-adsorbed Ni (111) surface at each temperature.

□-ray emitted from the hydrogen adsorbed Ni (111) surface at each temperature.

Figure 3 (a). LEED pattern of 1×1 (hydrogen-saturate) at 180K.

Figure 3 (b). LEED pattern of (2×2)-2H at 180K when after heat.

4. Conclusion

We have discussed the stability of the disorder phase of adsorbed hydrogens on Ni(111) surface, analyzing the positron annihilation γ -ray spectra emitted from the surface. It is suggested that there is a boundary line between the disorder phase and the gas phase in the high H-coverage region in the phase diagram.

5. References

[1] K Christmann, R J Behm, G Ertl, M A V Hove, and W H Weinberg 1979 *J. Chem Phys.* **70** 4168
[2] I Kanazawa, T Wada, Y Oishi, Y Terashima, K Fukutani, Y Murata, Y Ito, K Nozawa, F Komori 2005 *Nucl. Instrum. Methods Phys. Res. B* **232** 299
[3] Y Oishi, T Wada, I Kanazawa, K Fukutani, Y Murate, Y Ito, K Nozawa, K Komori 2005 *Appl. Surf. Sci.* **241** 119
[4] S Komagata, K Hirota, H Suzuki, M Osawa, S Arii, I Kanazawa, K Fukutani, K Nozawaand F Komori 2008 *Appl. Surf. Sci.* **255** 227
[5] E Domany and M Schick 1979 *Phys. Rev. B* **20** 3828
[6] R C Kittler and K H Bennemann 1979 *Solid state Common.* **32** 403
[7] K Nagai, Y Ohno, T Nakamura 1984 *Phys. Rev. B* **30** 1461
[8] E Domany, M Schick and J S Walker 1979 *Solid State Common.* **30** 331
[9] G Kresse and J Hafner 2000 *Surf. Sci.* **459** 287
[10] S Komagata, H Suzuki, K Hirota, I Kanazawa, K Fukutani, K Nozawa and F Komori 2010 *J. Phys.* **225** 012029
[11] S Komagata, K Hirota, S Arii, I Kanazawa, K Fukutani, K Nozawa, F Komori 2008 *Surf. Inter. Anal.* **40** 1713

Journal of Physics: Conference Series **262** (2011) 012033 doi:10.1088/1742-6596/262/1/012033

First results of *ab initio* simulations of scintillation detector characteristics

A Krille[1], M Nagl[2], U Vetter[2], H Hofsäss[2] and R Krause-Rehberg[1]

[1] Institute of Physics, Martin-Luther-University, von-Dankelmann-Platz 4, 06120 Halle (Saale), Germany

[2] 2. Physikalisches Institut, Georg-August-Universität Göttingen, Friedrich-Hund-Platz 1, D-37077 Göttingen, Germany

E-mail: arnold.krille@physik.uni-halle.de and mnagl@uni-goettingen.de

Abstract. We describe a new Monte-Carlo-simulation which models output signals of scintillator-photomultiplier tube (PMT) pairs . For the whole simulation from the initial photon-cascade within the scintillator to the final pulse as it is digitized at the output of the PMT only values from the scintillator's and PMT's technical manuals are used, no parameters need to be adapted. We find an excellent agreement between sampled and simulated signals. This allows to determine the influence of single parameters independent from the others.

With a special focus on positron lifetime as well as perturbed angular correlation spectrometers, which have high demands on the γ-photon incidence time determination's accuracy, the influence of the tts-parameter on the detector's overall time resolution was studied in detail.

1. Introduction

While positron lifetime spectroscopy (PLS) and perturbed angular correlation spectroscopy (PAC) are rather different methods of material research, they both rely on the time resolution of the used detectors. The PAC method may additionally (depending on the used isotopes) have rather high demands on energy resolution in order to separate gamma peaks close in energy often appearing in more complex decays. The PAC method allows for the analysis of local magnetic fields and electric field gradients at radioactive probe atoms in condensed matter and is mainly used for the characterization of metals[1] and semiconductors[2]. It is based on measuring correlations between gamma photons emitted from useful decay cascades which may show a modulated anisotropy due to the hyperfine interaction of the state dependent atom moments with the surrounding magnetic fields and electric field gradients[3]. PLS in contrast uses the statistically decaying lifetime of positrons in the sample to detect defects and voids in metals, crystals, semiconductors, polymers and meso-porous materials[4, 5, 6, 7].

Measuring energy and time of single γ-photons is usually accomplished with scintillator-photomultiplier-assemblies. These convert the incident photon into an electric pulse which carries time and energy information. Several years ago devices for digital processing of these signals became cheaper compared to their analog counter-parts.

Various groups have reported success of varying degree using digital equipment to extract the time- and energy-information [8, 9, 10]. However their results showed to be difficult to reproduce and seems to heavily depend on "hand selected" PMTs. Several attempts of simulating the pulses

Journal of Physics: Conference Series **262** (2011) 012033 doi:10.1088/1742-6596/262/1/012033

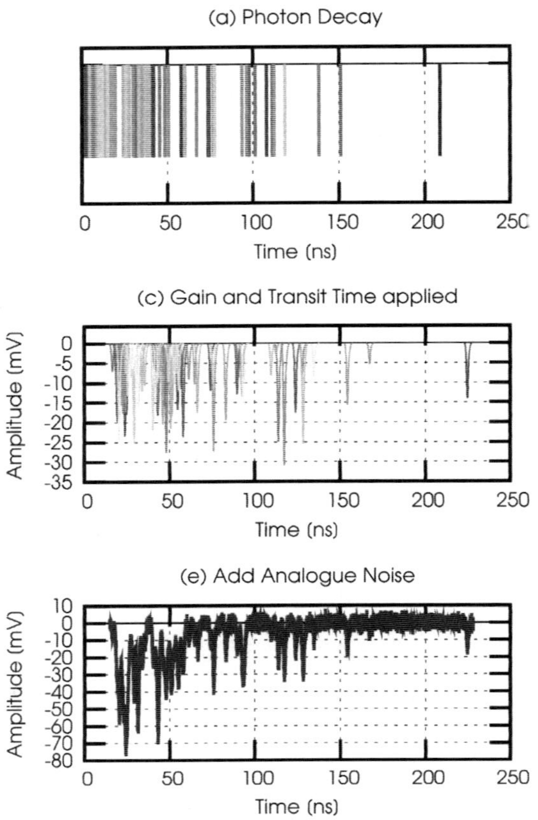

Figure 1. Simulation of the impact of a single gamma ray event explained graphically: The photons emitted from the scintillator (a) are convolved with the single-electron's anode signal (b), scaled and delayed to simulate gain and transit-time (c), all signals are summed (d), additive white gaussian noise is added (e) and finally the quantization of the digitizer applied (f).

coming from scintillator and photomultiplier where done but had deficiencies[11, 12]. They contained parameters that had to be "fitted to reality" and while they gave good results for one combination of scintillator, PMT and digital routine, the predictions for other combinations were out of scope. Additionally they did not reflect the physics of scintillators and photomultipliers.

2. The Simulation Method

When the γ-photon hits the scintillator it creates a photo electron which excites color centers. These color center excitations decay with a decay constant specific to the scintillation material and emit secondery photons in the visible or UV range. The total number of secondary photons corresponds to the incident γ-photon's energy. The secondary photons enter the PMT, hit its photo cathode and produce photo electrons which are then accelerated and multiplied in a dynode system with typically eight to twelve stages. The electrons from the last dynode (typically several millions per single photo electron from the photo cathode) finally hit the anode. By connecting a transimpedance amplifier to the anode this transient charge is measured as a short voltage pulse over time.

The complete process is simulated by first creating the poisson distributed random number specifying the number of photo electrons emerging on the PMT's cathode. The poisson

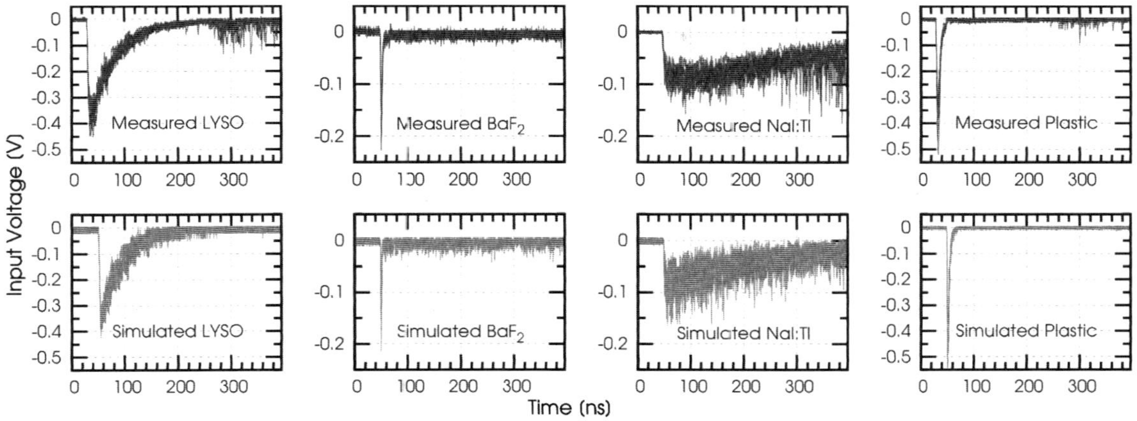

Figure 2. Comparing real (upper row) with simulated (lower row) signals for LYSO, BaF_2, NaI:Ti and plastic. PMT for both simulation and measurement was the Hamamatsu R3377. The simulated pulses match in shape, rise time, length, energy resolution and voltage range. Each panel shows 80 randomly selected signals.

distribution's λ value is given by the scintillators and PMT's overall efficiencies. Exponentially distributed random numbers are drawn from that distribution resembling the points in time each simulated secondary photon emerged from its color center's decay. Then each of these cathode electrons is convolved with a function describing the temporal distribution of the associated anode electrons. This function is approximated by two gaussian functions to account for the multiplication by the dynodes as both base-material and coating have an influence here. It approximates curves published by PMT vendors and is scaled by the PMT's rise time value according to the data-sheet. To account for the amplification (through each dynode's gain and the total number of dynodes) and transit-time (TT) as well as transit time spread (TTS), the pulses are delayed by TT plus a gaussian distributed random number simulating the TTS and scaled by a cascade of poisson distributed random numbers. Then all these single pulses are added up to form the pulse as it is measured as a voltage across the input resistor of the digitizer. To compare the simulation with digitally recorded signals, two more steps are necessary: Additive white gaussian noise is added in order to simulate the analog noise of the input amplifier and the signal is quantized to a defined number of bits and filtered to re-construct the relation of frequency to effective-number-of-bits in the digitizer. The figure 1 shows each step. A more concise description of the simulation method will be published separately.

Even though we only used parameters from the data sheets of scintillators, photomultipliers and digitizers, the resulting pulses are in excellent agreement with the measured signals. We compared the general shapes for $Lu_{1.8}Y_{0.2}SiO_5$:Ce (LYSO), BaF_2, NaI:Ti and plastic scintillators in figure 2. Energy distribution and energy resolution, rise times and lengths of the simulated pulses excellently match their real-world counterparts.

3. Results from the Simulation: Transit Time Spread

The transit time is the mean time difference between the emission of a single photo electron from the cathode and the arrival of the corresponding electron bunch at the anode. The *transit time spread* (TTS) is the spread of the transit time due to different trajectories, spread in the initial energies of the secondary photons, irregularities in the dynode material and shape as well as the static electric field.

In order to compare the simulations with real events, numerous signals were recorded with

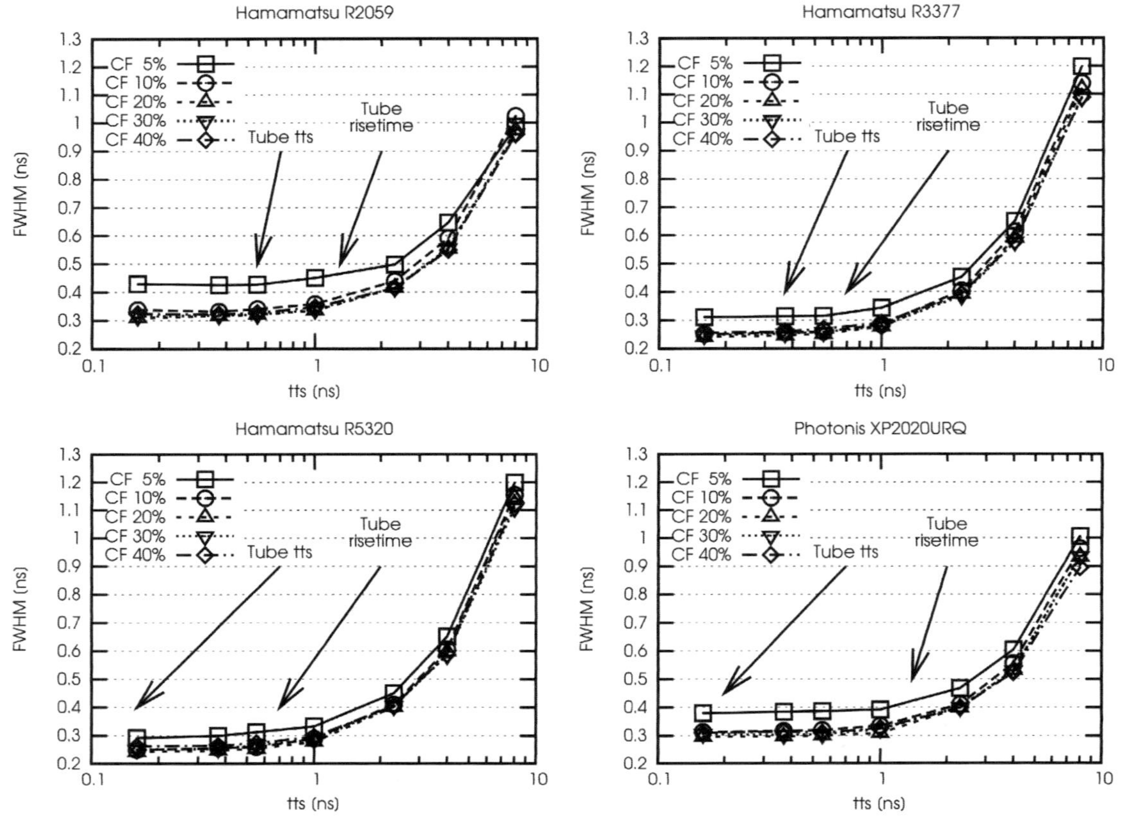

Figure 3. Simulation of variations of the transit time spreads (TTS) for several PMTs at different constant fraction trigger levels (CF). The original TTSs of the tubes are marked. Obviously the TTS has little impact on the time resolution for most PMTs unless the TTS exceeds the rise time.

digitizers featuring a resolution of 8 bit at a sampling rate of 4 GS/s. As input for the simulations we used the technical details of R2059, R3377, R5320 tubes from Hamamatsu and the well-known XP2020URQ tube from Photonis from the data sheets. The simulation was done using different values for the TTS in order to investigate the time resolution's dependency on this parameter. Figure 3 shows the results of these simulations. The comparison with the real TTS values of the R3377 and XP2020 PMTs show an excellent agreement. Also the time resolution of simulated signals is very close to the measured values. Remaining differences are in the range of fluctuations observable between tubes of the same type.

Obviously the TTS is not a limiting factor for the detector's time resolution. The graphs show very clearly that the time resolution is significantly influenced by the TTS only in the case where it is higher than the signals' rise time value.

4. Conclusion

Comparing the simulated signals to measured ones an excellent agreement is obtained. The major advantage of the presented simulation method is that it relies solely on parameters given in technical data sheets of scintillators (like [13]) and photomultipliers. The method successfully models the physics of scintillation detectors.

Very good agreement is found concerning time resolution, signal shape, rise time, pulse length, and energy resolution. Moreover it was shown that the signal rise time which depends on the scintillator's properties as well as the PMT's rise time value limits the achievable time resolution. Decreasing the transit time spread below the rise time value results in no significant improvement of the time resolution.

Acknowledgments

This work was supported by the German BMBF (Grant No. 05KK7NH1 and 05K10MG3).

Appendix A. References

[1] Recknagel E, Schatz G and Wichert T 1983 Hyperfine interactions of defects in metals *Hyperfine Interactions of Radioactive Nuclei* Topics in Current Physics ed Christiansen J (Springer) chap 4, pp 133–204
[2] Wichert T, Deicher M, Grbel G, Keller R, Schulz N and Skudlik H 1989 *Applied Physics A: Materials Science & Processing* **48** 59–85
[3] Frauenfelder H and Steffen R 1965 Angular distribution of nuclear radiation *Alpha, Beta and Gamma Ray Spectroscopy* ed Siegbahn K (North Holland, Amsterdam) p 997
[4] Krause-Rehberg R and Leipner H 2001 *Positrons in Solids* (Springer-Verlag)
[5] Mäkinen J, Corbel C, Hautojärvi P, Moser P and Pierre F 1989 *Phys. Rev. B* **39** 10162–10173
[6] Krause-Rehberg R, Brohl M, Leipner H S, Drost T, Polity A, Beyer U and Alexander H 1993 *Phys. Rev. B* **47** 13266–13276
[7] Dlubek G, Pionteck J, Yu Y, Thränert S, Elsayed M, Badawi E and Krause-Rehberg R 2008 *Macromolecular Chemistry and Physics* **209** 1920–1930
[8] Bečvář F 2007 *Nuclear Instruments and Methods in Physics Research B* **261** 871–874
[9] Saito H, Nagashima Y, Kurihara T and Hyodo T 2002 *Nuclear Instruments and Methods in Physics Research A* **487** 612–617
[10] Nissilä J, Rytsölä K, Aavikko R, Laakso A, Saarinen K and Hautojärvi P 2005 *Nuclear Instruments and Methods in Physics Research A* **538** 778–789
[11] Krille A, Krause-Rehberg R, Jungmann M, Becvar F and Brauer G 2008 *Applied Surface Science* **255** 93 – 95 ISSN 0169-4332 proceedings of the Eleventh International Workshop on Slow Positron Beam Techniques for Solids and Surfaces URL `http://www.sciencedirect.com/science/article/B6THY-4SJ2WPD-X/2/b8a99d9a23d5cb0c53955bf55523ea38`
[12] Krille A Digital positron lifetime: the influence of noise (IOP Publishing) proceedings of the PSD 08 workshop. To be published.
[13] Derenzo S E scintillators.lbl.gov URL `http://scintillators.lbl.gov/`

Journal of Physics: Conference Series **262** (2011) 012034 doi:10.1088/1742-6596/262/1/012034

NDT study of oxide dispersion strengthened steels

Vladimir Krsjak[1], Zoltan Szaraz[1], Jarmila Degmova[2] and Peter Hähner[1]

[1]Joint Research Centre, Institute for Energy, European Commission, P.O. Box 2, 1755 ZG Petten, Netherlands
[2]Slovak University of Technology, FEI, Ilkovicova 3, 81219 Bratislava, Slovakia

E-mail: vladimir.krsjak@ec.europa.eu

Abstract. Various oxide dispersion strengthened (ODS) steels (PM2000, MA956, ODM751 and ODS Eurofer) have been investigated using positron annihilation lifetime spectroscopy. Preliminary characterization of the vacancy type defects and the yttria nanoparticles are reported in this paper. The interpretation of the experimental data considers also the results of magnetic Barkhausen noise (BN) measurements as well as SEM and TEM investigations. Significant differences due to the presence of the yttria oxides (namely Y2O3) were observed in the studied materials in comparison to conventional ferritic/martensitic steels (e.g. Eurofer). Higher positron mean lifetime in ODS steels is, however, not only due to the presence of dispersoids, but also a result of vacancy agglomeration (clusters of 6-8 vacancies) which have been confirmed in recrystallised ODM751 and MA956 materials. On the other hand, positron trapping at dislocations was observed in the as-extruded ODS Eurofer in contrast to the other, recrystallised, materials where the high temperature treatment had led to the static recovery of these defects. It is suggested that some defects which are present (dislocations, vacancy clusters and dispersoids) and affect positron trapping in the materials are also important pinning sites for the magnetic domains. This is reflected by a shift of the signal peak in the BN spectra. The present complementary study provides more comprehensive information about materials microstructure and can support the interpretation of the physical/mechanical testing results (hardness, fracture mechanics etc.) obtained on these materials.

1. Introduction

The performance of currently available classes of structural materials qualified for various applications in present nuclear reactors is insufficient for many applications envisaged for next generation nuclear systems (so-called GenIV), which is why extensive R&D is being carried out into the development and qualification of new structural materials. Higher temperatures and fuel burn-up, longer service lifetimes, as well as new coolant environments represent important material issues for the safe and economic operation of GenIV reactor systems. Good radiation and corrosion resistance as well as improved creep properties in an extended range of temperatures have been achieved with ferritic steels strengthened by dispersion of nano-sized yttria particles. Nevertheless, the particular effects of the mechanical alloying process and of the specific service conditions (neutron flux and fluence, thermo-mechanical loads, chemical environment of the coolant etc.) on the performance of these oxide dispersion strengthened (ODS) steels need further qualification.

Positron annihilation lifetime spectroscopy (PALS) is a technique which is widely used in the field of radiation embrittlement studies [1, 2]. Its unique sensitivity to radiation induced defects provides information which cannot be obtained by any other technique in a non-destructive way. Published studies on the ODS steels show the capability of positron annihilation spectroscopy techniques to elucidate the nature, localization and thermal stability of the defects, which are responsible for the dimensional (swelling) and the mechanical properties stability under neutron irradiation [3].

Barkhausen Noise (BN) analysis is based on the inductive measurement of a noise-like signal, which is associated with the stochastic re-arrangement of magnetic domains, and generated when a varying magnetic field is applied to a ferromagnetic sample. This technique is known to be very sensitive to microstructural features such as precipitates and lattice defects in materials, because it probes the discontinuous motion of domain walls caused by the presence of any type of pinning centers [4]. Since many of these defects (dislocations, grain boundaries, voids etc.) are also possible trapping sites for positrons, we can expect a correlation to exist with the findings by PALS.

An application of both described techniques in a common study of iron-based model alloys has been published earlier [5, 6], but without specific interpretation of the results obtained. The present paper proposes a possible interpretation of particular characteristics obtained from the ODS steels under investigation and confirms the feasibility of the combination of these techniques for monitoring microstructural changes caused by service exposure (radiation and/or thermal treatment).

2. Experimental

Three commercial ferritic ODS steels (PM2000, MA956 and ODM751) and an ODS variant of ferritic-martensitic Eurofer steel were chosen for this complementary study. Nominal chemical compositions of the materials are compiled in the Table 1. The ODS alloys were produced by mechanical alloying, i.e., matrix powders were mixed and milled together with yttria particles to form solid solutions with uniform dispersion of oxide particles, and the mixtures were then consolidated using hot isostatic pressing and extrusion. The PM2000 and MA956 alloys were supplied as rods in the fully recrystallised condition for improved creep strength. The recrystallisation heat treatment resulted in a coarse columnar grain structure. The ODM751 alloy was received as tubes in recrystallised form as well, whereas the as-extruded ODS Eurofer alloy exhibited a fine-grained microstructure. The distribution and the sizes of the oxide particles were investigated by SEM and TEM. The diameter of the oxide particles was found to vary from 10 to 70 nm.

Table 1. Chemical composition of the studied materials in wt. % (Fe – balance).

Material	Manufacturer	Cr	Al	Ti	W	Mn	Mo	C	Y_2O_3
PM2000	Plansee	20	5.5	0.3	-	-	-	0.01	0.5
MA956	Incoloy	20	4.5	0.4	-	-	-	0.04	0.5
ODM 751	Dour Metal	16	4.5	0.6	-	-	1.5	0.01	0.5
ODS Eurofer	Plansee	8.9	-	-	1.1	0.4	-	0.07	0.2

Microstructural studies often face the problem of variations in the microstructure of different samples, due to not fully uniform fabrication, heat treatment or irradiation processes. An advantage of non destructive testing (NDT) resides in the fact that specific information can be provided by different experimental techniques applied on the same sample. This offers the opportunity to correlate measurements from different experimental techniques while making sure that the results refer to identical conditions of the material microstructure. With respect to this, the present sample size was chosen as 10x10x0.5 mm^3, which is a typical size for PALS, but also suitable for a number of other NDT techniques including BN measurements.

PALS measurements were performed in conventional fast-slow setup of the apparatus, with the use of BaF_2 scintillator detectors and Ortec electronics. The used ^{22}Na positron source with an activity of 3.6 MBq and encapsulated in kapton foil, enabled quick measurements (~ 4 hours) with reasonable FWHM (< 220ps with the used detectors) and source contribution (~ 18%). Analytical data processing was performed using the LT 9.0 program and two-components decomposition of the spectra.
Magnetic BN analyses have been performed using the commercial BN analyzer Stresstech Rollscan200 with an external magnetizing unit and pen-type sensor with ferrite core diameter of 3mm. Magnetizing fields were induced by an alternating current of 1A amplitude and frequency of 10Hz. The picked-up signals were amplified and filtered to discriminate frequencies below 500Hz. Signals were acquired and analyzed by ViewScan software from Stresstech. The BN parameters were determined by averaging 10 measurements (five in longitudinal direction and five in transverse direction to the material extrusion). To minimize the effect of residual magnetism, individual samples were demagnetized after each measurement.

3. Results and discussion

3.1. Positron annihilation lifetime spectroscopy

Two-component decomposition (Table 2) of the positron lifetime spectra was found to be most suitable for all materials studied. With this approach we can attribute the first component to the material bulk properties (reduced due to positron trapping at defects) while the second component comes from positrons trapped in defects. Higher τ_1 value in the as-extruded ODS Eurofer is attributed to the presence of dislocations, as high dislocation densities are common for the steels produced by mechanical alloying [7]. High temperature recrystallization leads to the recovered microstructure with large, almost dislocation free grains which results in shorter positron mean lifetime (MLT).

Table 2. PALS results from the ODS steels measurements.

Material	t_1 [ps]	I_1 [%]	t_2 [ps]	I_2 [%]	MLT [ps]	Annihilation rate in trapping sites [s^{-1}]	Trapping rate in defects [s^{-1}]
PM2000	79	45	226	55	160	4.40E+09	5.70E+09
MA956	78	47	246	53	167	4.10E+09	5.67E+09
ODM751	78	49	259	51	171	3.90E+09	5.44E+09
ODS Eurofer	129	43	241	57	193	4.10E+09	6.55E+09

The effect of the yttria particles on the positron lifetime can be seen in a considerably higher (> 10%) MLT values in comparison to conventional (non-ODS) ferritic/martensitic steels reported earlier [8]. The relatively small content of yttria oxides plays a role here due to significant positron affinity of yttrium (-5.31). Thus, for example, a positron would be trapped in a pure Y phase in an iron (positron affinity = -3.84) matrix with a binding energy of 1.5eV [9]. Positron lifetime in the Y_2O_3 bulk, experimentally reported as 239ps [10], together with the positron lifetime in small vacancy clusters, is described by the component τ_2. As can be seen in Fig.1, there is an overlap of the positron lifetimes in those defects, which is why the different trapping sites cannot be unambiguously distinguished. Nevertheless, some specific microstructural techniques (e.g. small angle neutron scattering, SANS) can provide additional information about the size, distribution and volume fraction of yttria nano-particles, which may then be used as an additional input, in order to reveal the information about density and type of vacancy clusters.

Figure 1. Measured positron lifetimes in open volume defects, together with theoretical values for different vacancy clusters (single vacancies, five- and eight- vacancy clusters, respectively, calculations for bcc Fe) and yttria oxide.

The highest value of the τ_2 component (259 ps) was found in the ODM751 steel, indicating the presence of vacancy agglomerations (6-7 vacancies) in the material. The range of the vacancy cluster sizes in the materials studied here (4-7 vacancies) is in agreement with published work of Xu et al [11], who observed similar features (4-6 vacancies) in nanocluster-strengthened ferritic steel and described the essential role of vacancies in the formation and stabilization of nanoclusters in this material. A quantitative analysis of the vacancy type defects, however, cannot be carried out without additional characterization of the yttria nanoparticles of the present ODS steels, e.g. by the SANS technique.

3.2. Magnetic Barkhausen noise measurements

The results of the BN spectra analysis are summarized in Table 3. Considering the values of signal amplitudes (signal strengths of the events described in terms of the RMS and the peak envelop height), and relative position of the peak (characteristic of the coercive field for domain wall motion), we can recognize three different types of signals. While ODS Eurofer and ODM751 steels show distinct behaviours, the other two materials, PM2000 and MA956, exhibit very similar features. We assume that this is due to the differences and commonalities in the chemical compositions of the materials (see Table 1).

Table 3. The results of the BN measurements on the ODS steels.

Material	RMS [a.u.]	Peak height [a.u.]	FWHM [ms]	Peak position [ms]
PM2000	3±0.2	6.6±0.3	21.5±1.6	4.5±0.2
MA956	3.1±0.4	6.3±0.6	16.9±1.8	4.9±0.3
ODM751	4.2±0.3	9.1±0.6	18.3±0.9	6.6±0.2
ODS Eurofer	2.8±0.2	6.2±0.3	18.5±1.9	9.9±0.2

According to theory and published works [12], one can expect a decreasing BN signal amplitude and a shift of the peak to the higher values of magnetizing field, when magnetic domain wall pinning is stronger. Although the results of the BN peak and signal RMS are scattered, the comparison of recrystallized (PM2000, MA956, ODM751) and non-recrystallized (ODS Eurofer) materials indicates agreement with this assumption as regards domain wall pinning by grain boundaries. However, the higher carbon content of the ODS Eurofer can also play a role in decreasing the signal amplitude of ODS Eurofer [12].

Recalling that a stronger pinning of the magnetic domain walls induced by the presence of defects in the microstructure, is revealed by a shift of the signal peak to higher values of the coercive magnetic fields (here recorded as a time shift of the peak position), the BN peak position can therefore serve as a quantitative feature of the defects acting as pinning sites for magnetic domains. It is natural to assume that some of these defects can act as trapping sites for positrons as well, and therefore a correlation with positron lifetime technique is suggested. The overall positron trapping kinetics in a material is characterised by the positron mean lifetime (MLT). As can be seen in the Fig.2, there is indeed a correlation between positron trapping and magnetic domain wall mobility.

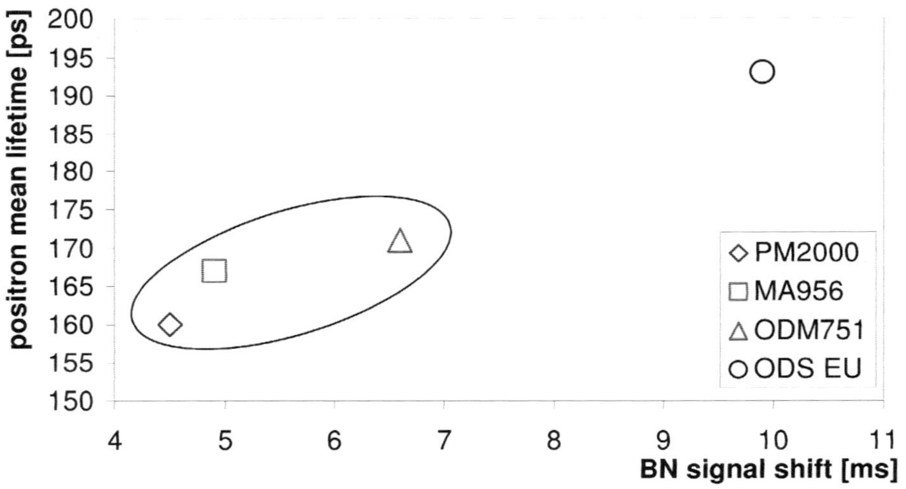

Figure 2. Correlation between positron mean lifetime and relative shift of the BN peak.

Both techniques, positron annihilation lifetime spectroscopy and magnetic Barkhausen noise analysis, show similar characteristics of the microstructural features of the individual ODS steels. Recrystallisation and the concomitant recovery of lattice defects influence the parameters measured by the distinct NDT techniques (See Fig.2.). We assume that the significant increase of the positron MLT and the shift of the BN signal towards higher magnetic fields in ODS Eurofer, as compared to the three recrystallised materials investigated, are both due to the higher dislocation density in that material. This can be seen in the Fig.1, where the τ_1 component related to the reduced bulk lifetime is considerably higher for ODS Eurofer, apparently due to positron trapping at dislocation lines. On the other hand, variations in the BN and PALS values of the recrystallised materials are supposed to be due to different sizes of vacancy clusters present in these materials.

4. Conclusions

Material defects (vacancy clusters, dislocations, grain boundaries yttria particles,) which act as positron trapping sites of different dimensions also contribute to the pinning of magnetic domains and impede the mobility of domain walls. In the present paper this situation has been explored by a complementary study combining the PALS and BN techniques, in order to identify a correlation between the distinct techniques. The experimental results provide evidence for the existence of that correlation, which is going to be further explored by studies of the thermal ageing behaviour of the ODS steels (α' precipitation at intermediate temperatures). More detailed information about size and density of defects will be available from on-going SANS experiments, which will provide the characteristics of the yttria nano-particles and the α' precipitates.

5. References

[1] Brauer G, Liszkay L, Molnar B, Krause R 1991 *Nucl. Eng. Design* **127** 47
[2] Slugen V, Segers D, de Bakker P M A, de Grave E, Magula V, van Hoecke T, van Waeyenberge B 1999 *J. Nucl. Mater.* **274** 273
[3] Ortega Y, de Castro V, Monge M A, Muñoz A, Leguey T and Pareja R 2008 *J Nucl. Mater.* **376** 222–228
[4] Yamaura S, Furuya Y and Watanabe T 2001 *Acta mater.* **49** 3019–3027.
[5] Debarberis L, Acosta B, Sevini F, Pirfo S, Hyde J M, Hutchings M T and Ortner S 2004 *NDT&E International* **37** 19–22
[6] Dobmann G, Debarberis L and Coste J F, 2001 *Nucl. Eng. Design* 206 363–374
[7] Klueh R L, Hashimoto N and Maziasz P J 2005 *Scripta Materialia* **53** 275-280
[8] Zeman A, Debarberis L, Kocik J, Slugen V and Keilova E 2007 *J Nucl. Mater.* **362** 259–267
[9] M.J. Alinger, S.C. Glade, B.D. Wirth, G.R. Odette, T. Toyama, Y. Nagai, M. Hasegawa, 2009 *Mat. Sci. and Eng.* **518** 150-157
[10] Damonte L C, Taylor M A, Desimoni J and Runco J 2007 *Rad. Phys. Chem.* **76** 248
[11] Xu J, Liu C T, Miller M K and Chen H 2009 *Phys. Rev. B* **79**, 020204
[12] Hug-Amalric A, Meilland P, Kleber X and Merlin J, Magnetic Barkhausen noise measurement potentialities for metallurgical transformations characterization in multi-phase High Strength steels, *proc. 9th Europ. Conf. on NDT*, September 2006, Berlin (Germany)

Journal of Physics: Conference Series **262** (2011) 012035 doi:10.1088/1742-6596/262/1/012035

Development of spin-polarized positron source using high energy proton beam

M Maekawa, Y Fukaya, Y Yabuuchi and A Kawasuso
Advanced Science Research Center, Japan Atomic Energy Agency
1233 Watanuki, Takasaki, Gunma, 370-1292, Japan

E-mail: maekawa.masaki@jaea.go.jp

Abstract. To obtain a highly spin polarized positron beam, ^{68}Ge isotope have been produced using a nuclear reaction of ^{69}Ga(p,2n)^{68}Ge. As target materials, we examined a metal form ^{69}Ga stable isotope and a GaN substrate. By 20 MeV proton irradiation, the production of ^{68}Ge source was confirmed in both targets. The production rates of ^{68}Ge were 0.16 and 0.53 MBq/μA/h for the metal Ga and GaN target, respectively. The spin polarizations of positrons emitted from ^{68}Ge was estimated to be approximately 50 to 70%.

1. Introduction

Using spin-polarized positrons, the excess spins in magnetic substances can be detected in positron annihilation measurements [1]. This will be useful in the studies of spin-electronics materials such as half-metals, diluted magnetic semiconductors and so on. Positrons emitted from β^+-decay nuclei are longitudinally polarized because of the parity non-conservation law. The longitudinal spin-polarization of positrons from β^+-emitters is given by $(v/c)\times(1+\cos\alpha)/2$, where, v is the positron speed, c is the light speed and α is the open angle from a specific direction. To obtain highly spin-polarized positrons, it is important to use β^+-emitters with higher endpoint energies. Sodium 22 source is widely used in the positron annihilation spectroscopy method. However, the maximum longitudinal spin polarization of positron emitted from sodium 22 is estimated to be 70 % from its endpoint energy [2]. For sufficient measurements with spin-polarized positrons, a strong enough source emitting highly polarized positrons is needed. For this purpose, we focus on a ^{68}Ge isotope. This isotope decays to ^{68}Ga with a half-life of 280 days. When ^{68}Ga decays with a half-life of 68 min, a positron (89 %) is emitted. The endpoint positron energy is 1.9 MeV and hence the theoretical maximum longitudinal spin polarization is approximately 94%. This radioisotope can be produced though the nuclear reaction of ^{69}Ga(p,2n)^{68}Ge. The threshold energy of this reaction is 11 MeV. By multiple irradiations, ^{68}Ge isotope for the generation of the slow positron beam can be accumulated. In this study, we report the production of ^{68}Ge by proton irradiation using some target materials.

Stable isotopes of Ga are ^{69}Ga(60.1 %) and ^{71}Ga(39.9 %). The cross sections of the production of ^{68}Ge isotope for ^{69}Ga and ^{71}Ga are 0.465×10^{-28} and 0.137×10^{-28} m^2, respectively. If isotope-separated ^{69}Ga was used as a target material, the production of ^{68}Ge is effective. During the high-energy proton irradiation, metal Ga is expected to melt because of the low melting point (29 °C). Therefore, to prevent the leakage of the melt Ga, a source structure to seal the melt Ga is required. When the gallium compound having a high melting point is used, such a difficulty can be avoided.

However the production efficiency of ^{68}Ge isotopes decreases because the content of ^{69}Ga decreases. In this study, we examined two types of targets; a metal form ^{69}Ga stable isotope and a GaN substrate.

2. Simulation

First we carried out a simulation to determine the optimum condition of proton irradiation to produce ^{68}Ge isotope by the Induced Radioactivity Analysis Code (IRAC) [3]. Figure 1(a) shows the depth profile of ^{68}Ge isotope at the proton energy of 20 and 25 MeV for the pure ^{69}Ga target. Total amounts of ^{68}Ge isotope were estimated to 1.1 MeV/μA/h and 2.3MeV/μA/h for the 20 and 25 MeV protons, respectively. Considering the transmission of positrons from the inside of target materials, more positrons are obtained in the case of the 20 MeV irradiation [2]. Figure 1(b) shows the depth profile of ^{68}Ge isotope for the GaN target with the proton beam of 20 MeV. Total amount of ^{68}Ge isotope was estimated to 0.56 MeV/μA/h. The total amount of ^{69}Ge for the metal ^{69}Ga target is approximately 2 times higher than that of GaN.

3. Experimental

Figure 2 shows the schematics of the source capsule. To avoid leakage of melted ^{69}Ga during irradiation, metal ^{69}Ga was put into a carbon tray (diameter:8mm, depth:0.5 mm or 1.0 mm) with a beryllium cap, which are rarely corroded by the Ga metal. The source capsule is finally capped by the Ti window. The thicknesses of Ga targets are 0.5 mm and 1 mm. These were irradiated with 20 MeV and 25 MeV protons, respectively, though the bdellium window. The thickness of GaN target is 0.475 mm. This target was directly irradiated with 20 MeV protons. Proton irradiation was carried out using the cyclotron at Japan Atomic Energy Agency. During the irradiation, the above targets were cooled by running water. The irradiation chamber was evacuated with a base pressure of 6×10^{-4} Pa.

After the irradiation, the amount of ^{68}Ge was measured using intensities of photopeaks of 511 keV and 1.07MeV. The former is from the annihilation of positrons emitted from ^{68}Ga. The later is derived from the decay of ^{68}Ge. The spin polarization of emitted positrons was estimated from the magnetic field dependence of *S*-parameter arising from two-gamma self-annihilation of positronium in a fused silica [4, 5]. The magnetic field was applied by SmCo permanent magnets. By changing the distance of pole gap, the strength of the magnetic field was controlled. Schematic of the measurement system is shown in figure 3.

4. Results and discussions

Figure 4 shows the estimated amounts of ^{68}Ge isotope for the different targets and irradiation energies.

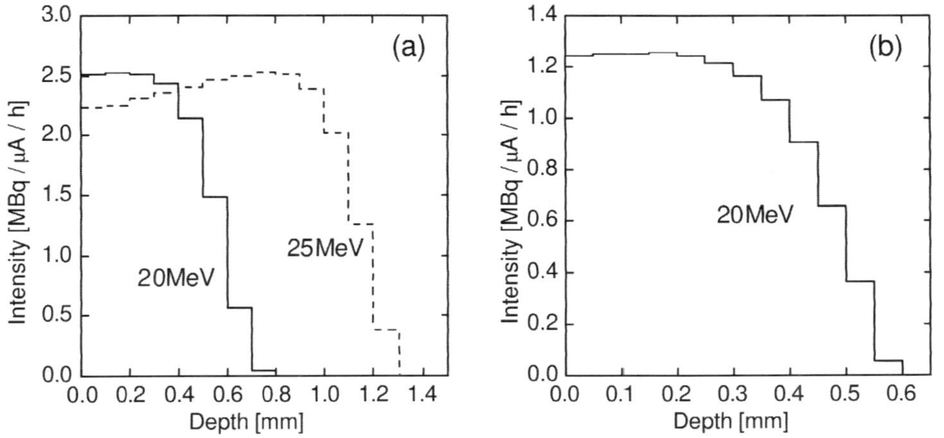

Figure 1. Simulation of depth profiles of ^{68}Ge source produced by proton irradiation. (a) Metal Ga target with energies of 20 and 25 MeV, (b) GaN target with energy of 20 MeV.

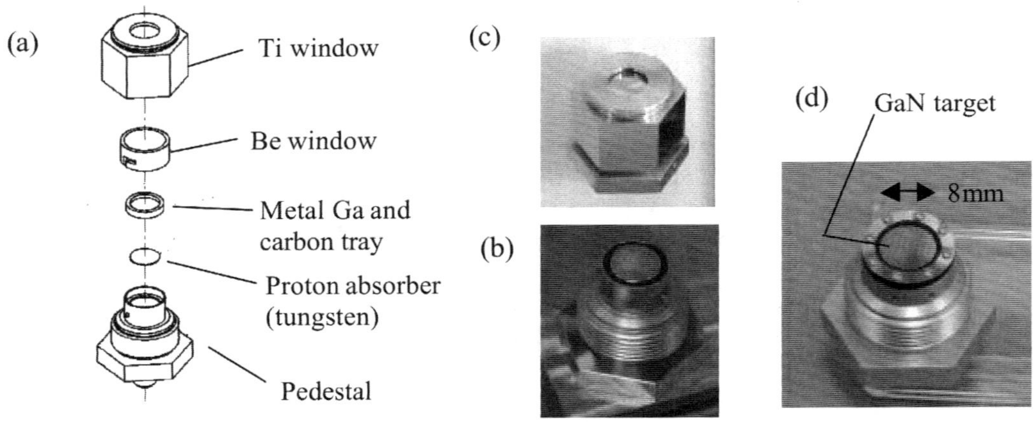

Figure 2 Schematics of the source capsule. (a) Structure of the capsule, (b) source capsule with metal Ga, (c) Ti window and (d) source capsule with GaN target.

When the metal ^{69}Ga targets were used, the production rates of ^{68}Ge were 0.16 and 0.14 MBq/μA/h for 20 and 25 MeV, respectively. These values are one order of magnitude lower than those expected from the simulation. After the irradiation, the metal Ga targets were melted and shrunk to spherical shapes of 2~3 mm in the capsule. Probably, only a part of protons were irradiated to the metal ^{69}Ga targets. In the case of the GaN target, no physical change was observed after the irradiation. The efficiency was estimated to be 0.53 MBq/μA/h. This value is in good agreement with the simulated value. Thus, the production of the ^{68}Ge isotope was confirmed in both targets, while an improvement to avoid the change of the shape due to the irradiation is needed in the case of metal targets.

Figure 5 shows the magnetic field dependences of S parameters from a fused silica obtained using the ^{68}Ge and ^{22}Na sources. As compared to the case of ^{22}Na source, a larger asymmetry is observed for the case of ^{68}Ge isotope suggesting a higher spin polarization. Theoretical curves for several spin polarizations are also shown. From these results, the spin polarizations of positrons emitted from ^{68}Ge and ^{22}Na were estimated to be 50 to 70 % and 25 to 30 %, respectively. Although the theoretical maximum spin polarization is 94%, the average spin polarization of positrons emitting to 2π direction

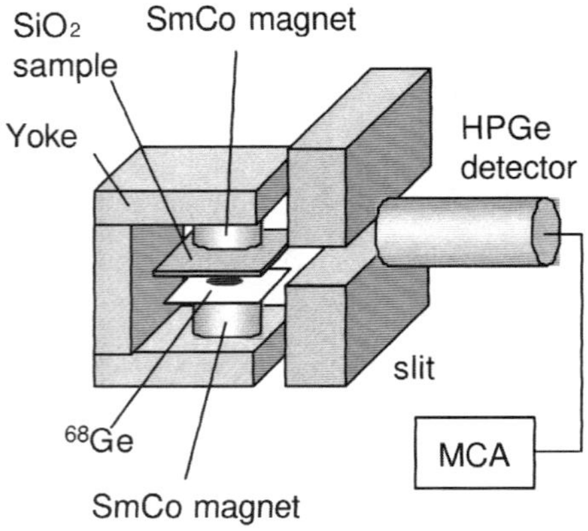

Figure 3. Schematics of the spin polarization measurement system.

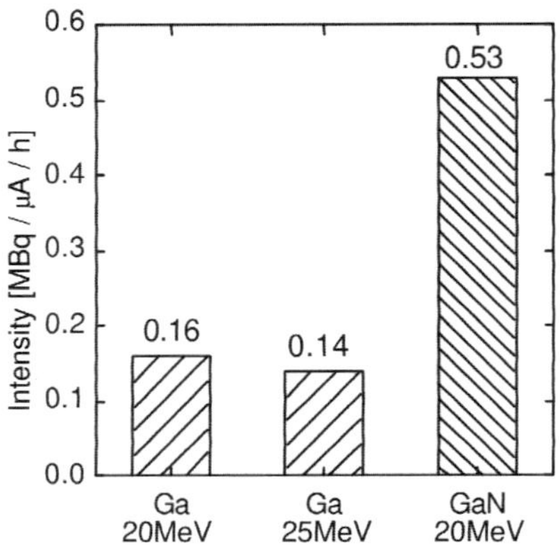

Figure 4. Amount of produced ^{68}Ge source with different targets and irradiation energies.

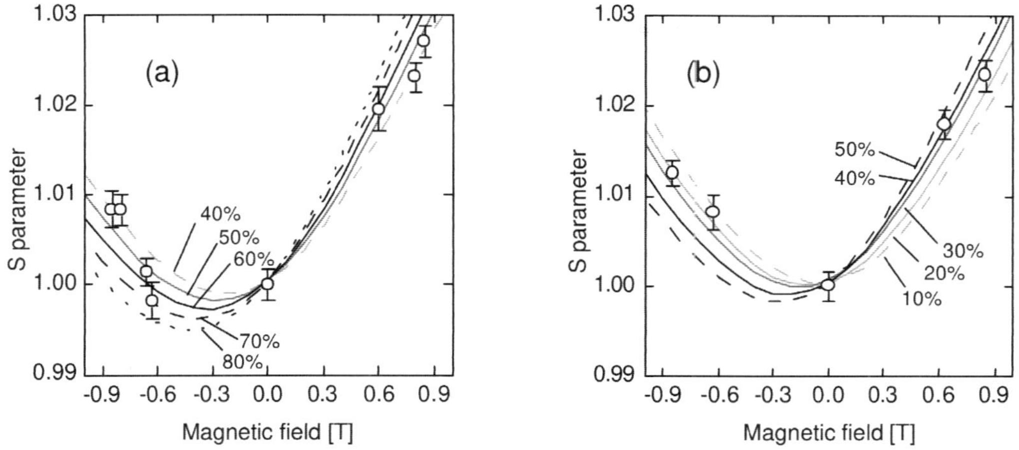

Figure 5. The magnetic field dependences of S parameters from a fused silica obtained using the (a)^{68}Ge and (b)^{22}Na sources. Theoretical curves for several spin polarizations are also shown.

become half of the theoretical value. In these measurements, the sample size was 20 mm square, the gap width of the pole piece was 3 mm and the diameter of the source was 5 mm. The open angle α was limited within 75 degree. From this geometrical reason, the measured spin polarization might have been decreased to 63 % of the maximum theoretical value $((1+\cos(75^{o})/2))$.

4. Summary

By the proton irradiation to the metal Ga and the GaN targets, production of ^{68}Ge was confirmed. Spin polarization of positrons emitted from the ^{68}Ge isotopes was found to be higher than that from ^{22}Na. The amount of ^{68}Ge isotope can be increased by multiple irradiations. For instance, ten successive proton irradiations with the proton current of 10 μA and irradiation duration of 10 hours could yield approximately 500 MBq activity which is a feasible intensity for generating a spin-polarized slow positron beam.

References

[1] Berko S 1967 Positron Annihilation in Ferromagnetic Solids, in: Stewart A T and Roellig L O (Eds.) *Positron Annihilation*, Academic Press pp. 61-80.
[2] Kawasuso A, Fukaya Y, Maekawa M and Yabuuchi A 2010 *J. Phys. Conf. Series* **225** 012028.
[3] Tanaka S, Fukuda M and Nishimura K 1997 *JAERI-Data-Code* **97-019** 1-91.
[4] Nagai Y, Nagashima Y, Kim J, Itoh Y and Hyodo T 2000 *Nucl. Instr. Methods* **B171** 199-203.
[5] Nagashima Y and Hyodo T 1990 *Phys. Rev.* **B 41** 3937-3941.

Study of interfaces in organic semiconductor heterojunctions

P Maheshwari[1], D Dutta[1], K Sudarshan[1], S K Sharma[1], S Samanta[2], A Singh[2], D K Aswal[2] and P K Pujari[1]

[1]Radiochemistry Division, [2]Technical Physics Division, Bhabha Atomic Research Centre, Mumbai 400 085, India.

E-mail: pujari@barc.gov.in

Abstract. The defect structure at the organic heterojunctions is studied using slow positron beam. The structural and electronic properties of heterojunctions are of technological and fundamental importance for understanding and optimization of electronic processes in organic devices. Interface trap centres play a significant role in the electrical conduction through the junctions. Depth dependent Doppler broadened annihilation measurements have been carried out in p- and n-type organic semiconductor thin films (30-80 nm) both single as well as multilayers grown on quartz substrate. The objective of the present study is to investigate the defect structure and to understand the behavior of positrons at the charged organic interfaces. Our result shows the sensitivity of positrons to the interfacial disorders that may be a convoluted effect of the presence of defects as well as the influence of the charge dipole in multilayers.

1. Introduction

Organic electronics has attracted worldwide attention due to the advantages of low-cost, ease of processing, compatibility with a wide range of substrates, large area and the ease of tailoring the material characteristics. The potential devices comprise of organic transistors [1], organic photovoltaic cells [2,3], organic light emitting diodes [4] and organic memory devices [5]. Organic heterojunctions are intrinsic to organic devices. Considerable work is focused on the heterojunctions towards understanding the physics of device operation and performance [6-9]. Many aspects of device operation are quite well understood. However, there is considerable knowledge gap in the understanding of structural and electronic properties of interfaces between organic and inorganic components which are important for understanding and optimization of electronic processes in these devices. Engineering of the interfacial properties to improve device performance by utilizing various surface modification schemes is another area of current interest.

In organic devices, energy level alignment and interfacial dipole has been observed at the interfaces due to charge transfer at the heterojunctions [10, 11]. The dependence of molecular orientation on charge transfer and interfacial dipole has also been reported [12, 13]. The molecular orientation near the interfaces, mainly governed by molecule-molecule and molecule-substrate interaction can be crucial for the device efficiency [14, 15]. Phthalocyanines are a class of low molecular weight organic molecules exhibiting semiconductor properties. They represent one of the most promising candidates for ordered organic thin films as these systems possess advantageous attributes such as thermal and chemical stability, and, excellent film growth characteristics [16]. The electronic and optical properties of organic semiconductor (OSC) thin films strongly depend on their structure and morphology [17].

The fabrication of heterojuctions involves controlled growth of OSC thin films. However, fabrication of hetero-multilayered structure is a challenge due to large anisotropy and relatively weak nature of interaction forces between organic molecules [16].

Beam based positron annihilation spectroscopy has been found to be a very sensitive technique to probe open volume defects present at the surfaces and interfaces, not possible by any other conventional defect characterization techniques. The unique repulsive force between positron and ion cores of the material makes positron ideal for probing defects, voids and free volumes. The presence of open volume defects/ free volume manifests as increase in lifetime and narrowing of the momentum distribution. The latter is often described by a shape parameter, namely Doppler broadening *S*-parameter defined as the ratio of the number of counts in the central part of 511 keV gamma line to the total number of counts under the peak. The increase in *S*-parameter corresponds to an increase in number or size of the defect or void.

The morphology of OSC thin films as a function of temperature and the type of substrate as well as the thickness of the films has been reported by us earlier [18]. In the present work, microstructure of the organic-organic as well as organic-inorganic (film-substrate) interfaces of the heterojunctions has been studied. Doppler broadened *S*-parameter as a function of positron energy is seen to be sensitive to the interfacial microstructure. We have observed for the first time, sharp contrast in *S*-parameter at the interfaces in organic heterojunctions indicating the presence of considerable disorder contrary to the reported ordered superstructure obtained from X-ray measurements.

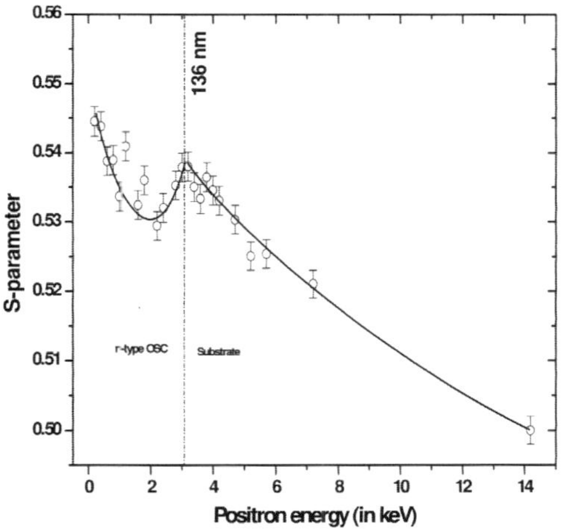

Figure 1. S-parameter for 80 nm n-type OSC film grown on quartz

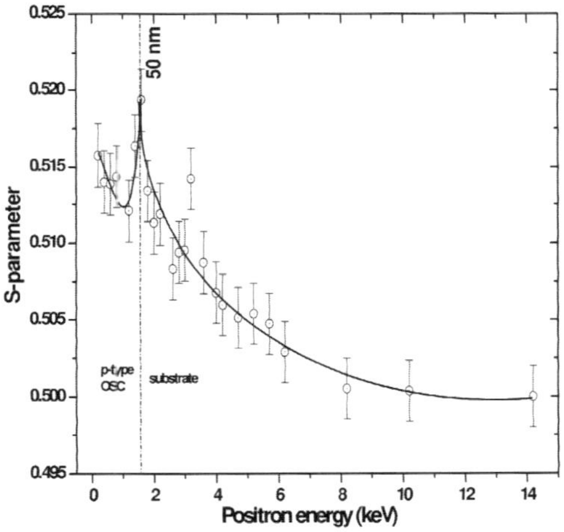

Figure 2. S-parameter for 80 nm p-type OSC film grown on quartz

2. Experiment

Single layer and multilayer structure of n-type (copper-hexadecafluoro-phthalocyanine; $F_{16}CuPc$) and p-type (cobalt-phthalocyanine; CoPc) OSC materials have been prepared on quartz substrate by thermal evaporation using Hind high Vac thermal evaporation system (model no-12A4T) under vacuum better than 10^{-6} mbar. The substrate was cleaned by sonication with trichloroethylene, acetone and methanol separately for 5 minutes each before the deposition. The films were deposited at the substrate temperature of 100°C and the deposition rate was 0.2 Å/sec. Single layer films of n and p-type material were 80 nm thick and the individual film thickness for p-n and n-p-n junction system was 60 nm and 30 nm, respectively. The thickness of the films was measured by quartz crystal monitor.

Depth dependent Doppler broadening measurements were carried out using slow positron beam at BARC in the energy range 200 eV-20 keV. A high efficiency high purity germanium detector with a resolution of 2 keV at 1332 keV photopeak of ^{60}Co was used for Doppler broadening measurements and spectra with 2×10^5 counts were acquired at each energy. The ratio of integral counts within ~ 2 keV energy windows centered at 511 keV and total photo peak area was used to evaluate the *S*-parameter.

3. Results and Discussion

The variation of *S*-parameter as a function of positron energy for the single layer (p and n-type OSC) and multilayer films (p-n and n-p-n junction) are shown in figures 1-4, with solid lines as eye-guides. In general, there is a decrease in *S*-parameter with beam energy followed by an increase at the interface (organic-organic/ organic-inorganic, as the case may be) and reaching the bulk value of quartz at higher energies. The most important observation is the enhancement of *S*-parameter at the interfaces.

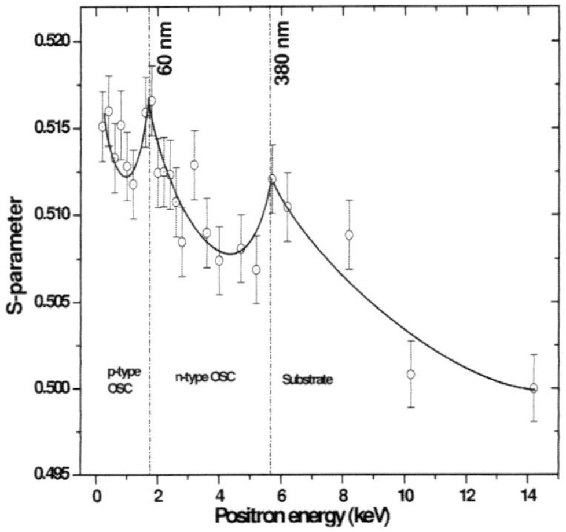

Figure 3. S-parameter for p-n Junction (each layer 60 nm)

Figure 1 and 2 shows the *S*-parameter variation as a function of depth for n-type and p-type OSC, respectively. The *S*-parameter variation for single layer n-type and p-type films of $F_{16}CuPc$ and CoPc on quartz substrate reflects an organic-inorganic (film-substrate) interface. The energy corresponding to the maximum of *S*-parameter value indicates the mean depth of the interface between the film and the quartz substrate as calculated from the range-energy relation given by, $\langle z \rangle = (40/\rho)*E^{\wedge}1.6$, where, ρ is the density of the material (1.7g/cm^3 for OSC and 2.32g/cm^3 for quartz) and E is the energy of positron in keV. The observed *S*-parameter profile indicates the sharp contrasting interfacial characteristics between the film and the substrate. The high value of *S*-parameter at the interface indicates the presence of open volume defects/ domains (trap centers) and reveals the sharp microstructure variation between organic and inorganic materials as indicated by the fall in *S*-parameter value thereafter. The epitaxial relation of film and the substrate is important not only in crystallographic sense but also for growth behavior. The organic molecules have large anisotropy and internal degrees of freedom due to low symmetry and extended size [16]. The effect of size difference of unit cells of organic molecules and the inorganic substrate as well as molecule-substrate interaction leads to translational and orientational domains, which are the sources of disorder [16]. The subsequent growth of the film is accomplished by the interplay of molecule-molecule and molecule-substrate interaction [19].

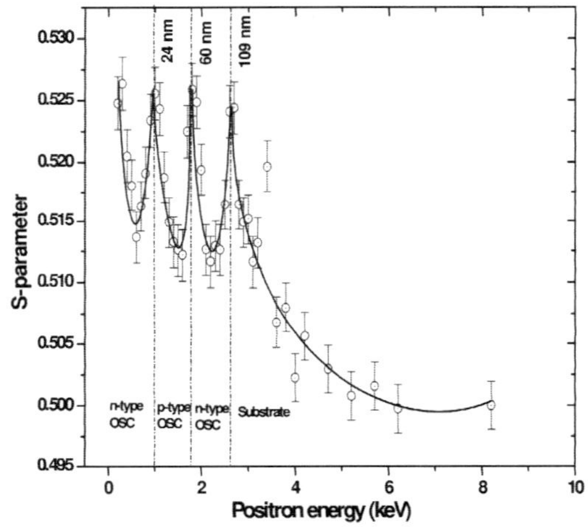

Figure 4. S-parameter for n-p-n junction (each layer 30 nm)

The organic-organic interfaces are distinctly clear in the case of multilayered structures; p-n and n-p-n heterojunctions. These interfaces experience weak van der Waals forces existing between the organic molecules. It has been reported in literature that these structures exhibit ordered arrangement of molecules across the heterojunction [20]. Also, energy level bending due to charge transfer at the heterojunction leads to the formation of charge accumulation layer and interfacial dipole [21]. However, the presence of defects/ open volumes/ domains of disorder are clearly revealed from our slow positron beam studies. Figure 3 shows *S*-parameter profile for p-n junction. This structure comprises of both an organic-organic (n and p-type OSC) and organic-inorganic (OSC-substrate) interface. A sharp contrast in *S*-parameter can be seen at the two interfaces corresponding to energies 1.8 keV and 5.7 kev respectively. Similar interfacial characteristics has also been observed for n-p-n junction represented by the sharp oscillatory behavior of *S*-parameter near to the interfaces between n-p and p-n type OSC materials (bulk film) as well as film-substrate as shown in figure 4. The increase in *S*-parameter value near the interface is the convoluted effect of defects arising due to the growth behavior of different type of organic molecules and the intrinsic electric field due to interfacial dipole at the interface. It is to be noted that for energies higher than 1.8 keV, the standard Makhovian profile for positron implantation shows wide (in comparison to film thickness) and smoothly varying form and should not lead to sharp interfaces (as reflected from oscillatory behavior of S-parameter). However, electric field at the interfaces due to charge accumulation might have modified the standard Makhovian profile. This must be taken into account to explain the sharp oscillations in S-parameter. The mean depths of these interfaces as calculated by positron implantation profile are observed to be shifted towards the larger depths. The shift in the depth appears to be due to the cumulative effect of increased diffusion of positrons on the substrate side, where diffusion length has been found to be high and leading to broadening of implantation profile. However, the role of electric field, especially in multilayers cannot be ruled out.

In conclusion, we have reported positron studies on the organic heterojunctions. A large degree of disorder or presence of open volume defects at the interfaces are seen which are believed to have ordered structure from X-ray studies. The observed contrast in the S-parameter at the interfaces makes the positron beam technique a potential characterization tool in organic electronics and, we believe, more detailed information on the heterojunction microstructure can be obtained through suitable modeling.

4. References

[1] Dodabalapur A, Katz H E, Torsi L and Haddon R C 1995 *Science* **269** 1560
[2] Tang C W and Vanslyke S A 1987 *App.l Phys. Lett.* **51** 913
[3] Peumans P and Forrest S R 2001 *Appl. Phys. Lett.* **79** 126
[4] Tang C W and Vanslyke S A 1986 *Appl. Phys. Lett.* **48** 153
[5] Ma L, Liu J. Pyo S and Yamg Y 2002 *App.l Phys. Lett.* **80** 362
[6] Akaike K, Kanai K, Ouchi Y and Seki K 2009 *Appl. Phys. Lett.* **95** 113306
[7] Tang J X, Zhou Y C, Liu Z T, Lee C S and Lee S T 2008 *Appl. Phys. Lett.* **93** 043512
[8] Liu Z T, Lo M F, Wang H B, Ng T W, Roy V A L, Lee C S and Lee S T 2009 *Appl. Phys. Lett.* **95** 093307
[9] Brumbach M, Placencia D and Armstrong N R 2008 *J. Phys. Chem. C* **112** 3142
[10] Haibo W, Liu Z, Ng T W, Lo M F, Lee C S and Lee S T 2010 *Appl .Phys. Lett.* **96** 173303
[11] Rajagopal A, Wu C I and Kahn A 1997 *J. Appl. Phys.* **83** 2649
[12] Chen W, Chen S, Huang H, Qi D C, Gao X Y and Wee A T S 2008 *Appl. Phys. Lett.* 92 063308
[13] Chen W, Chen S, Huang Y L, Huang H, Qi D C, Gao X Y, Ma J and Wee A T S 2009 *J. Appl. Phys.* **106** 064910
[14] Kowarik S, Gerlach A, Sellner S, Schreiber F, Cavalcanti L and Konovalov O 2006 *Phys. Rev. Lett.* **96** 125504
[15] Thayer G E, Sadowski J T, zu Heringdorf F M, Sakurai T and Tromp R M 2005 *Phys. Rev. Lett.* **95** 256106
[16] Brutting W, *Physics of Organic semiconductors* 2008 WILEY-VCH Verlag GmbH & Co. KGaA
[17] Ribic P R and Bratina G 2007 *J. Vac. Sci. Technol. B* **25** 1152
[18] Maheshwari P, Dutta D, Samanta S, Singh A, Aswal D K, Reddy A V R and Pujari P K 2009 *Phys. Stat. Solidi C* **6** 2589
[19] Debnath A K, Samanta S, Singh A, Aswal D K, Gupta S K, Yakhmi J V, Deshpande S K, Poswal A K and Surgers C 2008 *Physica E* **41** 154
[20] Akimichi H, Inoshita T, Hotta S, Noge S and Sakaki H 1993 *Appl.Phys. Lett.* **63** 3158
[21] Chen W, Chen S, Huang H, Qi D C, Gao X Y and Wee A T S 2008 *Appl. Phys. Lett.* **92** 063308

Journal of Physics: Conference Series **262** (2011) 012037 doi:10.1088/1742-6596/262/1/012037

o-Ps cooling for antihydrogen production

S Mariazzi[1] **and R S Brusa**
Dipartimento di Fisica, Università di Trento and INFN, Gruppo collegato di Trento, Via Sommarive 14, I-38123 Povo, Trento, Italy

e-mail: mariazzi@science.unitn.it

Abstract. AEgIS (Antimatter Experiment: Gravity, Interferometry, Spectroscopy) experiment in progress at CERN has planned to produce antihydrogen ($\overline{H}$) by using the resonant charge exchange process between orthopositronium (*o*-Ps) and antiproton ($\overline{p}$). Its primary goal is the realization of a cold $\overline{H}$ beam for the first measurement of the Earth's gravitational acceleration on antimatter. A demanding condition for large *o*-Ps - $\overline{p}$ cross section is the cooling of *o*-Ps at temperatures lower than at least 160 K. We will present recent results of *o*-Ps formation and cooling at cryogenic temperatures in a novel silicon positron/Ps converter in which oxidized nano-channels perpendicular to the surface were produced. The possibility to tune the nano-channels dimension to optimize *o*-Ps kinetic energy emission and to avoid *o*-Ps quantum confinement is discussed.

1. Introduction

Many $\overline{H}$ atoms were produced by the ATHENA [1] and successively by ATRAP [2] collaborations, by trapping and mixing cooled e^+ and $\overline{p}$. However, to perform precise spectroscopy and antimatter gravity measurements, cold $\overline{H}$ must be produced. One possibility to obtain cold $\overline{H}$ is the use of the charge exchange reaction $\overline{p} + o - Ps^* \rightarrow \overline{H}^* + e^-$, where the star indicates Rydberg states [3]. This process was verified experimentally by ATRAP obtaining Ps by charge exchange with excited Cesium atoms and then $\overline{H}$ by the above reaction [4]. AEgIS experiment aims to realize a $\overline{H}$ beam by the charge exchange process and to measure the Earth's local gravitational acceleration by observing the vertical displacement of the beam after crossing a classical Moirè deflectometer [5]. In AEgIS's approach, cold *o*-Ps will be produced implanting e^+ with an energy E in an e^+-Ps porous converter. In porous molecular materials, a large fraction of implanted e^+ can form *o*-Ps that is emitted into the pores with energy of few eV [6]. In the pores *o*-Ps can be cooled by collisions with the walls and, in presence of interconnected porosities towards the surface, it may escape into the vacuum [6]. To obtain a large charge exchange cross section, *o*-Ps must be cooled at temperature T < 160 K [7]. Several experiments have shown that the *o*-Ps kinetic energy is reduced close to the thermal energy in the pores [see Ref. 8]. In the case of a sample held at T < 300 K the quantum confinement of *o*-Ps in the pores plays a fundamental role setting a lower limit to the kinetic energy reachable by *o*-Ps [9, 10, 11]. Thus the possibility to tune shape and pores dimensions becomes essential. In this work we review and discuss recent results on *o*-Ps formation, cooling at cryogenic T and emission into the

1 To whom any correspondence should be addressed.

vacuum from a novel e^+-Ps converter composed of a regular array of tunable oxidized nano-channels (n-c) in Si [12, 13].

2. Experimental

Regular n-c perpendicular to the surface were obtained by electrochemical etching in Si p-type (100) [13]. A sample with n-c between 5-8 nm in diameter was synthesized by etching Si with 10 mA/cm^2 for 15 minutes (sample #A) [12]. Slightly smaller pores (4-7 nm) were obtained with an etching of 13 minutes (sample #0) [13]. After etching and cleaning, the inner surfaces of the n-c were oxidized for 2 h at 100°C in air. Two samples produced as sample #0 were subjected to two (sample #2) and four cycles (sample #4) of etching for 1 min and a re-oxidation in air. With this procedure the diameter of n-c is enlarged without changing the density of the n-c: the diameter was 8-14 nm in #2 and 14-20 nm in #4. The *o*-Ps formation and out-diffusion through n-c was analyzed by 2γ-3γ annihilation ratio (3γ-PAS) measurements [13]. With these measurements the fraction $F_{3\gamma}(E)$ of implanted e^+ forming *o*-Ps, was extracted. The mean e^+ implantation depth z was related to E through the equation $z = (40/\rho)E^{1.6}$ (z in nanometers when the material density ρ , 1.9 g/cm^3 for the etched silicon layer, and E are expressed in g/cm^3 and keV, respectively [13]). *o*-Ps perpendicular kinetic energy $E_\perp$ was obtained by a TOF apparatus. The time of flight t was measured as the difference between the stop signal (given by the *o*-Ps self-annihilation in flight after being escaped into the vacuum) and the start signal (given by secondary electrons emitted at the e^+ time arrival at the target). The background, due to prompt e^+ annihilation γ scattered into the stop detector, was estimated by measuring a sample without *o*-Ps formation and subtracted from the as-recorded TOF spectra. The spectra were then corrected for the *o*-Ps finite lifetime and *o*-Ps permanence in front to the stop detector (for details see Ref. 12). The time t was related to $E_\perp$ by $E_\perp = m_o(z_0/t)^2$, where m_o is the e^+ rest mass.

3. Results and Discussion

The possibility to tune the size of n-c by electrochemical etching allows tackling the two most important problems related to *o*-Ps cooling at cryogenic T: the *o*-Ps quantum confinement in the pores and the optimization of time spent by *o*-Ps in the n-c to maximize the cooling process and minimizing the pick-off. If n-c are modeled as an infinitive potential well with a quadratic base of side a and infinite length along the direction perpendicular to the surface of the sample the *o*-Ps ground state can be evaluated by $E_g = (\hbar^2\pi^2)/(2m_0a^2)$ and the corresponding lower T limit is $T = 2/3k_BE_g$, (k_B is the Boltzmann constant and $\hbar$ the reduced Plank constant). Thus if *o*-Ps has to be cooled at a certain T a target with appropriate n-c dimension has to be synthesized to overcome the quantum confinement. The minimum reachable T is 60-180 K in #0, 45-116 K in #A, 15-45 K in #2 and 7-15 K in #4. As in our TOF apparatus we were limited at 150 K minimum T [12] we chose, for a more detailed analysis, sample A# respect with samples #2 and #4 in which the cooling time is expected to be longer because of the reduced frequency of interaction with the walls due to the channel dimensions [9, 14]. Being limited to cool *o*-Ps at T ≥ 150 K, an evaluation of the thermalization time can be obtained by the *o*-Ps-free atoms interaction model [14]. In fact the thermal wavelength of *o*-Ps at these temperatures is around 3-4 nm smaller than the n-c size of samples A# [8]. Otherwise a quantum approach would be necessary [8, 9]. The calculation of the *o*-Ps energy loss points out that a complete thermalization at 150 K occurs in 11-15 ns for n-c with size in the 5-8 nm range when an effective mass for the atoms of the pore surface M = 25 a.m.u. is assumed. Thus for optimizing the *o*-Ps cooling at 150 K in sample #A, *o*-Ps has to remain in the n-c a time > 15 ns but not longer in order to reduce the pick-off effect. In Fig.1 the fraction $F_{3\gamma}(E)$ vs. e^+ implantation energy E measured at 300 K in all samples is shown. The $F_{3\gamma}(E)$ curves were fitted with a model based on the diffusion equation [13] above 2 keV in samples #A, #0 and #2 and above 5 keV in sample #4. Below these energies the data are altered by the

Journal of Physics: Conference Series **262** (2011) 012037 doi:10.1088/1742-6596/262/1/012037

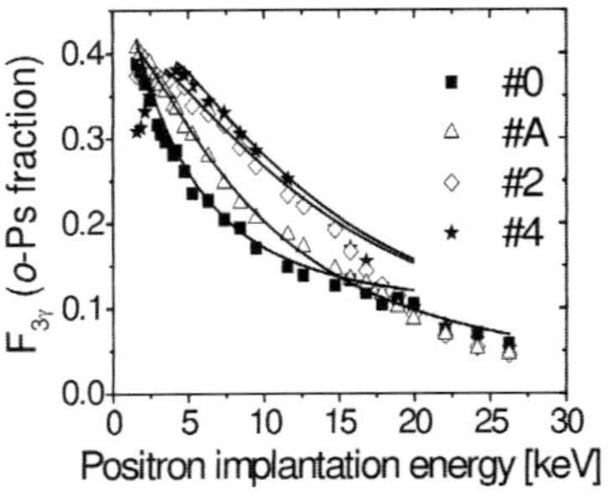

Figure 1. $F_{3\gamma}(E)$ as a function of e^+ implantation energy measured at 300 K. The continuous lines are best fits to the data using the *o*-Ps diffusion model of Ref. 13. Error bars are inside the symbols.

fraction of fast *o*-Ps not detected by HpGe detectors [13]. The *o*-Ps diffusion length results 300 ± 30 nm, 743 ± 15 nm, 1430 ± 140 nm and 1325 ± 125 nm in sample #0, #A, #2 and #4, respectively. The fitting gives also the *o*-Ps fraction annihilating via 3γ in the n-c and the fraction emitted into the vacuum after out-diffusion through the n-c. At 7 keV, in #0, around 7 % of implanted e^+ annihilate as *o*-Ps in the n-c while ~13 % reaches the vacuum as *o*-Ps. Always at 7 keV around 27 %, 31 % and 32 % of e^+ are emitted as *o*-Ps from sample #A, #2 and #4, respectively [12, 13]. In these samples the amount of *o*-Ps annihilating via 3γ in the n-c decreases to < 1%. The value of the diffusion coefficient in each sample can be estimated as $D_{o-Ps} = L^2_{o-Ps} / \tau$ where τ is the mean *o*-Ps lifetime in the considered open volumes that can be calculated by the rectangular extension of the Tao-Eldrup (RTE) model [6]. D_{o-Ps} values of 0.012±0.002, 0.069±0.003, 0.18±0.02, 0.13±0.02 cm^2/s are found for the samples #0, #A, #2, #4, respectively [12, 13].

In Fig.2 the mean *o*-Ps time of permanence in the n-c calculated from $\bar{t} = \bar{z}^2 /(2D_{o-Ps})$ [10] is shown. As previously calculated a complete thermalization at T = 150 K would be occurred in #A in

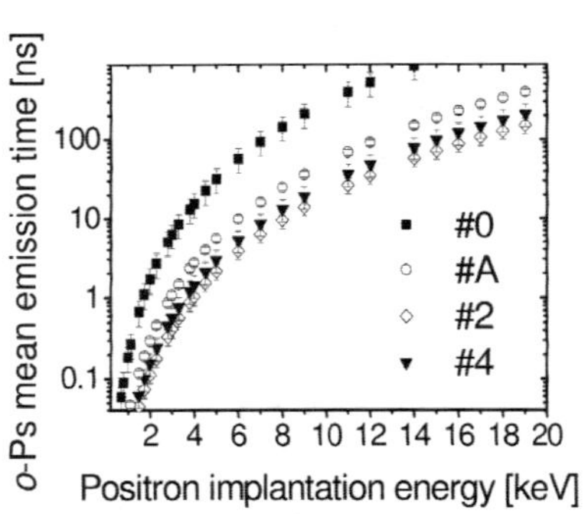

Figure 2. Mean *o*-Ps time of permanence in the n-c vs. e^+ implantation energy calculated according to Ref. [10].

<15 ns. If we compare these mean emission times with the estimated thermalization time at 150 K, we can observe that the *o*-Ps mean emission time in #A is shorter than the threshold of 11-15 ns up to e^+ implantation energy of around 7 keV where it reaches 16 ± 1 ns. Therefore, it is expected to have a fraction of thermalized *o*-Ps by implanting e^+ around this energy. It is interesting to note that the $\bar{t}$ at 7 keV in #0 becomes 90 ± 30 ns as a consequence of the lower D_{o-Ps} while in samples #2 and #4, where D_{o-Ps} is larger, it drops to 6 ± 1 ns and 9 ± 2 ns, respectively. The TOF spectrum measured at 150 K for sample A# is shown in the inset of Fig.3. In Fig.3 *o*-Ps energy spectrum obtained by multiplying the TOF spectrum by t^3 is shown in a semi-log scale. In this scale, the energy spectrum is well fitted below 0.25 eV by two lines corresponding to two Maxwellian beams at two different temperatures. The line with higher slope at lower energy is due to the distribution of e^+ thermalized at the sample T of 150 K. The presence of a second warmer distribution with a T of 1260 ± 15 K could be explained

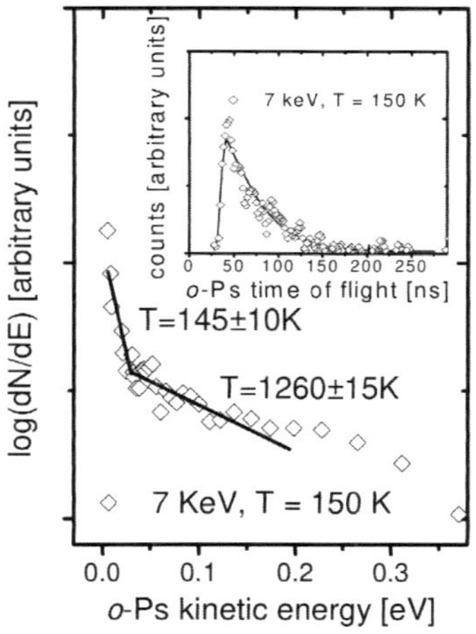

Figure 3. *o*-Ps energy spectrum obtained by multiplying by t^3 the data of the time of flight spectrum measured in sample #A (inset). The two-exponential fits points out the presence of two different Maxwellian beam distributions at the indicated T. The continuous line in the inset is a fit with an asymmetrical double sigmoidal function to guide the eye.

by the fact that in these n-c with high *o*-Ps diffusivity, *o*-Ps formed at a depth corresponding to a mean permanence time shorter than the estimated thermalization time of 11-15 ns does not thermalize before to be emitted into the vacuum [12]. The *o*-Ps fraction thermalized at 150 K can be calculated by integration of the tail area of the time of flight spectrum, i.e. from $t_M = z_0 / v_{th}$ to the end of the distribution where v_{th} is the *o*-Ps velocity corresponding to the sample T. This fraction is ~9 % of emitted *o*-Ps. The same result can be obtained by integration of the Maxwellian distribution corresponding to 150 K in the energy spectrum.

4. Conclusion

It was shown that a) high yield of *o*-Ps can be obtained from a oxidized n-c produced in a silicon matrix; b) cooling at T as lower as 150 K is possible thanks the tuning of the n-c size that allows to overcome the quantum confining of *o*-Ps. With the present apparatuses able to produce e^+ bunches of $2\text{x}10^8$-10^9 e^+, a beam of about $5\text{x}10^6$-10^7 *o*-Ps cooled atoms at 150 K could be obtained.

[1] Amoretti M, et al., (ATHENA Collaboration) 2002 *Nature* **419** 456
[2] Gabrielse G, et al., (ATRAP Collaboration) 2004 *Phys. Rev. Lett.* **93** 073401
[3] Humberston J W, Charlton M, Jacobsen F M and Deutch B I 1987 *J. Phys. B* **20** L25
[4] Storry C H, et al., (ATRAP Collaboration) 2004 *Phys. Rev. Lett.* **93** 263401
[5] Kellerbauer A, et al., (AEgIS Proto-Collaboration), 2008 *Nucl. Instrum. Methods Phys. Res. B* **266** 351
[6] Gidley D W, et al. in *New Directions in Antimatter Chemistry and Physics*, edited by Surko C M and Gianturco F A (Kluwer Academic, Dordrecht, 2001), p. 151.
[7] Proposal for AEgIS experiment at the CERN, SPSC-P-334-2007-017
[8] Brusa R S, and Dupasquier A, in *Physics with many positrons*, International School of Physics Enrico Fermi, Course CLXXIV edited by R.S. Brusa, A. Dupasquier, A.P. Mills
[9] Mariazzi S, Salemi A and Brusa R S 2008 *Phys. Rev. B* **78** 085428
[10] Cassidy D B, et al., 2010 *Phys. Rev. A* **81** 012715
[11] Crivelli P, et al., 2010 *Phys. Rev. A* **81** 052703
[12] Mariazzi S, Bettotti P and Brusa R S 2010 *Phys. Rev. Lett.* **104** 243401
[13] Mariazzi S, Bettotti P, S Larcheri, L Toniutti and Brusa R S 2010 *Phys .Rev. B* . **81**, 235418
[14] Nagashima Y, et al. 1995 *Phys. Rev. A* **52** 258

Journal of Physics: Conference Series **262** (2011) 012038 doi:10.1088/1742-6596/262/1/012038

Characterization of point defects in yttria stabilized zirconia single crystals

O Melikhova[1], J Cizek[1], J Kuriplach[1], I Prochazka[1], W Anwand[2], G Brauer[2] and D Grambole[3]

[1] Charles University in Prague, Faculty of Mathematics and Physics, V Holesovickach 2, CZ-18000 Prague 8, Czech Republic
[2] Institut für Strahlenphysik, Forschungszentrum Dresden-Rossendorf, PO Box 510119, D-01314 Germany
[3] Institut für Ionenstrahlphysik und Materialforschung, Forschungszentrum Dresden-Rossendorf, PO Box 510119, D-01314 Germany

E-mail: oksivmel@yahoo.com

Abstract. Characterization of point defects in a fully stabilized ZrO_2 + 9 mol.% Y_2O_3 single crystal with cubic structure was performed in this work. It was found that the crystal contains a high density of vacancy-like defects characterized by a lifetime of 175 ps. First principle theoretical calculations showed that this lifetime is comparable with lifetime of positrons trapped in zirconium vacancies associated with hydrogen. In particular, in the vicinity of the zirconium vacancy hydrogen forms an O-H bond with one of the nearest neighbour oxygen atoms. The calculated bond length is close to 1 Å. Using nuclear reaction analysis it was found that the hydrogen concentration in the crystal is 0.3 at.%. This amount of hydrogen is sufficient to form zirconium vacancy – hydrogen complexes capable of saturated positron trapping.

1. Introduction

Zirconia (ZrO_2) exhibits a high thermal stability and a low thermal conductivity together with good mechanical properties [1]. At temperatures above 1380 °C zirconia exists in a high temperature cubic phase, which transforms into a tetragonal one and subsequently into a monoclinic phase at lower temperatures. The high temperature cubic phase can be stabilized down to room temperature by addition more than 8 mol % of trivalent yttria (Y_2O_3) [2]. Such a system is then called yttria stabilized zirconia (YSZ). An embedment of Y_2O_3 into the ZrO_2 host lattice leads to a deviation from stoichiometry resulting in the creation of a high amount of native oxygen vacancies (V_O). Hence, vacancies play certainly a very important role in YSZ.

The purpose of this work is the characterization of vacancies in a high quality YSZ single crystal employing positron lifetime (LT) measurements and slow positron implantation spectroscopy (SPIS). In order to facilitate interpretation of experimental data, we performed also theoretical modeling of defects in YSZ by state-of-the-art *ab initio* electronic structure calculations including ionic relaxations.

2. Experimental

ZrO_2 + 9 mol.% Y_2O_3 single crystal with (100) orientation and dimensions 10 × 10 × 0.5 mm^3 was grown by skull melting method and supplied by Crystec GmbH, Berlin. Surface of the crystal was

Journal of Physics: Conference Series **262** (2011) 012038 doi:10.1088/1742-6596/262/1/012038

optically polished. A digital LT spectrometer [3] with an excellent time resolution of 145 ps (FWHM ^{22}Na) was employed for LT measurements. SPIS investigations were performed on the magnetically guided variable energy slow positron beam SPONSOR [4]. The energy of incident positrons can be varied in the range from 0.03 to 36 keV. Doppler broadening of annihilation profile was analyzed using the *S* parameter. The hydrogen concentration in the crystal was determined by nuclear reaction analysis (NRA) using the resonant nuclear reaction $^{15}N + {}^{1}H \rightarrow {}^{12}C + {}^{4}He + \gamma$ rays. The hydrogen concentration was determined at a sample depth of ~ 100 nm using ^{15}N ions accelerated to 6.64 MeV.

positron state	τ (ps)	E_B (eV)
Bulk	151	-
V_O	153	0.05
V_{Zr}	287	2.33
V_{Zr} + H	175	1.83
V_{Zr} + 2 H	172	0.88
V_{Zr} + 4 H	155	0.08
V_{Zr} + 6 H	150	0.00

Table 1. Results of *ab-initio* theoretical calculations of positron parameters: bulk positron lifetime for ZrO_2 crystal, positron lifetimes τ and binding energies E_B for various defects including O-vacancy (V_O), Zr-vacancy (V_{Zr}) and Zr vacancy associated with various number of hydrogen atoms (V_{Zr}+ *n*H).

3. Theoretical calculations

Ab initio calculations of atomic relaxations around defects were determined using the Vienna *ab initio* simulation package (VASP) [5] employing a generalized gradient approximation (GGA) projector augmented-wave (PAW) pseudopotential. *4s*, *4p*, *5s* and *4d* Zr electrons and *2s* and *2p* O and *1s* H electrons were included in the self-consistent calculations of electronic structure and ion relaxations. Defects were modeled in the cubic ZrO_2 (fluorite) lattice with the lattice parameter a = 5.08 Å [6] using 96 atom-based supercells employing 2×2×2 *k*-mesh sampling centered at the Γ point.

Positron lifetimes and positron binding energies to defects were calculated in the limit of vanishing positron density [7]. The self-consistent valence electron density calculated by VASP with added frozen core electrons was used in positron calculations. Hence charge transfer is *included* in the calculations of positron parameters. Since in our previous calculations [8] we found that the GGA scheme for electron-positron correlations formulated by Barbiellini et al. [9] when used with self-consistent electron density gives positron lifetimes which are in the best agreement with experiment, we used this scheme in all calculations in this paper.

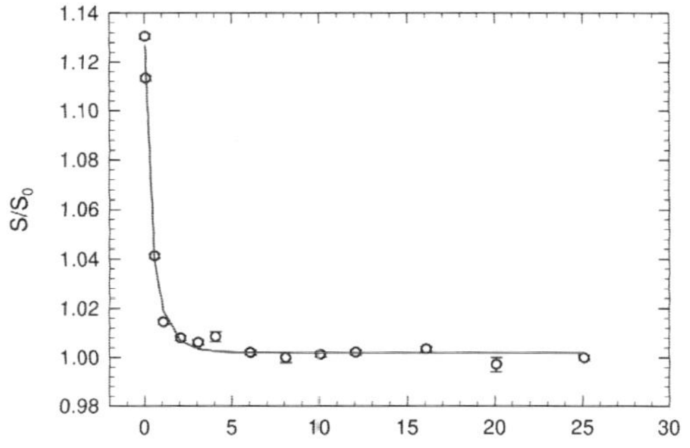

Figure 1. Dependence of the *S* parameter (open circles) on the positron energy the *E* for the ZrO_2 + 9 mol.%Y_2O_3 single crystal. The *S* parameter was normalized to the bulk value S_0 measured at *E* = 25 keV. The solid line shows a model curve fitted by the VEPFIT [10].

4. Results and discussion

LT measurements revealed that the YSZ crystal exhibits a single component spectrum (apart from the source component) with a lifetime of (175 ± 1) ps, which is remarkably longer than the calculated bulk lifetime in zirconia (see table 1). This testifies that YSZ crystal contains a high density of defects which cause saturated positron trapping. This is also supported also by SPIS results: the dependence of the *S* parameter on the positron energy is plotted in figure 1. The solid line in the figure is a model function calculated by the VEPFIT software [10] assuming a single layer model. A very short positron diffusion length $L_+ = (8.6 \pm 0.4)$ nm obtained from fitting gives a strong evidence that almost all positrons are trapped at defects. The concentration of defects can be estimated from SPIS results using the relation

$$c = \frac{1}{\nu \tau_B}\left(\frac{L_{B,+}^2}{L_+^2} - 1\right), \tag{1}$$

where τ_B and $L_{B,+}$ denote, respectively, the bulk positron lifetime and diffusion length in a perfect ZrO_2 lattice, while L_+ is the mean positron diffusion length measured in the sample. Here we used the calculated zirconia bulk lifetime τ_B = 151 ps (see table 1) and $L_{B,+} \approx 150$ nm, which is a typical positron diffusion length in similar materials [7]. Using the specific trapping rate $\nu \approx 10^{14}$ at. s^{-1}, which is a typical value for monovacancies in most solids [7], equation (1) gives the defect concentration in YSZ crystal $c \approx 2$ at.%. Despite of uncertainties of input quantities, it is clear that the YSZ crystal studied exhibits a very high concentration of defects comparable in order of magnitude with the concentration of substitutional Y atoms (6.2 at.%).

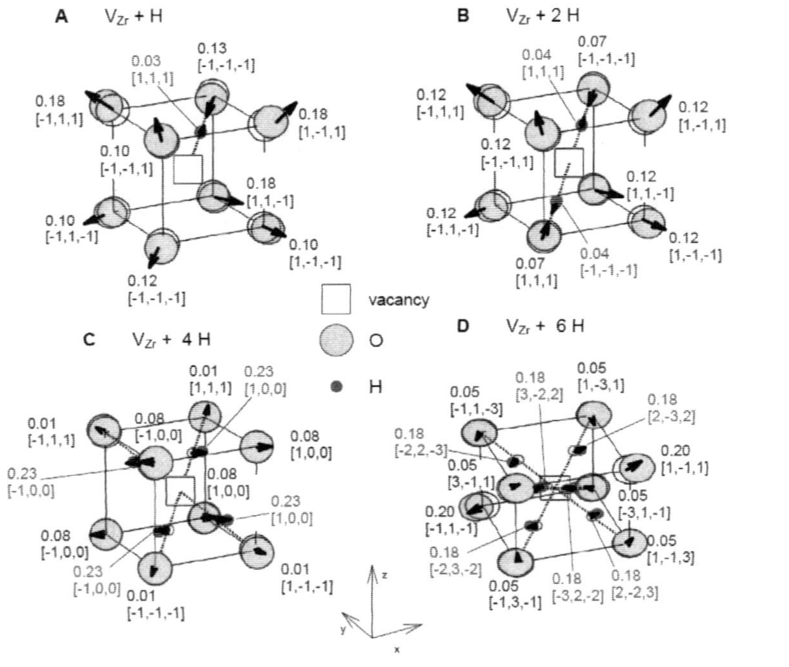

Figure 2. Relaxed geometries for V_{Zr} associated with various number of hydrogen atoms (V_{Zr} + *n*H). Relaxed positions of the nearest neighbour O atoms and trapped H atoms are shown by filled symbols, while open symbols show positions in the rigid lattice. The values in the figure are displacement magnitudes (in Å) and directions. Displacements of H atoms are given with respect to the site in the middle of a line connecting the vacancy and the nearest neighbour O atom.

Table 1 shows calculated positron lifetimes for a perfect ZrO_2 crystal and various defect configurations. Our previous calculations [8,11] showed that V_O and also its complexes with Y atoms are too shallow traps unable to confine positrons. Hence positron traps detected in the YSZ crystal should be defects containing the zirconium vacancy (V_{Zr}), which is a deep positron trap [8,11].

However, one can see in the table 1 that calculated lifetime of positrons trapped in V_{Zr} is longer than 175 ps determined in experiment. Note that Damonte et al. [12] found in ZrO_2 + 8 mol.% Y_2O_3 crystal a component with lifetime of 196 ps, i.e. again shorter lifetime than that calculated for V_{Zr}. A shortening of the positron lifetime may be caused by some impurity associated with V_{Zr} which reduces its free volume. When considering possible impurities, hydrogen is the most promising candidate. First, the presence of hydrogen can hardly be avoided in any method of crystal growth. Second, hydrogen is usually highly mobile in materials and can be relatively easily incorporated into the ZrO_2 structure due its tendency to bind with oxygen.

Indeed, nuclear reaction analysis (NRA) shows that the hydrogen concentration in the studied single crystal is (0.33 ± 0.03) at.% and hydrogen is in a 'tightly bound' state in the lattice. Although the hydrogen concentration determined by NRA is lower than the estimated concentration of V_{Zr}, it is still high enough to cause saturated positron trapping in $V_{Zr}+n$H complexes.

Figure 2 shows relaxed geometries of V_{Zr} complexes with hydrogen considered in this work. In general our calculations indicate that hydrogen is trapped in V_{Zr} and forms an O-H pair with one of the nearest neighbor O atoms. In case of multiple H atoms trapped in V_{Zr}, multiple O-H pairs are formed. Trapped H atoms are always displaced towards one of the nearest neighbor O anion, which relaxes slightly toward hydrogen. In V_{Zr} + 1H and V_{Zr} + 2H complexes H relaxes outward in the <111> direction, while O relaxes inward. The O anions which did not form O-H pairs relax outward. If more H atoms are trapped in V_{Zr}, then their mutual repulsion leads to displacements in a direction other than <111>. However, one can always observe the coupling of H and O atoms and formation of an O-H pair with bond length close to 1 Å.

Hydrogen associated with V_{Zr} causes a reduction of free volume and an increase of local electron density in vacancy. This is reflected by a remarkable shortening of the positron lifetime which can be seen in table 1. Note that V_{Zr} associated with more than 4 H atoms is not capable of positron trapping anymore. For this reason we stopped our calculations at V_{Zr} + 6H complex. The calculated lifetime of positrons trapped in V_{Zr} + 1H and V_{Zr} + 2H agree well with the lifetime measured on the YSZ crystal. Hence, our calculations suggest that YSZ crystal contains V_{Zr} associated with hydrogen.

5. Conclusions

Characterization of point defects in ZrO_2 + 9 mol.% Y_2O_3 single crystal was performed in this work. The crystal contains high density of defects related to V_{Zr}. First principles calculations indicate that V_{Zr} may be associated with hydrogen introduced into the sample unintentionally during the growing process. Nuclear reaction analysis revealed that the crystal indeed contains noticeable amount of hydrogen. Hence, hydrogen should be seriously considered as an important impurity in YSZ crystals.

Acknowledgments

This work was supported by the Academy of Science of the Czech Republic (project KJB101120906) and the Ministry of Schools, Youths and Sports of the Czech Republic (projects MSM 0021620834 and COST OC 165).

References

[1] Ryshkewitz E and Richardson D W 1985 *Oxide Ceramics: Physical Chemistry and Technology* (Hasskell, General Ceramics)
[2] Goff J P, Hayes W, Hull S, Hutchings M T and Clausen K N 1999 *Phys. Rev. B* **59** 14202
[3] Becvar F, Cizek J, Prochazka I 2008 *Appl. Surf. Sci.* **255** 111
[4] Anwand W, Kissener H-R and Brauer G 1995 *Acta Phys. Polon. A* **88** 7
[5] Kresse G and Hafner J 1994 *Phys. Rev. B* **49** 14251
[6] Howard C J, Hill R J and Reichert B E 1988 *Acta Crystallogr. Sect. B: Struct. Sci.* **44** 116
[7] Puska M J and Nieminen R M 1994 *Rev. Mod. Phys.* **66** 841
[8] Melikhova O, Kuriplach J, Cizek J, Prochazka I, Brauer G, Anwand W 2010 *J. Phys.: Conf. Series* **225** 012035
[9] Barbiellini B, Puska M J, Korhonen T, Harju A, Torsti T and Nieminen R M 1996 *Phys. Rev. B* **53** 16201
[10] van Veen A, Schut H, Clement M, de Nijs J, Kruseman A and Ijpma M 1995 *Appl. Surf. Sci.* **85** 216
[11] Cizek J, Melikhova O, Prochazka I, Kuriplach J, Kuzel R, Brauer G, Anwand W, Konstantinova T E and Danilenko I A 2010 *Phys. Rev. B* **81** 024116
[12] Damonte LC, Taylor MA, Desimoni J, Runco J 2007 *Rad. Phys. Chem.* **76** 248

Journal of Physics: Conference Series **262** (2011) 012039 doi:10.1088/1742-6596/262/1/012039

Quenched-in vacancies in Fe-Al alloys

O Melikhova[1], J Cizek[1], F Lukac[1], I Prochazka[1], J Kuriplach[1], W Anwand[2] and G Brauer[2]

[1] Charles University in Prague, Faculty of Mathematics and Physics, V Holesovickach 2, 180 00 Prague 8, Czech Republic
[2] Institut für Strahlenphysik, Forschungszentrum Dresden-Rossendorf, Postfach 510119, D-01314, Dresden, Germany

E-mail: oksivmel@yahoo.com

Abstract. Quenched-in vacancies in Fe-Al alloys with Al content ranging from 24 to 49 at.% were investigated employing two complementary techniques of positron annihilation: slow positron implantation spectroscopy and positron lifetime measurements. It was found that quenched alloys exhibit a very high concentration of vacancies. Although the free positron component cannot be resolved in positron lifetime spectrum in majority of samples, the concentration of quenched-in vacancies can be still determined from the positron diffusion length measured by a variable energy positron beam. The lowest concentration of vacancies was found in a stoichiometric (SM) Fe_3Al alloy. The concentration of defects increases with increasing degree of non-stoichiometry with respect to Fe_3Al, i.e. in alloys with under-SM and over-SM Al concentration. However, the increase in concentration of quenched-in defects is more pronounced in Al-rich alloys, i.e. alloys containing more than 25 at.% of Al.

1. Introduction

Intermetallic Fe-Al alloys are prospective structure materials for high-temperature applications. A special feature of Fe-Al alloys is an extremely high concentration of vacancies (up to several at.%) which can be created at elevated temperatures and relatively easily quenched down to room temperature [1]. Vacancies play a very important role in Fe-Al alloys and cause, for example, a strong hardening [2]. Thus, it is obvious that a detailed investigation of vacancies and vacancy-related defects in Fe-Al alloys is necessary for understanding of physical properties of these materials. Positron lifetime (LT) spectroscopy has been employed [3,4] for investigations of vacancies in Fe-Al alloys with Al content ranging from 25 at.% (Fe_3Al alloy with $D0_3$ structure) to 50 at.% (FeAl alloy with B2 structure). The lifetime of positrons trapped at defects varied in the range from 175 to 200 ps depending on Al content and thermal treatment [3,4]. This indicates that various configurations of vacancy-like defects are existing in Fe-Al alloys. There are still controversies among various authors in assignment of positron lifetimes to particular configurations of such defects.

The aim of this work is to characterise quenched-in defects in Fe-Al alloys with Al content (C(Al)) varying in the range from 24 to 49 at.%. Two complementary techniques of positron annihilation were employed: (i) LT spectroscopy which enables to identify defects and determine their concentrations and (ii) slow positron implantation spectroscopy (SPIS) which allows for an estimation of defect concentration using the positron diffusion length even in materials with very high defect densities.

Journal of Physics: Conference Series **262** (2011) 012039 doi:10.1088/1742-6596/262/1/012039

2. Experimental

In the present work we studied Fe-Al alloys with *C*(Al) = 24, 25, 28, 40, and 49 at.% prepared from pure Fe (99.99%) and Al (99.999%) by induction melting. The specimens were annealed at 1000°C for 1h in vacuum (10^{-3} mbar) encapsulated in silica glass ampoules. The annealing treatment was finished by quenching of the silica glass ampoules into water of room temperature.

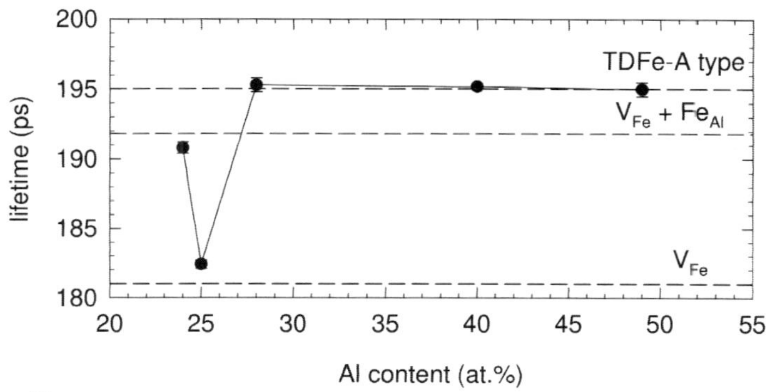

Figure 1. Lifetimes of positrons trapped at quenched-in defects in alloys with various Al content. Dashed lines shows theoretically calculated positron lifetimes for various defects, see text.

A $^{22}Na_2CO_3$ positron source with an activity of 1.2 MBq deposited on a 2 μm thick Mylar foil was used for LT studies. The source contribution consists of two weak components with lifetimes $\tau_2 \approx 368$ ps and $\tau_3 \approx$ 1.5 ns. A high resolution digital spectrometer [5] with a time resolution of 150 ps (FWHM, ^{22}Na) was employed. At least 10^7 annihilation events were accumulated in all LT spectra.

SPIS studies were performed on magnetically guided variable energy positron beam "SPONSOR" [6]. The energy of incident positrons was varied in the 0.03 to 36 keV range. The Doppler broadening of the annihilation line was measured by an HPGe detector with an energy resolution of (1.09 ± 0.01) keV at 511 keV and evaluated in terms of the *S* parameter. The *S* parameter values in this paper were normalized to the bulk *S* parameter for the Fe_3Al sample measured at a positron energy of 36 keV.

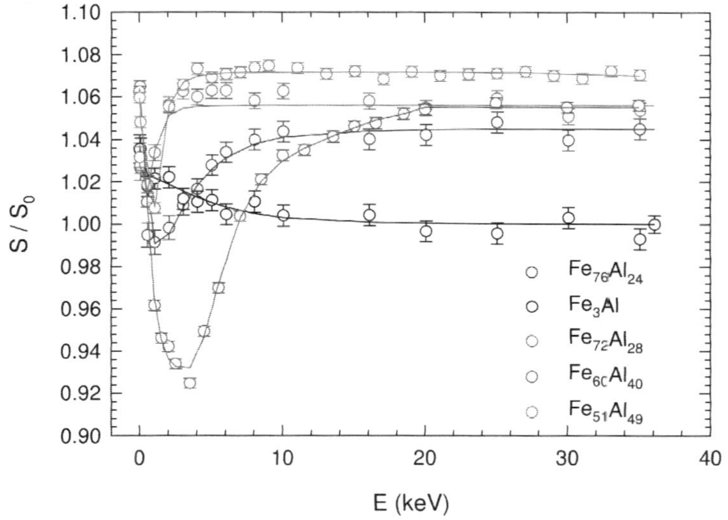

Figure 2. Dependence of the *S* parameter on the energy *E* of incident positrons. *S* parameter values are normalized to the Fe_3Al bulk value S_0 measured at *E* = 36 keV. Solid lines show fit by the VEPFIT software.

3. Results and discussion

The SM Fe_3Al alloy exhibits a two component LT spectrum. The shorter component with the lifetime τ_1 = (27 ± 9) ps and intensity I_1 = (10.5 ± 0.6) % represents a contribution of free positrons. The longer component with the lifetime τ_2 = (182.4 ± 0.3) ps and intensity I_2 = (89.5 ± 0.8) % can be attributed to positrons trapped at quenched-in defects. The lifetime τ_2 agrees well with the calculated lifetime of positrons trapped in Fe-vacancy (V_{Fe}) surrounded by 4 Fe and 4 Al nearest neighbours [7]. The quantity $\tau_f = (I_1 / \tau_1 + I_2 / \tau_2)^{-1} = 114$ ps agrees well with the bulk positron lifetime of 112 ps measured on annealed $Fe_{76.3}Al_{23.7}$ [8]. This testifies that the Fe_3Al sample contains a single type of uniformly distributed traps and their concentration can be calculated using the two state simple trapping model

$$c = \frac{1}{\nu}\frac{I_2}{I_1}\left(\frac{1}{\tau_f} - \frac{1}{\tau_2}\right). \qquad (1)$$

Here $\nu = 4 \times 10^{14}$ at. s^{-1} is the specific positron trapping rate for V_{Fe} [8]. From equation (1) one obtains that the concentration of V_{Fe} in Fe_3Al sample is $c = (5 \pm 1) \times 10^{-3}$ at.%.

All non-SM alloys exhibit single component LT spectra with the lifetime in the range 191–195 ps which obviously represents a contribution of positrons trapped at quenched-in defects. Hence, non-SM alloys have a very high density of defects ($c > 2 \times 10^{-2}$ at.% [7]), which causes saturated positron trapping, i.e. the free positron component is too short and small to be resolved. Figure 1 shows the lifetime of positrons trapped at defects plotted as a function of C(Al). In the $Fe_{76}Al_{24}$ alloy, a lifetime of 191 ps is detected, which agrees with the calculated lifetime of positrons trapped in V_{Fe} associated with an Fe antisite atom (V_{Fe}+Fe_{Al}). On the other hand, Al-rich alloys (C(Al) > 25 at.%) are characterized by a longer lifetime of 195 ps which is close to the calculated lifetime of a triple defect consisting of two V_{Fe} aligned in the [100] direction and connected by an Fe_{Al} atom (TDFe-A type) [3].

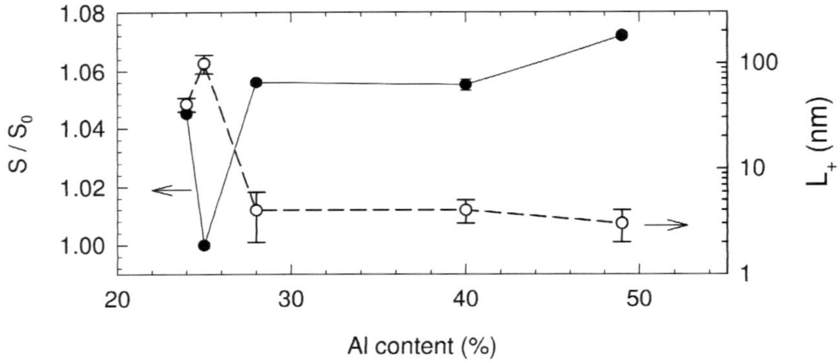

Figure 3. Bulk parameters obtained from fitting of $S(E)$ curves measured on Fe-Al alloys with various Al content: S parameter (full circles), positron diffusion length L_+ (open circles).

The dependence of the S parameter on the incident positron energy is plotted in figure 2. At very low energies $E \approx 30$ eV almost all positrons annihilate on the surface. A drop of S at positron energies $E < 10$ keV is due to a contribution of positrons annihilating in an oxide layer with a thickness of 10-40 nm formed on specimens during annealing. Almost no influence of the oxide layer on the $S(E)$ curve was found in the Fe_3Al alloy. Solid lines in figure 2 show results of fitting of the experimental points using the VEPFIT software package [9]. A two-layer model consisting of (i) an oxide surface layer and (ii) the bulk region were assumed in fitting with the exception of the Fe_3Al sample where no oxide layer was considered. The S parameter and positron diffusion length L_+ for the bulk region obtained from fitting are plotted in figure 3. The SM Fe_3Al alloy exhibits the lowest S parameter and the longest positron diffusion length. Non-SM Fe-Al alloys have significantly higher S parameters and shorter positron diffusion lengths indicating higher concentrations of quenched-in vacancy-like defects. The concentration of defects can be calculated from SPIS results using expression

$$c = \frac{1}{\nu\tau_B}\left(\frac{L_{B,+}^2}{L_+^2} - 1\right), \qquad (2)$$

where $L_{B,+} \approx 180$ nm [7] is the positron diffusion length in a defect-free Fe-Al alloy and L_+ is the positron diffusion length measured in the studied sample. It has been demonstrated [7,10] that L_+ can be reliably determined even in samples with a very high concentration of defects. Figure 4 shows the dependence of the concentration of quenched-in vacancy-like defects found using equation (2) on C(Al). Note that for triple defects we used $\nu = 8 \times 10^{14}$ at. s^{-1} since their open volume is roughly twice that of V_{Fe}. The concentration of quenched-in vacancies in Fe_3Al calculated from LT data using equation (1) is also shown in the figure and agrees well with that calculated from SPIS data. One can see in figure 4 that the SM Fe_3Al alloy exhibits the lowest concentration of quenched-in vacancies among the studied samples. Non-SM alloys feature a significantly increased c both for the under-SM composition ($Fe_{76}Al_{24}$) and for the over-SM one (C(Al) > 25 at.%).

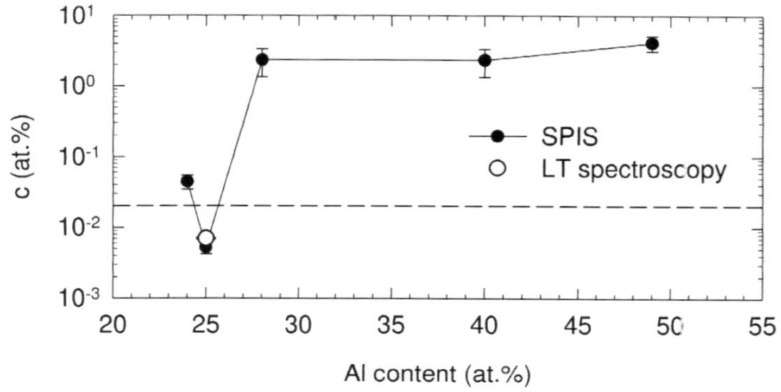

Figure 4. Concentration of vacancy-like defects deduced from SPIS data (full points) and LT data (open circles). The dashed line shows the maximum density of defects which can be determined by LT spectroscopy.

4. Conclusions

Open-volume defects in Fe-Al alloys quenched from 1000°C were identified as Fe-vacancies in the Fe_3Al alloy, Fe-vacancies associated with Fe_{Al} atoms in the alloy with C(Al) < 25 at.% and triple defects consisting of an Fe-divacancy plus an Fe_{Al} atom in alloys with C(Al) > 25 at.%. It was demonstrated that the concentration of defects can be reliably determined from SPIS results even in cases when the free positron component cannot be resolved in LT spectra. The concentration of vacancies in order of 10^{-2} at.% was found in the Fe_3Al alloy. Roughly 10 times higher concentration was found in the alloy with C(Al) < 25 at.%. Almost three orders of magnitude higher concentration of defects approaching 10 at.% was determined in Al-rich alloys (C(Al) > 25 at.%).

Acknowledgement

This work was supported by the Czech Scientific Foundation (project GA106/08/P133), the Academy of Science of Czech Republic (project KAN300100801) and the Ministry of Education, Youths and Sports of the Czech Republic through the research plan MSM 0021620834.

References

[1] Sautho G 1995 *Intermetallics* (Weinheim: VCH)
[2] Chang Y A, Pike L M, Liu C T, Bilbrey A R and Stone D S 1993 *Intermetallics* **1** 107
[3] de Diego N, Plazaola F, Jiménez J A and Serna J 2005 *Acta Mater.* **53** 163
[4] Haraguchi T, Hori F, Oshima R and Kogachi M 2001 *Intermetallics* **9** 763.
[5] Becvar F, Cizek J, Prochazka I 2008 *Appl. Surf. Sci.* **255** 111
[6] Anwand W, Kissener H-R and Brauer G 1995 *Acta Physica Polonica A* **88** 7
[7] Melikhova O, Cizek J, Kuriplach J, Prochazka I, Cieslar M, Anwand W and Brauer G 2010 *Intermetallics* **18** 592
[8] Schaefer H-E, Würschum R, Sob M, Zak T, Yu W Z, Eckert W and Banhart F 1990 *Phys. Rev. B* **41** 11869
[9] van Veen A, Schut H, Clement M, de Nijs J, Kruseman A and Ijpma M. 1995 *Appl. Surf. Sci.* **85** 216
[10] Krause-Rehberg R, Bondarenko V, Thiele E, Klemm R and Schell N 2005 *Nucl. Instr. Meth. B* **240** 719

Journal of Physics: Conference Series **262** (2011) 012040 doi:10.1088/1742-6596/262/1/012040

The role of positron annihilation lifetime studies and nuclear sensors for characterising porous materials

E Mume[1,2], A Uedono[3], G Mizunaga[3], D E Lynch[4] and S V Smith[1]

[1]Centres for Antimatter-Matter Studies (CAMS) at the Australian Nuclear Science and Technology Organisation (ANSTO) and the [2]Australian National University, [3]Institute of Applied Physics, University of Tsukuba, Tsukuba, Japan, [4]Exilica Ltd, UK

E-mail: Suzanne.smith@ansto.gov.au

Abstract. A series of nuclear sensors were designed to assess the chemistry within the nanopores of a porous material. The nuclear sensors of varying size, charge, and hydrophobicity were exposed to hollow silica shells (HSS) at varying pH. Uptake and release kinetics were studied over a 24 h period at room temperature. Preliminary study indicate positively charged nuclear sensors were selectively and rapidly (within 10 min) absorbed by the HSS at pH 7 to 9. PALS showed there were two types of pores (1.7 and 0.7 nm) present. The data suggest the nuclear sensors sit within the larger pore of the HSS. Both PALS and nuclear sensors are required to obtain an accurate insight into the nanoporosity of the hollow silica shells.

1. Introduction

A breakthrough in the synthesis of mesoporous silica materials with controlled particle size, morphology and porosity has enabled new applications in medicine, environment and industry.[1-2] The silica matrices are particularly attractive because of their chemical stability and biocompatibility. They can be readily internalised by plant and animal cells when functionalized and show no significant cytotoxicity.[3]

For many mesoporous silica nanoparticles, absorption on the surface of the pores can be controlled by size or morphology of the pores.[4] However it is difficult to predict under what condition the cargo molecules can be loaded and released. Small Angle Neutron Scattering (SANS), X-Ray Spectroscopy (SAXS), nitrogen gas adsorption-desorption, Scanning Electron Microscopy (SEM), Transmission Electron Microscopy (TEM), Brunauer Emmett Teller (BET) measurements and Positron Annihilation Lifetime Spectroscopy (PALS) all provide information about porosity in materials.[5] However, these tools provide little information about the chemistry within the nanopores of the silica shells. Understanding how to optimise the number of available pores and determining the physical properties within these pores continues to challenge the field. This study reports a preliminary investigation on how we use positron annihilation lifetime spectroscopy (PALS) and nuclear sensors (radiotracers) (see Figure 1) to probe the chemistry in these pores.

Journal of Physics: Conference Series **262** (2011) 012040 doi:10.1088/1742-6596/262/1/012040

2. Preparation

2.1. Preparation of HSS

The HSS were prepared as described previously.[6] Poly(1-methylpyrrol-2-ylsquaraine) (PMPS) were prepared by refluxing equimolar amounts of 1-methylpyrrole and squaric acid in butan-1-ol for 18 h. The PMPS beads were filtered off and then transferred to a crucible and held at 660°C for up to 3 h, after which time the sample was removed and cooled to ambient temperature.

2.2. Synthesis of Nuclear Sensors (NS)

The hexaaza cages ligand (Figure 1a) that consist of 6 secondary nitrogen donor groups that have an ability to encapsulate metal ions were synthesised in a similar method to that described in the literature.[7] The Ligands were radiolabelled with either Cu^{2+} or Co^{2+} metal ion solutions doped with $^{57}Co^{2+}$ or $^{64}Cu^{2+}$ radioisotopes, respectively. A typical procedure involved exposing equimolar amounts of the Ligand (dota, diamsar, or sarar-NCS) at 1.5 x 10^{-2} M with the radioisotope solution in sodium phosphate buffer (PBS) pH 7. The complexation reactions were monitored at 23 °C by instant thin layer chromatography (ITLC-SG) until complete (> 95%).

2.3. Binding of NS to HSS

A typical procedure involved mixing 10 mg (accurately weighed in triplicate) of HSS in 1480 μL of appropriate buffer (pH 3-9) at 23 °C. The resulting mixture was then vortexed for 5 s and left to agitate for 5 min. The HSS were then exposed to either free $^{57/nat}Co^{2+}$ or $^{54/nat}Cu^{2+}$ (1x10^{-3} M) or $^{57/nat}$Co-Ligand (1x10^{-4} M) in 20 μL. Uptake of the metal ions and $^{57/nat}$Co-Ligand were monitored until equilibrium was reached. The radioactivity associated with the HSS was measured using the gamma counter. The moles of nuclear sensors associated with the HSS were calculated in the following manner, A= B x % activity associated with HSS/weigh (mg) of HSS. Where A = mole of metal ion or metal complex bound to HSS per milligrams and B = initial moles of metal ion or metal complex.

2.4. Co^{2+}-doped HSS

HSS (300 mg) was swelled in 0.1 M potassium dihydrogen phosphate (PBS) solution (5 mL) for 5 min. Cobalt(II) chloride (0.82g) in PBS was added to the suspension and agitated for 24 h at 23 °C. The Co^{2+} doped HSS was isolated after centrifugation (1000 rpm for 1 min) washed with Milli-Q water (3 x 1 mL) and then dried in air.

2.5. PALS

Positron annihilation lifetime spectroscopy (PALS) was used to determine the microporosity within the HSS. The positron lifetimes were analysed using the Laplace inversion technique and fitted assuming three annihilation modes or components [i.e. free positron : β^+; *para*-positronium: *p*-Ps; *ortho*-positronium : *o*-Ps]. The third component, associated with the pick-off annihilation or *o*-Ps is used to determine the lifetime and relative intensities of the micropores. Eqn (1) shows the relationship between the lifetime spectrum of positrons, $S_{LT}(t)$ and the lifetime components. The observed spectra were analysed with a time resolution of about 170 ps (full-width-at-half-maximum, FWHM) by using the RESOLUTION computer program.[8]

$$S_{LT}(t)=\Sigma\, \lambda_i\, I_i \exp(-\lambda_i\, t), \quad (1)$$

where λ_i and I_i are respectively the i-th components of the annihilation rate of positrons and its intensity. The lifetime of the positrons, τ_i, is given by $1/\lambda_i$.

3. Results and Discussion

The HSS were prepared by the overcoating of sacrifical polymeric template particles with silicon precursor followed by thermal calcination in a furnace at 660 °C.[6] Ligands were synthesized in a similar manner to that described elsewhere [7] or purchased from a commercial supplier. Each ligand was then radiolabelled with $^{57/nat}Co$ at a metal ion (M) to ligand ratio of 1:1 at pH 7 at 23 °C and micromolar concentrations as previously described. ITLC was used to ensure the complexation of the $^{57/nat}Co$ was complete (>95%) [see Figure 2]. Uncomplexed ^{57}Co and ^{64}Cu were used to monitor the absorption of Co^{2+} and Cu^{2+}, respectively. HSS were exposed to each NS in buffered solution at pH 3 to 9. Solutions were sampled at varying time intervals (up to 24 h) to determine unbound NS. Figure 3 show the similar binding profiles for Cu^{2+} and Co^{2+} with significantly higher absorption (1.5 higher) for the former. The rate of absorption for the Cu^{2+} ion was also considerably faster than those for the Co^{2+} ion, reaching their maximum absorption within 10 minutes and 90 minutes, respectively.

The uptake of NS (Co-dota, Co-diamsar and Co-sarar-NCS) by the HSS was also monitored over time (up to 24 h) or until equilibrium was reached at pH 7. Figure 4 compares the binding of a series of NS (Co-dota, Co-diamsar and Co-sarar-NCS) to the HSS at pH 7. The data show that subtle changes in the structure has a significant effect on the binding properties of the HSS. The positively charged NS [the free metals, metal-ligand complexes] of varying size are preferably taken up by the HSS at higher pHs. Notably, negatively charged NS (Co-dota) was weakly absorbed compared to the positively charged NS (Figure 4), suggesting the pores of HSS are negatively charged at high pH.

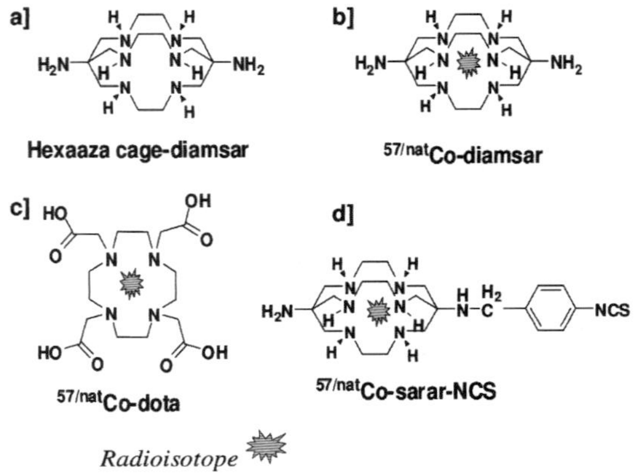

Figure 1. Ligands were radiolabelled with $^{57/nat}Co^{2+}$ to form the nuclear.

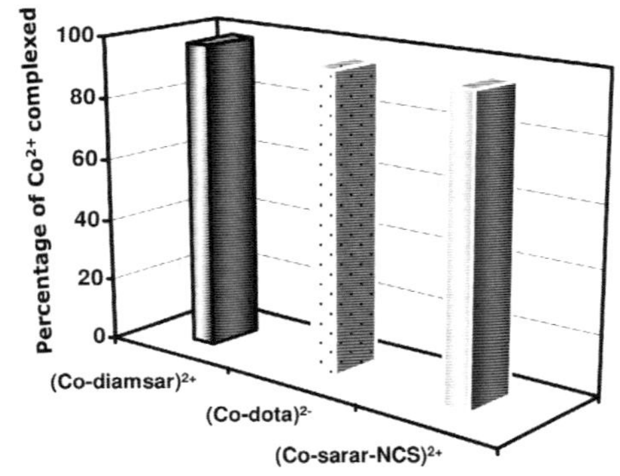

Figure 2. Percent of $^{57/nat}Co^{2+}$ complexed by dota, diamsar and sarar-NCS after 1 h.[a]
[a] $[M^{2+}] = 10^{-3}$ M; Buffer : 0.1M PBS pH 7; Temp = 23°C; Time = 1 h.

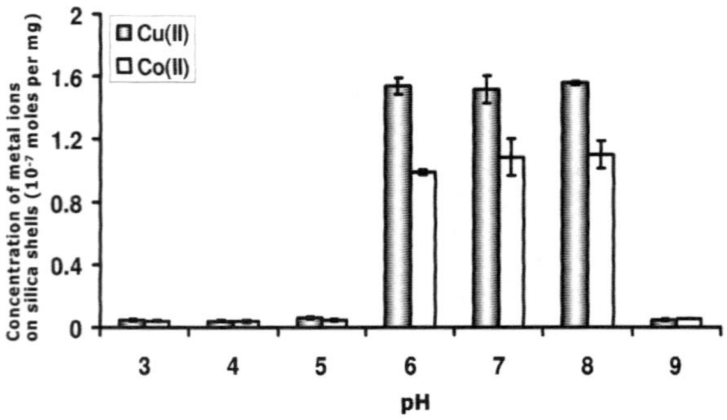

Figure 3. The effect of pH on metal binding properties of HSS at ambient temperature.[a]
[a] $[M^{2+}] = 10^{-3}$ mol/L; silica shell = 10 mg; Vol = 1.5 mL; Time = 10 min; Temp = 23°C; number of samples (n)=3.

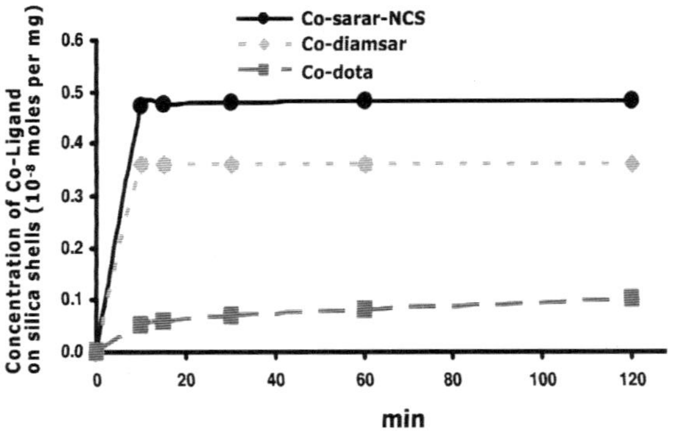

Figure 4. Moles of $[M\text{-}Ligand]^{2+}$ absorbed into HSS at pH 7 over time.[a]
[a] $[M\text{-}Ligand] = 10^{-4}$ mol/L; silica shell = 10 mg; Vol = 1.5 mL; Temp = 23°C; number of samples (n)=3.

In order to gain insight where the NS are located in the HSS, the HSS was loaded with Co(II) at physiological pH (7) and its PALS spectrum compared with that of the undoped HSS. Data were fitted using eqn (1) and results are summarized in Figure 5. Four lifetime components [τ_1, τ_2, τ_3 and τ_4] were present in the HSS sample and only three components were refined for the PALS spectrum of the doped with Co^{2+} HSS. The τ_3 and τ_4 components were fitted using the Tao-Eldrup model to determine the pore diameters.[9] Two pores sizes were evident in the HSS sample, a large pore 1.7 nm related to the τ_4 and a smaller pore at 0.6-0.7 nm defined by τ_3. The loss of the τ_4 component suggests the metal ion is trapped in this pore. Conversely the pores with diameter of 0.6-0.7 nm or smaller do not contribute to the metal trapping.

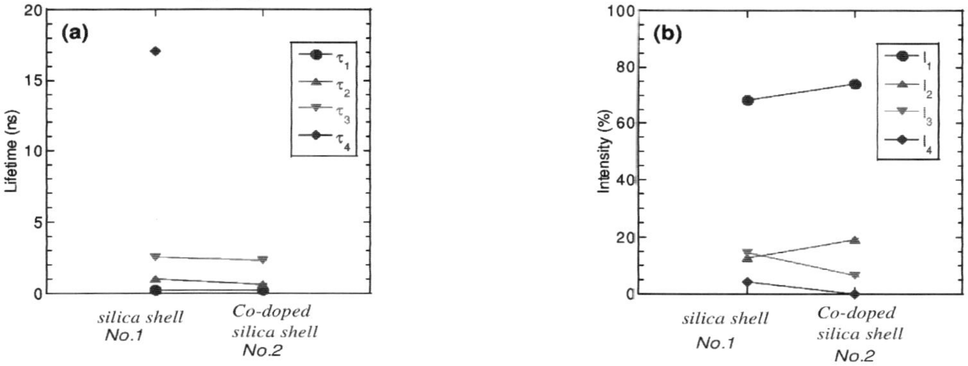

Figure 5. The lifetime and relative intensities of nanopores in the HSS.

4. Conclusions

Nuclear sensors are novel non-destructive tools for studying porosity of materials. Because of their high sensitivity (up to 10^{-5} ppb) only small quantities (e.g 10 milligrams) of material are required for analysis. Tailoring the shape, charge and hydrophobicity of the nuclear sensors provides information on the chemical reactivity of the pores within a porous material. PALS can be used to provide information on where the molecules might be absorbed in a material. Collectively, PALS and nuclear sensors have proven to be valuable tools to assess the chemistry and availability of micropores in materials. Further work will involve broadening the range of NS to gain a deeper understanding of the relationship between structure and function of novel porous materials.

Acknowledgements

We thank the ARC Centre of Excellence for Antimatter-Matter Studies

References

[1] Huh S, Wiench J W, Yoo J C, Pruski M, Lin V S Y 2003 *Chem. Mater.* **15** 424
[2] Trewyn B G, Whitman C M, Lin V S Y 2004 *Nano Lett.* **4** 2139
[3] Radu D R, Lai C Y, Jeftinija K, Rowe E W, Jeftinija S, Lin V S Y 2004 *J. Am. Chem. Soc.* ***126*** 13 216
[4] Vallet-Regi M, Ramila A, del Real R P, Perez-Pariente J 2001 *Chem. Mater.* **13** 308
[5] Smith S V 2009 *J. Phys. Conf. Ser.* **185** 012044
[6] Lynch D E, Nawaz Y and Bostrom T 2005 *Langmuir* **21** 6572
[7] Di Bartolo N M, Sargeson A M, Donlevy T M and Smith S V 2001 *J. Chem. Soc. Dalton Trans.* 2303
[8] Saito H, Nagashima Y, Kurihara T and Hyodo T 2002 *Nucl. Inst. Method A.* **487** 612
[9] Kirkegaard P, Eldrup M, Mogensen O E and Pedersen N J 1981 *Comput. Phys. Comm.* **23** 307

*Abbreviations: 1,4,7,10-tetraazacyclododecane-1,4,7,10-tetraacetic acid (dota); 3,6,10,13,16,19-hexaazabicyclo[6.6.6]eicosane-1,8-diamine (diamsar); N^1-(4-isothiocyanatobenzyl)-3,6,10,13,16,19-hexaaza-bicyclo[6,6,6]icosane-1,8-diamine (sarar-NCS); Metal complexes of these ligands will be denoted as (M-ligand), e.g. ((Co-diamsar).

Journal of Physics: Conference Series **262** (2011) 012041 doi:10.1088/1742-6596/262/1/012041

Towards the production of an energy-tunable positronium beam using Ps$^-$ photodetachment technique

Yasuyuki Nagashima, Koji Michishio, Takayuki Tachibana and Hiroki Terabe

Department of Physics, Tokyo University of Science, 1-3 Kagurazaka, Shinjuku-ku, Tokyo 162-8601, Japan

ynaga@rs.kagu.tus.ac.jp

Abstract. The production of an energy-tunable positronium beam by photodetachment of positronium negative ions, which was proposed many years ago, has become feasible by the recent development of efficient formation of the ions and successful photodetachment. Here we shall describe our current efforts directed towards the production of a positronium beam using this technique.

1. Introduction

It is well known that positronium (Ps) atoms can be formed and emitted with a maximum energy of a few eV when keV-energy positrons strike a solid target in vacuum. Ps can also be formed in gases, liquids and powders. However, once Ps atoms are formed, it is impossible to accelerate them to the required energy using an electric field.

Energy-tunable Ps beams have been developed using charge exchange of positrons with a low-density gas and thin foils. Ps formation by the reaction,

$$e^+ + M \rightarrow \mathrm{Ps} + M^+,$$

of a positron with a molecule, M, is a significant channel at intermediate energies [1,2]. The Ps kinetic energy, T_{Ps}, is related to that of the e^+ kinetic energy, T_{e^+}, by

$$T_{Ps} = T_{e^+} - E_I + E_{Ps}$$

where E_I is the ionization energy of the target gas. E_{Ps} is the Ps binding energy, i.e.,

$$E_{Ps} = 6.8\mathrm{eV}/n^2$$

and n is the principal quantum number of the Ps. This technique has been used for the observation of Ps specular reflection from the surface of LiF [3] and the measurement of Ps-gas scattering cross sections in the energy range below 100eV [2,4].

Ps atoms with kinetic energies from 10 to 500 eV have also been formed by the partial transmission of keV-energy positrons through a 50-Å carbon film in vacuum [5]. The Ps energy distribution

$$dN = (10^{-3}\,\mathrm{eV})^{1/2} E^{-3/2} dE\, d\Omega$$

has been obtained.

It has been proposed that the production of an energy-tunable monoenergetic Ps beam might be feasible by the photodetachment of accelerated positronium negative ions (Ps$^-$) [2,6,7]. Recently, we observed the efficient emission of Ps$^-$ formed from thermalized positrons impinging on tungsten, molybdenum, tantalum and iron surfaces coated with alkali metals [8-12]. Furthermore, we have succeeded in the photodetachment of Ps^- for the first time [13]. We now aim to realise the production of energy-tunable Ps beams using these techniques.

2. Efficient production of Ps^-

It is well known that electronic surface states are sensitive to adsorbates on the surfaces. The lowering of electron work functions by the deposition of an overlayer of alkali metal on a metal surface enables efficient emission of Ps^- from the surfaces.

The Ps^- work function, the energy required to emit Ps^- from a metal surface, is

$$\phi_{Ps^-} = \phi_+ + 2\phi_- - E_{Ps^-},$$

where ϕ_+ and ϕ_- are the positron and electron work functions, respectively, and E_{Ps^-} is the Ps^- binding energy, which has been calculated theoretically to be 7.13eV [14, 15]. When the surface is coated with an alkali metal, the Ps^- work function depends on the effect of the adsorbate dipole, D, as

$$\phi_{Ps^-}' = \phi_{Ps^-} - D.$$

The value of D for Cs on a tungsten surface is reported to be about 3eV for 2.2×10^{14} atoms /cm^2 [16]. This indicates that Ps^- emission is more feasible in this case than from the uncoated surfaces.

In 2008, we observed the efficient emission of Ps^- from a Cs coated W(100) surface [8]. The mean emission efficiency (the fraction of incident slow positrons yielding Ps^-) over a 3000s interval following Cs deposition was 1.25%, which is two orders of magnitude higher than that obtained using carbon thin films [6] and clean tungsten surfaces [17]. However, the efficiency decreased and was almost constant at about 0.1% after 60000s. The decrease may have been due to the accumulation of residual molecules in the target chamber onto the target surface or the structural change of the Cs layer. Recently, we observed that the Ps^- emission efficiency from Na coated tungsten is higher than that from the Cs coated surface, even though the effect of the adsorbate dipole for Na is smaller than that for Cs. Furthermore, the decrease of the emission efficiency was slower than that for the Cs coated surface and the efficiency was still higher than 0.5% after 3 days [12].

3. Photodetachment of Ps^-

The Ps^- photodetachment process has been studied theoretically and cross sections have been calculated [18,19]. Recently, we have succeeded in the first observation of the photodetachment of Ps^- in the slow positron beam facility in KEK, Tsukuba, Japan [13].

The positrons generated at the beam dump of an electron Linac with a repetition rate of 50Hz were collected and transported to the Na coated polycrystalline tungsten target [20]. The width of the positron pulse at the target position was 12ns. The Ps^- ions emitted from the target were accelerated and the Doppler shifted γ-rays were detected using Ge detectors. In order to lower the background due to the pile-up of the γ-rays produced in a short period of time, excess counts were reduced by surrounding the detectors in lead shielding with a slit to allow the passage of the γ-rays. Two detectors were used to obtain high statistics.

In order to photodetach the Ps^-, a Q-switched Nd: YAG pulsed laser of 12ns width was used. The wave length was 1064nm and the power was 1J/pulse. The repetition frequency was half of that for the slow positron beam and the data was acquired for laser on/off repeatedly.

Figure 1 shows the obtained γ-ray spectra. The peak attributed to Ps^- annihilation decreased upon laser irradiation. This shows that Ps^- ions were converted to ortho-Ps atoms, which self-annihilate into 3γ-rays, by the photodetachment.

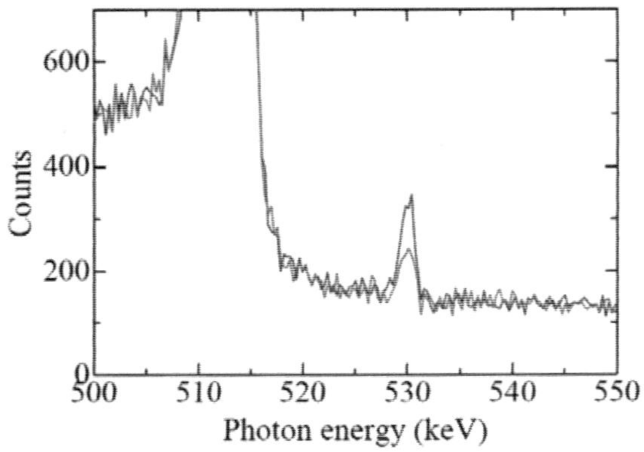

Figure 1. The γ-ray energy spectra for laser on (red) and laser off (blue).

4. Production of an energy-tunable Ps beam by the photodetachment of Ps$^-$

The success of the Ps$^-$ photodetachment technique opens the door to a new era of the production of energy-tunable Ps beams. Figure 2 shows the conceptual design of such a beam. The pulsed slow positron beam is transported along the curved magnetic field on to the Na coated tungsten surface. The emitted Ps$^-$ ions are accelerated and photodetached using a high power Q-switched Nd: YAG pulsed laser. Irradiation with a 230nm laser, where Feshbach resonance is predicted theoretically [18,19], might enable us to perform the photodetachment more efficiently. However, using a 1064nm laser will result in a beam with smaller angular divergence. The Ps beam is transported through the biased grids to eliminate charged particles such as electrons and positrons, and impinges onto the target.

Using the Ps beam, the investigation of Ps passing through thin foils suggested by Mills in 1989 [21] may be possible. Observation of coherent resonance excitation [22] of Ps may also be feasible if we can prepare a spatially periodic field with optimum conditions.

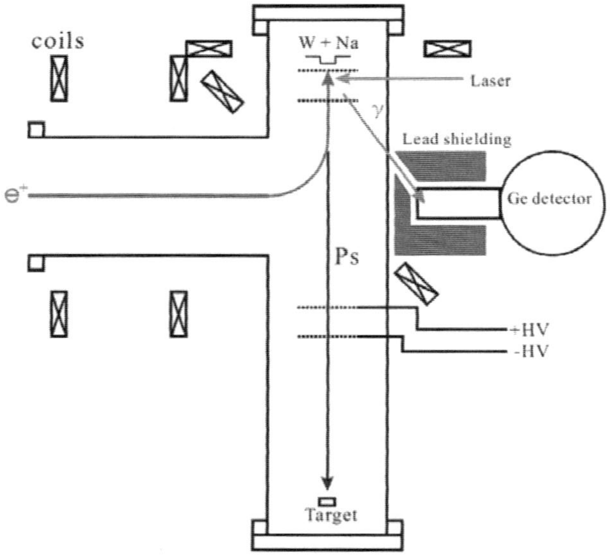

Figure 2. Schematic diagram of the proposed energy-tunable Ps beam apparatus.

Acknowledgments
We thank Professor A. P. Mills, Jr., Professor A. Yagishita, Professor A. Igarashi, Professor T. Kuga, Dr. K. Wada, Professor T. Hyodo and Professor T. Azuma for valuable discussions and their encouragement. We also thank Dr. S. Jinno, Professor M. Fujinami, Professor T. Shidara and Professor S. Ohsawa for their encouragement. We would like to thank the KEK staff for their generous support. This work was supported by the Reimei Research Promotion project (Japan Atomic Energy Agency) and the Matsuo Foundation.

References

[1] Brown B L 1985 *Bull. Ann. Phys. Soc.* **30** 30; *Positron annihilation* 1985 ed R M Singru and P C Jain (World Scientific, Singapore) p. 328
[2] Laricchia G, Armitage S, Kövér Á and Murtagh D J *Advances in Atomic, Molecular, and Optical Physics* **56** 2008 ed E Arimondo, P R Berman and C C Lin (Elsevier, Amsterdam) p.1
[3] Weber M H, Tang S, Berko S, Brown B L, Canter K F, Lynn K G, Mills A P, Roellig L O and Viescas A J 1988 *Phys. Rev. Lett.* **61** 2542
[4] Zafar N, Laricchia G, Charlton M and Garner A 1996 *Phys. Rev. Lett.* **76** 1595
[5] Mills A P and Crane W S 1985 *Phys. Rev. A* **31** 593
[6] Mills A P 1981 *Phys. Rev. Lett.* **46** 717
[7] Laricchia G *Positron Spectroscopy of Solids* 1995 ed A Dupasquier and A P Mills Jr (IOS, Ameterdam) p.401
[8] Nagashima Y, Hakodate T, Miyamoto A and Michishio K 2008 *New J. Phys.* **10** 123029
[9] Nagashima Y, Hakodate T, Miyamoto A, Michishio K and Terabe H 2009 *Phys. Status Solidi C* **6** 2291
[10] Nagashima Y, Hakodate T, Miyamoto A, Michishio K and Terabe H 2009 *J. Phys. Conf. Series* **194** 012039
[11] Michishio K, Tachibana T, Terabe H, Miyamoto A and Nagashima Y 2010 *J. Phys. Conf. Series* **199** 012003
[12] Terabe H. Michishio K, Tachibana T and Nagashima Y in this volume
[13] Michishio K, Tachibana T, Wada K, Hyodo T, Kuga T, Igarashi A, Yagishita A and Nagashima Y in preparation
[14] Frost A A, Inokuti M and Lowe J P 1964 *J. Chem, Phys. 41* 482
[15] Drake G W F and Grigorescu M 2005 *J. Phys. B: At. Mol. Opt. Phys.* **38** 3377
[16] See, for example, Kiejna A and Wojciechowski K F 1981 *Prog. Surf. Sci.* **11** 293
[17] Nagashima Y and Sakai T 2006 *New J. Phys.* **8** 319
[18] Ward S and Humbersron J 1987 *J. Phys. B: At. Mol. Phys.* **20** 127
[19] Igarashi A, Toshima N and Shimamura I 2000 *New J. Phys.* **2** 17.1
[20] Tachibana T, Michishio K, Terabe H, Wada K, Hyodo T, Kurihara T, Yagishita A and Nagashima Y *Nucl. Instr. and Meth. in Phys. Res. A*, in press
[21] Mills A P, Friedman P G and Zuckerman D M *Annihilation in Gases and Galaxies* 1989 ed R J Drackman (NASA Conference Publication 3058) p. 213
[22] See, for example, Okorokov V V 1965 *JETP Lett.* **2** 111

Journal of Physics: Conference Series **262** (2011) 012042 doi:10.1088/1742-6596/262/1/012042

Measurements of the total cross sections for positron scattering from He using an electrostatic high-brightness slow positron beam apparatus

Kazuaki Nagumo[a,1], Yuna Nitta[a], Masamitsu Hoshino[b], Hiroshi Tanaka[b] and Yasuyuki Nagashima[a,2]

[a] Department of Physics, Tokyo University of Science, 1-3 Kagurazaka, Shinjuku-ku, Tokyo 162-8601, Japan

[b] Department of Physics, Sophia University, 7-1 Kioicho, Chiyoda-ku, Tokyo 102-8554, Japan

E-mail; [1] kazu_nagumo@yahoo.co.jp, [2] ynaga@rs.kagu.tus.ac.jp

Abstract. We have measured the total cross sections for positron scattering from He using an electrostatic high-brightness slow positron beam apparatus. Our results are in reasonable agreement with the data by Jaduszliwer *et al* in the energy region below 15 eV, which have been obtained under the influence of the lower axial magnetic field than other groups.

1. Introduction

There have been considerable efforts devoted to positron scattering studies, in particular to the measurements of the total cross sections (TCS) [1][2]. Since the 1970s, several groups have developed magnetically confined or electrostatic slow positron beam systems for the positron scattering studies. Most of the TCS measurements have been performed using the magnetically confined beams in order to transport the slow positrons efficiently.

The values of TCS, σ_{tot}, for positron scattering from several gases have been derived by the Beer-Lambert law, in which positrons scattered by target molecules are considered as having been removed from the incident beam. The TCS is obtained from the relationship,

$$\sigma_{tot} = -\frac{k_B T}{pL} \ln \frac{I}{I_0} \quad (1)$$

where I and I_0 are the transmitted beam intensities with and without gas of pressure p and temperature T in the cell, respectively. L is the length of the gas cell and k_B is the Boltzmann's constant. Since the forward scattered positrons are constrained to follow helical trajectories by the magnetic field, the transmitted intensities with gas in the cell, I, is overestimated. This effect has been corrected, for example, using calculated differential cross sections [1].

In the present work, we have measured the TCS for He using an electrostatic slow positron beam apparatus, where the effect of the forward scattered positrons is smaller than that under the influence of magnetic fields.

2. Experimental Procedure

The apparatus used was an electrostatic high-brightness slow positron beam system [3][4]. Positrons from a ^{22}Na source of ~250 MBq activity were firstly moderated with a tungsten mesh moderator. The slow positrons emitted from the moderator were transported by several electrostatic lens elements. Then, the slow positrons were focused onto a remoderator at the energy of 5 keV with a diameter of approximately 2-3 mm for brightness enhancement. The remoderator was a tungsten (100) single crystal foil and used in the reflection geometry. Positrons reemitted from the remoderator were deflected and energy selected by an electrostatic hemispherical energy analyzer and then reached a 30.0 mm long gas cell, which had entrance and exit apertures of 2 mm diameter. Finally, the survived positrons through the cell were detected by the Channeltron. In order to prevent the effect of stray magnetic fields, the apparatus was covered with μ-metal shielding.

The pressure of the target gas was determined using an absolute capacitance manometer at the center of the cell. The gas pressure was maintained such that the total positron scattering probability was less than or equal to 20%. The resolution of the slow positron beam for the TCS measurements was about 250 meV (FWHM) [4]. The positron beam intensity was 1.5 e^+/s and 0.04 e^+/s at 100 eV and 5 eV incident energy, respectively.

3. Results and Discussion

The obtained values of TCS for He without corrections for the positron forward scattering were shown in Fig. 1 together with the prior experimental results [5]-[13] and the recent theoretical results [14][15].

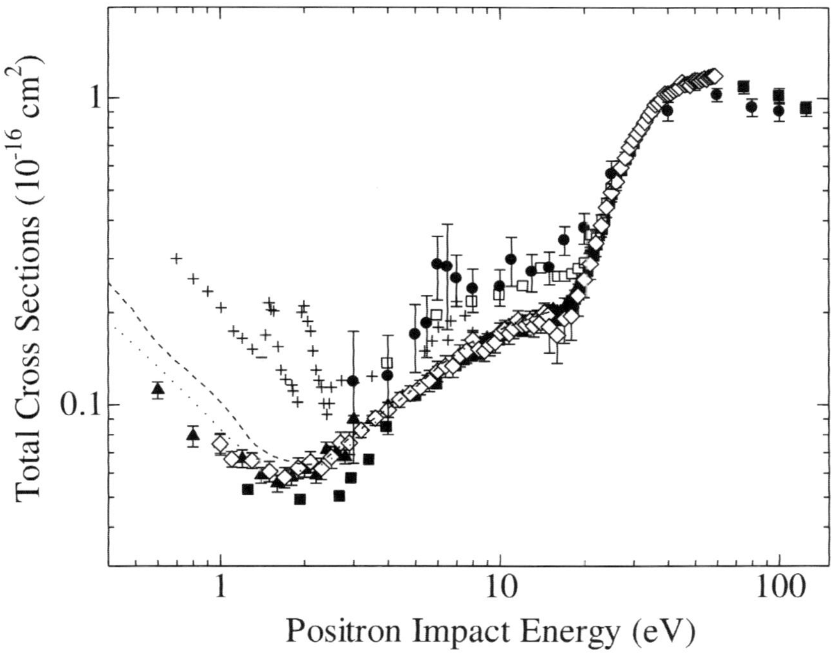

Figure 1. Total Cross Sections for positron scattering from He. Experiment: ●, Present results without corrections for the positron forward scattering; □, Jaduszliwer et al [5]-[7]; ■, Detroit group [8][9]; ▲, Mizogawa et al [10]; +, Karwasz et al [11]; ◇, ANU group [12][13]. Theory: ------, Reeth *et al* [14]; ······, Wu *et al* [15].

The present results are in reasonable agreement with the data by Jaduszliwer *et al* [5]-[7] in the region below 15 eV, which have been obtained under the influence of the lower axial magnetic field (2.75 G) than other groups. However, our results are higher than the measurements by Mizogawa *et al* [10] and ANU group [12][13], which agree well with the theoretical values of Reeth *et al* [14] and Wu *et al* [15], in the region between 4 and 20 eV. Although there should be some causes for the discrepancies, the reason is unknown.

We are presently working on the TCS measurements for Ne. The data will be published soon.

Acknowledgments
We are indebted to Wataru Suda, Daiki Tomida and Toshihide Hakodate for their excellent technical support. We thank Professor Stephen Buckman, Dr James Sullivan, Dr Casten Makochekanwa and Professor Gaetana Laricchia for useful discussions concerning the development of the apparatus and the preliminary measurements. We are also grateful to the Sophia University Techno Center Staff for their help with the engineering of our apparatus. This work is supported by the Matsuo Foundation.

References

[1] Charlton M and Humberston J 2001 *Positron Physics* (Cambridge: Cambridge University Press)
[2] Surko C M, Gribakin G F and Buckman S J 2005 *J. Phys B* **38** R57
[3] Suda W, Nagumo K, Tomida D, Hoshino M, Nagashima Y and Tanaka H 2008 *Nucl. Instrum. Methods B* **266** 416
[4] Nagumo K, Nitta Y, Hoshino M, Tanaka H, Nagashima Y 2010 *J. Phys: Conference Series* **199** 012024
[5] Jaduszliwer B, Keever W C, Paul D A L 1971 *Can. J. Phys.* **50** 1414
[6] Jaduszliwer B and Paul D A L 1973 *Can. J. Phys.* **51** 1565
[7] Jaduszliwer B, Nakashima A, Paul D A L 1974 *Can. J. Phys.* **53** 962
[8] Stein T S, Kauppila W E, Pol V, Smart J H, Jesion G 1978 *Phys. Rev. A* **17** 5
[9] Kauppila W E, Stein T S, Smart J H, Dababneh M S, Ho Y K, Sowning J P, Pol V 1981 *Phys. Rev. A* **24** 2
[10] Mizogawa T, Nakayama Y, Kawaratani T and Tosaki M 1985 *Phys. Rev. A* **31** 2171
[11] Karwasz G P, Pliszka D, Brusa R S, Perazzolli C 2005 *Acta Physica Polomoca A* **107** 666
[12] Sullivan J P, Makochekanwa C, Jones A, Caradonna P, Buckman S J 2008 *J. Phys. B* **41** 081001
[13] Caradonna P, Jones A, Mokochekanwa C, Slaughter D S, Sullivan J P, Buckman S J, Bray I, Fursa D V 2009 *Phys. Rev. A* **80** 032710
[14] Reeth P V and Humberston J W 1999 *J. Phys. B* **32** 3651
[15] Wu H, Bray I, Fursa D V, Stelbovics A 2004 *J. Phys. B* **37** L1

Journal of Physics: Conference Series **262** (2011) 012043 doi:10.1088/1742-6596/262/1/012043

Development of a dedicated superconducting accelerator for positron production

B E O'Rourke[1], N Oshima[1], R Kuroda[1], R Suzuki[1], T Ohdaira[1], A Kinomura[1], N Hayashizaki[2], E Minehara[3], H Yamauchi[4], Y Fukamizu[4], M Shikibu[4], T Kawamoto[4] and Y Minehara[4]

[1] National Institute of Advanced Industrial Science and Technology (AIST), AIST-Central 2, 1-1-1 Umezono, Tsukuba, Ibaraki 305 8568, Japan

[2] Tokyo Institute of Technology, Research Laboratory for Nuclear Reactors,2-12-1, Ookayama Meguro-ku, Tokyo 152-8550, Japan

[3] The Wakasa Wan Energy Research Centre, 64-52-1 Nagatani, Tsuruga, Fukui 941-0821, Japan

[4] TIME Corporation, 73-48 Numata Nishi-machi Ohara, Mihara, Hiroshima 729-0473, Japan

E-mail: `brian-orourke@aist.go.jp`

Abstract. We report on the current status of a project to develop a dedicated superconducting accelerator for slow positron production at AIST. Two 500 MHz, 5 cell cavities will form the basis of the new accelerator. Initial set-up and preliminary design activities are reported.

1. Introduction

The advanced defect characterization research group at AIST is currently generating positrons with a LINAC accelerator and has been at the forefront of variable energy positron based materials characterization using techniques such as positron annihilation lifetime spectroscopy (PALS) [1]. Recently a positron based microbeam analysis system has been developed which has dramatically increased the spatial resolution of PALS measurements [2]. However, this increased resolution comes at a cost of increased time required to analyze a sample as the microbeam needs to be scanned across the sample. The efficiency of such experiments is strongly influenced by the positron intensity at the target. Our group is now planning to install a new superconducting accelerator which will be dedicated to positron production. In this contribution we report on the proposed advantages of the superconducting accelerator and the current progress of the project.

2. Current Accelerator and Positron Beamline at AIST

At present positrons are produced at AIST via a normal-conducting LINAC. The current facilities at the AIST LINAC are shown in figure 1. Electrons from the LINAC at an energy of 70 MeV are directed on to a water cooled tantalum converter and tungsten moderator which produces a beam of slow positrons.

As positrons are only produced when the electron beam is on target, initially the positron beam has the same pulse structure as the accelerator. Typically the LINAC operates with 1 μs pulses at a repetition rate of 100 pps (pulses per second). In order to avoid saturation of detectors during experiments these pulses are stretched in a linear storage section. For PALS

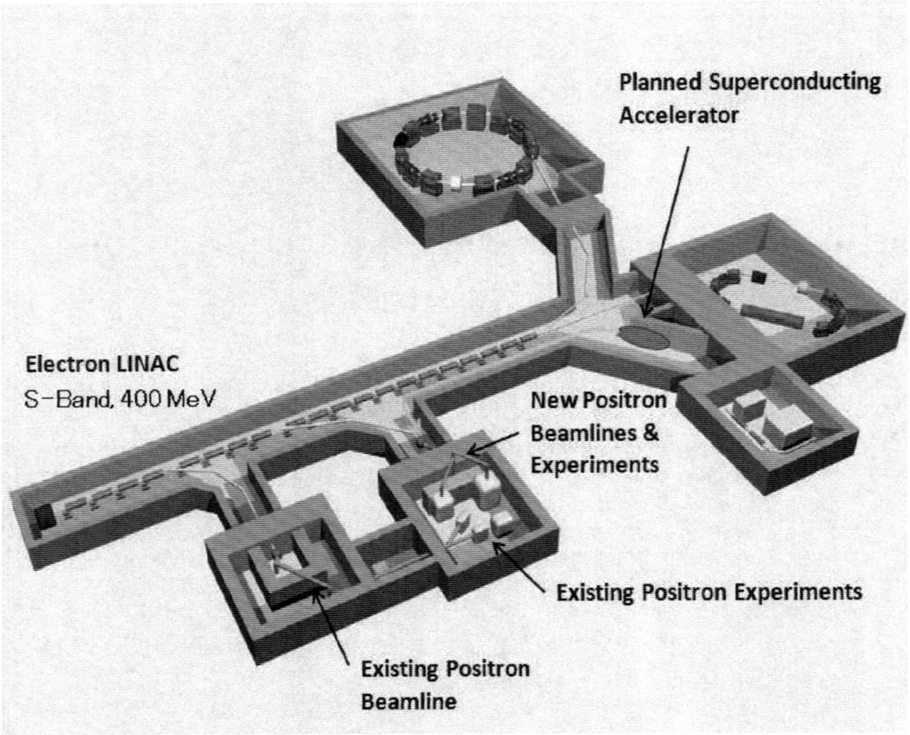

Figure 1. Current and planned facilities at AIST.

and positron auger electron spectroscopy (PAES) measurements these stretched pulses must then be chopped at high frequency and the short pulses bunched as described in reference [3]. It is clear that this low pulse rate from the LINAC is not suited to such experiments. A higher pulse rate would eliminate the need for such manipulation of the slow positron beam and would substantially increase its transport efficiency.

3. Superconducting Accelerator

In order to improve the situation outlined above it was decided to develop a dedicated accelerator for positron production. This proposed accelerator should be optimized for high current/high pulse rate operation and for this reason it was decided to use superconducting cavities. Our group at AIST has obtained two superconducting accelerator cavities (SCA), originally manufactured by Siemens in Germany and successfully operated since the early 1990's at the JAERI free electron laser facility [4]. The 5 cell niobium cavities operate at a frequency of 500 MHz and have a total length of 1.5 m. The maximum accelerating gradient is approximately 5 MV/m in pulsed mode and 3 MV/m for continuous operation. In pulsed mode (duty cycle around 3%) one module can therefore accelerate up to 7.5 MeV. For positron production the efficiency is reduced at low energy and an electron energy of more than 10 MeV is required. An initial single pass set-up with both cavities in series should provide a sufficiently high energy to demonstate positron porduction but for higher efficiency we envisage adding a re-circulating beamline to enable multi-pass, i.e. high-energy, operation.

A diagram of one of the 5 cell cavity modules is shown in figure 2 (both modules are identical). The modules are composed of a central superconducting chamber manufactured from high purity

Journal of Physics: Conference Series **262** (2011) 012043 doi:10.1088/1742-6596/262/1/012043

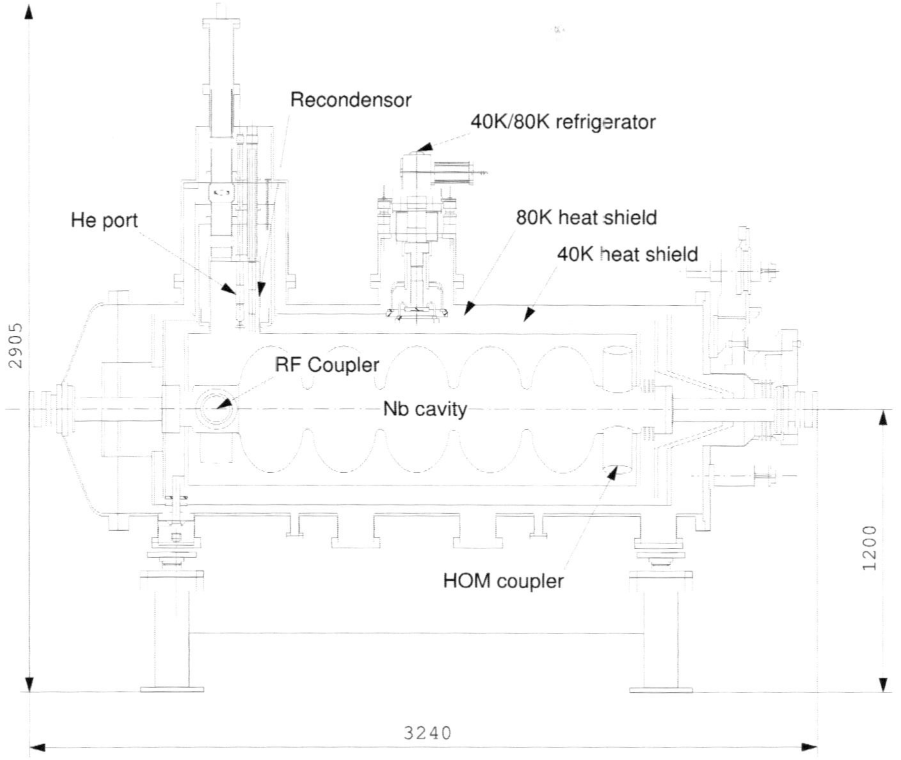

Figure 2. 5-cell superconducting accelerator module. Dimensions in mm.

niobium. This chamber is surrounded by a liquid helium tank with a capacity of approximately 400 l. The liquid He tank is further surrounded by two shield layers which are cooled by a separate cryostat and operate at 40 K and 80 K respectively. Finally these shield layers are thermally isolated from the external vessel by an outer vacuum layer.

4. Cavity Commissioning

Since the cavities are not new but were opearated successfully at JAERI for more than 15 years it is necessary to test the cavity performace. We have performed several tests on one of the modules in order to confirm it's current condition. These tests can be divided into two categories; i) vacuum and ii) refrigeration.

4.1. Vacuum Tests

The integrity of both the internal cavity vacuum and the external isolation vacuum were confirmed. In the case of the internal cavity great care must be taken to avoid introducing contamination onto the cavity walls which can affect the superconducting characteristics, leading to local heating and reduced performance. To this end we performed this test pumping under clean room conditions and ensured that when backfilling with N_2 contaminants were removed from the gas stream before entering the cavity.

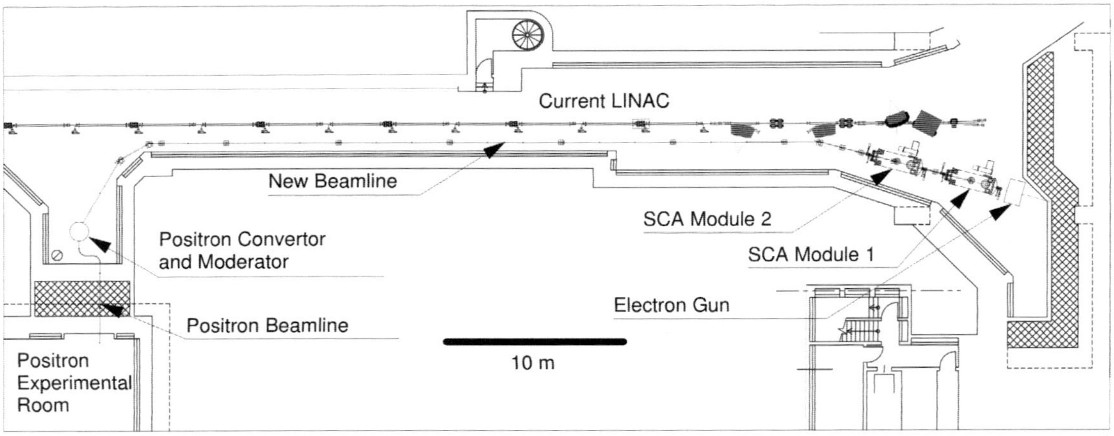

Figure 3. Layout of the proposed single-pass beamline.

4.2. Refrigeration Tests

In order to confirm the operation of the cryostat and heat exchangers a test cooling of the cavity with liquid He was performed. By measuring the temperature at various locations inside the module using internal platinum resistance temperature detectors the operation of both cryostats was monitored and after preliminary cooling with liquid N_2 it was possible to introduce liquid He. Efficient collection of liquid He confirmed that both the 4 K and shield refrigeration systems were working properly.

5. Current Status

After testing the operational capacity of one of the modules as described above this module was brought to AIST and lowered into the basement LINAC accelerator hall. It is planned to test and introduce the second cavity later this year. Figure 3 shows a proposed layout for the new beamline including both modules. The proposed beamline runs beside the current LINAC and can be constructed without affecting its operation. At present we are procceding with simulations of the full accelerator and positron beamline.

6. Conclusion

The advanced defect characterization group at AIST is currently developing a new accelerator dedicated to positron production. Two superconducting cavities have already been obtained and are under test. We have confirmed the vacuum and thermal integrity of one of the cavities and installed it in the AIST LINAC hall. A full beamline design is currently underway and we hope to complete the project within the next several years.

References

[1] Oshima N, Suzuki R, Ohdaira T, Kinomura A, Narumi T, Uedono A and Fujinami M 2009 *Rad. Phys. Chem.* **78** 1096
[2] Oshima N, Suzuki R, Ohdaira T, Kinomura A, Narumi T, Uedono A and Fujinami M 2009 *App. Phys. Lett.* **94** 094196
[3] Suzuki R, Kobayashi Y, Mikado T, Ohgaki H, Chiwaki M, Yamazaki T and Tomimasu T 1991 *Jpn. J. App. Phys.* **30** L532
[4] Kikuzawa N, Minehara E, Sawamura M, Nagai N, Takao M, Sugimoto M, Ohkubo M, Sasabe J, Suzuki Y and Kawarasaki Y 1993 *Nucl. Instrum. Meth. A* **331** 276

Journal of Physics: Conference Series **262** (2011) 012044 doi:10.1088/1742-6596/262/1/012044

Imaging of the distribution of average positron lifetimes by using a positron probe microanalyzer

N Oshima[1], R Suzuki[1], T Ohdaira[1], A Kinomura[1], S Kubota[2], H Watanabe[2], K Tenjinbayashi[2], A Uedono[2] and M Fujinami[3]

[1] National Institute of Advanced Industrial Science and Technology (AIST), 1-1-1 Umezono, Tsukuba, Ibaraki 305-8568, Japan

[2] Institute of Applied Physics, University of Tsukuba. Tsukuba, Ibaraki, 305-8573, Japan

[3] Department of Applied Chemistry, Chiba University, 1-33 Yayoi, Inage, Chiba 263-8552, Japan

E-mail: nagayasu-oshima@aist.go.jp

Abstract. A positron probe microanalyzer (PPMA) is a positron lifetime measurement system with a positron microbeam. The PPMA is used for obtaining average positron lifetime distribution images by scanning the positron injection position on a sample. Methods of data acquisition and analysis for obtaining average positron lifetimes by the PPMA are reported.

1. Introduction

Recently a positron lifetime measurement system with an intense positron microbeam generated by an electron accelerator was developed [1]. Therefore, positron annihilation lifetime spectroscopy (PALS) can be applied with a relatively good spatial resolution of about a few tens of micrometers. This apparatus is called a positron probe microanalyzer (PPMA) or a scanning positron microprobe (SPM). The advantage of the PPMA is that it can obtain two or three dimensional positron lifetime images (maps) using the scanned lateral injection positions (x, y) and the implantation depths (z) of the focused beam, which allow the defect distributions to be visually evaluated [1]. In PALS, normally more than 10^6 positron annihilation events must be recorded to obtain the complete lifetime spectrum. These spectra are then usually analyzed by the least squared fitting method using a model function consisting of a sum of a few decaying exponentials convoluted with the time resolution function [2]. The counting rate of positron annihilation events using the PPMA is $1 - 3 \times 10^3$ s^{-1} which is more than one order of magnitude larger than that obtained by other methods. However, lifetime measurements at many points (10^3-10^4 pixels) are required in order to obtain positron lifetime images. If 10^6 annihilation events are recorded at each point (like ordinary PALS) for an image consisting of 10^3-10^4 pixels, the experiment would take more than one week. This long time is practically unacceptable. On the other hand, average positron lifetimes can be obtained to an accuracy of several ps by recording a relatively low number of annihilation events, about 10^3-10^4. Therefore, images of average positron lifetime distributions can be obtained within practically reasonable times like 1 hour to 1 day. In this paper, methods of data acquisition and analysis for obtaining average positron lifetime images by the PPMA are reported. Statistical and systematic errors of the average lifetime are also discussed.

Journal of Physics: Conference Series **262** (2011) 012044 doi:10.1088/1742-6596/262/1/012044

2. Basic mathematical formulas

In general, positron lifetime spectra P and average positron lifetime $\overline{\tau}$ in materials are given by the following analytical formulas,

$$P(t) = \sum_i (I_i/\tau_i)\exp(-t/\tau_i) \quad \text{for } t \geq 0,$$
$$= 0 \quad \text{for } t < 0. \tag{1}$$
$$\overline{\tau} = \sum_i I_i\tau_i = \int_{-\infty}^{\infty} tP(t)dt \tag{2}$$

Here, τ_i is the positron lifetime of the i-th decaying component and I_i is its intensity ($\sum_i I_i = 1$). The experimentally obtained spectrum P_R differs from the analytical description given in equation (1) by a convolution with the time resolution function R, i.e.

$$P_R(t) = \int_{-\infty}^{\infty} P(t-t')\,R(t')dt' \tag{3}$$

where $\int_{-\infty}^{\infty} R(t)dt = 1$. The average lifetime of the distribution P_R is given by,

$$\overline{\tau}_R = \int_{-\infty}^{\infty} t\,P_R(t)dt \tag{4}$$

If R is given by a single Gaussian ($R(t) = (1/\sqrt{\pi}\sigma)\exp\left(-\left(t/\sqrt{2}\sigma\right)^2\right)$), the following relation can be proved numerically $\overline{\tau}_R = \overline{\tau}$ for $\tau_i > 0.2\sigma$. The PPMA can measure positron lifetimes with a typical resolution characterized by the full width at half maximum, $FWHM = 2.35\sigma \approx 200$ ps. This implies that average positron lifetimes can be obtained using formula (4) without deconvolution of R if $\overline{\tau} >\sim 30$ ps.

In the laboratory, positron annihilation spectra are normally obtained through digitalized recorders, e.g. micro channel analyzers (MCA), where the event number N_k is recorded in channel k. Then, average lifetimes are obtained by,

$$\overline{\tau}_{ex} = \sum_k t_k N_k/N_0 \tag{5}$$

Here, the constant background is negligibly small. The range of k under consideration should be large enough to include the whole spectrum shape. The total event number N_0 is constant ($N_0 = \sum_k N_k = const$) and is preset beforehand by the experimenter. t_k is the time corresponding to channel k and is determined by experimental calibration. The statistical error of $\overline{\tau}_{ex}$ is given by

$$(\delta\overline{\tau}_{ex})^2 \approx \sum_k \left(\sqrt{N_k}\partial\overline{\tau}_{ex}\big/\partial N_k\right)^2 = \sum_k \left(\sqrt{N_k}t_k\big/N_0\right)^2 \tag{6}$$

The time width corresponding to the channel width of the MCA is normally much shorter than the $FWHM$ time resolution, and hence, the following approximations are obtained,

$$\overline{\tau}_{ex} \approx \overline{\tau}_R \tag{7}$$
$$\delta\overline{\tau}_{ex} \approx N_0^{-1/2}\left(\int_{-\infty}^{\infty} t^2 P_R(t)dt\right)^{1/2} \tag{8}$$

3. Data acquisition and analysis using a scanning positron microbeam

3.1. Experimental setup

The experimental setup around a sample holder in the PPMA is shown in figure 1. Positron lifetime spectra were measured as the distribution of the time difference between the trigger signal from a beam pulsing system and the detection timing signal of an annihilation γ-ray from the detector. Three samples, pure iron (Fe), yttria-stabilized zirconia (YSZ) and Kapton, were mounted on the same holder and placed on an $x-y$ translation system [1]. The YSZ sample was $\delta L \approx 1$ mm thinner than the Fe and Kapton samples.

3.2. Analysis of positron lifetime spectra with traditional method

Positron lifetimes were measured with a beam of energy $E = 25$ keV and diameter of 50 micrometers. The counting rate of annihilation events was $\sim 1 \times 10^3$ s^{-1}. Measured lifetime spectra ($N_0 = 10^6$) were analyzed by the fitting program PALSfit [3]. Fitting results with average positron lifetimes $\overline{\tau}_{ex-0}$ are listed in table 1. The time resolution of the PPMA determined by the PALSfit program was $FWHM \approx 190$ ps.

3.3. Data acquisition and analysis of positron lifetime spectra for scanning experiment

Positron lifetime spectra of YSZ, Kapton, Fe and Fe_{2nd} (Fe_{2nd} was measured at a different point on the same Fe sample) were measured in sequence (see figure 1) and $\overline{\tau}_{ex}$ obtained using formula (6). This cycle was iterated 50 times to check the time stability. The second Fe position, Fe_{2nd}, was used for calibration and the recorded number of annihilation events at Fe_{2nd} was three times larger than those for YSZ, Kapton and Fe, i.e. ($N_0(Fe_{2nd}) = 3 \times N_0$(Fe, Kapton, YSZ)). The measurement was repeated with $N_0 = 1000$, 2000, 5000, and 10000. A larger N_0 for Fe_{2nd} was used to reduce systematic errors as discussed below. A plot of $\overline{\tau}_{ex}$(Fe, YSZ, Kapton) as a function of cycle number, j, in the case of $N_0 = 5000$ is shown in figure 2(a). The values of $\overline{\tau}_{ex-0}$ given in table 1 are also shown by arrows on the right side in figure 2. Absolute values of $\overline{\tau}_{ex}$(Fe, YSZ, Kapton) were calibrated as the average $\overline{\tau}_{ex}$(Fe) for the first five results becomes consistent with $\overline{\tau}_{ex-0}$(Fe). At early cycle numbers j in figure 2(a), $\overline{\tau}_{ex}$(Kapton) is consistent with $\overline{\tau}_{ex-0}$(Kapton) while $\overline{\tau}_{ex}$(YSZ) is inconsistent with $\overline{\tau}_{ex-0}$(YSZ) by about 10 ps. This inconsistency can be accounted for by considering the difference of the time of flight (TOF) of positrons from the pulsing system to the surface of the sample. The TOF for the YSZ sample was larger than that for Fe and YSZ by $\delta TOF = \delta L/(2E/m)^{1/2} = 10.6$ ps. In addition, it is clearly seen that $\overline{\tau}_{ex}$ decreased with cycle number j (elapsed time). This drift is mainly caused by the change of room temperature with time. In order to compensate for this drift, absolute values of average (Fe_{2nd}) measured at j and $j+1$ were calibrated to the value of $\overline{\tau}_{ex-0}(Fe_{2nd})$ and then the $\overline{\tau}_{ex}$(Fe, YSZ, Kapton) measured at j were calibrated to this modified $\overline{\tau}_{ex}(Fe_{2nd})$ value. The systematic error due to this calibration is only 10% of $\delta\overline{\tau}_{ex}$(Fe, YSZ, Kapton). After calibration using $\overline{\tau}_{ex}(Fe_{2nd})$ and correction of δTOF, $\overline{\tau}_{ex}$ for all samples are consistent with $\overline{\tau}_{ex-0}$ and were stabilized in time as shown in figure 2(b). The standard deviations of $\overline{\tau}_{ex}$, $\delta\overline{\tau}_{ex}$ were 4.1, 4.6 and 5.9 ps for Fe, YSZ and Kapton in the case of $N_0 = 5000$. $\overline{\tau}_{ex}$ and $\delta\overline{\tau}_{ex}$ as a function of N_0 are in shown figure 3. The size of $\delta\overline{\tau}_{ex}$ can be explained well by formula (8), where experimentally obtained spectra ($N_0 = 10^6$) were used to obtian $P_R(t)$, is proportional to $N_0^{-1/2}$ (see solid lines). These experimental results were almost consistent with results of Monte Carlo simulations [4].

As an example of the above analysis, figure 4 shows an average positron lifetime image of a ductile-fractured iron sample. Figure 4(a) shows an image obtained by a scanning electron microscope (SEM) of the same sample. The PPMA image (figure 4(b)) consisting of about 8×10^3 pixels ($N_0 = 5000$ for each pixel) was successfully measured in a practically acceptable time (about one day).

Table 1. Lifetimes τ_i and their intensities I_i obtained by deconvolution of resolution function R from the lifetime spectra ($N_0 = 10^6$) using PALSfit are listed. $\overline{\tau}_{ex-0}$ are the average positron lifetimes.

Sample	τ_1 (ns)	I_1 (%)	τ_2 (ns)	I_2 (%)	$\overline{\tau}_{ex-0}$ (ns)
Fe (Fe_{2nd})	0.109	90.6	0.324	9.4	0.135
YSZ	0.186	100.0	-	-	0.186
Kapton	0.387	99.8	3.1	0.2	0.392

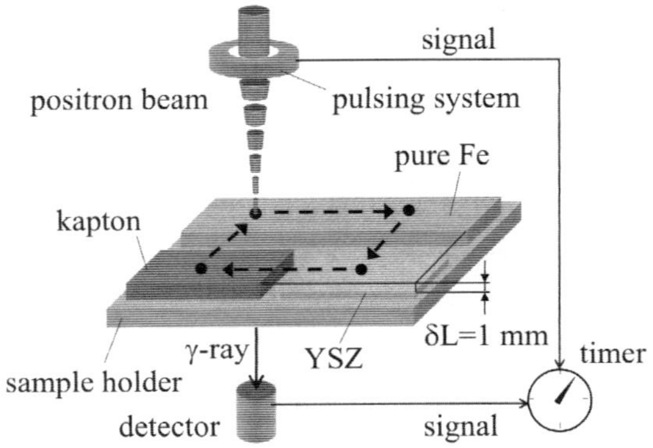

Figure 1. Schematic drawing of measurement procedure with the PPMA. Samples are mounted on the same holder and placed on an $x - y$ translation system.

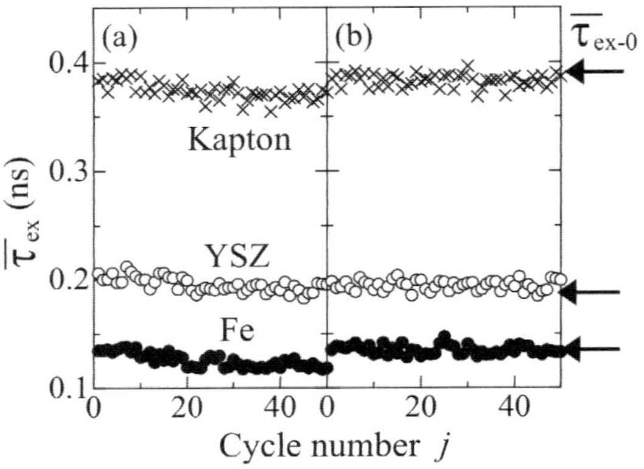

Figure 2. Fifty cycle measurements of $\overline{\tau}_{ex}$ in case of $N_0 = 5000$ for (a) before and (b) after improvement of statistical error and corrections of δTOF. $\overline{\tau}_{ex}$ in (b) were consistent with $\overline{\tau}_{ex-0}$ for all samples.

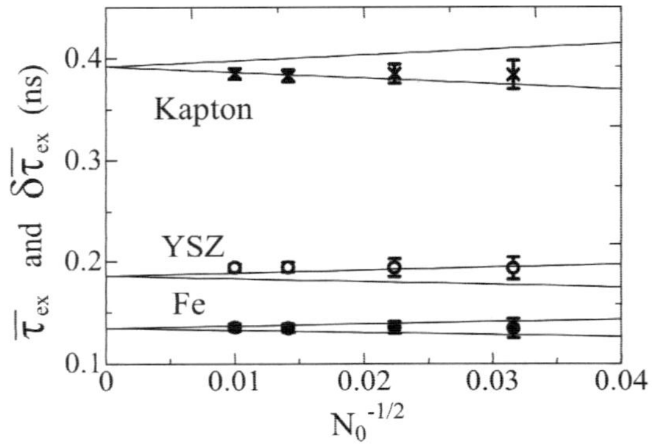

Figure 3. Dependence of $\overline{\tau}_{ex}$ and $\delta\overline{\tau}_{ex}$ on N_0. The $\delta\overline{\tau}_{ex}$ are shown as error bars at each plot of $\overline{\tau}_{ex}$. The size of $\delta\overline{\tau}_{ex}$ were well explained by formula (8) (solid lines).

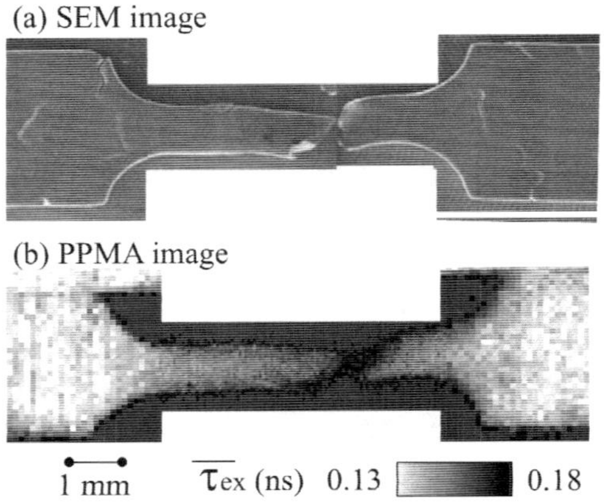

Figure 4. Image of a ductile-fractured iron sample obtained by (a) the SEM and (b) the PPMA. The PPMA image, consisting of about 8×10^3 pixels ($N_0 = 5000$ for each pixel), was measured in about one day.

4. Summary

Data acquisition techniques and analysis for obtaining average positron lifetime at each image using the PPMA are discussed. Using these methods average positron lifetime $\overline{\tau}_{ex}$ at each point were measured with an accuracy of around several ps by recording only $N_0 = 10^3 - 10^4$ positron annihilation events. The PPMA can therefore obtain average positron lifetime distribution images consisting of $10^3 - 10^4$ pixels within practically reasonable times.

References

[1] Oshima N, Suzuki R, Ohdaira T, Kinomura A, Narumi T, Uedono A, Fujinami M 2009 *Appl. Phys. Lett.* **94** 194104
[2] Krause-Rehberg R and Leipner H S 1999 *Positron Annihilation in Semiconductors, Solid-State Sciences* Vol. 127 (Berlin: Springer)
[3] Olsen J V, Kirkegaard P, Pedersen N J, Eldlup M 2007 *Phys. Stat. Sol. (c).* **4** 4004
[4] Suzuki R, Mikado T, Ohgaki H, Chiwaki M, Yamazaki Y 1995 *Mat. Sci. Forum.* **175-178** 217

Journal of Physics: Conference Series **262** (2011) 012045 doi:10.1088/1742-6596/262/1/012045

Microstructural studies of poly (perfluorosulfonic acid) membrane doped with silver nanoparticles using positron annihilation spectroscopy

Pushkar N Patil, K Sudarshan, S K Sharma, D Dutta, P Maheshwari and P K Pujari[1]

Radiochemistry Division, Bhabha Atomic Research Centre, Mumbai – 400085

[1]pujari@barc.gov.in

Abstract. The Nafion -117 (H^+ form) was converted to Ag^+ form by ion exchange. The silver nanoparticles were formed in the membrane by reduction of Ag^+ (of Ag^+ form membrane) using an ionic reducing agent, $NaBH_4$, and a neutral reducing agent, formamide. The free volume hole sizes in the nanoparticle doped and undoped Nafion were measured by positron annihilation lifetime spectroscopy (PALS). The surface modifications in the membrane by incorporation of nanoparticles were investigated using slow positron beam. The changes in the bulk free volume of Nafion with presence of nanoparticles were found to be nominal. In contrast, the surface microstructure appears to have undergone significant change.

1. Introduction

Poly (perfluorosulfonic acid) membrane (Nafion®) is one of the most studied synthetic membranes for various applications [1,2]. The Nafion-117 consists of polytetrafluoroethylene backbone and perfluoro ether side chains terminated by $SO_3^-H^+$ group. The technological importance of the Nafion is due to its selective permeability to cations, low resistance to current flow and excellent chemical, mechanical and thermal stability. The various properties of Nafion are dependent on its microstructure. The free volume properties of Nafion were shown to be strongly correlated to the diffusion of gases through the membranes [3]. Incorporation of nanoparticles into the membranes has been shown to change the properties of the membranes drastically [4]. The properties of nanoparticles in the matrix are strongly related to their size, shape and their spatial distribution in the matrix. In this regard, Nafion has received much attention as a host for variety of nanoparticles [5,6]. The incorporation of metal oxide nanoparticles in the Nafion was shown to improve proton conductivity and water retention properties [6]. The Nafion stabilized nanoparticles were also used as sensors [7,8]. Although there are many studies on the correlations of the physicochemical properties with the free volume structure of Nafion [3,9,10], the microstructural changes that may occur with incorporation of nanoparticles in the membrane have not been reported so far. The positron annihilation spectroscopy (PAS) is the powerful tool to study the free volume hole properties in the membranes [11]. The variable energy slow positron beam is used to probe the microstructure of the surfaces of thin films.

In the present work, Nafion-117 in H^+ form is converted to Ag^+ form by ion exchange method. The Ag^+ ions were reduced, to form nanoparticles in the membrane, using either sodium borohydride ($NaBH_4$) or formamide. The resulting microstructural changes were studied using PAS.

2. Experimental

2.1. Sample preparation

Nafion-117 ion exchange membrane having thickness of 178 μm, IEC of 0.91 mmol/g and density 1.95 g/cc was purchased from Aldrich. The H^+ form of the Nafion was equilibrated with 0.5 M NaCl solution for 30 h to convert to the Na^+ form. It was further converted to Ag^+ form by immersing in 0.25 M $AgNO_3$ for 48 h [12]. The silver nanoparticles were formed in the Nafion membrane by reduction of Ag^+. For this purpose Ag^+ form of the membranes were immersed in 0.66 M $NaBH_4$ or formamide for 30 mins, with continuous stirring. The membrane samples were washed with boiled water to remove any unwanted residue.

2.2. Positron annihilation studies

The PALS measurements were performed at room temperature using a fast-fast coincidence system consisting of two BaF_2 scintillation detectors. The time resolution of the spectrometer was 350 ps. For PALS measurements, 20 Ci source of ^{22}Na deposited on 8 m Kapton film was sandwiched between stack of Nafion membrane with four pieces on each side and wrapped in an aluminum foil. Total area under each lifetime spectra was about 10^6. The positron lifetime spectra were fitted using PATFIT-88. The continuous distribution of lifetime was obtained using CONTIN. The free volume hole radii were calculated from o-Ps lifetimes. The details of the data analysis are given elsewhere [13].

Depth dependent Doppler broadening measurements of the surface of the membrane were carried out using a slow positron beam at BARC, in the energy range 200 eV- 10 keV. A 30% relative efficiency high purity germanium detector with a resolution of 2 keV at 1332 keV photo peak of ^{60}Co was used for Doppler broadening measurements and spectra with 2×10^5 counts were acquired at each energy. The ratio of the counts within ~ 2.0 keV energy window centered at 511keV and total photo peak area was used to evaluate the *S*-parameter. The $3\gamma/2\gamma$ ratio or relative Ps yield was estimated as the ratio of total counts in the 375-480 keV region of the annihilation gamma-spectra to counts under the 511 keV peak.

3. Results and discussion

The two reducing agents used in the present study to reduce Ag^+ in the membrane to silver nanoparticles are $NaBH_4$ and formamide. The Nafion is a cation exchange membrane and does not allow negative ions to permeate into it. When borohydride (BH_4^-) is used as a reducing agent, Ag^+ ions at the ion exchange sites are replaced with Na^+ ions as represented below,

$$[-R-SO_3^-Ag^+]_{mem}+[NaBH_4]_{aq}\rightarrow[-R-SO_3^-Na^+]_{mem}+[BH_4^-]_{aq}+[Ag^+]_{mem}$$

and the replaced Ag^+ ions moving out of the membrane are reduced to Ag nanoparticles near the surface regions of the membrane by BH_4^- as

$$[BH_4^-]_{aq}+[Ag^+]_{mem}+3H_2O\rightarrow[H_3BO_3]_{aq}+[Ag^o]_{mem}+\tfrac{7}{2}H_2(g)$$

Thus, the use of borohydride as reducing agent renders the Nafion membrane in the Na^+ form with nanoparticles concentrated near the surface regions extending up to ~15 μm [12] from the surface. Formamide, on the other hand, is neutral and it can invade the matrix of the Nafion. The reduction of silver ions in the membrane by formamide can be represented as

$$2[-R-SO_3^-Ag^+]_{mem}+[HCONH_2]_{aq}+[H_2O]_{mem}\rightarrow 2[-R-SO_3^-H^+]_{mem}+[NH_2COOH]_{aq}+2[Ag^0]_{mem}$$

The carbamic acid decomposes $NH_2COOH\rightarrow CO_2(g)+NH_3$. Thus, the use of formamide as reducing agent gives the Nafion with homogeneously distributed nanoparticles within the membrane and leaves the backbone in the H^+ form. Further details about the spatial distributions of silver nanoparticles in Nafion with preparation method are described by Kumar et al. [14]. The Nafion membranes with silver nanoparticles produced using $NaBH_4$ and formamide as reducing agents are referred to as Na^+/Ag and H^+/Ag, respectively, while the Nafion without nanoparticles are referred to with cation present in the ion exchange sites. The formation of nanoparticles in these membranes was confirmed by XRD measurements and the size of the nanoparticles estimated using Debye-Scherrer formula were found to be 15±5 nm and 10±3 nm respectively. The results are in agreement with the sizes reported in the literature [14].

The o-Ps lifetimes and intensities in the Nafion samples with and without nanoparticles are shown in figure 1. The pore radius distributions calculated from the lifetime distributions are given in figure 2. The o-Ps lifetime in the H^+ form of Nafion is in reasonable agreement with the reported values [5,9]. The Nafion in Na^+ form showed marginally lower o-Ps lifetime than the H^+ form. The Ps lifetimes in the Nafion were shown to be sensitive to water retention in the membrane [9]. The shorter o-Ps lifetime in the Na^+ exchanged Nafion than the H^+ form might be due to less retention of water in the former because of the larger ionic size of Na^+. The Ag^+ form showed significantly lower o-Ps lifetime and intensity. The Ag^+ ions are known to be good scavengers of electron and hence inhibit the positronium formation [15]. The changes in the o-Ps lifetimes and intensities in Ag^+ loaded membrane might be due to chemical effects of Ag^+ rather than the free volume changes. The changes in the distribution of bulk o-Ps lifetimes or pore radii distributions in the membrane on introduction of nanoparticles are found to be marginal (Figure 2). However, the lowering in the intensity of o-Ps is significant and well beyond the statistical uncertainties (Figure 1). This shows that the nanoparticles are blocking some of the free volume holes in the Nafion membrane, though this fraction seems to be very small.

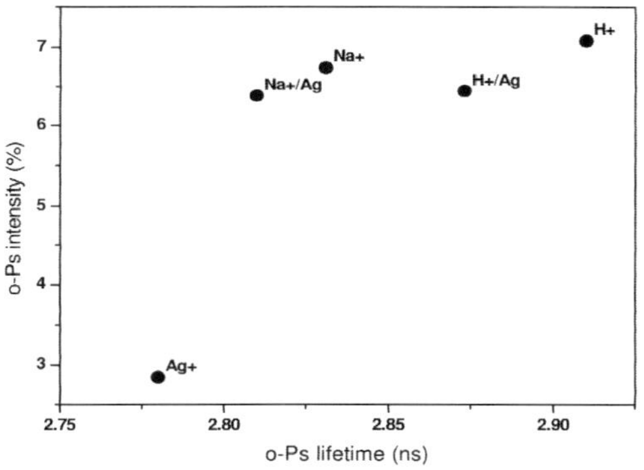

Figure 1. The o-Ps lifetimes and intensities in Nafion-117 samples

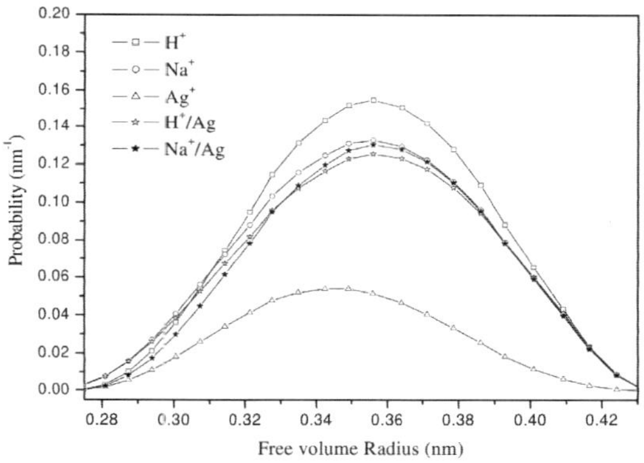

Figure 2. The free volume hole radius distributions in Nafion-117 samples

The slow positron beam studies have been carried out to understand the microstructural changes on the surface of the membrane. The energy dependent *S*-parameter in the membranes is shown in Figure 3. The mean implantation depth of the positrons in the film is also shown in the figure. The implantation depth (in μm) is calculated as $0.040\times E^{1.6}/\rho$, where E is the positron energy in keV and ρ is the density of the membrane in g/cc. The density differences that can arise due to incorporation of nanoparticles are ignored here in calculating the mean implantation depth. It can be seen from the figure that the maximum depth probed is ~1μm. Hence, the probing region only corresponds to a small part of surface nanoparticles even in the borohydride reduced samples where nanoparticles are spread over ~15μm [12]. It is interesting to note that the surface *S*-parameters in the nanoparticles doped membranes (in both H^+ and Na^+) are significantly higher than their counter parts without nanoparticles. The relative 3γ/2γ ratio in the surface of the membranes in the nanoparticles doped samples is also higher than the undoped samples as shown in figure 4, with exception of the Ag^+ form of Nafion. The higher *S*-parameter and higher Ps yield (3γ/2γ ratio) in the surface layers of the nanoparticles doped membranes suggest an increase in the free volumes in these regions contrary to the average bulk. The higher *S*-parameter in the surface of Ag^+ form of membrane with lower Ps yield might be due to formation of a metallic layer of silver on the surface of the membrane when the sample is stored in air for a longer duration. The possible increase in the free volumes or open porosity near the surface regions might be due to the release of gaseous byproducts during reduction of Ag^+ to silver nanoparticles.

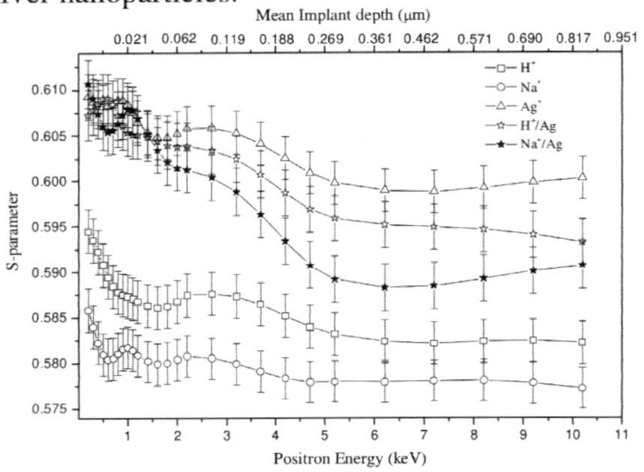

Figure 3. The *S*-parameter as a function of positron energy and mean implantation depth.

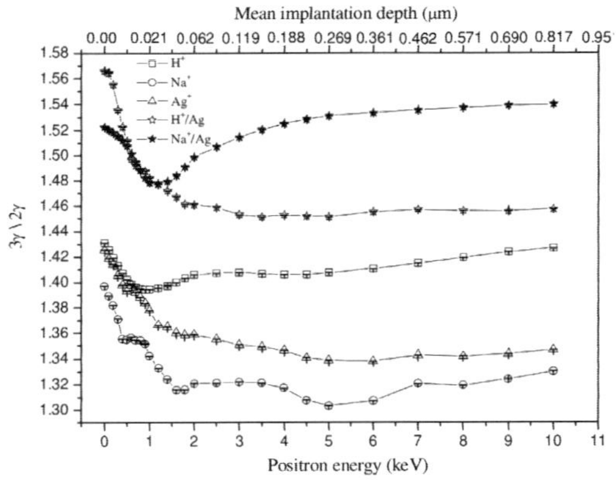

Figure 3. The 3γ/2γ ratio as a function of positron energy and mean implantation depth.

4. Conclusion

The microstructural changes in the Nafion-117 membrane with the formation of silver nanoparticles have been monitored using conventional positron annihilation lifetime spectroscopy and slow positron beam. The studies indicate reduction in the number of free volume holes without much change in the average hole radii with formation of nanoparticles in both H^+ and Na^+ forms of Nafion. However, the surface of the Nafion membrane undergoes significant microstructural changes i.e. becomes more porous on formation of nanoparticles as indicated by high Ps yield and high *S*-parameter. This is probably caused by release of gaseous byproducts during the reduction of Ag^+ ions to nanoparticles.

5. References

[1] Heitner-wirguin C, 1996. *J. Membr. Sci.* **120**, 1-33.
[2] Mauritz KA, Moore RB, 2004. *Chem. Rev.* **104**, 4535-4585.
[3] Mohamed HFM, Kobayashi Y, Kuroda CS, and Ohira A, 2010. *J. Phys: Conf. Ser.* **225**, 012038.
[4] Giannelis EP, 1996. *Adv. Mater.* **8**, 29-35.
[5] Ladewig BP, Knott RB, Hill AJ, Riches JD, White JW, Martin DJ, Diniz da Costa JC, and Gao Qing Lu, 2007. *Chem. Mater.* **19**, 2372-2381.
[6] Ke Li, Gongbo Ye, Jingjing Pan, Haining Zhang and Mu Pan, 2010. *J. Membr. Sci.* **347**, 26-31.
[7] Sujie Xing, He Xu, Junshui Chen, Guoyue Shi, Litong Jin, 2010. *J. Electroanal. Chem.* (In press)
[8] Keith B. Male, Sabahudin Hrapovic, Yali Liu, Dashan Wang, John HT Luong, 2004. *Anal. Chim. Acta.* **516**, 35-41.
[9] Sodaye HS, Pujari PK, Goswami A, and Manohar SB, 1997. *J. Polym. Sci. B: Polym. Phys.* **35**, 771-776.
[10] Shibahara Y, Sodaye HS, Akiyama Y, Nishijima S, Honda Y, Isoyama G, Tagawa S, 2010. *J. Power sources,* **195**, 5934-5937.
[11] Pethrick RA, 1997. *Prog. Polym. Sci.,* **22**, 1-47.
[12] Sachdeva A, Sodaye S, Pandey AK, and Goswami A, 2006. *Anal. Chem.* **78**, 7169-7174.
[13] Patil PN, Kathi S, Dutta D, Pujari PK, 2010. *Polym. Bull.,* **65**, 577-587.
[14] Kumar R, Pandey AK, Dhara S, Mishra NL, Ramagiri SV, Bellare JR, and Goswami A, 2010. *J. Membr. Sci.* **352**, 247-254.
[15] Ache HJ, Wild RE, Bartal LJ, 1977. *J. Phys. Chem.* **81**, 941-943.

Journal of Physics: Conference Series **262** (2011) 012046 doi:10.1088/1742-6596/262/1/012046

Positrons in gas filled traps and their transport in molecular gases

Z Lj Petrović [1], A Banković [1], S Marjanović [1], M Šuvakov[1], S Dujko[1,2,3], G Malović[1], R D White[3] and S J Buckman[4]

[1]Institute of Physics, University of Belgrade, Pregrevica 118, POB 68, Zemun, Serbia

[2]Centre for Mathematics and Informatics, POB 94079, 1090 GB Amsterdam, The Netherlands

[3]ARC Centre for Antimatter-Matter Studies, James Cook University, Townsville 4810, QLD, Australia

[4]ARC Centre for Antimatter-Matter Studies, The Australian National University, Canberra, ACT, 0200, Australia

zoran@ipb.ac.rs

Abstract. In this paper we give a review of two recent developments in positron transport, calculation of transport coefficients for a relatively complete set of collision cross sections for water vapour and for application of they Monte Carlo technique to model gas filled subexcitation positron traps such as Penning Malmberg Surko (Surko) trap. Calculated transport coefficients, very much like those for argon and other molecular gases show several new kinetic phenomena. The most important is the negative differential conductivity (NDC) for the bulk drift velocity when the flux drift velocity shows no sign of NDC. These results in water vapour are similar to the results in argon or hydrogen. The same technique that has been used for positron (and previously electron) transport may be applied to model development of particles in a Surko trap. We have provided calculation of the ensemble of positrons in the trap from an initial beam like distribution to the fully thermalised distribution. This model, however, does not include plasma effects (interaction between charged particles) and may be applied for lower positron densities.

1. Introduction

It has recently become possible to model positron transport in gaseous media based on the measured scattering cross sections. The efforts of groups in San Diego [1-2], Trento [3], London [4], Detroit [5] Swansea [6] and recently Canberra [7,8] led to measurements of a wide range of data which cover all relevant momentum and energy transfer processes as well as the processes that change the number of particles (annihilation and positronium formation). Sometimes, however, one needs to complete the set with theoretical results which are also becoming reasonably abundant [9-11], preferably those results based on theories that were tested against experiments on other similar targets. In any case sufficiently detailed and complete sets are now available for argon, hydrogen, nitrogen and few more relevant gases.

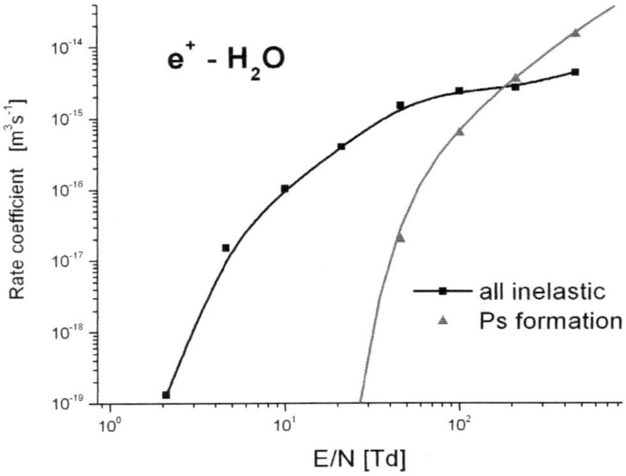

Figure 1. Rate coefficients for positrons in water vapour: Ps formation and inelastic losses.

Having in mind that the principal constituent of a human body is H_2O, we attempted to compile a set of cross sections that could be used to model behaviour of positrons in a human body. While a combination of experimental data and calculations was sufficient with some extrapolations to form the set, we have taken advantage of the recent measurements at the ANU in Canberra to provide two of the most important cross sections [8], the total scattering and the positronium formation cross section. It is important to note that all of the relevant processes, while providing a good balance and effective behaviour of the positron swarms, cannot be tested against the measured transport data yet as such experiments (with two exceptions [12,13]) are not available at the moment in a sufficiently wide range of normalized electric fields (E/N).

In our previous reviews of preliminary results we presented the first results for positron transport including the data for argon and some aspects of transport in molecular gases and in the crossed magnetic and electric fields [14]. Finally we have dealt with thermalization of positronium and transport in liquids which are associated with applications of positron transport [15]. Final results have been published for argon [16] and partly for nitrogen [18] with a special analysis of the negative differential conductivity (NDC) induced by positronium formation [16].

In this paper we present new developments with transport of positrons in water vapour and first attempts to model low density positron traps in gases.

2. Positron transport in water vapour

Before addressing the results we should perhaps discuss whether one has sufficient motivation for swarm (transport) studies of positrons and how those could be justified by applications.

The time scales of possible swarm experiments in gases that are at a high enough pressure are very short and yet positrons survive long enough to allow us determination of transport coefficients [12,13]. A review of past, existing and possible future swarm experiments with the discussion how to overcome formidable difficulties facing studies of positron swarms was given by Charlton [19]. It is important to note that a swarm system should not be viewed as a competition with traps in efficiency of storing swarms. Absence of multiplication (as compared to electrons) does not allow us to form stable ensembles of particles, so for any measurement the price is in the loss of particles. Such an experiment would thus provide transport data allowing us to adjust the desired properties by electric and magnetic fields- see accompanying paper [20], model some of the applications and also giving us a possibility to test and benchmark the calculations of the models of such applications. This is especially important as it would be very difficult to base all the comparisons on predicted trajectories and some averaged, accurately measured and calculated, data are needed. Finally such experiments would provide a natural occurrence of some of the predicted and yet to be predicted kinetic phenomena driven by non-conservative processes, Ps formation in this case.

Transport coefficients in water vapour calculated by a standard Monte Carlo procedure [16] show similar features as in argon and hydrogen. In other words, while mean energy is similar to what was observed for electrons, the drift velocity shows the duality of drift velocities, with NDC unlike anything seen for electrons (although the effect was predicted for electrons [21] in a model gas). The ***flux*** drift velocity increases gradually throughout the range of electric fields covered and at the same

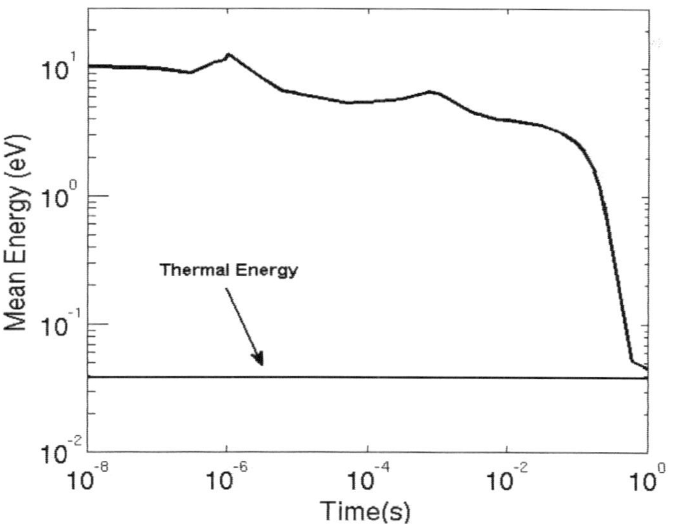

Figure 2. Temporal development of the mean kinetic energy of positrons in Surko trap.

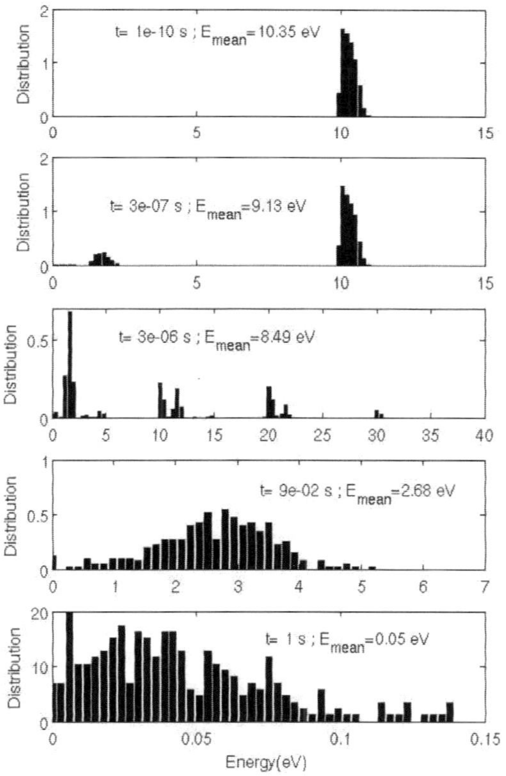

Figure 3. Temporal development of the energy distribution function (probability of finding particles in a given energy bin) of positrons in a trap.

time the ***bulk*** drift velocity shows a very pronounced NDC with a maximum difference between the two that is greater than an order of magnitude. With the exception of hydrogen there are no experiments confirming such drift velocities. Our good experience with electron swarms teaches us that our predictions are quite accurate but one would prefer to have the transport data to normalize the entire sets of cross sections.

One example of the data shown here in Fig. 1 is the rate for Ps formation compared to other elastic and inelastic processes.

3. Monte Carlo modelling of a Penning Malmberg Surko trap

The initial development and analysis of a Surko trap was done on the basis of assumed beam like properties of particles in the trap. Yet particles in such traps start from a beam like distribution and thermalize to an isotropic Maxwellian distribution at room temperature. In that respect the Surko trap is similar to the Frank Hertz experiment that will have its 100th anniversary in 2014. Very much like in the case of the Frank Hertz experiment [22] we believe that proper modeling of the Surko trap may be performed by swarm techniques that may provide description of a non-hydrodynamic particle kinetics (when the swarm is not in equilibrium with the local field but may depend on the spatial or temporal history).

In Figure 2 we show temporal development of the mean energy of positrons in trap. At the same time we show energy distributions at several times after the particles have been introduced to the trap. Unlike the real trap where particles are fed into the trap continuously here we introduce the whole batch together in order to follow the temporal development.

While providing the qualitative physical picture of the trap that is already available, the modeling also does so in a quantitative manner thus allowing comparisons between different gases, different fields, analysis of a possible benefit of added axial or radial electric fields.

Acknowledgements
This work was funded by the MNTRS project 141025, COE Nonequilibrium processes in plasmas and environmental science, ARC Centre for Antimatter-Matter Studies and the International Science Linkages Program. Authors are grateful to J. Marler, W. Tattersall, J. Sullivan and C. Makochekanwa for collaboration and advice and to G. Garcia, C. Surko and M. Charlton for many useful comments.

References

[1] Marler JP and Surko CM 2005 *Phys. Rev. A* **72** 062702
[2] Sullivan JP, Gilbert SJ, Marler JP, Greaves RG, Buckman SJ and Surko CM 2002 Phys. Rev. A 66 042708
Marler JP, Sullivan JP and Surko CM 2005 *Phys. Rev. A* **71** 022701
[3] Zecca A, Chiari L, Sarkar A, Lima MAP, Bettega MHF, Nixon KL and Brunger MJ 2008 *Phys. Rev.* A **78** 042707
Zecca A, Sanyal D, Chakrabarti M and Brunger MJ 2006 *J. Phys. B: At. Mol. Opt. Phys.* **39** 1597
[4] Laricchia G, Van Reeth P, Szluinska M and Moxom J 2002 *J. Phys B.* **35** 2525
[5] Kauppila WE, Stein TS, Smart JH, Dababneh MS, Ho YK, Downing JP and Pol V 1981 *Phys. Rev.* A **24** 725
Hoffman KR, Dababneh MS, Hsieh YF, Kauppila WE, Pol V, Smart JH, Stein TS 1982 *Phys. Rev.* A **25** 1393
[6] Charlton M 1985 *Rep. Prog. Phys.* **48** 737
Charlton M and Humberston JW 2001 *Positron Physics* (Cambridge University Press).
[7] Sullivan J P, Makochekanwa C, Jones A, Caradonna P and Buckman S J, 2008 J. Phys. B: At. Mol. Opt. Phys. **41** 081001
[8] Makochekanwa C, Bankovic A, Tattersall W, Jones A, CaradonnaP, Slaughter D.S, Nixon K, Brunger M.J., Petrovic ZLj, Sullivan J.P. and Buckman S.J., 2009 New Journal of Physics **11** 103036
[9] McEachran R P 2006 Private communication
[10] Gilmore S, Blackwood JE and Walters HRJ 2004 *Nucl. Instr. Meth. B* **221** 129
[11] F.A. Gianturco, T. Mukherjee, Phys. Rev. A 55 (1997) 1044.
[12] Bose N, Paul D. A. L., and Tsai J. S., 1981 J. Phys. B 14, L227.
[13] Charlton, M 1985 J. Phys. B 18, L667
[14] Marler JP, Petrović ZLj, Banković A, Dujko S, Šuvakov M, Malović G and Buckman SJ 2009 *Phys. Plasmas*
[15] Sullivan J P, Buckman S J, Jones A, Caradonna P, Makochekanwa C, Slaughter D, Petrović Z Lj,. Banković A, Dujko S, Marler J P and White R D 2009 Journal of Physics: Conference Series **162** (2009) 012002
[16] Šuvakov M, Petrović ZLj, Marler JP, Buckman SJ, Robson RE and Malović G 2008 *New J. Phys.* **10 No 5** 053034
[17] Banković A, Petrović ZLj, Robson RE, Marler JP, Dujko S and Malović G 2008 Nucl. Instrum.Methods B
[18] Banković A, Marler JP, Šuvakov M, Malović G and Petrović ZLj 2008 *Nuclear Inst. Methods, B* **266** 462
[19] Charlton M, *Journal of Physics: Conference Series* **162** (2009) 012003
[20] Banković A, Dujko S , White RD, Malović G , Buckman SJ and Petrović ZLj, this issue
[21] Vrhovac SB and Petrović ZLj 1996 *Phys. Rev. E* **53** 4012
[22] Dujko S, White R D and Petrović Z Lj, 2008 J. Phys. D: Appl. Phys. 41 025206

Journal of Physics: Conference Series **262** (2011) 012047 doi:10.1088/1742-6596/262/1/012047

Kinematically complete experiments for positron-impact ionization of helium atoms at the NEPOMUC facility

T Pflüger*, M Holzwarth*, A Senftleben*, X Ren*, A Dorn*, J Ullrich*, L R Hargreaves‡, B Lohmann‡, D S Slaughter†1, J P Sullivan†, J C Lower† and S J Buckman†

*Max-Planck-Institute for Nuclear Physics, Heidelberg, Germany
‡Centre for Antimatter - Matter Studies, University of Adelaide, Adelaide, Australia
†Centre for Antimatter - Matter Studies, Australian National University, Canberra, Australia

E-mail: A.Dorn@mpi-hd.mpg.de

Abstract. Positron impact ionization of helium is studied with fully resolved momentum vectors of all continuum particles. An imaging multi-particle momentum spectrometer (reaction microscope) detecting all final state particles over the full solid angle was applied. This apparatus was connected to the NEPOMUC facility delivering intense positron beams tuneable over a large energy range with good beam quality. At 80 eV impact energy about 5000 triple coincidence events were collected. Cross sections as function of the longitudinal particle momentum show strong differences compared to respective electron impact ionization data most likely originating from the reversed post collision interaction in both cases. Calculations using the 3 Coulomb wavefunction method show clear discrepancies from the experimental results.

1. Introduction

The scientific and technological uses of low energy positrons are numerous including the formation of neutral antihydrogen atoms in order to test quantum electro dynamics (QED), fundamental symmetries in nature (the CPT theorem) or the action of gravity on antimatter. Examples for technological applications of positrons are positron emission tomography (PET) for material characterization or the study of metabolic processes, plasma diagnostics and new ways to study clusters and nanoparticles. All applications depend on the quantitative understanding of the basic interactions of positrons with matter. The project presented here aims to study ionizing collisions of positrons with light atoms (H, He) or molecules (H_2), representing the most fundamental dynamical few-body systems, in unprecedented detail. Insight into the dominant dynamical mechanisms involved and benchmark data for theory are urgently required since, while nearly exact ab initio calculations are available for static bound states of quantum mechanical few-body systems, dynamical reactions have proven to be much more difficult to be solved. Even for the simplest non-trivial system, the three-body problem as it is realized in low energy electron impact single ionization of hydrogen, accurate theoretical calculations have become available only within the recent years [1]. More complex systems, e.g., involving multi-electron targets as helium still pose severe problems for theory [2] and can be treated only in an approximate way. The most stringent test of calculations and the most detailed insight into the reaction is obtained by kinematically complete experiments in which the momentum vectors of all continuum particles are determined. For electron impact, such so-called (e,2e) experiments became possible very early, beginning from 1969, and a large part of our present knowledge on ionization of atoms by charged particle impact is based on these studies.

Collisions involving positrons are expected to be significantly different from analogous collisions involving electrons in particular in the low energy range up to ≈ 100 eV providing new tests of our understanding of basic atomic physics [3] (for a recent review see [4]). Reasons are the absence of the Pauli exclusion principle and therefore of the exchange interaction characteristic of electrons, the added richness of the positronium channel and the repulsive short-range positron-atom interaction, in contrast to the attractive electron-atom interaction. Experimental data in particular for differential cross sections are scarce. For ionization, so far, mainly integrated cross sections were measured and fully differential cross section data exist only for one distinguished collision geometry where both, the ionized target electron and the positron projectile are emitted into the forward direction after the collision [5, 6]. Reasons are the lack of available intense positron beam sources and the low efficiency of conventional experimental coincidence spectrometers. Recently, first results from highly efficient multi-particle imaging spectrometers were presented at higher impact energy [7]. The project described here aims to set-up a multi-particle imaging spectrometer (reaction microscope) with large phase space acceptance, improved momentum resolution and high efficiency. Here we report on the results of a first pilot beam time at the NEPOMUC positron beam facility at the neutron research reactor FRMII in Garching.

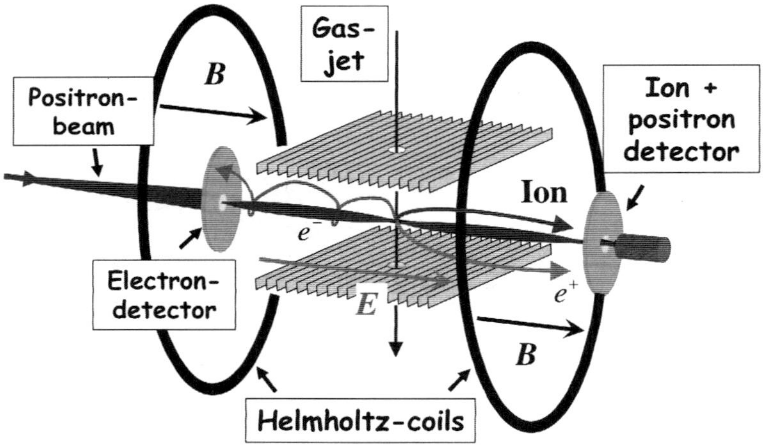

Figure 1. Scheme of the apparatus.

2. Experimental technique

2.1. The reaction microscope

A sketch of the apparatus is shown in Figure 1. This reaction microscope is normally used for electron impact ionization studies [8] and has been adapted for the use of a positron projectile beam. A collimated projectile beam crosses a target gas jet producing a well defined reaction volume of mm size. The reaction microscope technique [9] is based on the projection of all charged final state particles produced in an ionization reaction by means of homogenous electric and magnetic fields parallel to the projectile beam axis onto microchannel plate (MCP) detectors. The scattered positrons and the recoiling ions are registered by the detector in forward direction with respect to the incoming beam and the ejected electron by the backwards detector. The momentum vectors which the particles obtained in the collision then are calculated from their times-of-flight (TOF) and detection positions. As indicated in Figure 1, the cyclotron trajectory of the electron in the 6 Gauß magnetic spectrometer field serves to transversally confine the trajectories of the final state electron to the sensitive area of the detectors which is 8 cm in diameter. Furthermore, it guides and focuses the incoming projectile beam into the target gas jet and subsequently into a central bore of the MCPs of the forward detector. The target gas beam is produced by supersonic expansion of the gas with 20 bar stagnation pressure through a 30 μm nozzle into the vacuum. A two stage collimation and differential pumping using subsequent skimmers finally delivers a gas beam of 2 mm cross section with 10^{12} $1/cm^3$ density in the collision point of the main chamber at a background pressure of 10^{-8} mbar. The supersonic expansion delivers an internal cold target with a temperature of roughly 1 K. This is required to provide a low target atom momentum spread and, therefore, a momentum resolution of about 0.1 – 0.2 a.u. for the ions produced in collisions with the positrons.

2.2. Attachment to the NEPOMUC facility

The spectrometer was attached to the open beam port of the NEPOMUC positron beam facility at the research reactor FRMII [10]. Here, nominally a flux of 10^7 positrons per second with 20 eV kinetic energy is delivered. The beam diameter is 2 mm and it is guided in a 60 Gauß solenoidal magnetic field. In order to transfer the beam into the reaction microscope an adiabatic field transition stage was used along which the field smoothly decreased over a distance of 60 cm from 60 Gauß to 6 Gauß. For optimal conditions the B-field reduction by a factor of 10 gives rise to the increase of the positron beam cross section area by the same factor. Therefore, a subsequent collimation with an aperture to the original beam size of 2 mm should result in a positron flux of 10^6 1/s. In reality only about $2{\cdot}10^5$ 1/s were obtained using a collimation aperture of 5 mm measured with the positron MCP detector of the spectrometer. The reasons could not be unambiguously clarified during the beam time and have to be investigated in the future. There could be imperfections in the magnetic field transition stage or in the beam transportation line geometry which was passing by the reaction microscope rather close and therefore the B-fields from the spectrometer and the beam line interfered. During the experiment a triple coincidence rate for He ionization of roughly 1/120s and altogether 5000 events within the available beam time were obtained.

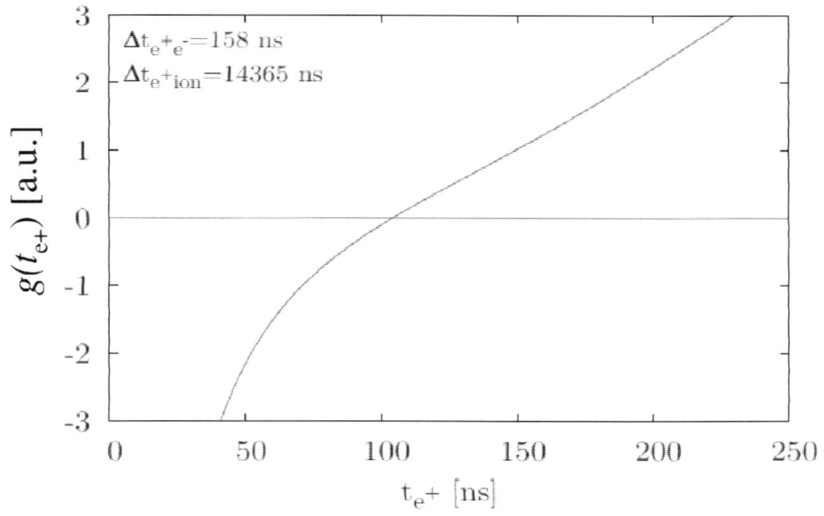

Figure 2. The dependence of the functional given in equation (1) from the positron TOF for particular relative TOFs of the particles shown in the diagram

2.3. Particle time of flight determination

In order to calculate the initial momentum of the charged particles produced in a collision their TOF to the MCP detector is required. Since the projectile beam is not pulsed the collision time is not provided directly. For high projectile beam energies in the range of several hundred eV the timing signal of the scattered and detected positron is a good time reference for the collision time since in most cases the relative energy loss in the collision is small compared to absolute positron energy and its TOF, therefore, in a good approximation is a constant. At low energies as it is the case here the scattered positron's TOF varies considerably from collision to collision and this is a signature of the full three body complexity of the reaction we are interested in.

Experimentally measured quantities are the relative TOF of the particles with respect to each other, i.e. the time difference between the positron and the electron Δt_{e+e-} and between the positron and the recoiling ion Δt_{e+He+}. If we take the positron TOF t_{e+} as an unknown variable we can express the longitudinal momena of all particles to be a function of this quantity:

$$k_{e+}^{\parallel} = f(t_{e+})$$
$$k_{e-}^{\parallel} = f(t_{e-}) = f(t_{e+} + \Delta t_{e+e-})$$
$$k_{He+}^{\parallel} = f(t_{He+}) = f(t_{e+} + \Delta t_{e+He+})$$

Now we make use of momentum conservation requiring that the sum of all final state longitudinal momenta is equal to the incoming projectile momentum assuming that the target atom is standing still initially.

$$k_{Sum}^{\parallel}(t_{e+}) - k_0^{\parallel} = g(t_{e+}) = 0 \quad (1)$$

Thus, finding the root of the functional $g(t_{e+})$ provides the required positron TOF t_{e+} from which the other particles' TOFs t_{e-} and t_{He+} can be obtained. As an example in Figure 2 $g(t_{e+})$ is plotted as function of t_{He+} for particular relative TOFs Δt_{e+e-} and Δt_{e+He+}. The accuracy of this procedure is about 0.1 – 0.2 a.u. in longitudinal momentum, i.e. the longitudinal momentum resolution of the spectrometer.

3. Results

Altogether about 5000 triple coincidence events were collected during a data accumulation time of roughly 5 days. This number is not sufficient for the extraction of fully differential cross sections requiring the selection of a small subset of these data by fixing, e.g., the projectile scattering angle, an ejected electron solid angle element with respect to the projectile scattering plane and its kinetic energy within certain constraints. In order to do so about one to two orders of magnitude more events are required. Therefore, here partially integrated data are presented. In Figure 3 (top diagram) the cross section is plotted as function of the longitudinal momentum of the scattered positron as well as function of the ejected electron. For comparison a respective cross section for electron impact ionization at slightly higher impact energy of 100 eV is also shown in Figure 3 (bottom diagram). Since in this case ejected and scattered electrons cannot be distinguished only one curve is plotted showing also a sharp peak at high forward momentum which can be assigned to forward scattered projectiles and a broad maximum around zero momentum corresponding to ionized electrons. For positron impact, the situation is clear since the projectile and the ejected electron can be distinguished. As expected a clear forward emission of the positron is found with a rather sharp cut-off close to the maximum momentum at 2.1 a.u. corresponding to a forward scattered positron which has lost just the He ionization energy. Interestingly also the emitted electrons are strongly forward emitted with the cross section maximum close to 1 a.u. This is in strong contrast to the results obtained for electron

Journal of Physics: Conference Series **262** (2011) 012047 doi:10.1088/1742-6596/262/1/012047

impact ionization where the maximum of the emitted electrons is found at zero momentum. A likely explanation for this behaviour is the different post collision interaction (PCI) in both cases. For positron impact, the emitted electron is attracted by the scattered positron and dragged forward. For electron impact the ionized electron is repelled by the scattered projectile. As a result, the projectile peak is also broader for positron impact, since there is additional energy loss due to PCI while the contrary is true for electron impact.

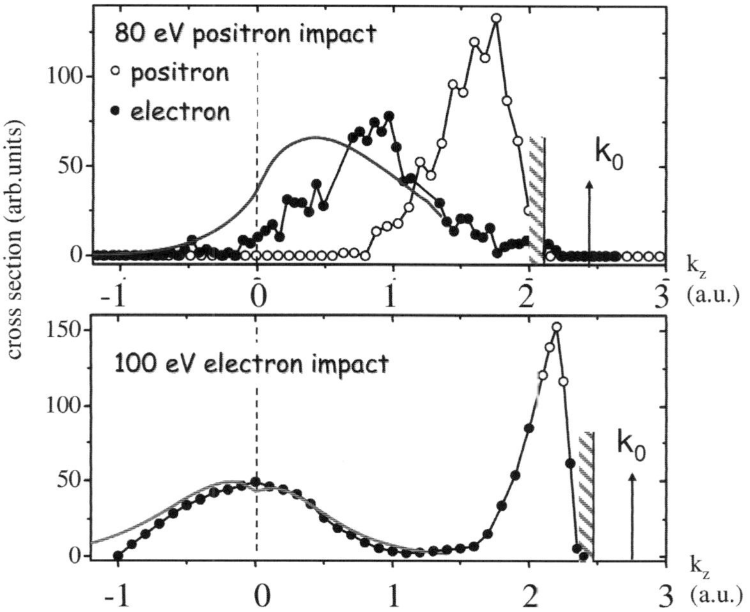

Figure 3. Positron (top) and electron (bottom) impact single ionization cross sections as a function of the longitudinal momentum of one final state continuum particle. This is the scattered positron or the ejected electron in case of positron impact and one free electron in case of electron impact. The plotted curves are 3C-calculations.

The measurements are compared with calculations using the 3-Coulomb wavefunction method (3C) which takes into account the Coulomb interactions within each two-body subsystem in the final state continuum. While this wavefunction is not an exact solution of the three-body problem at small inter-particle distances it has proven to yield reliable fully differential cross sections for electron impact ionization in the energy range down to about 100 eV [8]. Indeed, the calculation compares rather well with experiment in case of electron impact ionization showing its cross section maximum also close to zero momentum and a width slightly larger than experimentally observed. For positron impact, there is a significant deviation of the cross section maximum which is shifted to lower momentum by almost 0.5 a.u. The origin of this discrepancy is not clear and might indicate difficulties of the model to describe the increased interaction of two forward emitted Coulomb particles.

4. Conclusions

In summary we have demonstrated the feasibility of kinematically complete experiments for low energy positron impact ionization of atoms using a reaction microscope in combination with a magnetically guided continuous positron beam. For 80 eV impact energy, triple coincidences of all final state particles were recorded allowing the reconstruction of their absolute times-of-flight by exploiting momentum conservation. As an example, cross sections as function of the longitudinal momentum of the scattered projectile and the ejected electron were presented and compared with respective electron impact ionization data and calculations using the 3C model. Significant differences of the emitted electron distribution were found between positron and electron impact ionization and assigned to the reversed PCI in both cases. The origin of the discrepancy found in comparison with 3C theory is not clear so far. Here a respective calculation within a different theoretical approach would be helpful. In future fully differential cross sections will be obtained using longer data accumulation times or higher positron beam intensities combining a reaction microscope with the Australian Positron Beamline Facility [11].

Acknowledgements
The authors gratefully acknowledge the strong support from the NEPOMUC group headed by Christoph Hugenschmidt.

5. References

[1] Rescigno T N *et al.*, Science **286**, 2474 (1999); Bray I, *Phys. Rev. Lett.* **89**, 273201 (2002)
[2] Schulz M *et al.*, Nature **422**, 48 (2003)
[3] Della Picca R *et al.*, *Nucl. Instr. Meth. B* **247**, 52 (2006)
[4] Surko C M, Gribakin G F, Buckmann S J, *J. Phys. B* **38**, R57 (2005)
[5] Arcidiacono C, Kövér Á, and Laricchia G, *Phys. Rev. Lett.* **95**, 223202 (2005)
[6] Kövér Á and Laricchia G, *Phys. Rev. Lett.* **80**, 5309 (1998)
[7] de Lucio O G, Otranto S,Olson R E, DuBois R D, *Phys. Rev. Lett.* **104**, 163201 (2010)
[8] Dürr M, Dimopoulou C, Dorn A, Najjari B, Bray I, Fursa D V, Zhangjin Chen, Madison D H, Bartschat K and Ullrich J, *J. Phys. B* **39**, 4097 (2006)
[9] Ullrich J, Moshammer R, Dorn A, Dörner R, Schmidt L Ph H and Schmidt-Böcking H, *Rep. Prog. Phys.* **66**, 1463 (2003)
[10] Hugenschmidt C, Löwe B, Mayer J, Piochacz C, Pikart P, Repper R, Stadlbauer M and Schreckenbach K, *Nucl. Instr. Meth.* A **593**, 616 (2008)
[11] Sullivan J P, Jones A, Caradonna P, Makochekanwa C and Buckman S J, *Rev. Sci. Inst.* **79,** 113105 (2008)

Journal of Physics: Conference Series **262** (2011) 012048 doi:10.1088/1742-6596/262/1/012048

A study of porosity of synthetic polymer nanoparticles using PALS

B Pham[1], P Guagliardo[2], J Williams[2], S Samarin[2] and S V Smith[1]
[1]Centre for Antimatter-Matter Studies, Australian Nuclear Science and Technology Organisation (ANSTO) NSW 2232, Australia
[2]Centre for Antimatter-Matter Studies, School of Physics, University of Western Australia, WA 6009, Australia

E-mail: binh.pham@ansto.gov.au and svs@ansto.gov.au

Abstract. Positron annihilation lifetime spectroscopy (PALS) has been used to study the free volume in dry synthetic polymer nanoparticles of various sizes. A series of poly(styrene/divinyl benzene) particles with diameters in the range of 100 to 500 nm were synthesized and then carefully chemically treated using the sulfonation process, to increase their porosity. The particles were characterised by Scanning Electron Microscopy (SEM), light scattering and PALS. Light scattering gave larger size for the treated particles, reflecting the hydration effect and therefore the increase in porosity. PALS spectra of untreated and treated particles gave four and three life-time components, respectively. Analysis by PAScual version 1.3.0 program indicated there was a reduction in the intensity and the type of the micropores in the treated particles. The data suggest PALS is a sensitive tool for detecting changes in microporosity in particles. The conflicting results obtained for light scattering compared to PALS for chemically treated particles is difficult to resolve and suggests sample preparation of polymeric materials for PALS is the critical factor.

1. Introduction

The development of drug delivery vehicles and imaging agents using polymeric materials, especially polymer nanoparticles, continues to grow [1-2]. In these systems the free volume sites are important for controlling the uptake, storage, separation and release of the molecules of interest. Thus, in order to optimise their performance, extensive characterisation of their free volume is essential.

Positron annihilation lifetime spectroscopy (PALS) was used to study the microporosity of polymers and polymeric blends/composites [3-7]. PALS is particularly useful as a qualitative tool for measuring the effect of processes on the porosity in materials. However the question whether the sensitivity and resolution of the positron annihilation measurements is affected by the particle size has not been answered. A preliminary investigation into the effect of particle size on the analysis of porosity in synthetic polymer nanoparticles using PALS is presented. A series of monodispersed polymer particles (ranging from 100 to 500 nm in diameter) was synthesized, treated by sulfonation. PALS was used to determine the porosity of particles before and after the treatment.

[1] To whom any correspondence should be addressed.

2. Experimental procedure

2.1. Synthesis of polymer nanoparticles:

Polymer particles of various diameters (100, 200 and 530 nm by SEM) were synthesized by emulsion polymerisations using the chemicals summarised in Table 1. Reagents were weighed into a round bottom flask. The flask was capped and deoxygenated with nitrogen gas for 15 minutes, and then immersed in a 70°C oil bath with constant stirring for at least 20 hours. The product was dried in a fume hood overnight and then in an 80°C oven for 48 hours to remove water and any free monomer(s).

Table 1 - Recipes and sizes of polymer particles from the emulsion polymerisations

	Particles 1 (100 nm in diameter)	*Particles 2 (200 nm in diameter)*	*Particles 3 (530 nm in diameter)*
Styrene (g)	15.042	23.677	10.080
Divinyl Benzene (g)	0.308	0.378	0.511
Sodium Dodecyl Sulphate (g)	0.115	0.029	0.00
Initiator V-501 (g)	0.116	0.146	0.146
NaOH (g)	0.038	0.045	0.073
Deionised Water (g)	62.210	72.184	40.615

2.2. Modification of porosity for the polymer particles:

A portion of the dry polymer particles was then chemically treated using the sulfonation process from Holboke *et al.* to increase the porosity of the polymer particles [8]. Typically, 0.02 g of silver sulphate was added to concentrated sulphuric acid (15 mL) while stirring and warmed to 90°C. To this mixture was slowly added 1 g of dry polymer particles. The resultant suspension was heated at 90°C for 3 hours with gentle stirring. The mixture was then poured into a 6 M sulphuric acid solution at room temperature (100 ml) with gentle stirring. The particles were centrifuged at 5000 rpm and washed thoroughly with deionised water. The resultant product was dried in an 80°C oven for 48 hours.

2.3. Characterisation of particles:

2.3.1. Scanning Electron Microscopy: (ZEISS, ULTRA *Plus*, 10kV energy): polymer particles were deposited on a 200 mesh carbon coated copper grid. Micrographs of the particles before and after sulfonation are illustrated in Figures 1 and 2.

2.3.2. Light scattering: Zetasizer (ZetaNano, Malvern) was used to measure size distributions of polymer particle suspensions before and after the sulfonation. Data are summarised in Table 2.

2.3.3. PALS: the instrument consists of a fast-fast spectrometer with detectors comprising of a BC418 plastic scintillator coupled to a Burle 8850 photomultiplier tube. A 30 μCi ^{22}NaCl source was used for the collecting spectra. The time resolution of the instrument was determined using a high purity annealed nickel to be approximately 200 ps. Each spectrum has at least 2×10^6 counts and was analysed by PAScual version 1.3.0 [9]. Data are given in Table 3.

3. Results and discussion

SEM images in Figure 1 show that the polymer particles obtained from the emulsion polymerisations (untreated particles) are monodispersed and have mean diameters of 100, 200 and 530 nm. The particles after the sulfonation (treated particles) (Figure 2-B) retain their original shape and size (Figure 2-A). The SEM technique measures particles in their dry form (i.e. free of water and unreacted monomers).

The light scattering measures the diameters of the particles suspended in the aqueous solution. It is important to note that particles become hydrated in the aqueous solution and if there is any unreacted monomer it can influence the radii obtained. As expected particle diameters obtained by the light scattering were consistently higher than those given by SEM (see Table 2). For the Particles 1 and 3, diameters given by the light scattering appear to be increased by 1.3 to 1.5 fold, while the variation in the Particles 2 is only by 1.03 fold. This is mostly likely due to the presence of unreacted monomer remaining in the former particles.

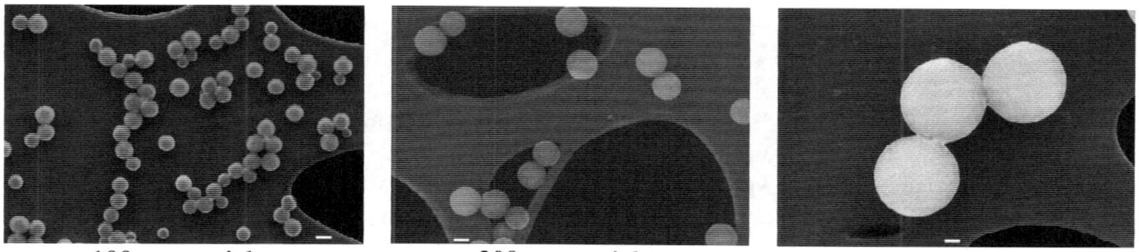

100 nm particles 200 nm particles 530 nm particles

Figure 1 - SEM micrographs of untreated particles. Scale bar: 100 nm

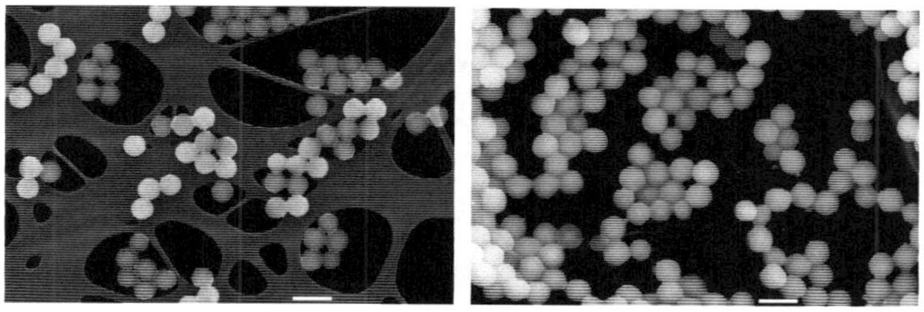

A B

Figure 2: SEM micrographs for 530 nm particles: untreated (A) and treated (B). Scale bar: 1 μm

The light scattering data revealed that the diameter of treated particles increased by 2 fold compared to untreated particles. Such an increase confirmed the treatment by sulfonation was successful (See Table 2).

Table 2 – Summary of the colour and average diameter of untreated and treated particles

Particles	*Colour*		*Diameter by SEM of dry particles (nm)*	*Diameter by Light Scattering of hydrated particles (nm)*	
	Untreated	*Treated*		*Untreated*	*Treated*
1	White	Light brown	100	153	320
2	White	Grey	200	216	-
3	White	Creamy grey	500	665	1329

Prior to lifetime measurements the samples were heated at 80°C (close to the glass transition temperature of polystyrene of 100°C) for 24 hours to remove remaining moisture. The o-Ps lifetimes and intensities are given in Table 3.

Table 3: Ortho-positronium lifetimes and intensity for particles

Particles	**D (nm)**	**τ_3 (ns)**	**I_3 (%)**	**τ_4 (ns)**	**I_4 (%)**	**d_3 (nm)**	**d_4 (nm)**
1 untreated	100	2.03 (0.01)	30.7 (0.1)	29.4 (0.1)	1.5 (0.1)	0.65	2.54
2 untreated	200	2.01 (0.01)	33.1 (0.1)	15.0 (0.9)	0.8 (0.1)	0.65	1.75
1 treated	100	1.58 (0.01)	11.6 (0.1)	-	-	0.56	-
2 treated	200	1.69 (0.01)	11.8 (0.7)	-	-	0.58	-
3 treated	530	1.57 (0.01)	10.2 (0.8)	-	-	0.55	-

particle diameter (D), o-Ps lifetimes (τ), intensities (I) and pore diameters (d)
Colours of particles: white (untreated), light brown (1, treated), grey (2, treated) and creamy grey (3, treated)

Spectra of 100 nm and 200nm untreated particles could be deconvoluted into four discrete components. Two of these components (less than 0.5 ns) are associated with p-Ps and free positron annihilation, as well as annihilations in the source material and supporting foil. The longer components (more than 1 ns) are of higher interest for the present study. They are associated with o-Ps pick-off annihilation in the free volume of the polymeric particles. The τ_3 values (approximately 2 ns, using the RTE model [10]) for untreated Particles 1 and 2 correspond to a pore diameter d_3 of 0.65 nm. The intensity of this component (I_3) exceeds 30%, indicating that these materials contain significant nanoporosity. Their lifetime τ_4 components (29.4 and 15 ns, respectively) correspond to larger pores with diameters d_4 of 2.54 and 1.75 nm, respectively. However, this component could be due to annihilations in the space between the particles or at free surfaces and the chemical and preparation processes for forming these particles are unlikely to generate to distinct type of pores.

For the treated particles the spectra could be deconvoluted into only three discrete components. Very close values for lifetimes τ_3 observed for treated Particles 1 and 3 correspond to a pore diameter of 0.56 nm. A slightly larger τ_3 of 1.69 ns for treated Particles 2 corresponds to a slightly larger pore diameter of 0.58 nm. I_3 values for these samples were in the range of 10.2-11.8%.

The light scattering measurements suggest that the microporosity of these particles is increased after sulfonation. However there was a conflict between the results given by light scattering and by PALS. Measurements by light scattering indicated the hydration of treated particles increased due to an increase in the porosity. In contrast, the PALS results suggest the treated particles have significantly lower free volume in comparison to the treated particles. The τ_3 values indicate that the average pore diameter of the treated particles decreased by around 1 Å. The intensity of this component also decreased significantly.

While the particles were washed extensively after the sulfonation process, their colour changed from white to cream and brown as the particles decreased in size. Hence the PALS results could reflect the presence of silver ions inside the particles. The different observations from light scattering and PALS could also be due to the sample preparation and drying processes deployed for particles prior to PALS measurements. As the particles have polymer chains at low cross-linker (divinylbenzene) concentration, drying at 80°C (close to their glass transition temperature) could have resulted in the collapsing in the nanostructure of the particles.

Future experiments will be directed toward the effect of sample preparation (i.e. drying conditions) and cross-linking agent used to synthesize the polymer particles on PALS analysis.

4. Conclusion

PALS was proved to be a sensitive technique to measure the free volume in the synthetic polymer particles of a range of particle sizes. The results show that the untreated polymer particles have comparable τ_3 values but there is a significant difference in respective τ_4 values. The size of the nanopores for the treated particles is similar but there is no evidence of τ_4. There appears to be an effect of particles size on the PALS analysis, however the results are inconclusive and further work is warranted. The light scattering data indicate an increase in porosity for treated particles however PALS show a reduction in free space. Future experiments will investigate the effects of particle size and sample preparation on the PALS measurements.

5. References

[1] Palko, H A et al. 2010 *Inhalation Toxicology* **22** 657
[2] Garcia I et al. R 2010 *Nanomedicine* **5** 777
[3] Kluin J E et al. R 1993 *Macromolecules* **26** 1853
[4] Jean Y C 1990 *Microchem. J.* **42** 72
[5] Zaydouri I and Grivet M 2009 *Radiation Physics and Chemistry*, **78**
[6] Zhou W, Wang J, Gong Zh, Gong J, Qi N and Wang B 2009 *Appl. Phys. Lett.* **94**,
[7] Peters R M, Paramo J A, Quarles C A and Strzhemechny Y M 2009 *AIP Conf. Proc.* 1099, 965
[8] Holboke A E and Pinnell R P 1989 *J. Chem. Educ.* **66** 613
[9] Pascual-Izarra C et al 2009 *Nucl. Instr. Meth. Phys. Res. A* **603** 456
[10] Dull T L et al. 2001 *J. Phys. Chem. B* **105** 4657

Journal of Physics: Conference Series **262** (2011) 012049 doi:10.1088/1742-6596/262/1/012049

The experimental determination of the phase space distribution of a positron beam

C Piochacz and C Hugenschmidt
Technische Universität München, Forschungsneutronenquelle FRM II, Lichtenbergstr. 1, D-85747 Garching
E-mail: christian.piochacz@frm2.tum.de

Abstract. The NEPOMUC positron beam facility hosts four successfully operated positron instruments which cover the fields of solid state, surface and fundamental physics. Two further setups are under construction and a variety of positron experiments were accomplished at the open beam port. The success of almost all experiments till now has been dependent on the quality of the positron beam provided by NEPOMUC. Therefore, the source, the beamline, and some key components as e.g. the remoderator and the beam switch are the subject of continuous optimizations and improvements. A crucial part of our efforts is a meaningful characterization of the positron beam. The most important measure of the quality of a positron beam is the phase space volume occupied by the beam particles and hence, a novel device was developed whose measurement concept is presented here.

1. Introduction

The NEutron induced POsitron source Munich (NEPOMUC) provides the world's most intense positron beam, and by installing the NEPOMUC remoderator a high-brightness beam became available. This unique beam is used routineously for several experiments and in many of them, the positron beam has to be focused onto the sample in order to allow laterally resolved measurements [1].

2. The phase space concept

The minimal focus which can be achieved depends first of all on the size of the phase space volume (PSV) which is occupied by the beam particles. Only if the spot gets very small, the aberrations of the used lenses get dominant. In general, the phase space is spanned by 3 space and 3 momentum coordinates and has hence 6 dimensions. Due to this construction, the state of one positron of a beam at a certain time is entirely described by a point in the phase space. Consequently, a positron beam—which is an ensemble of positrons—can be described as a set of points in the phase space or as a volume which enclosures these points. The important theorem of Liouville states, that this 6-dimensional PSV keeps constant under the influence of conservative forces. By regarding symmetries and approximations the 6-dimensional phase space can be divided in subspaces in such a way, that the Liouville theorem applies also for each PSV in these subspaces. E.g. in rotational symmetric systems, where only paraxial rays are regarded, the four lateral components $(x, p_\mathrm{x}, y, p_\mathrm{y})$ of the 6 dimensional space can be substituted by a two dimensional subspace spanned by the distance from the optical axis r and by the conjugated canonical momentum P_r, which equals the physical momentum p_r in the electrostatic case. In this representation of the phase space, the volume which includes a certain amount of the beam intensity has a simple 2 dimensional geometric shape as e.g. an ellipse.

If aberrations are neglected lenses shear the transversal phase space volume along the P_r axis, and drift sections lead to a shearing along the r axis. The aberrations of lenses lead to a non-linear transformation of the phase space volume and hence to a perturbation of it as shown for spherical aberrations in Fig. 1. In addition, this figure helps to understand, why (second order) aberrations lead to a blur of the geometric focus and why not the *real* PSV but the *effective* PSV is the determining parameter of the minimal spot size. Since the effective PSV is the envelope of the real PSV, the effective one is always larger. Not only the classical aberrations of lenses but also imperfections in longitudinal magnetic fields, which are used for adiabatic beam guidance, lead to nonlinear perturbations of the PSV. Especially in the case of long beam lines as at the NEPOMUC facility, were the solenoidal magnetic field has to be interrupted several times for e.g. pumping and beam diagnostic, the effective PSV can increase considerably.

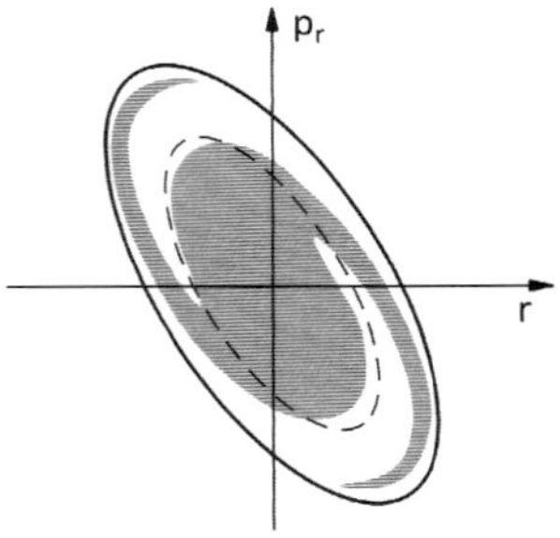

Figure 1. Due to the aberrations of lenses and due to imperfect magnetic, adiabatic beam transport the initial PSV (dashed ellipse) gets perturbed. Because of the Liouville theorem, the perturbed PSV (gray shaded area) must have the equal size as the initial volume, however, the volume of the envelope, i.e. the effective PSV is remarkable larger.

3. Measurement of the PSV

Until now, at NEPOMUC and its instruments the lateral PSV has been only estimated by two different methods. The first is based on measuring the beam diameter and the longitudinal energy spread within the magnetic guiding field. Due to energy conservation the transversal energy spread should be in the same order of the longitudinal energy spread and hence the transversal PSV can be estimated. By the other approach the smallest attainable spot which can be achieved with a rotational symmetric focusing lens is measured. By knowing the details of the imaging and the aberrations of the lens, the PSV can be calculated in principle. However, some of the imaging or lens probabilities have to be gained e.g. by simulations or have to be estimated. Hence, both approaches may lead only to rough values and moreover, they provide only an absolute value of the PSV and no information of the intensity distribution within the phase space.

In order to overcome this a novel method for the determination of the intensity distribution of a positron beam within the lateral phase space was developed. The experimental setup consists of a magnetic field termination to extract the positrons from the magnetic guiding field, a mask, which divides the beam in several sub-beams, and a lateral resolving intensity sensitive detection system made of a MCP, a scintillation screen and a CCD camera (see Fig. 2). After the magnetic field termination and after the mask the beam diverges free, where the diverging angle is determined by the ratio of the transversal and longitudinal momentum. As the longitudinal momentum can be tuned separately in front and behind the mask a homogeneous illumination and a proper imaging of the mask pattern onto the MCP can be ensured. The mask selects certain radii r_i of the lateral phase space (r, p_r). Due to the transversal momentum p_r^i the image of these radii r_i' are enlarged. Hence, (the averaged) transversal momentum p_r^i of one sub-beam

Journal of Physics: Conference Series **262** (2011) 012049 doi:10.1088/1742-6596/262/1/012049

can be determined by measuring the growth $(r_i' - r_i)$. With the obtained set of (r_i, p_{r}^i) and the relative intensity I of the subbeams, the orientation and the intensity distribution along the main axis of the PSV can be plotted (see Fig. 3).

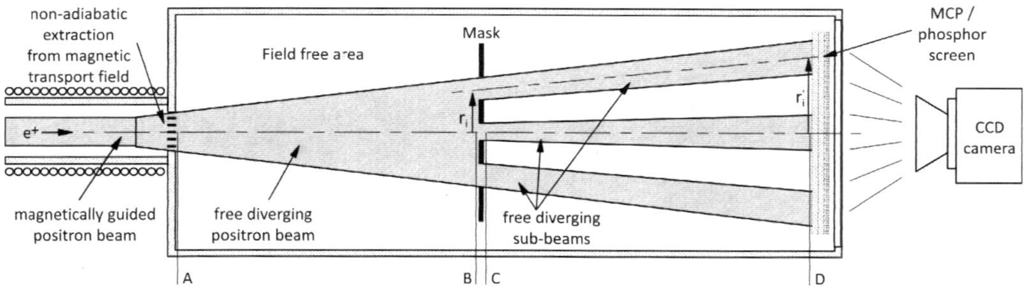

Figure 2. Sketch of the setup for the determination of the transversal phase space distribution of a positron beam.

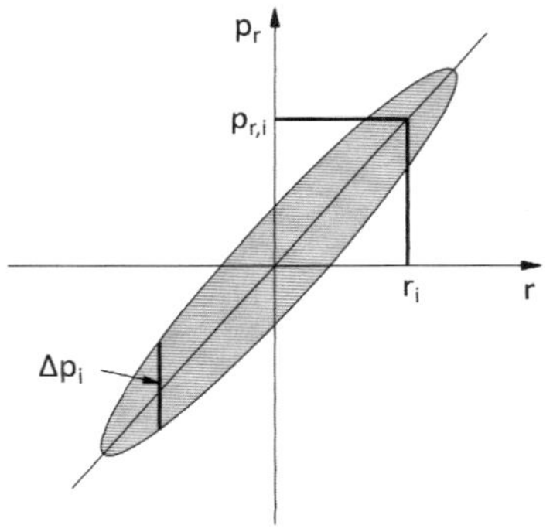

Figure 3. The orientation and the intensity distribution along the main axis is given by a set of $(r_{\mathrm{i}}, p_{\mathrm{i}}, I)$. The r_{i} are given by the mask of the setup, the p_{i} is proportional to the enlargement of the r_{i} during the drift and the relative intensity I can be measured with the CCD camera. For the ascertainment of the width Δp_{i} of the PSV the radial blur of the subbeams has to be analyzed as described in the text.

In order to determine the extension of the phase space volume, the radial intensity distribution of the subbeams has to be regarded. By considering a hypothetical phase space distribution which is constant up to a maximum radius and up to a maximal radial momentum, the radial intensity distribution at the MCP can be deduced from simple geometric considerations (see left column of Fig. 4). Due to the drift the initial PSV gets sheered along the r-axis. Hence, the rectangular PSV is mapped to a parallelogram. The mask selects from this enlarged beam well defined areas with certain radii and hence from the PSV slices with a width equivalent to the hole diameters of the mask. Since after the mask a second drift follows, these slices are also sheered. Because the measured radial intensity distribution of the beams is the projection of the PSV onto the radial axis, the radial intensity distributions of the subbeams have to be isosceles trapezia[1] This means, the subbeams appear with blurs at the MCP and the width of these blurs depends on the ratio of the transversal to the longitudinal momentum of the subbeams and the length of the second drift.

A more realistic radial PSV distribution and the according radial intensity distribution are given on the right side of Fig. 4. In this case numerical simulations are necessary to reveal the exact shape of the radial intensity distributions but as shown in the last sub-figure, the characteristic slopes at the borders are still observable.

[1] This does not hold for subbeams which stem from the border of the primary beam. In this case the shape might be an irregular pentagon

Journal of Physics: Conference Series **262** (2011) 012049 doi:10.1088/1742-6596/262/1/012049

4. Conclusion

In this work the setup of a novel device is presented and by numerical calculations it is demonstrated that not only simple but also non-uniform PSV distributions can be measured. An implementation of the setup was constructed and very first experiments were successfully performed. However, for the measurement of the PSV of the NEPOMUC beam further experiments are necessary.

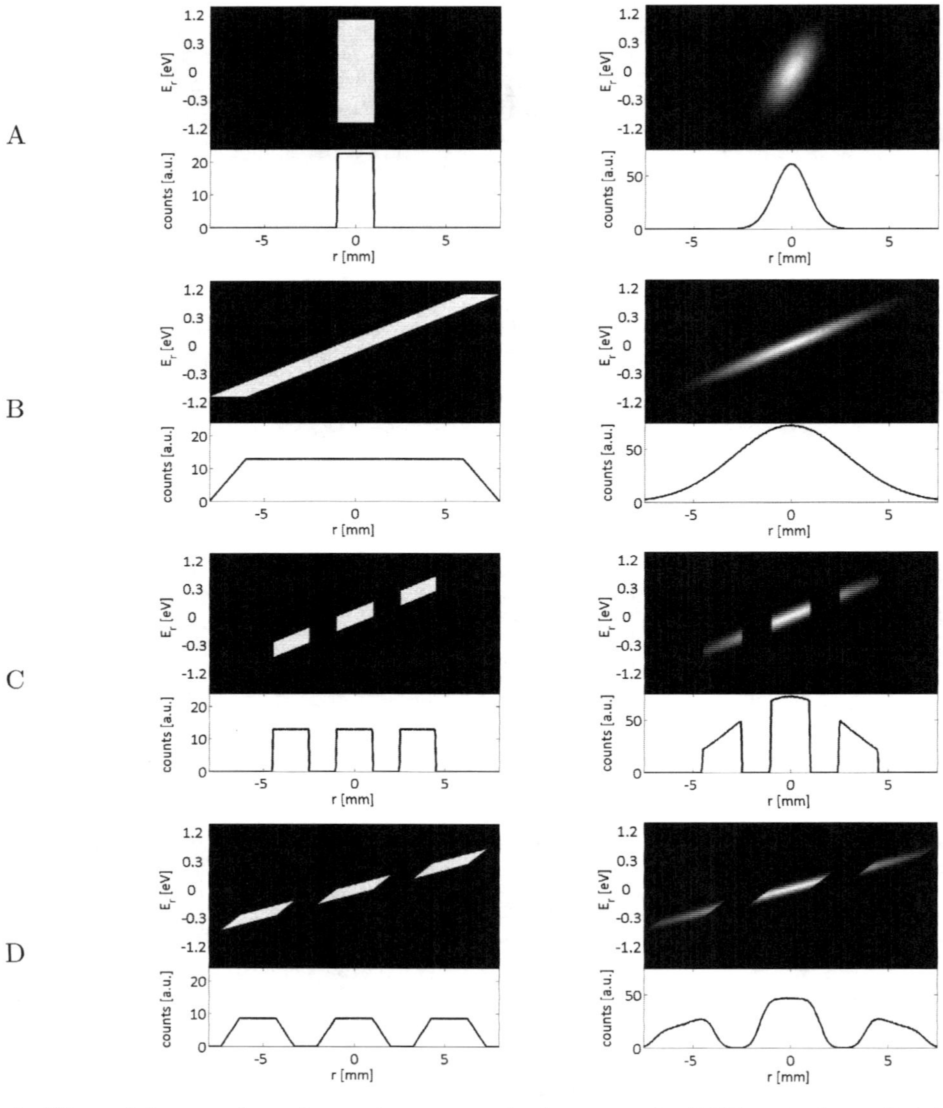

Figure 4. The subfigures show from the top to the bottom the PSV and the radial intensity distribution at the positions A, B, C, D marked in Fig. 2. On the left side a rectangular and homogenous filled PSV is supposed and on the right side the simulation of a more realistic 2-dimensional Gaussian distribution is shown.

References

[1] C. Hugenschmidt et al., In this proceedings

Journal of Physics: Conference Series **262** (2011) 012050 doi:10.1088/1742-6596/262/1/012050

Hydrogen-induced surface modifications in ZnO single crystals

I. Prochazka[1], J Cizek[1], W Anwand[2], G Brauer[2], D Grambole[3] and H Schmidt[3]

[1] Charles University in Prague, Faculty of Mathematics and Physics, V Holesovickach 2, CZ-18000 Prague 8, Czech Republic

[2] Institut für Strahlenphysik, Forschungszentrum Dresden-Rossendorf, PO Box 510119, D-01314 Germany

[3] Institut für Ionenstrahlphysik und Materialforschung, Forschungszentrum Dresden-Rossendorf, PO Box 510119, D-01314 Germany

E-mail: ivan.prochazka@mff.cuni.cz

Abstract. Surface changes in ZnO single crystals electrochemically doped with hydrogen were investigated in this work using slow positron implantation spectroscopy (SPIS) combined with atomic force microscopy (AFM) and optical microscopy. It was found that hexagonally shaped pyramids were formed on the surface of hydrogen-loaded crystals. The formation of these pyramids can be explained by hydrogen-induced plastic deformation realized by a slip in the [0001] direction. Such a picture is supported (i) by AFM where steps of a height comparable with the *c*-lattice parameter were found at the base of the pyramids, and (ii) by SPIS which revealed a defected subsurface layer, formed by the hydrogen-induced plastic deformation and exhibiting an enhanced concentration of open-volume defects in hydrogen-loaded crystals.

1. Introduction

ZnO is a wide band gap semiconductor with a variety of applications including UV light emitting diodes, optoelectronic devices, and gas sensors. Due to a progress in crystal growth, high quality single crystals are nowadays available. Properties of ZnO crystals are strongly influenced by lattice defects. A detailed characterization of intrinsic defects and impurities in ZnO crystals is, therefore, very important. Defect studies of hydrothermally grown ZnO single crystals revealed that hydrogen is the most important impurity in ZnO crystals [1]. It was found that hydrogen is coupled with zinc vacancies and vacancy-hydrogen complexes are formed. Moreover, it was shown that a high amount of hydrogen can be introduced into a ZnO crystal by electrochemical loading [2]. In this work, we performed detailed investigations of surface modifications in a hydrogen-loaded ZnO crystal.

2. Experimental

A hydrothermally grown ZnO (0001) oriented single crystal (MaTecK GmbH) with O-terminated surface was investigated. The crystal was covered by a 20 nm thick Pd over-layer deposited by cold cathode beam sputtering. The Pd over-layer acts as a catalyst for hydrogen permeation into ZnO and prevents hydrogen losses due to the formation of H_2 molecules on the surface [3]. The specimens were electrochemically doped with hydrogen in a cell filled with an 1:1 mixture of H_3PO_4 and glycerine. Hydrogen charging was performed at a constant current of 0.3 mA for 24 h using a Pt counter-electrode and the loaded specimen acting thereby as cathode. The hydrogen concentration introduced into the sample was determined at a depth of ~ 100 nm by nuclear reaction analysis (NRA), employing the resonant nuclear reaction $^{15}N + {}^{1}H \rightarrow {}^{12}C + {}^{4}He + \gamma$ rays [4] induced by 6.64 MeV ^{15}N ions. Slow positron implantation spectroscopy (SPIS) investigations were performed on a magnetically guided slow positron beam "SPONSOR" [5]. The energies of incident positrons were selected in the range from 0.03 to 36 keV. Doppler broadened annihilation profiles were analyzed in terms of the S parameters. Atomic force microscopy (AFM) measurements were carried out in the tapping mode using a Dimension DI 3100 microscope (Veeco Instruments). The surface of the specimens was examined also by a metallographic optical microscope (OM) type Arsenal AM-2T.

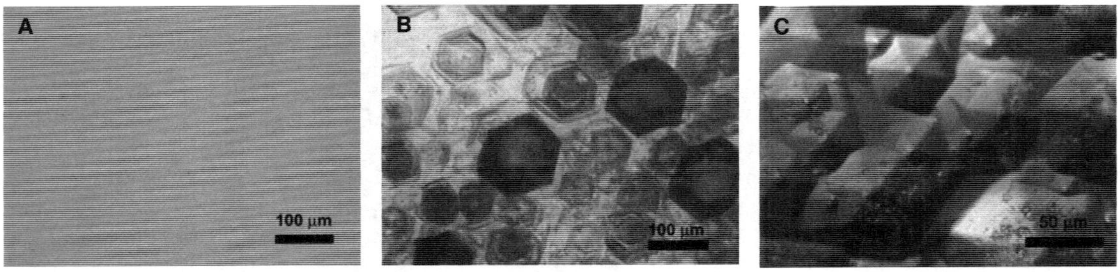

Figure 1 OM images of ZnO crystal surface. (A) virgin sample (reflected light), (B) hydrogen-loaded sample (loaded side, reflected light), (C) hydrogen-loaded sample (loaded side, glazing incidence light).

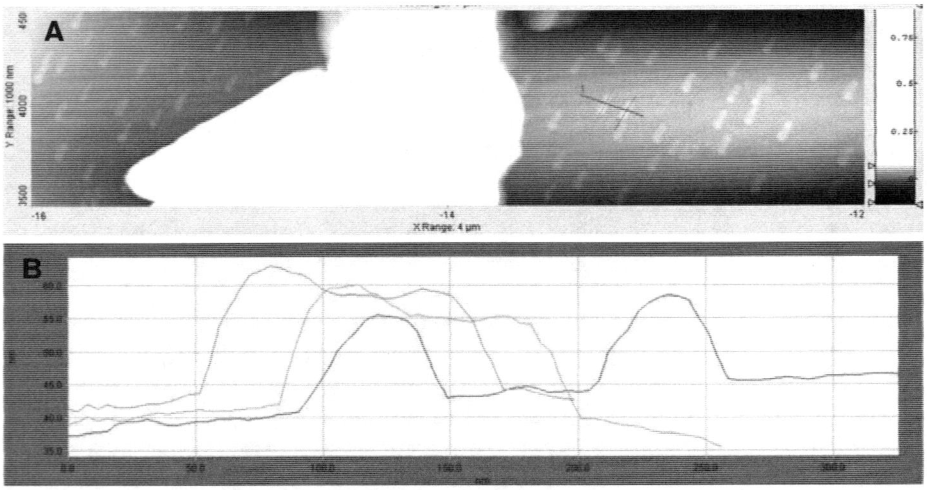

Figure 2. (A) AFM image of hydrogen loaded crystal surface showing a lot of crystallites formed by hydrogen loading, (B) height profile taken across the lines 1,2,3 shown in the upper panel.

3. Results and discussion

NRA data on the hydrogen-loaded crystal revealed a hydrogen concentration of (44.3 ± 0.1) at.-% on the loaded side (covered with Pd cap) and (17.2 ± 0.1) at.-% on the backside. These concentrations considerably exceed that of (0.3 ± 0.1) at.% found [1] in the virgin (unloaded) crystal. Hence NRA confirmed a high amount of hydrogen being introduced into the crystal by electrochemical loading.

Fig. 1 shows OM images of the sample surface. The virgin crystal exhibits just a flat polished surface (Fig. 1A). However, on the hydrogen-loaded crystal we observed a lot of hexagonal pyramids shown in Fig. 1B. Interestingly, all pyramids have exactly the same orientation with respect to the crystal. Observations in glazing incidence light (Fig. 1C) revealed that the pyramids are grown out of the sample. The formation of pyramids can be explained by lattice expansion due to hydrogen occupying the bond-centered lattice sites and forming O-H bonds parallel with the *c*-axis [6]. When the hydrogen-induced stresses exceed the yield stress, plastic deformation takes place via a slip in [0001] direction.

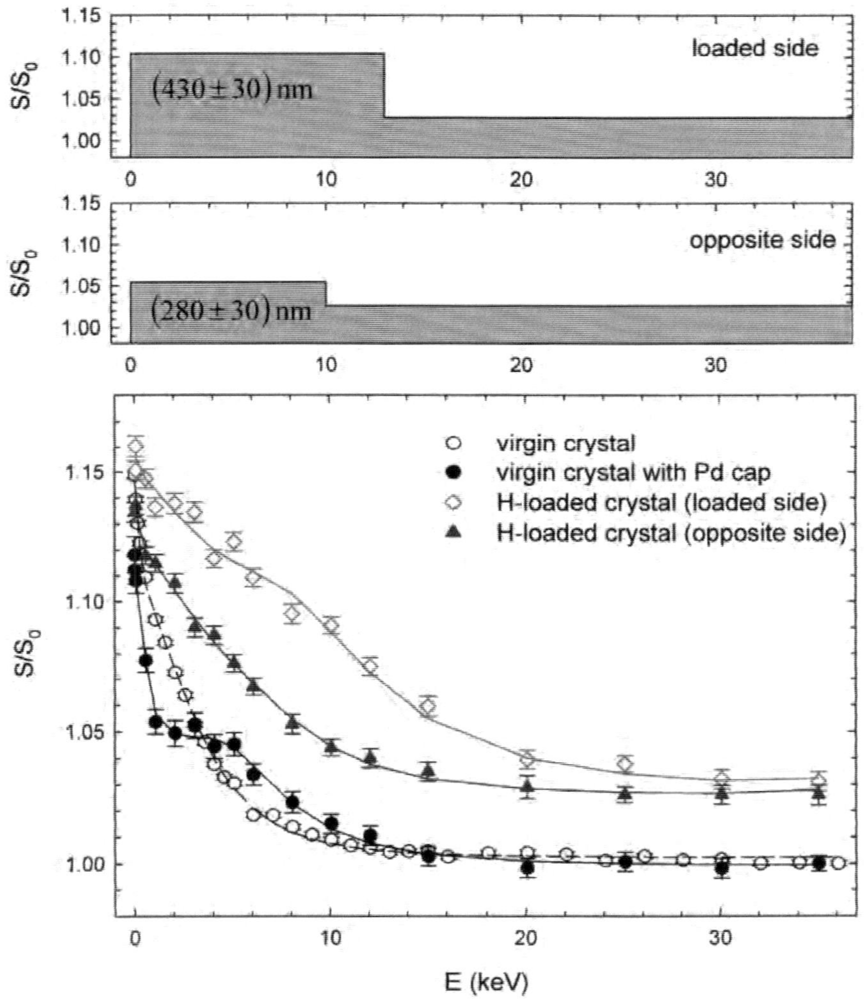

Figure 3. Dependence of the S parameter on positron energy E for the virgin crystal, the virgin crystal covered with Pd cap, and the hydrogen loaded crystal measured on the loaded and the opposite side. All S parameters are normalized to the value S_0 measured on the virgin crystal at E = 35 keV. Solid lines show fit by VEPFIT code [7]. A single layer model was assumed for the virgin crystal. Two layers, including the Pd cap and the ZnO bulk, were used for the virgin crystal covered with the Pd cap. Two layers, involving the defect-rich subsurface layer and the ZnO bulk, were considered for the hydrogen-loaded crystal (the hydrogen-loaded and the opposite side) and are shown in the upper panels. Thicknesses obtained for the defect-rich subsurface layer are given in the upper panels, too.

An AFM image of the hydrogen-loaded crystal is shown in Fig. 2. A lot of tiny crystallites introduced by hydrogen loading can be seen in Fig. 2A. Height profiles of two selected crystallites are plotted in

Journal of Physics: Conference Series **262** (2011) 012050 doi:10.1088/1742-6596/262/1/012050

Fig. 2B. Fig. 2A shows that all nanoscopic bulges have a common orientation with the crystal. Hence AFM testifies that hydrogen loading leads to the formation of nanoscopic features on the surface which may be regarded as nuclei of the micrometer size pyramids observed by OM. This is further supported by the fact that steps with a height comparable to the ZnO *c*-lattice constant were observed by AFM on the slope of these pyramids (Fig. 2B).
SPIS results are plotted in Fig. 3 as dependences of S parameter on positron energy. The solid lines are model curves fitted by the VEPFIT software [7]. Using a single layer model, a relatively short positron diffusion length of $L_+ = (46 \pm 1)$ nm was determined for the virgin ZnO crystal. Hence, the virgin crystal contains a high density of positron traps. As discussed in detail in Ref. 1 these traps are related to Zn-vacancies. The S(E) curve for the virgin crystal covered with the Pd layer shows a knob around the energy of 2 keV caused by positrons annihilating inside the Pd layer. However, the S parameter for the ZnO bulk is found to remain unchanged. A positron diffusion length of $L_+ = (47 \pm 2)$ nm obtained by VEPFIT is virtually the same as in the uncovered virgin sample. The S(E) curves measured on the hydrogen loaded crystal (the loaded and the backsides, respectively) are plotted in Fig. 3 as well. One can see from the figure that hydrogen loading leads to a large increase of S parameter in the sub-surface layer corresponding to the positron energy range 1-10 keV and also to an increase of the S parameter for the bulk ZnO region. Hence, two layers involving (i) the sub-surface layer with a very high concentration of defects, and (ii) the bulk ZnO layer were assumed in the VEPFIT analyses of S(E) curves for the loaded crystal. This model is shown schematically in the upper panels in Fig. 3. The formation of a defected sub-surface layer in the hydrogen-loaded crystal strongly supports the picture of hydrogen-induced plastic deformation. Moreover, one can see in Fig. 3 that the S parameter in the defect-rich sub-surface layer is higher on the loaded side of the crystal and the sub-surface layer here also extends more into the bulk of the crystal. An increase of S parameter in the ZnO bulk region indicates that new defects were introduced also into the bulk. However, seemingly the highest concentration of defects was created in the sub-surface layers.

4. Conclusions

A high hydrogen concentration can be introduced into a ZnO crystal by electrochemical loading. Surface studies by OM and AFM revealed the formation of nanometer size bulges in hydrogen–loaded samples which further grow into micrometer size hexagonal pyramids. SPIS investigations revealed an overall increase of the defect density in a hydrogen-loaded crystal and the formation of a defect-rich sub surface layer. These results indicate that structure modifications of hydrogen-loaded samples are likely caused by a slip in [0001] direction caused by stresses arising from a hydrogen-induced lattice expansion.

5. References

[1] Brauer G, Anwand W, Grambole D, Grenzer J, Skorupa W, Cizek J, Kuriplach J, Prochazka I, Ling C C, So C K, Schulz D and Klimm D 2009 *Phys. Rev. B* **79** 115212
[2] Cizek J, Zaludova N, Vlach M, Danis S, Kuriplach J, Prochazka I, Brauer G, Anwand W, Grambole D, Skorupa W, Gemma R, Kirchheim R and Pundt A 2008 *J. Appl. Phys.* **103** 053508
[3] Kirchheim R 1988 *Prog. Mater. Sci.* **32** 261
[4] Lanford W A 1995 *Handbook of Modern Ion Beam Materials Analysis*, ed R Tesmer and M Nastasi (Pittsburg: Materials Research Society) p 193
[5] Anwand W, Kissener H-R and Brauer G 1995 *Acta Phys. Polon. A* **88** 7
[6] Van de Walle C G 2000 *Phys. Rev. Lett.* **85** 1012
[7] van Veen A, Schut H, Clement M, de Nijs J, Kruseman A and Ijpma M 1995 *Appl. Surf. Sci.* **85** 216

Acknowledgments

This work was supported by the Academy of Science of the Czech Republic (project KAN300100801) and the Ministry of Schools, Youths and Sports of the Czech Republic (project MSM 0021620834).

Vacancy identification in Co^+ doped rutile TiO_2 crystal with positron annihilation spectroscopy

X B Qin[1], P Zhang[1, 2], L H Liang[1, 3], B Z Zhao[1], R S Yu[1], B Y Wang[1] and W M Wu[3]

1. Key Laboratory of Nuclear Analysis Techniques, Institute of High Energy Physics, Chinese Academy of Sciences, Beijing 100049, China.

2. School of Physical Engineering, Zhengzhou University, Zhengzhou, 450001, China.

3. College of Physical Science and Technology, Guangxi University, Nanning 530004, China.

qinxb@ihep.ac.cn (Qin) and wangboy@iehp.ac.cn (Wang B.Y.)

Abstract. Co-doped rutile TiO_2 films were synthesized by ion implantation. Variable energy positron annihilation Doppler broadening spectroscopy and coincidence Doppler broadening measurements were performed for identification of the vacancies. A newly formed type of vacancy can be concluded by the S-W plot and the CDB results indicated that the oxygen vacancy (V_O) complex Ti-Co-V_O and/or Ti-V_O are formed with Co ions implantation and the vacancy concentration is increased with increase of dopant dose.

1. Introduction

Since the observation of room temperature ferromagnetism (RT-FM) in cobalt doped TiO_2 crystal, the diluted magnetic semiconductors (DMS) have led to intense interest in these materials due to their potential use in spintronic devices, which make use of both the spin and charge of electrons [1-5]. The origin of ferromagnetism is nowadays a hot topic and is not clearly understood [1-8]. The strong interaction between Co ions and O vacancy was found to play a key role in the occurrence and stability of FM [3, 9-12]. Calculations indicated that Ti^{3+} ions with one 3*d* electrons are usually generated in slightly reduced TiO_2 [12, 13]. However, experimental identification of the oxygen vacancy complex has remained unresolved. There is thus a great need to experimentally clarify the oxygen vacancy complex.

Positron annihilation spectroscopy (PAS) is a powerful technique for evaluating vacancy-type defects [14-18]. Thermalized positrons in solids get trapped at vacancies because of the missed positive charge of the ion cores and finally annihilated with surrounding electrons. At vacancies, positron-electron momentum distribution is narrowed due to the reduced electron density. Meanwhile, the core electron momentum distribution can be used to identify the atoms around the vacancies. Coincidence detection of broadened annihilation photons (CDB) can eliminate random background and yield three orders of magnitude better signal-to-noise ratio in Doppler broadening measurements, which enables us to observe the high-momentum annihilations with core electrons specific to each element [19-21]. Thus, CDB measurements are expected to provide additional information about the atoms surrounding the annihilation site. For this end, we prepared a series of Co^+ ion implanted rutile TiO_2 and characterized the vacancies inside them by PAS method.

2. Experimental

Five rutile TiO_2 (110) crystals with thickness of 500 um were purchased from Shanghai institute of optics and fine mechanics, Chinese academy of sciences (CAS). Samples 1# to 4# were implanted with the dose of 1.25×10^{16} ions/cm^2, 2×10^{16} ions/cm^2, 3×10^{16} ions/cm^2 and 4×10^{16} ions/cm^2 at institute of semiconductors, CAS. The implanting energy was fixed to be 100 keV for all samples with the mean implanting depth about 40nm which can be simulated by SRIM code. Also an un-doped sample marked 0# was prepared for comparison.

The Doppler broadening annihilation radiation at 511keV peak was recorded by a HPGe detector. The line width is characterized by S and W parameters. The S parameter is defined as the ratio of the gamma-ray counts in the central part of the 511 keV peak (510.24–511.76 keV) to that in the entire peak (504.2–517.8 keV). W parameter was defined as the ratio of the summed gamma-ray counts in the range of 513.6–517.8 keV and 504.2–508.4 keV to the total number of counts in the entire peak.

For CDB measurement, the energies of annihilating γ ray pairs (denoted by E_1 and E_2) are simultaneously registered in the two detectors located at nearly 180° to each other. The difference of the energies of the two γ rays is given by cPL and the total energy $E_t=E_1+E_2$ is given by $2m_0c^2$ (neglecting the thermal energies and chemical potentials), where PL is the longitudinal momentum component of the positron-electron pair along the direction of the γ ray emission, c is the speed of light and m_0 is the electron rest mass. Selecting the coincidence event under the condition of $2m_0c^2$-2.4keV< E_t < $2m_0c^2$+2.4keV, each spectrum is collected for more than 38 hours to get eight million counts. The single-detector Doppler broadening and CDB measurements were performed at Beijing intense slow positron beam.

3. Results and Discussion

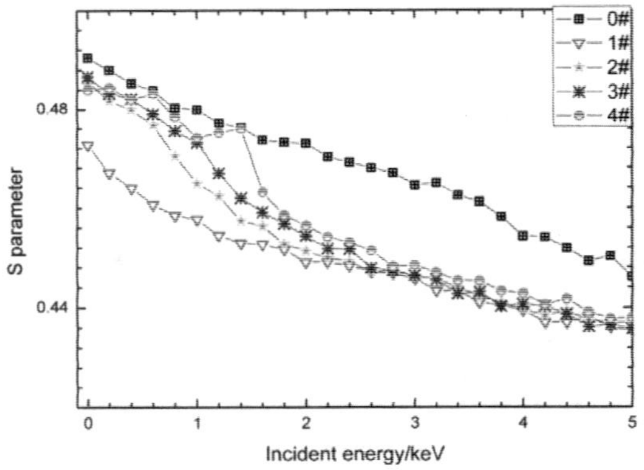

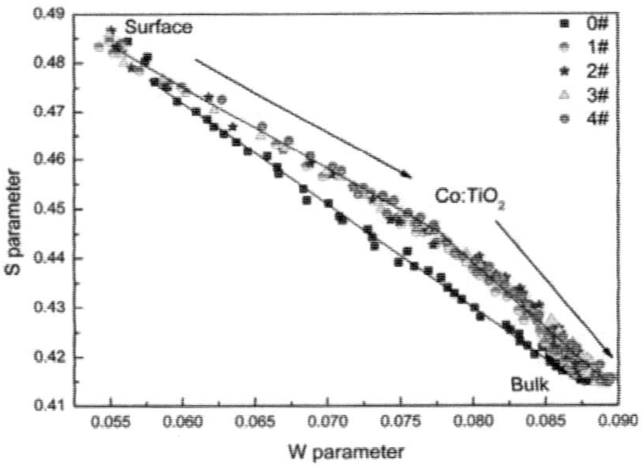

Fig.1 S parameters as a function of incident positron energy. Fig. 2 S-W plot of all samples.

The S parameters of all samples as a function of positron incident energy from zero to 10 keV are shown in figure 1. The S parameters decreased gradually from sample surface until deep inside the bulk. When Co^+ was implanted, the S parameter decreased sharply at implanting zone and slowly increased with increase of implanting dose. It's easy to conclude that more vacancies are formed during implantation with increase of dopant dose corresponding to the increase of S parameters. Sharp decrease of S parameters after implantation implies a conversion of annihilation mechanism, which can also be concluded by the S-W plot of all samples as shown in figure 2. S-W plots are useful in detecting the change in the nature of positron trapping defects, since the slope of the plots is a fingerprint of a specific vacancy [22]. For un-doped sample 0#, the S-W parameter changed linearly from surface to the bulk which means a single annihilation mechanism, but for samples 1# to 4#, double kinds of annihilation mechanism can be found and one of them at (0.0077,0.4457) is believed to be induced by implantation.

The dominant vacancy-type defects in CoTiO films include V_{Ti}, V_O, complex involving V_{Ti} or V_O, and other impurity. CDB measurements were expected to provide additional information about the atoms surrounding the annihilation site and the results are shown in figure 3. Sample 0#, 1#, 3# and annealed Ti bulk were measured with incident positron energy fixed at 2 keV with the mean depth of 28.5 nm experiencely calculated by $d_{(nm)}=40E^{1.6}{}_{(keV)} / \rho_{(g/cm3)}$, which is smaller than SRIM results[21]. The surface effect should be surely considered, but there would be much defects generated with increase of dopant dose which traps positrons more effectively. In order to clarify the type of vacancies induced by dopant, ratio curves were made via dividing each measured CDB spectrum by that of sample 0#. Three different zones can be obviously distinguished along the momentum. At zone about $(0\sim3)\times10^{-3}$ m_0c, positrons annihilated with valence electrons. Zone $(3\sim10)\times10^{-3}$ m_0c was believed to be induced by the Ti vacancies where positrons annihilated with $2p$ electrons of oxygen [16], because the ratio decreased due to the absence of oxygen for Ti sample. But for samples 1# and 3#, the ratio corresponding to this zone increased since much more Ti vacancies were induced by ion implantation, and also a partial of positrons will be annihilated on constituent oxygen atoms. The prompt peak at $(10\sim18)\times10^{-3}$ m_0c should be induced by 3d electrons of Ti and Co[20]. With similar electron structure except for the $3d$ electrons (Ti: *1s2 2s2 2p6 3s2 3p6 3d2 4s2*; Co: *1s2 2s2 2p6 3s2 3p6 3d7 4s2*), Co has five more atomic nucleus so that its $3d$ electron momentum is larger than that of Ti, this leads to the broadening of the prompt peak centered at ~ 15×10^{-3} m_0c compared with that of Ti sample in figure 3. It is interesting that the Ti and substitutional Co at the Ti sites enhanced positron annihilation although they are positively charged, thereby non-trapping centers for positrons. Considering the fact that there are excess Ti $3d$ electrons around the oxygen vacancy upon ion implantation, the above vacancy complex should be weakly negatively charged. So, these results imply the presence of Ti-Co-V_O or Ti-V_o complex.

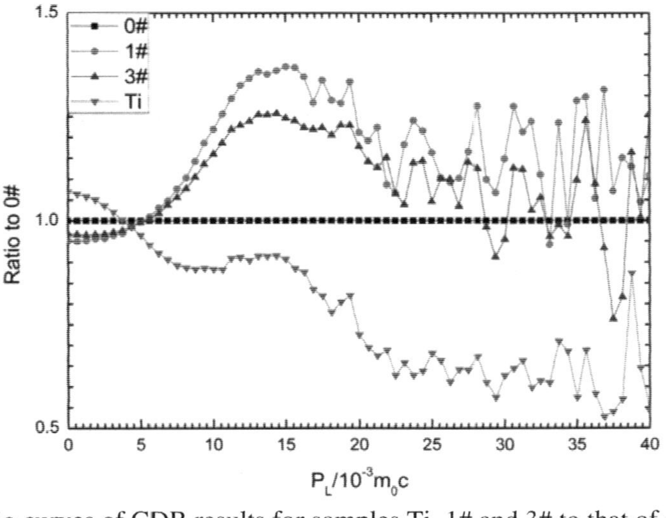

Fig.3. Ratio curves of CDB results for samples Ti, 1# and 3# to that of sample 0#.

4. Conclusion

In summary, we have studied the Co ion implanted rutile TiO_2 samples by positron annihilation Doppler broadening spectroscopy and CDB measurements with variable energy slow positron beam. The S parameters show that the concentration of vacancies is increased with increase of dopant dose and a new type of annihilation mechanism was distinguished by S-W plot. With CDB measurements, we identified the existence of Ti-Vo and/or Ti-Co-Vo vacancy complex because of the existence of excess Ti $3d$ electrons around the oxygen vacancy.

References

[1] Akdogan N, Nefedov A, Zabel H, esterholt K, Becker H W, Somsen C, Gok S, Bashir A, Khaibullin R, and Tagirov L, 2009 *J. Phys. D: Appl. Phys* **42** 115005
[2] Jia C W, Xie E Q, Zhao J G, Duan H G, and Zhang Y Z, 2007 *Mater. Sci. Eng., B* **140** 10
[3] Kim D, Hong J, Park Y R, and Kim K J, 2009 *J. Phys.: Condens. Matter.* **21** 195405
[4] Mi W B, Jiang E Y, and Bai H L, 2009 *J. Magn. Magn. Mater.* **321** 2472
[5] Park M S, Kwon S K, and Min B I, 2002 *Phys. Rev. B* **65** 161201
[6] Shinde S R, Ogale S B, Higgins J S, Zheng H, Millis A J, Kulkarni V N, Ramesh R, Greene R L, and Venkatesan T, 2004 *Phys. Rev. Lett.* **92** 166601
[7] Sullivan J M, and Erwin S C, 2003 *Phys. Rev. B.* **67** 144415
[8] Yang Z, Liu G, and Wu R, 2003 *Phys. Rev. B* **67** 060402
[9] Chambers S A, Droubay T, Wang C M, Lea AS, Folks L, Deline V, and Anders S, 2003 *Appl. Phys. Lett.* **82** 1257
[10] Jaffe J E, Droubay T C, and Chambers S A, 2005 *J. Appl. Phys.* **97** 073908
[11] Uedono A, Ishibashi S, Keller S, Moe C, Cantu P, Katona T M, Kamber D S, Wu Y, Letts E, Newman S A, Nakamura S, Speck J S, Mishra U K, DenBaars S P, Onuma T, and Chichibu S F, 2009 *J. Appl. Phys.* **105** 054501
[12] Yan W, Sun Z, Pan Z, Liu Q, Yao T, Wu Z, Song C, Zeng F, Xie Y, Hu T, and Wei S, 2009 *Appl. Phys. Lett.* **94** 042508
[13] Takenaka H, and Singh D J, 2007 *Phys. Rev. B* **75** 241102
[14] Alatalo M, Barbiellini B, Hakala M, Kauppinen H, Korhonen T, Puska M J, Saarinen K, Hautojarvi P, and Nieminen R M, 1996 *Phys. Rev. B* **54** 2397
[15] Coleman P G, Burrows C P, Mahapatra R, and Wright N G, 2007 *J. Appl. Phys.* **102** 014106
[16] Kuriplach J, Brauer G, Anwand W, Nicht E M, Coleman P G, and Wagner N, 1999 *Phys. Rev. B* **59** 1948
[17] Puska M J, and Nieminen R M, 1994 *Rev. Modern Phys.* **66** 841
[18] Takenaka H, and Singh D J, 2008 *Phys. Rev. B* **77** 155132
[19] Brauer G, Becvar F, Anwand W, and Skorupa W, 2006 *Appl. Surf. Sci.* **252** 3368
[20] Nagai Y, Tang Z, and Hasegawa M, 2000 *Radiat. Phys. Chem.* **58** 737
[21] Yang X, Xia H, Qin X, Li W, Dai Y, Liu X, Zhao M, Xia Y, Yan S, and Wang B, 2009 *Carbon* **47** 1399
[22] Yang X L, Zhu W X, Wang C D, Fang H, Yu T J, Yang Z J, Zhang G Y, Qin X B, Yu R S, and Wang B Y, 2009 *Appl. Phys. Lett.* **94** 151907

Journal of Physics: Conference Series **262** (2011) 012052 doi:10.1088/1742-6596/262/1/012052

The generation of deformation damage during fatigue loading in Al-Cu alloy studied by the Doppler Broadening technique

M L Ruiz-Ripoll[1,2], H Schut[1], N H van Dijk[1], R C Alderliesten[2], S van der Zwaag[2] and R Benedictus[2]

[1] Delft University of Technology, faculty of Applied Sciences, Mekelweg 15 NL-2629 JB, Delft, The Netherlands.
[2] Delft University of Technology, faculty of Aerospace Engineering, Kluyverweg 1 NL-2629 HS, Delft, The Netherlands.

E-mail: M.L.RuizRipoll@tudelft.nl

Abstract. We have investigated the evolution of damage during fatigue loading in Al–Cu 2024-T3 alloys using the positron annihilation Doppler Broadening (DB) technique. This technique enables us to monitor in a non destructive way, at the atomic and vacancy level, the formation of deformation defects and their interaction with solute atoms at selected stages of fatigue testing. The changes in the S and W Doppler Broadening parameters are linked to the changes in fatigue behavior at lower stress levels. The material was tested under constant amplitude fatigue loading at four different stress levels and DB tests were conducted at selected stages of fatigue lifetime. The results are compared to those obtained during static tensile tests.

1. Introduction

Plastic deformation of precipitation hardened aluminium alloys leads to the displacement of dislocations through the matrix containing a dense network of precipitates. With increasing deformation levels the defects link to form voids which initiate nano sized cracks that subsequently grow and ultimately lead to failure of the material [1-2]. In order to observe the evolution of these defects and to monitor the damage process directly, the positron annihilation Doppler Broadening (DB) technique has been used. Positrons are effectively trapped at defects associated with open volume defects such as vacancies, vacancy clusters and misfit location at the precipitate/matrix interface or inside solute aggregates with higher positron affinity than the host material. Upon trapping a positron annihilates with an electron into two 511 keV photons. Because of the momentum of the electron the measured energy of the photon is shifted by an amount of $\Delta E = p_L c/2$, with c the speed of light and p_L the longitudinal component of the positron-electron momentum along the direction of emission of the photons. In an annihilation spectrum these shifts lead to a broadening of the 511 keV photo peak which is quantified by the S- and W-parameters [1]. S and W reflect annihilations with valence and core electrons, respectively, and are therefore ideal to monitor the generation and evolution of deformation induced defects and the chemical environment at the positron trapping sites. We have used these properties to investigate the role of Cu solute atoms on the evolution of damage and the effect of intermittent annealing periods on the fatigue lifetime [2-6].

2. Experimental

Commercial aluminium 2024-T3 (AlCu4Mg1) sheet metal has been used as base material. The material has been solution treated at 495 °C then cold worked and finally aged at RT. The useful length of the dog-bone shaped specimens was 120 mm with 1.2 mm thickness. The total length was 240 mm and the radius of transition was 185 mm (k_t=1.1). The Al alloy contains Cu (3.80-4.90 at.%), Mg (1.20-1.80 at.%), Mn (0.30-0.90 at.%), Fe (0.50 at.%), Si (0.50 at.%), Zn (0.25 at.%) and other constituents with concentrations below 0.15 at.%.

Two different routes for mechanical loading of the samples have been followed. The first one consists of continuous constant amplitude fatigue tests at selected stress levels until failure (referred to as T3-C). The second route combines fatigue testing interrupted by annealing periods (referred to as T3-A). The T3-A sequence comprises of:

1. Fatigue loading for 50, 100, 200 and 400 kcycles depending on the maximum stress σ_m applied. These were σ_1=0.97, σ_2=0.89, σ_3=0.81 and σ_4=0.76 of the yield strength of 345 MPa [7]. The stress ratio ($R = \sigma_{min}/\sigma_{max}$) was in all cases 0.1.
2. A Doppler Broadening measurement to monitor the effect of the mechanical loading.
3. An annealing period during which the material is annealed at 65 °C for 24h (in air) and thereafter quenched in water at room temperature.
4. A second DB measurement to investigate the effect of the annealing during step 3.
5. Repetition of the above steps until failure.

The fatigue tests were conducted using a servo hydraulic testing machine with a 60 kN load capacity and a steel alloy grip based on ASTM E606 [8] at 25 Hz frequency.

The original condition of each sample was established before the loading with an additional DB test. To avoid further natural aging effects the specimens were stored at -20 °C after each fatigue sequence and DB test. The DB experiments were carried out using a ^{22}Na source sealed between two 7 μm thick Kapton foils, sandwiched between two identical specimens. Two HPGe detectors were used to detect both annihilation photons in time coincidence with an overall energy resolution of 0.9 keV at 511 keV. The S parameter is calculated as the ratio of the counts registered in a fixed central momentum window ($|p_{//}| < 3.5\times10^{-3}$ m_oc) to the total number of counts in the photo peak. This choice of the momentum window makes the S parameter sensitive to annihilations with low momentum valence electrons. Similarly, the W parameter is obtained from the high momentum regions (W_{left} and W_{right} with $10\times10^{-3}m_oc< |p_{//}| < 26\times10^{-3}m_oc$) and accounts for annihilations with high momentum core electrons. All S and W data presented in this paper are normalized to the S and W values of pure, well annealed aluminium.

3. Results

In figure 1 the number of cycles to failure as a function of the maximum load (σ_m) is shown in an S-N plot. It can be seen that the results follow the trend of a typical Wöhler curve. In the case where fatigue and annealing periods are alternated (T3-A), unlike the results presented by Lumley [2] no improvements in life times were observed with respect to those for the continuously loaded sample (T3-C), but the amount of scatter in lifetime seemed to be smaller, in agreement with his earlier observations. In figure 2 the Doppler-broadening data are shown together with those of the as prepared and the failed T3-C material. Also shown are the lines connecting the S,W points for pure Al, Cu and Mg and an aluminium vacancy (V_{Al}). In the S-W plot all points are located on a line parallel to the Al-Cu line intersecting the Al–V_{Al} line near the S,W point associated with the vacancy. Clearly, the observed trends are indicative for the generation of defects associated with (uncommitted) Cu solute atoms. Figure 3 focuses on the results obtained for samples exposed to the lowest stress level (σ_4). Here we noted that during the first half of the lifetime (<400 kcycles) the S,W points shift into the direction of a Cu rich positron trapping site, both after cycling and the subsequent annealing. Only in case the number of cycles approaches the failure regime the defects that are generated have a more vacancy type character. This trend is also observed at the lower stress level σ_3. For higher stress levels σ_1 and σ_2 the early failure of the samples did not provide enough data points to draw conclusions.

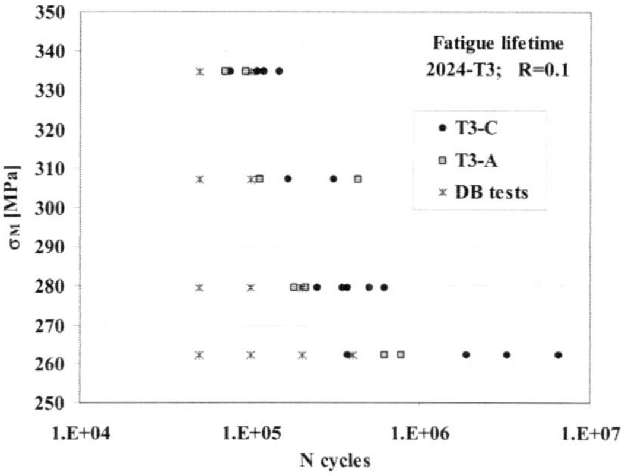

Figure 1. Lifetime results for R=0.1 for the T3-C and T3-A material. The number of cycles after which the DB tests were carried out are marked by asterisks.

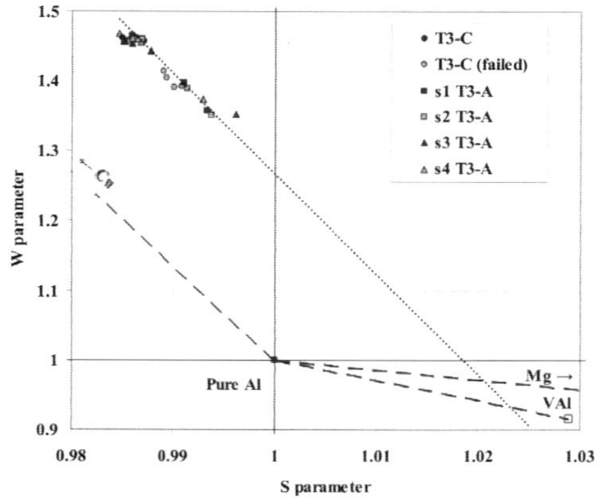

Figure 2. S-W plot obtained from Doppler Broadening tests. Note: The S,W point for Cu and Mg lie outside the range of axis.

4. Discussion

The fatigue loading results are compared with the ones obtained by Hautakangas et al. [1,5] for stepwise static tensile loading of T3 and under aged (UA) alloys obtained by solutionizing T3 samples at 495 °C for 30 minutes in N_2 gas followed by quenching in ice water and subsequent artificial ageing at 195 °C for 5 minutes. The static deformation of T3 material only shows an increase of S and decrease of W for strain levels above 2.3 % (see figure 4) corresponding to stress levels above the yield strength. Clearly, open volume defects are created. Ageing at RT after the deformation step shows no further change in S and W indicating that the defects are stable [5]. For the UA material, in which a significant amount of Cu is still in solution, static deformation (to 3% strain) again results in the creation of defects, but in this case ageing at RT leads to formation of additional Cu precipitates.

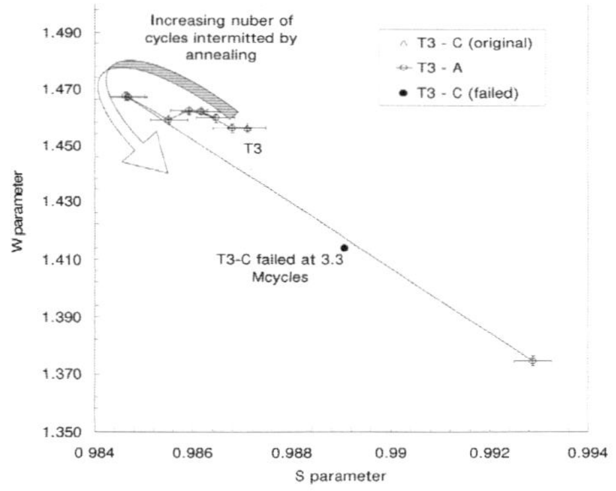

Figure 3. Experimental S-W paths obtained for σ_4 stress levels.

In the present fatigue experiments on T3 material the maximum stress level applied stays below the yield point and hence only minor changes in S and W parameters are expected. However, for samples tested at low stress levels a trend is visible where the data shifts into the direction of the S,W point for Cu. Since RT ageing can not explain this behavior we conclude that during cyclic loading the defects and dislocations together with the limited amount of Cu present (of the order of 0.1%) have formed additional positrons trapping sites. This process does not involve all the defects generated during the fatigue testing because a subsequent annealing treatment at 65 °C moves the S,W point further into the direction of positron trapping sites associated with Cu. Only in case the loading has resulted in sample failure the data shifts into the high S, low W direction, as the plastic zone surrounding the crack tip must have passed through the sample volume probed.

In order to compare the results of the static (figure 4) and fatigue loading experiments (figure 2) we define $D=\sqrt{(S-S_o)^2+(W-W_o)^2}$ as the distance of an S,W point to the S_o,W_o point of the as prepared T3 sample. In figure 5 D is plotted as a function of the calculated real stresses. Cyclic loaded samples that were not broken all have a value of D close to zero irrespective of the load level. This is not unexpected since the fatigue stress levels are in the elastic region and damage evolution is less wide spread. As pointed out earlier, in broken samples local plastic deformation in a region around the crack, which is probed in the PA test, has occurred. For theses samples D has the same magnitude as the strained ones. However, as the amount of local deformation can not be quantified uniquely, care should be taken in quantifying the degree of damage accumulation and interpreting differences between samples.

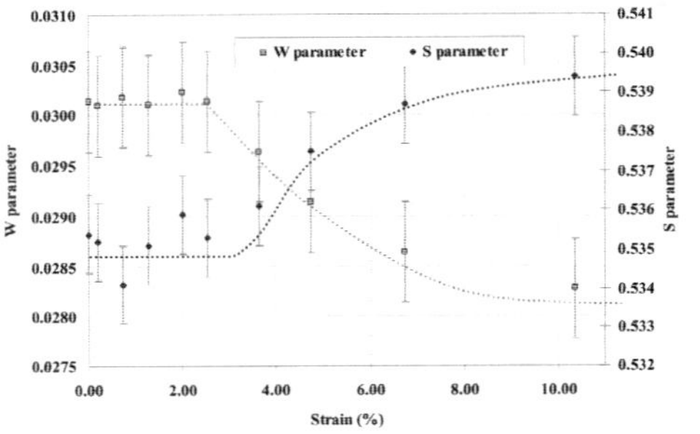

Figure 4. S and W parameters as a function of the total strain. Data from [5].

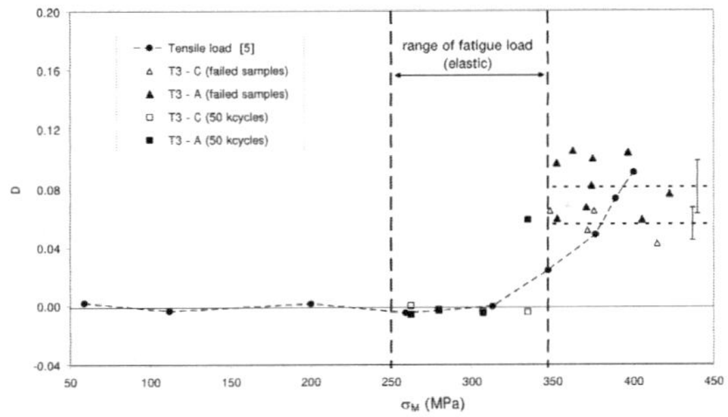

Figure 5. Comparison between static and dynamic loads in terms of maximum real stresses and D.

5. Conclusion

While the damage accumulation of AA2024 T3 material during plastic straining has been demonstrated convincingly, only weak PA spectrum changes have been recorded for samples loaded dynamically. More detailed PA measurements are therefore required to measure fatigue damage evolution and its disappearance by intermittent low temperature annealing treatments.

References

[1] Hautakangas S, Schut H, van der Zwaag S, Rivera Diaz del Castillo P E J and van Dijk N H. 2007 Positron annihilation spectroscopy as a tool to develop self healing in aluminium alloys *Phys. Status Solidi (c)* **4 -10** 3469-72.

[2] Lumley R N, O'Donell R G, Polmear I J and Griffiths J R 2005 Enhanced fatigue resistance by underageing and Al-Cu-Mg-Ag alloy *Mater. Forum* **29** 256-61.

[3] Lumley R N, Polmear I J and Morton A J 2005 Development of mechanical properties during secondary aging in aluminium alloys *Mater. Sci. Tech. Ser.* **21** 1025-32.

[4] Lumley R N, Morton A J and Polmear I J 2002 Enhanced creep performance in an Al-Cu-Mg-Ag alloy through underageing *Acta Mater.* **50** 3597-608.

[5] Hautakangas S, Schut H, van der Zwaag S, Rivera Diaz del Castillo P E J and van Dijk N H. 2008 Self healing of deformation damage in underaged Al-Cu-Mg alloys *Scripta Mater.* **58** 719-22.

[6] Dupasquier A, Kögel G and Somoza A 2004 Studies of light alloys by positron annihilation techniques *Acta Mater.* **52** 4707–26.

[7] Annual Book of ASTM Standards 03.01.2007. ASTM E8M-08.

[8] Annual Book of ASTM Standards 03.01.2005. ASTM E606-92.

The authors acknowledge the financial support of the IOP Self Healing Materials (the Netherlands) in the form of project SHM08736, "Development of Self Healing in precipitation hardened and dispersoid strengthened aluminium alloys and determination of their fatigue properties".

Journal of Physics: Conference Series **262** (2011) 012053 doi:10.1088/1742-6596/262/1/012053

Positron annihilation lifetime measurements of He-ion-irradiated Fe using pulsed positron beam

K Sato[1], A Kinomura[2], T Omura[3], Q Xu[1], T Yoshiie[1], R Kasada[3], A Kimura[3] and K Morishita[3]
[1] Research Reactor Institute, Kyoto University, Kumatori-cho, Osaka 590-0494, Japan
[2] National Institute of Advanced Industrial Science and Technology, Tsukuba, Ibaraki 305-8568, Japan
[3] Institute of Advanced Energy, Kyoto University, Gokasho, Uji, Kyoto 611-0011, Japan

E-mail: ksato@rri.kyoto-u.ac.jp

Abstract. The positron annihilation lifetimes of He-ion-irradiated Fe samples were measured by a pulsed positron beam technique. The relative thickness of the surface oxidation layer was measured by Auger microprobe measurements. Unirradiated samples with different surface oxidation layer thicknesses had very similar second lifetime component intensities, indicating that the surface oxidation layer thickness does not affect positron lifetime spectra. In lifetime measurements using 10-keV positrons, the second lifetime component decreased monotonically with increasing irradiation dose. The mean lifetime in the 10-keV measurements was longer than that in 15-keV measurements (which correspond to the peak region of vacancy production); this is thought to be due to vacancy clusters absorbing He atoms.

1. Introduction

In fission and fusion reactors, large amounts of He are introduced into materials by either direct implantation or indirect means such as nuclear transmutation reactions. Because of its extremely low solubility and high mobility in materials, He tends to strongly agglomerate into bubbles, which deteriorates the mechanical properties of materials [1–3]. Therefore, it is critical to investigate the interaction between He and irradiation-induced defects, especially vacancies. Sugano *et al.* did not observe He bubbles in transmission electron microscopy images of α-Fe samples that been irradiated with 8 keV He ions at room temperature and subsequently annealed at temperatures up to 673 K, but they did observe bubbles in samples that had been annealed at 773 K [4]. To study the growth of He bubbles, it is essential to obtain information about the growth of small vacancy clusters and the interaction between small vacancy clusters and He atoms, which cannot be observed by transmission electron microscopy.

Positron annihilation spectroscopy can detect vacancy clusters and even single vacancies. Positron annihilation spectroscopy using slow-positron beam techniques is very effective for investigating the near-surface microstructure of ion-irradiated materials. Fujinami *et al.* performed a positron beam study of He–vacancy complexes in He-ion-implanted Si and detected the formation of $He–V_2$ complexes [5].

In the present study, we investigate the effect of the He irradiation dose on the formation of vacancy clusters in Fe by performing positron annihilation lifetime (PAL) measurements using a pulsed positron beam. Since oxidation layers readily form on the surfaces of pure Fe samples, the effect of the surface oxidation layer (SOL) on PAL is also discussed.

2. Experimental procedure

Samples were prepared from 99.99% pure Fe that was cold rolled into approximately 0.l-mm-thick sheets. Each sheet was cut into 15 × 15 mm^2 pieces. All samples were mechanically polished using an abrasive coating with 1-μm-diameter particles. They were then annealed at 1073 K for 2 h in H_2 flow. He ion irradiation was performed at the National Institute of Advanced Industrial Science and Technology (AIST), Japan. The irradiation doses of the incident 150-keV He ions were 1×10^{18}, 1×10^{19}, and 1×10^{20} He^+/m^2. The irradiation was performed at room temperature. Calculations using SRIM code [6] gave a displacement damage peak of approximately 400 nm and a He penetration peak depth of about 450 nm. PAL measurements were performed using a variable-energy slow-positron beamline at the AIST linac [7]. The time resolution of the PAL measurements was about 270 ps. 10- and 15-keV positrons were implanted. Positron stopping profiles were calculated using the method described in Refs. 8–10. The average stopping depths of 10- and 15-keV positrons were 212 nm and 408 nm, respectively. Lifetime spectra were decomposed using the PALSfit program [11]. The relative thickness of the SOL was analyzed by a field-emission Auger microprobe (Jeol, JAMP-9500FV) at the Institute of Advanced Energy, Kyoto University, which utilized a 10-keV electron beam (current: 10 nA) to produce Auger electrons and a 1-keV Ar beam (current: 610 nA) for ion etching. The etching time per cycle was 2 s and the etching rate was 10.1 nm/min for Si. No data was obtained for Fe, so only the relative distribution of O atoms from the surface was obtained.

3. Results and discussion

3.1. Effect of surface oxidation layer on PAL

Table 1 lists the PALs of unirradiated samples measured by positrons with penetration energies of 10 and 15 keV. τ_3 probably originates from the background signal due to reemitted positrons or from the experimental apparatus. τ_1 is the lifetime of the matrix. The lifetimes of mechanically polished samples do not differ much from the lifetime of 106–110 ps obtained by conventional methods [12,13]. τ_1 of samples chemically polished with hydrogen fluoride is longer than the matrix lifetime (about 110 ps). This sample did not have a mirror surface, which may be one reason why it has a longer lifetime than the matrix. τ_2 is probably the lifetime of the SOL or the annihilation lifetime of positrons that penetrate inside the sample and then diffuse to the surface. The oxidation layer contains large voids whose lifetimes are about 500 ps [14].

Figure 1 shows the depth distribution of O atoms from the surface in unirradiated Fe. The horizontal axis denotes the ion etching time and the vertical axis corresponds to the amount of O atoms. The number of O atoms decreases with increasing depth. The chemically polished sample has a thinner SOL than the mechanically polished sample. If positrons directly implanted in the oxidation layer influence the positron lifetime, its intensity would vary with its thickness; however, all three treatments gave very similar intensities for τ_2. Therefore, τ_2 is thought to mainly originate from positrons diffusing from the bulk to the oxide interface. In the irradiated samples, positrons are trapped at defects and thus may not diffuse so easily. Consequently, we ignored the effect of SOL when we considered the results for the irradiated samples.

3.2. PAL measurements

Tables 2 and 3 give the PALs of He-ion-irradiated samples with positron penetration energies of 10 and 15 keV, respectively. In all cases, long lifetimes of about 1 ns are obtained for τ_3; they are due to the background signal caused by reemitted positrons or the experimental apparatus. In the 10-keV measurements, τ_1 remains constant and τ_2 decreases monotonically with increasing irradiation dose. In the 15-keV measurements, τ_1 increases for 10^{20} He^+/m^2 irradiation and τ_2 decreases initially and then increases with increasing irradiation dose.

Journal of Physics: Conference Series **262** (2011) 012053 doi:10.1088/1742-6596/262/1/012053

Table 1. PALs of unirradiated samples measured by positrons with penetration energies of 10 and 15 keV. MP: mechanical polishing; CP: chemical polishing; SS: soft sheet; HS: hard sheet. "–" indicates that it was not possible to decompose the spectra into three components.

Positron energy	Surface treatment	τ_1 (ps)	I_1 (%)	τ_2 (ps)	I_2 (%)	τ_3 (ps)	I_3 (%)
10 keV	MP with SS	107 ± 6	68 ± 1	394 ± 3	31 ± 1	1141 ± 44	1 ± 0.1
10 keV	MP with HS	110 ± 1	70 ± 1	407 ± 1	30 ± 1	–	–
10 keV	CP	124 ± 1	71 ± 1	351 ± 2	29 ± 1	–	–
15 keV	MP with SS	115 ± 1	84 ± 1	403 ± 3	15 ± 1	1023 ± 35	1 ± 0.1

Table 2. PALs of He-ion-irradiated samples with a positron penetration energy of 10 keV. τ_{av} was calculated using $(\tau_1 I_1+\tau_2 I_2)/(I_1+I_2)$.

Irradiation dose	τ_{av} (ps)	τ_1 (ps)	I_1 (%)	τ_2 (ps)	I_2 (%)	τ_3 (ps)	I_3 (%)
10^{18} He^+/m^2	228	138 ± 10	58 ± 1	355 ± 4	41 ± 1	1456 ± 64	1 ± 0.1
10^{19} He^+/m^2	223	148 ± 9	58 ± 2	330 ± 6	41 ± 2	1132 ± 39	1 ± 0.1
10^{20} He^+/m^2	224	148 ± 11	53 ± 3	311 ± 6	46 ± 2	935 ± 25	1 ± 0.1

Table 3. PALs of He-ion-irradiated samples with a positron penetration energy of 15 keV. τ_{av} was calculated using $(\tau_1 I_1+\tau_2 I_2)/(I_1+I_2)$.

Irradiation dose	τ_{av} (ps)	τ_1 (ps)	I_1 (%)	τ_2 (ps)	I_2 (%)	τ_3 (ps)	I_3 (%)
10^{18} He^+/m^2	195	142 ± 3	66 ± 2	301 ± 5	33 ± 2	1053 ± 32	1 ± 0.1
10^{19} He^+/m^2	192	145 ± 6	59 ± 4	262 ± 7	40 ± 4	978 ± 24	1 ± 0.1
10^{20} He^+/m^2	221	186 ± 4	74 ± 4	326 ± 12	25 ± 4	944 ± 37	1 ± 0.1

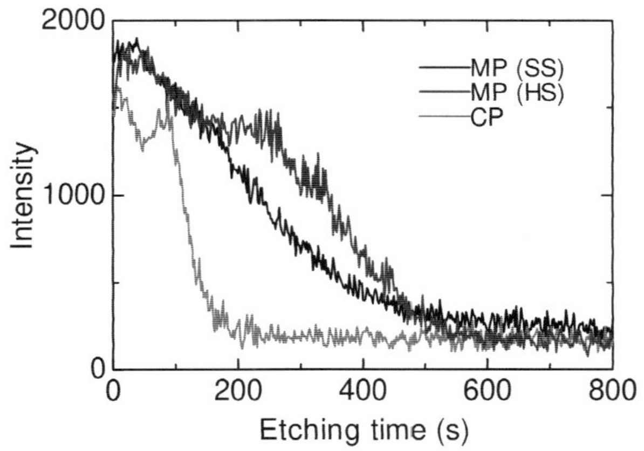

Figure 1. The depth distribution of O atoms from the surface in unirradiated Fe. MP: mechanically polished; CP: chemically polished; SS: soft sheet; HS: hard sheet.

In this study, the peak region of the vacancies and He atoms was measured at a positron penetration energy of 15 keV and the peak height of the generated vacancies was about three times higher than that at a depth of 210 nm (average penetration depth of 10-keV measurements). Troev *et al.* calculated PALS for vacancy clusters containing He atoms in Fe [15]. The positron lifetime of He bubbles decreases with increasing number of He atoms in vacancy clusters. The lifetime of vacancy clusters composed of 12 vacancies (V_{12}) with no He atoms is about 310 ps, whereas the lifetime of V_{12} containing 9 He atoms is about 200 ps. τ_1 was about 140 ps. It is the mixed lifetime of single vacancy–He atoms complexes (V–He_n) and dislocation loops. τ_2 (262–355 ps) is the lifetime for vacancy cluster–He atoms complexes (V_m–He_n).

When τ_{av} was measured for the same irradiation dose and at a different positron penetration energy; τ_{av} for 15-keV measurements is shorter than that of 10-keV measurements despite the high amount of vacancies produced. For 10^{18} and 10^{19} He^+/m^2 irradiation, τ_2 of 15-keV measurements is also shorter than that of 10-keV measurements. This is because the amount of He atoms increased in the defect peak region so that more He atoms were absorbed by vacancy clusters.

In 10-keV measurements, vacancy clusters are formed by low dose irradiation and the vacancy clusters absorb He atoms by high dose irradiation. Consequently, the mean lifetime decreases with increasing irradiation dose. In 15-keV measurements, vacancy clusters also absorb He atoms. However, the production rate of vacancies is high in the defect peak region, and the growth speed of vacancy clusters exceeds the reduced lifetime caused due to the absorption of He atoms. These phenomena need to be demonstrated by performing simulations based on rate theory.

4. Concluding remarks

The PAL of Fe irradiated with He ions was measured by a pulsed positron beam technique. The relative thickness of the SOL was obtained using an Auger microprobe. The SOL thickness did not influence the lifetime spectra. PAL measurements detected the growth of vacancy clusters and V_n–He_m complexes. The PAL decreases as the number of He atoms vacancy clusters that are absorbed increases. In the next step, the isochronal annealing behaviour of PAL will be investigated using a pulsed positron beam.

Acknowledgements

This study was partially financially supported by the Budget for Nuclear Research of the Ministry of Education, Culture, Sports, Science and Technology, Japan, based on the screening and counselling by the Atomic Energy Commission. A part of this work was conducted under the Collaboration Program of the Laboratory for Complex Energy Processes, Institute of Advanced Energy, Kyoto University.

References

[1] Trinkaus H and Singh B N 2003 *J. Nucl. Mater.* **323** 229
[2] Chen J, Jung P and Trinkaus H 2000 *Phys. Rev.*B **61** 12923
[3] Xu Q, Yoshida N and Yoshiie T 2007 *J. Nucl. Mater.* **367–370** 806
[4] Sugano R, Morishita K, Kimura A, Iwakiri H and Yoshida N 2004 *J. Nucl. Mater.* **329–333** 942
[5] Fujinami M, Miyagoe T, Sawada T, Suzuki R, Ohdaira T and Akahane T 2003 *Physica* B **340–342** 724
[6] Ziegler J F, Biersack J P and Littmark U 1985 *The Stopping and Range of Ions in Solids* (New York: Pergamon)
[7] Suzuki R, Ohdaira T and Mikado T 2000 *Radiat. Phys. Chem.* **58** 603
[8] Nieminen R M and Oliva J 1980 *Phys. Rev.* B **22** 2226
[9] Valkealahti S and Nieminen R M 1984 *Appl. Phys.* A **35** 51
[10] Vehanen A, Saarinen K, Hautojärvi P and Huomo H 1987 *Phys. Rev.* B **35** 4606
[11] Olsen J V, Kirkegaard P, Pedersen N J and Eldrup M 2007 *Phys. Stat. Sol.* C **4** 4004
[12] Hautojarvi P, Judin T, Vehanen A, Yli-Kauppila J, Johansson J, Verdone J and Moser P 1979 *Sol. Stat. Commun.* **29** 855
[13] Onitsuka T, Takenaka M, Kuramoto E, Nagai Y and Hasegawa M 2002 *Phys. Rev.* B **65** 012204
[14] Wu UC, Chen YQ, Wang B, Wang SJ, Jean YC, Suzuki R and Ohdaira T 2006 *Appl. Surf. Sci.* **252** 3274
[15] Troev T, Popov E, Staikov P and Nankov N 2009 *Phys. Stat. Sol.* C **6** 2373

Journal of Physics: Conference Series **262** (2011) 012054 doi:10.1088/1742-6596/262/1/012054

Understanding the effect of nanoporosity on optimizing the performance of self-healing materials for anti-corrosion applications

S Sellaiyan[1,2], S V Smith[1], A E Hughes[3], A Miller[4], D R Jenkins[5] and A Uedono[6]

Centres for Antimatter-Matter Studies (CAMS) at the [1]Australian Nuclear Science and Technology Organisation (ANSTO) and the [2]Australian National University.
[3]CSIRO Future Manufacturing Flagship, Clayton South MDC 3169.
[4]CSIRO Mathematical and Information Sciences, Glen Osmond, SA 5064.
[5]CSIRO Mathematical and Information Sciences, North Ryde NSW 1670.
[6]Institute of Applied Physics, University of Tsukuba, Ibaraki 305-8573, Japan

E-mail: svs@ansto.gov.au

Abstract. The chromate-based epoxy primer film was prepared on glass and metal surfaces using various graded doctor blades. The quality and consistency of the films was assessed using scanning electron microscopy and the porosity within the film characterized by positron annihilation lifetime spectroscopy. The positron lifetime (τ) distribution for the epoxy polymer matrix was resolved using the CONTIN program. The free volume was found from the *ortho*-positronium component. The optimum thickness for the films was established for future structure/activity studies.

1. Introduction

Materials that have the ability to heal themselves are known as self-healing. They have the ability to restore the material to its original set of properties after thermal or mechanical damage. They can be metals, ceramics, polymers or their composites. One well known class of inhibited paints is the chromate based primers, which are still used in the most severe environments to protect a wide range of metallic substrates [1,2]. These types of coating systems are currently employed to protect a wide range of engineered structures from corrosion [3]. In the aerospace industry, alloys are selected for their superior ratio of weight to mechanical performance. However, they have poor corrosion resistance and films must be developed to protect them. These protective films are generally paint based incorporating chemical inhibitors that have the ability to prevent or slow corrosive processes. The inhibitor contains a leachable component (anti-corrosive pigment) that is released in order to prevent the corrosive reactions. The free volumes or voids are important properties to incorporate into the films for the inhibitor to work efficiently. We are interested to understand how porosity influences the ability of these films to repair themselves.

Positron annihilation lifetime spectroscopy (PALS) technique is a powerful characterization tool for the study of free volume sizes and free volume fraction in polymeric materials. The utility of the

[1] Correspondence to S. V. Smith, Centre for Antimatter-Matter Studies (CAMS), Australian Nuclear Science and Technology Organisation, Locked Bag 2001, Kirrawee DC, NSW 2232, Australia.

positron in the study of polymers is enhanced by the fact that the positron can capture an electron and form a bound system called positronium. In polymers, the long lifetime component represents the formation of *ortho*-positronium, *o*-Ps, and it is this component that gives information on free volume characteristics within a material. Madani *et al.* [4] used PALS to characterize both dry and water-saturated epoxy systems incorporating a selection of non-chromate corrosion inhibitors. The same authors have studied voids at the interface between a $SrCrO_4$ inhibitor phase and an epoxy matrix using PALS [5]. While Cao *et al.* [6] examined primer/topcoat systems using Doppler broadening of energy spectroscopy (DBES), the mean implantation depth of the positrons was insufficient to provide any information about the primer layer.

The present study involves the use of PALS to determine the free volume or voids within the chromium doped primer (CP) paint films and establish the optimum thickness for future leaching studies.

2. Experimental

To the epikote 828 (1 g) (Epoxy resin produced from bisphenol A and epichlorohydrin, Resolution Performance Products) was added strontium chromate (0.5 g) followed by hardener 92133 (2.2 ml) (ANAC Advanced Coatings). The mixture was thoroughly mixed and then added to glass and Ni plate using a spatula. A range of films of various thicknesses (15, 22, 30 and 50 μm) on glass surface and a three layers thick film (100 μm thickness of each layer) on a Ni surface was prepared using various graded doctor blades.

The films were left to dry overnight and then characterized by Scanning Electron Microscope (SEM) and PALS. Morphology of chromium inhibited primer film were studied after platinum sputter coating with a scanning electron microscopy (ZEISS ultra plus SEM) at 10 kV accelerated voltage and 4 mm working distance. The lifetime spectra of positrons were measured using a conventional lifetime system (^{22}Na source, 10 μCi) with BaF_2 scintillators attached to H3378 (Hamamatsu Photonics) photomultiplier tubes and digital oscilloscope [7]. Each spectrum contained about 6 x10^6 counts.

3. Result and Discussion

Many trials were performed to obtain a consistent and uniform coating of the chromium based primer. The weight percentage of substituents was optimized to epikote (29.2%), strontium chromate (14.6%) and hardener (56.2%). The mixture was then left (5-10 mins) at room temperature to thicken prior to coating on the substrate. A typical SEM of films produced is illustrated in Figure 1.

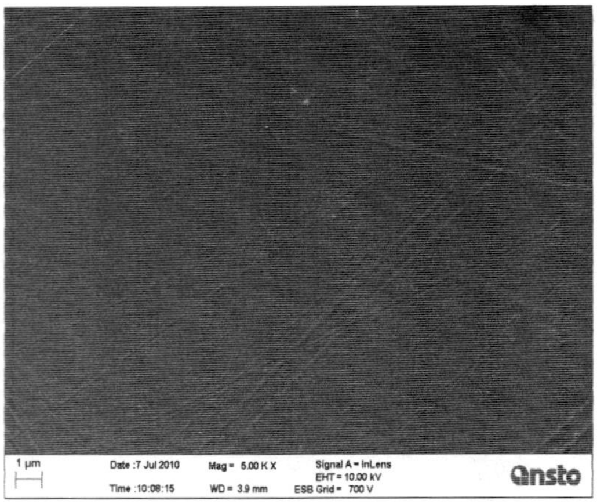

Figure 1. SEM Micrographs of CP films at 5000 X magnification

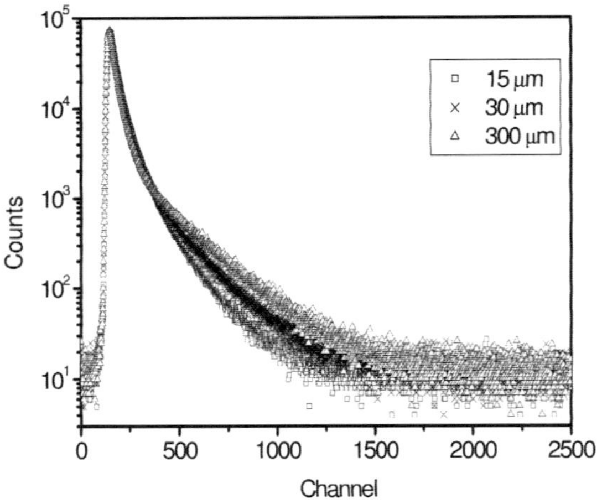

Figure 2. Positron lifetime spectra of 15, 30 and 300 μm thick CP films

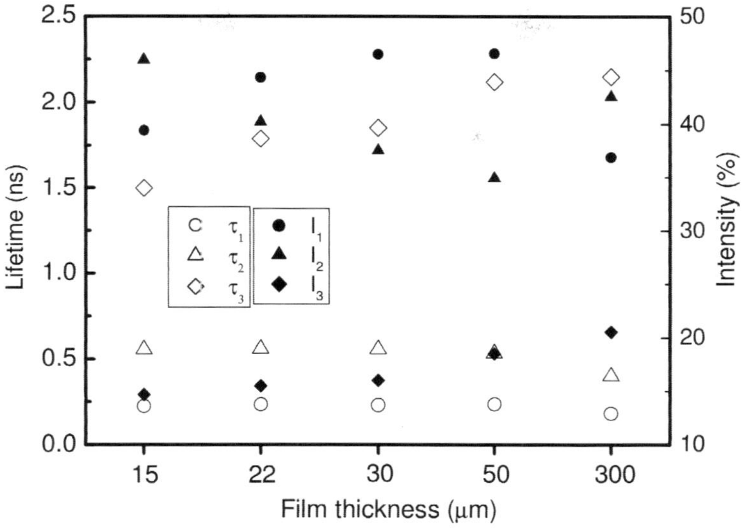

Figure 3. Positron Lifetime and relative intensity of 15, 22, 30, 50 and 300 μm thick CP films

Positron annihilation lifetime experiments were carried out on CP films deposited on glass plates (15, 22, 30 and 50 μm thick) and on Ni plates (300 μm thick). For the current study three different lifetime components were refined; τ_1, τ_2 and τ_3, with relative intensities of I_1, I_2 and I_3 (proportional to the probability of each process). Figure 2 shows typical lifetime spectra of the chromium inhibiter primer films. After fitting the data using the CONTIN code [8], we observe three lifetime components and their intensity as shown in Figure 3. The first lifetime (τ_1) is the weight average between the lifetime of *para*-Ps (*p*-Ps) and the relatively short lifetime of positrons in the polymer, the second lifetime (τ_2) is the intermediate lifetime of the free positrons prior to annihilation by an electron and the longest lifetime (τ_3) is due to the *o*-Ps.

For CP films with thickness of 15-50 μm on glass, a proportion of the positrons annihilate in the substrate. The lifetime of this *o*-Ps annihilates component in the substrate was established to be 0.960±0.006 ns with an intensity of 29±1%. The τ_3 value appeared to increase with increasing film thickness. This is because a higher proportion of *o*-Ps annihilated in the film instead of the substrate. For CP film with 50 μm thicknesses, however, the τ_3 value seems to be saturated. The lifetime of positrons in polymer films were found to be ~2.1 ns. However because of the additional positron component from the glass substrate further work was conducted using nickel as the substrate.

For CP film (3 x 100 μm thickness deposited on nickel substrate) almost all positrons were found to annihilate in the CP film. Therefore the τ_3 value observed could confidently be used to estimate the size of free volumes in the film. However as the film was multi-layered it is difficult to assess if the porosity occurs within each layer or between the layers. For the multi-layer film the τ_3 value was optimised (2.12±0.01 ns) using equation (1) to obtain mean free volume hole radius, *R*, according to Tao 1972 and Eldrup *et at.* 1981 [9,10].

$$\tau_3 = \frac{1}{2}\left[1 - \frac{R}{R_0} + \frac{1}{2\pi}\sin\left(\frac{2\pi R}{R_0}\right)\right]^{-1} \quad (1)$$

where τ_3 (o-Ps lifetime) and *R* (hole radius) are expressed in ns and nm, respectively.
and $R_0 = R + \Delta R$ where ΔR is a constant for an electron layer thickness (0.166 nm).
From this τ_3 value the mean free volume pore radius *R* was estimated to be 0.29 nm. The free volume size (V_f) was then calculated (i.e. $V_f = 4/3\pi R^3$) to be 0.11 nm^3 within the film.

Dull *et al.* [11] have extended Tao-Eldrup model to calculate the rectangular pores from *o*-Ps lifetime. Further analysis of current data using the Dull model to determine free volume as a rectangular pore gave a similar R value of 0.30 nm.

The free volume hole distribution can be calculated for *o*-Ps lifetime using equation (2).

$$S_{LT}(t) = \int_0^{\infty} \lambda\alpha(\lambda)e^{-\lambda t}d\lambda \tag{2}$$

where λ (the annihilation rate) is the reciprocal of lifetime τ with an annihilation probability-density function (PDF) λα.

The measured lifetime spectra were analyzed with a time resolution (full width at half-maximum) of approximately 170 ps by a numerical Laplace inversion technique [8]. Figure 4 gives a typical positron lifetime (τ) with PDF distributions for the 300 μm thick CP film. The three peaks correlate to *para*-positronium (*p*-Ps), positron, and *o*-Ps annihilations components, respectively. The *o*-Ps lifetimes values were ~2.1 ns and are consistent with analysis reported above.

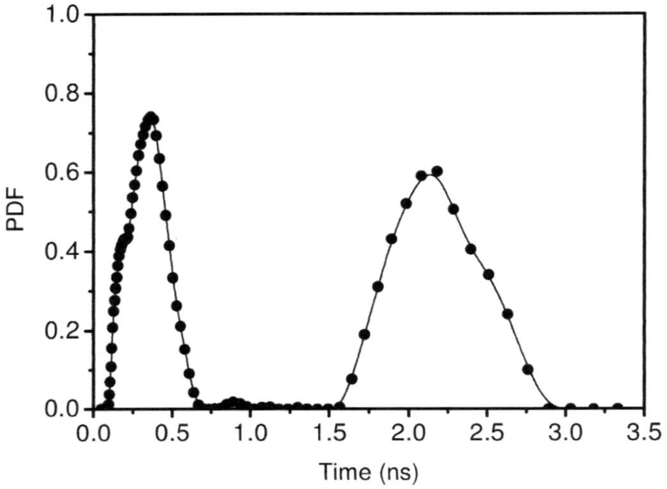

Figure 4. Positron Lifetime with PDF of 300 μm thick CP film

4. Conclusion

The chromate inhibited primer films were coated on glass and Ni substrate and analysed by PALS to determine optimum condition for future leaching studies. SEM show the films are consistent. The *o*-Ps component (τ_3=2.12 ns) from the PALS data were fitted to determine pore shape. Optimum thickness for CP films was found to be above 50 μm thickness.

References

[1] Scholes F H, Furman S A, Hughes A E, Nikpour T, Wright N, Curtis P R , Macrae C M, Intem S, and Hill A J 2006 *Prog. Org. Coat.* **56** 23
[2] Furman S A, Scholes F H, Hughes A E, and Lau D 2006 *Prog. Org. Coat.* **56** 33
[3] Ghosh S K 2009 *Self-healing Material: Fundementals, Design Strategies, and Applications* (Wiley-VCH Verlag GmbH & Co.KGaA)
[4] Madani M M, Miron R R, and Granata R D 1997 *J. Coat. Technol.* **69** 45
[5] Madani M M, Miron R R, and Granata R D 1999 (ANTEC 99 New York) p 2202
[6] Cao H, Zhang R, Sundar C S, Yuan J P, He Y, Sreczki T C, Jean Y C, and Nielsen B 1998 *Macromolecules* **31** 6627
[7] Saito H and Hyodo T 2003 *Radiat. Phys. Chem.* **68** 431
[8] Gregory R B 1991 *Nucl. Instrum. Methods Phys. Res. Sect. A* **303** 496
[9] Tao S J 1972 *J. Chem. Phys.* **56** 5499
[10] Eldrup M, Lightbody D, Sherwood J N 1981 *Chem. Phys.* **63** 51
[11] Dull T L, Frieze W E, Gidley D W, Sun J N, and Yee A F 2001 *J. Phys. Chem. B* **105** 4657

Journal of Physics: Conference Series **262** (2011) 012055 doi:10.1088/1742-6596/262/1/012055

The role of nuclear sensors and positrons for engineering nano and microtechnologies

Suzanne V Smith[1]

[1]Node Director, ARC COE in Anti-matter Matter Studies, Australian Nuclear Science and Technology Organisation, PO Box 1, Menai, 2234 NSW, Australia

E-mail: suzanne.smith@ansto.gov.au

Abstract. A sustainable nano-manufacturing future relies on optimisation of the design and synthetic approach, detailed understanding of structure/properties relationships and the ability to measure a products impact in the environment. This article outlines how bench-top PALS and nuclear techniques can be used in the routine analysis of a wide range of nanomaterials. Traditionally used in the semiconductor industry, PALS has proven to be useful not only in measuring porosity in polymeric materials but also in the monitoring of milling processes used to produce natural fibre powders. Nuclear sensors (radiotracers), designed to probe charge, size and hydrophilicity of nanomaterials, are used to evaluate the connectivity (availability) of these pores for interaction with media. Together they provide valuable information on structure/properties relationship of nanomaterials and insight into how the design of a material can be optimised. Furthermore, the highly sensitive nuclear sensors can be adapted for monitoring the impact of nanomaterials *in vivo* and the environment.

1. Background

Today, a successful product requires the analysis of its life-cycle and the potential side-effects of a design solution before it enters the market. Our communities expect our manufacturers to display philosophies that encourage the creation of new technologies to enhance our quality of life at a reduced cost, but also reduce the negative impacts on our environment.

Advances in micro and nanotechnologies have the potential to be wide reaching, influencing industry sectors such as healthcare, agrifood, transport, energy, materials and information and communications technologies. However, there are issues that continue to challenge the field such as the ability to;

1. predict and accurately measure performance of these materials
2. develop cost-effective scale up of their synthesis and
3. assess or forecast potential risk of these materials.

The community and regulatory bodies continue to demand a better understanding of the impact of these materials in their direct, manufacturing and ecological environments. Furthermore, there is high expectation that the next generation of products will have "zero" waste.

[1]To whom any correspondence should be addressed.

Many manufacturers are actively pursuing chemical processing strategies that align with the philosophy of Green Chemistry. Here the creator of a product chooses chemicals or chemical processes to actively reduce or eliminate negative impacts on the environment and waste. Such strategies have been common practice in the field of nuclear chemistry for many years and are summarised in Figure 1. The article will illustrate how we are adapting the practices in the nuclear chemistry and radiopharmaceutical field to aid the design, optimise the production, and evaluate the impact of nano and microtechnologies in their environment.

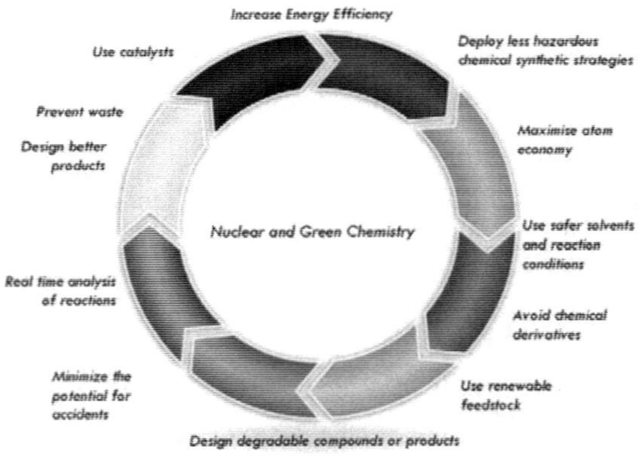

Figure 1. Philosophies of green and nuclear chemistry align.

2. Nuclear techniques

Nuclear techniques are the application of radioisotopes and detection systems for probing questions at the atomic and molecular level. They are widely deployed in the medical field, where radiotracers or radiopharmaceuticals are designed to specifically locate a disease site for imaging or treat it with radiation. Most of the radioisotopes used in radiopharmaceuticals are short-lived (i.e. hours to days). Their synthesis must be simple in design, 100% reliable, and chemically and energy efficient. Imaging is possible at extremely low concentration (10^{-10} to 10^{-12} M) with the intent of keeping toxicity to a minimum.

3. Engineering nanomaterials

As the nanoparticles decrease in size, the ratio of surface area to volume increases, often resulting in changes in their electronic properties (e.g. quantum dots) or reactivity (e.g. catalysts). Properties change drastically from the bulk material to small clusters of atoms [see Figure 2a and 2b).[1] Particles of the same chemical composition but different size and shape can have dramatically varied performance and interaction with biological systems.[2] The change in the available atoms on the surface of a particle is the rationale for such changes in behaviour.

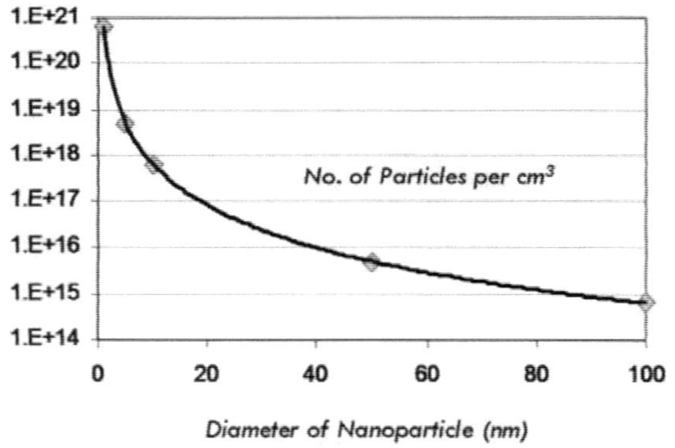

Figure 2a Number of particles in the bulk increase per volume as the diameter decreases.

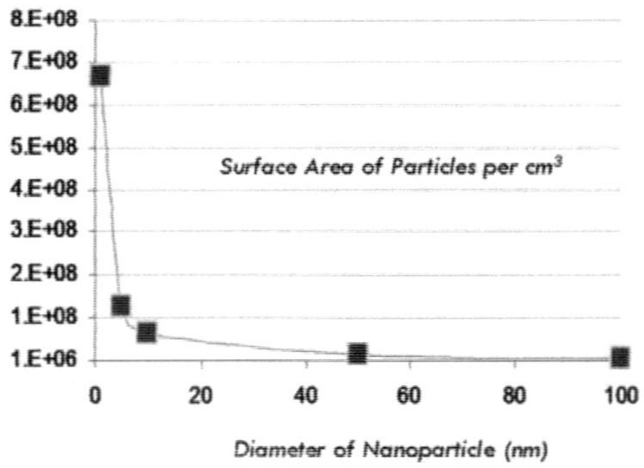

Figure 2b The surface area of the materials increases dramatically as the particles decrease in size.

The strategies to synthesize nanomaterials (see Figure 3) are generally described as *Top Down and Bottom Up* approaches. *Top-down* involves the grinding or milling of materials into small particles and the Bottom up requires the building of layers on a surface, atom by atom. *Top-down* approach, often termed as atomization, uses less chemicals but involves physical processes, such as milling to break down materials into smaller particles. The challenge here is to maintain the overall distinct molecular structure but generate micro and nanoparticles of the bulk material which will improve the overall properties of the bulk material such as diffusion and/or surface reactivity. The bottom-up approach can involve the generation of defects or vacancies in materials such as those found in wasteforms or semi-conductors. These defects and voids act to compensate for charge deficiencies and therefore provide greater stability. Induced microporosity (free volume) can improve the strength of a material such as synthetic bone, or it can also weaken a film whose role is to protect the underlying surface from the environment.

4. Porosity and PALS

The sizes of the holes in materials can be readily determined using a number of conventional techniques, such as transmission electron microscopy (TEM), BET surface analysis, and small angle neutron and X-ray scattering (SANS and SAXS). However, extensive sample manipulation, vacuum conditions, high skill level and limited availability prevents their routine use. Recent advances in detectors, digital oscilloscopes and software analysis for positron annihilation techniques is providing new opportunities for the bench-top Positron Annihilation Lifetime Spectroscopy systems (PALS).

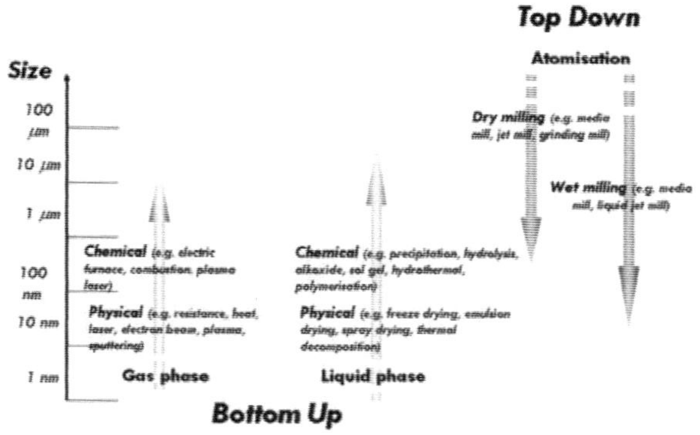

Figure 3. Summary of *Bottom Up* and *Top Down* approaches to the synthesis of nanomaterials

Classically used in the measurement of vacancies in semi-conductors and wasteforms, PALS is becoming a powerful tool for the analysis of soft matter. Requiring minimal sample preparation it can be used to analyse vacancies from defects (0.1 nm) to voids (up to 10 nm) in a wide array of materials. More recently it has been used to measure the porosity of polymeric materials and we are now interested to evaluate its potential role for assessing biological materials such as wool and silk powders.

Wool and silk fibres are traditionally used in the textile industry, but fibres too short for weaving go to waste. Increasing demand for renewable products has encouraged scientists to explore new applications for these fibres in areas such as wound dressing, constituents, cosmetics and pharmaceuticals. Re-crystallisation and milling processes have been investigated for the production of smaller fibrous particles, with the former more attractive for translation to large scale production.

Generating a fibrous powder from silk and wool fibres can dramatically enhance features such as rate and selectivity of metal binding properties.[3,4] PALS was used to evaluate the change in microporosity of fibres compared to jet-milled particles. PALS showed no change in size or incidence of micropores in natural fibres compared to milled samples. Solution studies investigating the binding of heavy metal ions confirmed total amount metal ion absorbed per mg of material had not changed. However a dramatic increase in the rate of metal ion absorbed was demonstrated.

5. Nuclear Sensors

A library of nuclear sensors (radiotracers) is under development to probe the charge, size and hydrophobicity of meso and micropores in materials. They are designed around the hexa aza cage compound, *sarar*, originally designed for use in ^{64}Cu labelling of antibodies and peptides for Positron Emission Tomography (PET).[5,6] The hexa aza cage can be synthesized via a metal template synthesis and readily modified to change the size and charge of the ligands (Figure 4).[7] The hexa aza cages can stably bind quantitatively to both long and short lived radioisotopes (^{57}Co and ^{64}Cu, respectively) and therefore can be used to monitor a range of processes in materials. Attaching a linker group to the hexa

Journal of Physics: Conference Series **262** (2011) 012055 doi:10.1088/1742-6596/262/1/012055

aza cages allows one to monitor the reactive groups on the surfaces of films and particles.[8] While PALS can provide information on the porosity within a material, it cannot indicate if the pores are connected or non-connected. Hence the nuclear sensors provide information on not only if the pores are available but also the charge within those pores. As the nuclear sensors can be detected at extremely low concentrations (up to 10^{-5} ppb), only small quantities (milligrams) of the nanomaterials are required for analysis. Furthermore the materials under investigation can be evaluated under conditions relevant to intended application providing more accurate information on the materials' properties. Highly adaptable for high-through-put applications, the nuclear sensors are currently being evaluated for on-line monitoring of the synthesis and stability of materials.

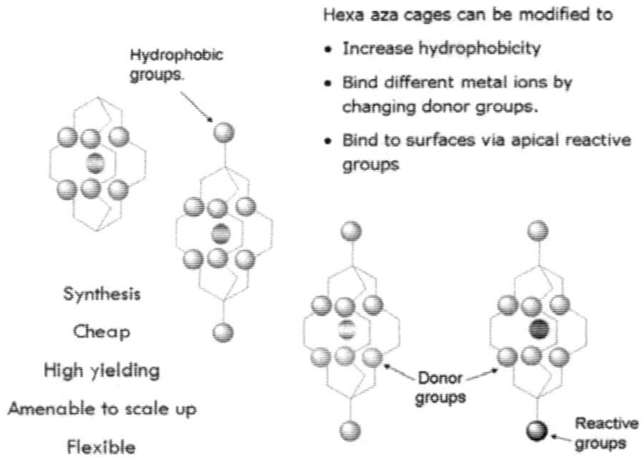

Figure 4 Schematic of the design of hexa aza cages

6. Application of PALS and Nuclear Sensors

We have used both PALS and nuclear sensors to provide insight into the relationship between structure and function of a wide range of materials and also for the optimisation of their synthesis and quality control of batch to batch production of nanomaterials. A selection of studies conducted in our laboratory will be described here.

Recent advances in the synthesis of mesoporous silica materials with controlled particle size, morphology and porosity has enabled their application in medicine, environment and industry. Amorphous silica is difficult to analyse by conventional techniques and understanding how to optimise the number of available pores and determining the physical properties within these pores continues to challenge the field.

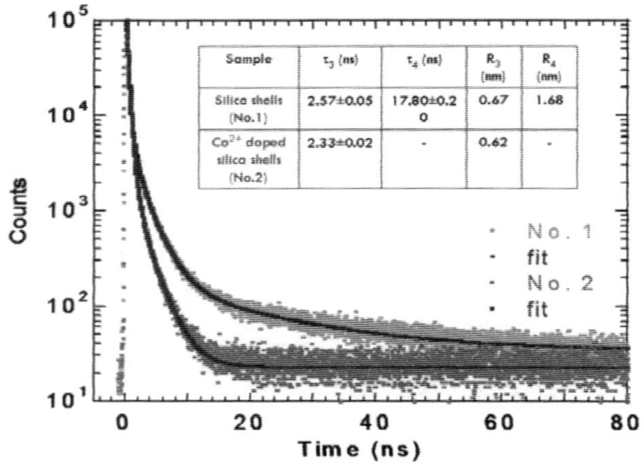

Figure 5. A comparison of PALS spectra for free and Co(II) doped hollow silica shells.

Hollow silica shells synthesized via a polymer template and high temperatures were studied using PALS and nuclear sensors.[9] Studies with nuclear sensors, outlined in detail elsewhere by Mume *et al* in these proceedings, show the porous silica preferentially bound positively charged molecules under basic conditions. In addition the micropores of the hollow silica shells preferred hydrophobic molecules over hydrophilic molecules. A PALS investigation into the microporosity of hollow silica shells, revealed that there were two types of micropores present (see Figure 5). When the hollow silica shells were doped with Co(II) ions, PALS showed there was an absence of large pores indicating that the metal ions were housed within the larges pores of the hollow silica shells.

Clay particles are another interesting class of materials, readily deployed in a range of applications from cosmetics and pharmaceuticals, to clean up agents for harmful oxyanions and for the efficient dispersion of pesticides. They are found in nature or can be easily synthesized. They have a high aspect ratio and their interlayer spacing can be readily manipulated by metal ion substitution. The interlayer spacings and morphology of the clay particles are usually determined using XRD and TEM, however small platelet size, lack of long-range order and sample preparation can interfere with full analysis. As these characteristics do not interfere with PALS it was investigated as an alternative tool for the analysis of a series of natural and synthetic clay particles.[10] A difference in the PALS spectra was clearly evident and the o-Ps lifetimes, their intensities and the pore shape (spherical and square channel geometries) were analysed. The data clearly showed that PALS was able to detect differences in interlayer spacing of the clays and was sensitive to changes as a result of metal ion doping. Further work will explore the sensitivity of PALS to changes in metal ion doping of clays and its value as a routine analysis tool for clay materials.

The alloys developed for the aerospace industry are selected for their superior weight to mechanical performance ratio. However, because of their poor corrosion resistance they must be coated with films to protect them. These self-healing films are generally paint based, incorporating chemical inhibitors that have the ability to prevent or slow corrosive processes. The free volumes or voids are important properties to incorporate into the films for the inhibitor to work efficiently. We are developing methods to analyze the relationship between the voids and the stability of these films and using this information to model diffusion properties of these materials.

Nuclear sensors or radioactive analogues of the inhibitors are synthesized and incorporated into the films. The films are exposed to corrosive environments and the release of the radioactivity or inhibitor can be monitored by gamma detection. Production of the films to a consistent level of porosity is an important element of this work. Slow positron beamline techniques are ideal for depth profiling porosity in films. However it is important to establish the effect of density of the films and the doping of the inhibitor on the annihilation of positrons. Initial studies have shown the potential PALS has for not only characterizing a range of anti-corrosion films but also for the selection of films suitable for slow positron beam analysis.[11]

7. Conclusion

PALS and nuclear sensors are ideal tools for understanding the relationship between porosity and performance of a wide range of materials from hard to soft matter, and including biological and composite materials. They are ideal for optimising materials synthesis and proving to be useful for the quality control or assessment of batch to batch reliability of production processes. Both techniques require minimal sample preparation and therefore are more likely to provide information relevant to sample applications. Extremely sensitive and independent of media, the nuclear sensors have the potential for high-throughput screening and on-line monitoring of materials.

Acknowledgments

The author wishes to acknowledge financial support from Australian Research Council Centre of Excellence in Antimatter Matter Studies and contribution from students, post-doctorates and collaborators from Deakin University, University of Western Australia and Australian National University in Australia and University of Tsukuba in Japan and Exilica Ltd. UK.

Abbreviations

Because the IUPAC names for the ligands are long and complicated, those ligands described in this paper have been abbreviated as follows:

sarar = 1-*N*-(4-aminobenzyl)-3,6,10,13,16,19-hexaazabicyclo[6.6.6]eicosane-1,8-diamine

References

[1] Service R F 2004 *Science* **304**(5678) 1732-4

[2] Service R F 2005 *Environmental Health Prospectus* **113** 823-39

[3] Naik R, Wen G, Dharmaprakash M S, Hureau S, Uedono A, Wang X, Liu X, Cookson P G and Smith S V 2010 *J. Appl. Poly. Sci.* **115** 1642-50

[4] Rajkhowa R, Naik R, Wang L, Smith S V and Wang X 2010 *J. Appl. Poly. Sci.* accepted

[5] Smith S V 2007 Molecular imaging with copper-64 in the drug discovery and development arena, *Expert Opin. Drug Discovery* **2** 659-72

[6] Di Bartolo N M, Sargeson A M, Donlevy T M and Smith S V 2001 Synthesis of a new cage ligand, sarar, and its complexation with selected transition metal ions for potential use in radioimaging *Journal of the Chem. Soc. Dalton Trans.* 2303-9

[7] Smith S V 2008 Sarar technology for the application of copper-64 in biology and materials science *Q. J. Nucl. Med. Mol. Imaging* **52** 193-202

[8] Huang S, Joso R, Fuchs A, Barner L and Smith S V 2008 Probing solid-liquid interfaces using radiotracer technology: Characterization of functionalized microspheres *Chem. Mat.* **20** 5375-80

[9] Mume E, Uedono A, Mizunaga G, Lynch D E and Smith S V 2010 The role of positron annihilation lifetime studies and nuclear sensors for characterising porous materials *J. Phys.: Conf. Ser.* accepted

[10] Fong N, Guagliardo P, Williams J, Musumeci A, Martin D and Smith S V 2010 Clay particles - potential of positron annihilation lifetime spectroscopy (PALS) for studying interlayer spacing *J. Phys.: Conf. Ser.* accepted

[11] Selvakumar S, Smith S V, Hughes A E, Miller A, Jenkins D R and Uedono A 2010 Understanding the effect of nanoporosity on optimizing the performance of self-healing materials for anti-corrosion applications *J. Phys.: Conf. Ser.* accepted

Low energy positron interactions with rare gas atoms: threshold features and benchmark cross sections

James Sullivan[1], Peter Caradonna[1], Adric Jones[1], Josh Machacek[1], Casten Makochekanwa[1], Robert McEachran[1], Dennis Mueller[2] and Stephen Buckman[1]

ARC Centre for Antimatter-Matter Studies,
[1]The Australian National University, Canberra ACT 0200 Australia
[2]The University of North Texas, Denton, Texas, USA

Email: Stephen.buckman@anu.edu.au

Abstract. Using a high resolution (~60 meV FWHM) positron beam from a Surko buffer-gas trap, we have measured grand total cross sections as well as total elastic, positronium formation and electronic excitation cross sections in the rare gases. Strong threshold (Wigner) cusps are observed in the elastic scattering channel in all the rare gases at the opening of the positronium formation channel. We present low energy elastic scattering measurements for neon which show excellent agreement with recent theory and potentially establish this system as a 'benchmark'.

1. Introduction

The advent of high-resolution, high-flux positron beams from buffer-gas traps has the potential to unravel many new processes and features in low energy positron scattering from atoms and molecules. The past ten years, since these beams have been applied to atomic physics research, has seen the first state-resolved vibrational and electronic excitation measurements, the first absolute differential cross section measurements and insights into the long-standing challenge of explaining the anomalously high annihilation rates for positrons on some molecules.

In this paper we present recent measurements of positron interactions with rare gas atoms and focus on several new aspects of the collision process that have been revealed with the high resolution available from the buffer-gas trap beam. In measurements of the energy dependence of elastic scattering we observe prominent cusp-like features which are centred around the threshold for positronium formation, E_i-6.8 eV, where E_i is the first ionization potential of the atom concerned. In other measurements of elastic scattering, we show a comparison between the measured total elastic cross section for neon and recent theoretical work.

2. Experimental aspects

Space does not permit a detailed exposition of the experimental apparatus and techniques that have been used here. They have been described previously in a number of papers [1,2] and we will only briefly discuss the relevant points here. Positrons from a radioactive ^{22}Na source are moderated, trapped and cooled in a buffer-gas trap before being released from the trap in a pulsed beam. Each pulse is a few μs in width and contains on the order of a 1000 positrons. The energy of the pulsed beam can be readily varied between 0-60 eV, the repetition rate is typically 200 Hz, and the energy width typically 50-60 meV, full width at half maximum. The positron beam is radially confined by a strong, ~500 Gauss, axial magnetic field.

The positrons exit the trap and are passed through a scattering cell containing the gas under study. Those transmitted through the scattering cell are energy analysed by a retarding potential analyzer. The spectrum of transmitted positrons as a function of retarding potential can then be used to obtain the scattering cross sections in a manner which has been discussed previously [2]. In the present work we present data for two recent examples of experimental investigations with this apparatus – 'benchmark' total elastic scattering for neon, and structures in the elastic cross section in the energy region near the threshold for positronium formation in the rare gases. Some of this work has been recently reported [3].

3. Results and Discussion

3.1. Threshold cusps in the rare gases

Threshold (Wigner) cusps have been observed in many scattering processes involving electrons and have been predicted for positron scattering at the threshold of the Ps formation channel. The initial prediction was made by Campeanu *et al.* [4] and there have been several experimental tests and observations over the years [eg. 5-8] as well as a semi-empirical prediction for the rare gases [9].

In the present investigations elastic scattering measurements in each of the rare gases He-Xe have revealed significant features in the cross section at the threshold for the opening of the positronium (Ps) channel. Examples of these structures are shown in figures 1(a) and (b) for He and Ar, respectively.

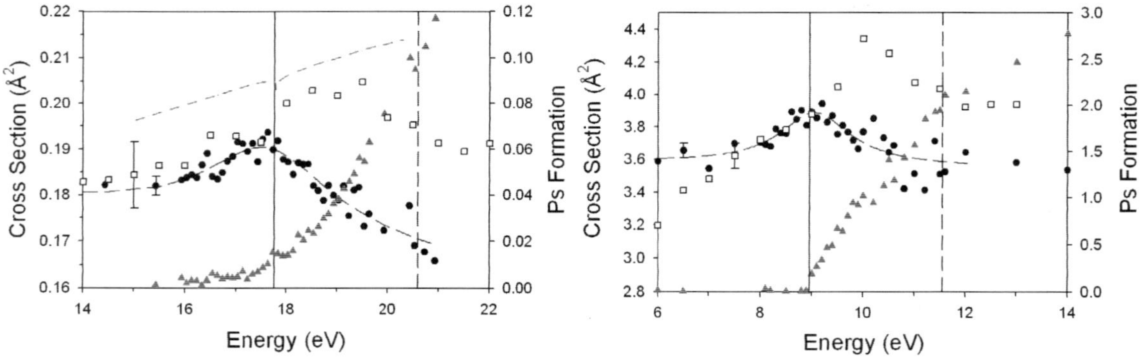

Figure 1. Total elastic and positronium formation cross sections for He (left) and Ar (right). (●) present elastic cross section, the long dashed line is a fitted curve to the data; (□) result of [7]; (- - -) theoretical result of [10] for He; (▲) Ps formation cross section. The vertical lines represent the Ps threshold (solid) and first excited state threshold (dashed).

Broad cusp features are observed in both He and Ar, with the elastic cross section rising just before the Ps formation threshold, and then falling off above this energy across the Ore gap – the energy gap between the Ps formation threshold and the first excited state of the atom. In He and Ar these features represent about a 10% perturbation on the elastic cross section. In the other rare gases, which we do not show due to space limitations, the magnitude of the feature varies somewhat, with the strongest presence being observed for Xe. Also shown in figure 1 are the recently published results of Coleman *et al.* [6,7] and the results of a variational calculation of the elastic scattering cross section due to van Reeth and Humberston [9] for He.

There are some similarities and several significant differences between the present work and the recent measurements of Coleman and co-workers. For both He and Ne there is reasonable agreement between the two measurements with both observing cusp-like behaviour in the elastic cross section, although the result of [7] indicates a peak in the elastic cross section which is about 1 eV above the Ps threshold. For the heavier gases, Ar, Kr, Xe, Coleman and co-workers characterize their observation as an upward step in the elastic cross section which occurs above the Ps threshold and which is quite different to that observed in the present work, particularly in the case of Xe. A full description of these recent results can be found in Jones *et al.* [3] and Jay and Coleman [7].

3.2. Elastic positron-neon scattering

The other example that we provide of recent measurements is for the total elastic cross section for neon at energies up to the Ps formation threshold. One rationale for these measurements was the broad spread of absolute values that are observed in both previous experimental and theoretical investigations of this cross section. This is demonstrated in Figure 2 where we show a selection of results from previous experiment and calculation, indicating a spread in cross section values of 50% or more at some energies below the Ps formation threshold, where only the elastic scattering channel is open.

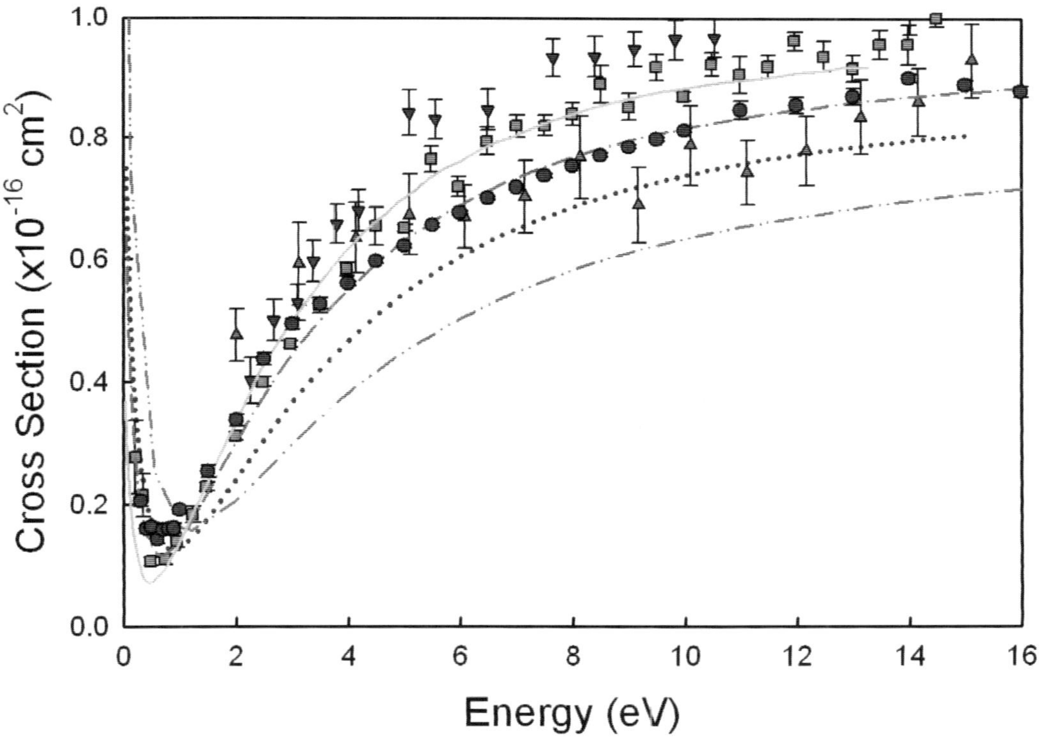

Figure 2: Total elastic cross section for positron scattering from neon. Several experimental measurements are displayed: (●) Present result, (▲) Canter *et al.* [11], (▼) Charlton *et al.* [12], (■) Stein *et al.* [13], as well as a number of theoretical calculations: ( ) Dzuba *et al.* [14], (· · · · ·) McEachran *et al.* [15], (— · —) McEachran (this work), (— · · —) Nakanishi and Schraeder [16]

Journal of Physics: Conference Series **262** (2011) 012056 doi:10.1088/1742-6596/262/1/012056

The present results, which have high statistical and absolute accuracy, show best agreement with the measurements of Canter *et al.* between 2-14 eV, and are in excellent agreement with the recent polarized orbital calculation of McEachran through the entire energy range. Whilst all experiments and theory show the same general features of the cross section, the present result is in essentially perfect agreement with the polarized orbital calculation in particular. Note that the present experimental results have also been corrected for forward scattering effects which, for the present experimental circumstances, result in corrections of more than 20% below 0.7 eV, but less than 1% above 9 eV.

4. Conclusions

The essential features of a trap-based positron beam – high energy resolution, high efficiency and statistical accuracy - are demonstrated with measurements of elastic scattering in the rare gases. We have identified strong channel coupling features in the elastic scattering channel, which appear as a result of the strong opening of the Ps formation channel, and manifest in the formation of Wigner cusps. The features observed in the lighter gases are similar in form to those observed by Coleman and colleagues, with some differences in energy dependence evident. In the heavier gases, however, we see quite different behaviour in the elastic cross section than is observed by Coleman et al. One could only speculate as to the source of these discrepancies. However it is clear that both experiments observe evidence of coupling between the positronium formation and elastic scattering channels at or near the energy of the Ps formation threshold.

Acknowledgements

We gratefully acknowledge the support of the Australian Research Council's Centre of Excellence Program.

References

[1] J. P. Sullivan, A. Jones, P. Caradonna, C. Makochekanwa and S. J. Buckman, Rev. Sci. Instrum. **79** 113105 (2008)
[2] J. P. Sullivan, S. J. Gilbert, J. P. Marler, R.G. Greaves, S. J. Buckman, and C. M. Surko, Phys. Rev. A **66** 042708 (2002)
[3] A.C. Jones, P. Caradonna, C. Makochekanwa, D.S. Slaughter, R.P. McEachran, J.R. Machacek, J.P. Sullivan and S.J. Buckman Phys. Rev. Lett. **105** 073201 (2010)
[4] R. I. Campeanu, D. Fromme, G. Kruset, R. P. McEachran, L. A. Parcell, W. Raith, G. Sinapius, and A. D. Stauffer, J. Phys. B **20** 3557 (1987)
[5] P. G. Coleman, K. A. Johnston, A. M. G. Cox, A. Goodyear, and M. Charlton J. Phys. B **25** L585 (1992)
[6] P. G. Coleman, N. Cheesman, and E. R. Lowry Phys. Rev. Lett. **102** 173201 (2009)
[7] P.M. Jay and P.G. Coleman Phys. Rev. A**82** 012701 (2010)
[8] P. Caradonna, A.C. Jones, C. Makochekanwa, D.S. Slaughter, J.P. Sullivan, S.J. Buckman, I. Bray and D. Fursa Phys Rev. A**80** 032710 (2009)
[9] W. E. Meyerhof and G. Laricchia J. Phys. B **30** 2221 (1997)
[10] P. Van Reeth and J.W. Humberston J. Phys. B **32** L103 (1999)
[11] K. F. Canter, P.G. Coleman, T. C. Griffith, and G. R. Heyland Appl. Phys. A **3**, 249 (1974)
[12] M. Charlton, G. Laricchia, T. C. Griffith, G. L. Wright, and G. R. Heyland J. Phys. B **17** 4945 (1984)
[13] T. S. Stein, W. E. Kauppila, V. Pol, J. H. Smart, and G. Jesion Phys. Rev. A **17** 1600 (1978)
[14] V. A. Dzuba, V. V. Flambaum, G. F. Gribakin, and W. A. King, J. Phys. B **29** 3151 (1996)
[15] R. P. McEachran, A. G. Ryman, and A. D. Stauffer, J. Phys. B **11** 551 (1978)
[16] H. Nakanishi and D. M. Schrader, Phys. Rev. A **34** 1823 (1986)

Journal of Physics: Conference Series **262** (2011) 012057 doi:10.1088/1742-6596/262/1/012057

Simulations of pulses in a buffer gas positron trap

W Tattersall[1,2], R D White[2], R E Robson[2], J P Sullivan[1], S J Buckman[1]
[1]ARC Centre for Antimatter-Matter Studies, Research School of Physics and Engineering, The Australian National University, Canberra, ACT, Australia
[2]ARC Centre for Antimatter-Matter Studies, School of Engineering and Physical Sciences, James Cook University, Townsville, QLD, Australia
E-mail: `wade.tattersall@anu.edu.au`

Abstract. In this study we simulate positron transport properties for various configurations of the gases and electric fields used in the Australian Positron Beamline Facility positron trap, which is based on the Surko buffer-gas trap. In an attempt to further improve the time and energy resolution of the trap and thus the associated scattering experiments, we apply a Monte-Carlo simulation procedure to a variety of possible configurations of the dumping stage of the trap.

1. Introduction

Studies of low-energy positron interactions with matter require generation of a beam with high energy-resolution, as achieved for example in the Surko trap [1, 2, 3] through thermalisation of positrons in a mixture of buffer gases. A similar trap has been developed for the Australian Positron Beamline Facility (APBF), located at the Australian National University (ANU), though the resolution of the beam has hitherto been adjusted on an empirical basis only. Moreover, the time-dependent effects of the 'dump' stage of the trap have yet to be investigated, something which is crucial for achieving optimal energy and time resolution of the final beam. The present paper addresses these effects using Monte Carlo simulation, which though a well known technique in general [4, 5, 6], requires measures to treat the space- and time-dependent electric fields which exist during the dump stage of the trap. This paper will describe progress directed to that end.

2. Simulation procedure

The simulation is a typical Monte Carlo track structure code, operating in a similar manner to, but developed independently of previous simulations [7]. Electric fields derive from sharp axial changes in potential ('potential walls'), effectively introducing a second form of elastic 'collision' for a positron. Temporal effects are accounted for by allowing the height of the potentials to vary with time. However motion between the potential walls is essentially free and kinetic energy is conserved, and it is not necessary to employ a null-collision method [4].

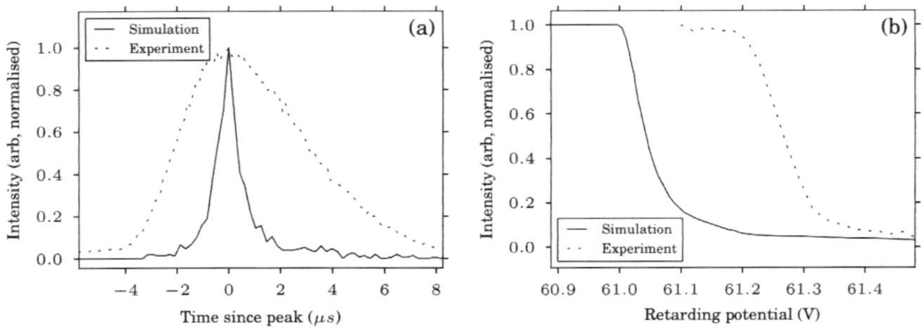

Figure 1. Comparison between experiment and simulation (a) pulse time profiles and (b) energy cut-offs. All simulation parameters are equal to those known for the experiment. Total gas pressure is 1.1 mTorr, ramping profile is logarithmic, starting at -4 V and increasing to 62 V. The final potential wall is set to 61 V. Positrons are initially thermalised to 300 K.

3. Simulation of the dumping stage

3.1. Simulation Parameters

Our simulation of the ANU beamline trap begins with positrons that have been cooled and located within the dumping electrode - the loading and cooling phases of the trap are not simulated – and the average positron energy equals that of the buffer gas, a mixture of N_2 and CF_4, at 23°C and at experimental pressures (see details in Figure 1). Collision cross-sections are a combination of experimental and theoretical values [8]. A virtual multi-channel plate detector (MCP) was placed two metres from the dumping electrode, just as the real MCP is in the experiment. The potentials were assumed to have infinitesimal edge widths, and a fast slew rate. The energy and time at which each particle reaches the detector is recorded. In this way, the experimental pulse shape and energy cut-off curves may be simulated, with the latter determining the energies of the positrons that strike the MCP.

In the actual experiment, a retarding potential analyser (RPA) is used to resolve positron energies. As the RPA potential increases, the positron counts must monotonically decrease since the RPA is blocking an increasing proportion of the beam. This is reflected in our simulation in terms of a cumulative distribution of positron energies.

Comparisons between experiment and simulation of the pulse shape and energy cut-off curves are shown in figure 1. The agreement is better for energy cut-offs, though there is a shift of approximately 0.2V, which may be indicative of a mis-match between the calibrations of the RPA and the final electrode of the trap, or due in part to other factors, such as contact potential differences. It should be noted that this energy cutoff is used for calibrating the experiment every time a set of data is measured, so the energy drift is known and the analysis of the results compensates for it. All in all, we would claim only a qualitative agreement with experiment, but we nevertheless feel that even these initial results help provide good physical insight and provide a useful guide for optimal tuning of experimental parameters, as outlined below.

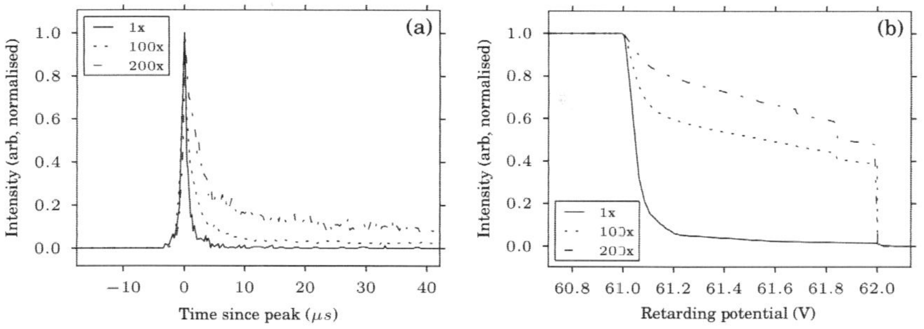

Figure 2. Comparison between different gas pressures for (a) pulse time profiles and (b) energy cut-offs. Total gas pressures are 1x, 100x and 200x those used in the experiment. All other parameters are as in Figure 1.

3.2. Effect of trapping gas density

The simulation was repeated with different gas pressures, ranging from no gas to a pressure two hundred times greater than that of the experiment. Increasing pressure broadens the time and energy distribution of the resultant pulse, something which is to be expected, as collisions with gas molecules can either increase or decrease the time that the positron remains in the dumping stage of the trap.

One point of interest is that when the positrons and gas molecules are at the temperature and pressure used in the experiment, the total collision frequency is of the order of 10^5 Hz, meaning that there are on the order of 10 gas collisions per positron during the dumping time of the trap (which is 650 μs), which has a negligible effect on any of the properties of the produced pulse.

Comparisons of the pulse time profiles and energy cut-offs are shown in figure 2. For purposes of experiment, a lower gas density is desirable, but this must be balanced against the rate at which thermalisation occurs during the cooling phase. The stepped nature of the energy cutoff is due to vibrational and rotational excitations within the area above the final wall. The positrons have a relatively high energy in this region because the dumping electrode potential is higher than that of the wall electrode, and the frequency of collisions with gas molecules becomes significant.

3.3. Effect of initial positron energy distribution

As explained earlier, our simulation begins with positrons in thermal equilibrium with the background gas, at room temperature, which corresponds to an experiment where the positrons have been allowed to cool for a sufficient time. We now look at the effect of increasing the average initial energy, from thermal energy up to a completely uncooled beam that retains an isotropic velocity spread corresponding to 1.5 eV.

Since higher energies result in a higher collision frequency, gas molecule interactions have a larger impact on the results of the simulation in precisely the same manner as in the earlier gas density variations. To avoid conflating these interaction effects with the effects of faster positrons colliding with the walls in the time-dependent potential, the gas density was set to zero.

Figure 3 shows a comparison of pulse time profiles and energy cut-offs. A lower initial positron temperature results in less variability in the energies and times at which the positrons are detected. However, the slower velocities of the positrons mean that they travel across the dumping potential relatively slowly, and hence have less frequent opportunities to escape the trap, which can lead to a larger energy spread. This is discussed further below.

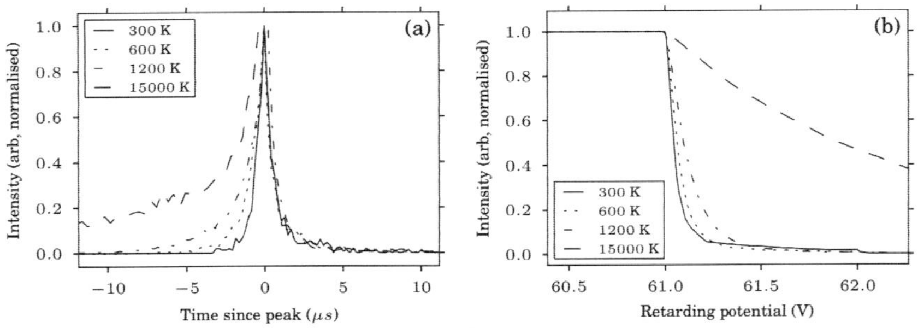

Figure 3. Comparison between different initial positron energies for (a) pulse time profiles and (b) energy cut-offs. Temperatures refer to equivalent isotropic velocity distribution for positrons with three degrees of freedom. No gas was included. All other parameters are as in Figure 1.

3.4. Effect of dump ramp profile

At room temperature, the average energy of a positron is given by $\epsilon = \frac{3}{2}kT \approx 0.0388\,\mathrm{eV}$, with a corresponding average speed of $v = \sqrt{\frac{2\epsilon}{m_e}} \approx 1.168\times10^5\,\mathrm{m}s^{-1}$. The average velocity component in the direction of the electrodes is then $v_z = v/\sqrt{3} \approx 6.743\times10^4 \mathrm{m}s^{-1}$. The width of the dumping electrode is 6 cm, so the positron on average bounces from wall to wall at a frequency of 1.1 MHz, meaning that during a typical dumping time of 650 μs, the positron has approximately 325 opportunities to leave the trap. If the dumping potential exceeds the potential required to eject a positron from the trap in between two of these opportunities, then the positron will pick up excess energy, leading to a broadened energy distribution and slightly higher average energy. The details depend on the rate at which the dumping function changes in this period. Ideally, the dumping potential should stop precisely when it has reached the same level as the wall potential, but this is not practical due to fluctuations in the potentials set by the amplifiers.

In the ANU experiment, three different types of functions have been used to control the dumping electrode to varying effect. The instant dump consists of switching the electrode from low to high as quickly as the equipment allows (the slew rate is approximately 30 V/μs). The result would be a pulse of positrons that would have an energy distribution of the same width, but shifted up by an amount corresponding to the potential of the final electrode, were it not for the transport time effect mentioned above. This effect results in 90% of the positrons remaining in the dumping electrode until the end of the ramp, rendering the potential of the final electrode inconsequential, and setting the beam energy to that of the top of the dumping ramp. In addition, every positron will gain sufficient potential to leave the trap within a few microseconds, leading to a high time resolution.

The bilinear ramp style consists of two linear ramps that share a common centre point. The logarithmic ramp initially increases quickly, but only gradually approaches the final value, thereby reducing the energy variation between positrons located in different parts of the dumping electrode when the dump potential passes the wall potential. The pulse shapes and energy cutoffs for each of these dumping functions are provided in figure 4.

3.5. Other Variables

Several other possible variables were investigated, and found to have little effect on the resultant pulse. Thus, for example, a spatial gradient in the electric potentials has little effect, other than to increase the collision rates with gas molecules, resulting in a longer time within the trap and additional signal loss due to positronium formation and annihilation. On the other hand, modifying the ratio of gases has a similar effect to modifying the gas density.

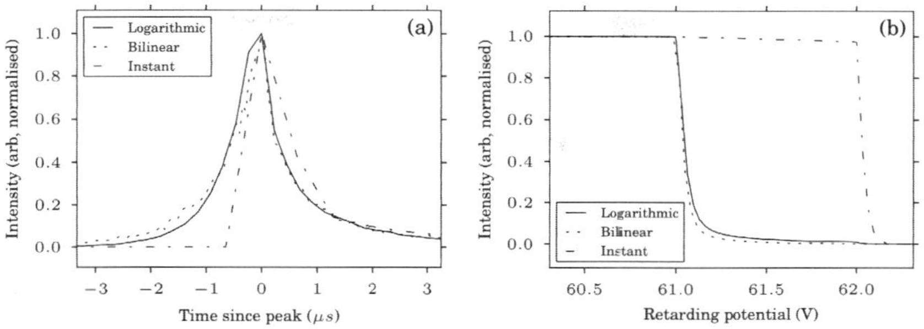

Figure 4. Comparison between different ramping functions for (a) pulse time profiles and (b) energy cut-offs. Ramping functions are as described in section 3.4, no gas was included, all other parameters are as in Figure 1.

4. Concluding Remarks

We have addressed key questions relating to optimising the operation of the traps used in the ANU beamlines. In future work we will improve the accuracy of the simulation to achieve better agreement with experiment, and consider the effect of altering a wider range of experimental variables. This will allow experimatalists to achieve better time and energy resolution in experiment and ultimately more precise data.

Acknowledgments

The authors are grateful to the Australian Research Council for support under the Centres of Excellence Program, and to experimentalists J. Machacek and A. Jones for data and discussions on the topics related to this work.

[1] J P Sullivan, A Jones, P Caradonna, C Makochekanwa, and S J Buckman. A positron trap and beam apparatus for atomic and molecular scattering experiments. *Review of Scientific Instruments*, 79(11):113105, 2008.

[2] A Jones, P Caradonna, C Makochekanwa, D Slaughter, D Mueller, J P Sullivan, and S J Buckman. High resolution positron interactions. *Journal of Physics: Conference Series*, 194(1):012033, 2009.

[3] C M Surko, G F Gribakin, and S J Buckman. Low-energy positron interactions with atoms and molecules. *Journal of Physics B: Atomic and Molecular Physics*, 38(6):R57–R126, March 2005.

[4] H R Skullerud. The stochastic computer simulation of ion motion in a gas subjected to a constant electric field. *Journal of Physics D: Applied Physics*, 1(11):1567, 1968.

[5] C Champion and C LeLoirec. Positron follow-up in liquid water: I. a new monte carlo track-structure code. *Physics in Medicine and Biology*, 51:1707–1723, April 2006.

[6] S Dujko, R D White, K F Ness, Z Lj Petrović, and R E Robson. Non-conservative electron transport in CF_4 in electric and magnetic fields crossed at arbitrary angles. *Journal of Physics D: Applied Physics*, 39(22):4788, 2006.

[7] M Šuvakov, Z Lj Petrović, J P Marler, S J Buckman, R E Robson, and G Malović. Monte carlo simulation of non-conservative positron transport in pure argon. *New Journal of Physics*, 10(5):053034, 2008.

[8] A Banković, J P Marler, M Šuvakov, G Malović, and Z L Petrović. Transport coefficients for positron swarms in nitrogen. *Nuclear Instruments and Methods in Physics Research B*, 266:462–465, February 2008.

Journal of Physics: Conference Series **262** (2011) 012059 doi:10.1088/1742-6596/262/1/012059

Positron annihilation study of defects in electron-irradiated single crystal zinc oxide

C K To[1], B Yang[1], C D Beling[1], S Fung[1], C C Ling[1] and M Gong[2]

[1] *Department of Physics, The University of Hong Kong, Hong Kong, People's Republic of China*

[2] *Department of Physics, Sichuan University, Chengdu, People's Republic of China*

Email: sfung@hkucc.hku.hk, edwardto04@yahoo.com.hk

Abstract. Pressurized melt grown zinc oxide (ZnO) single crystals purchased from Cermet Inc. were irradiated by 2MeV electrons with fluence of $6x10^{17}cm^{-2}$. Isochronal annealing from 100 ℃-800℃ was performed on the crystals under argon and air ambience. Variable Energy Doppler Broadening Spectroscopy (VEDBS) was carried out on both the as-grown and the irradiated samples at each annealing step. The migration, agglomeration and annealing of grown-in and irradiated-introduced defects were studied. It was observed that the grown-in vacancy-type defects concentration decreased at 300°C and 600 °C. For the irradiated sample annealed in argon, the positron trapping vacancy-type defect concentration decreased at 300°C and 600°C. Further annealing the as-grown and irradiated samples in argon increased the S parameter further. For the irradiated sample annealed in air, the vacancy-type defect concentration decreases at 300°C and 700°C.

1. Introduction

ZnO is a II-VI semiconductor possessing a direct wide band gap of ~3.3eV. It has a large exciton binding energy of 60meV. It is a potential candidate for making novel chemical sensors, piezoelectric transducers, short wavelength optoelectronic devices and spintronic devices[1-3]. The presence of defects strongly degrades the optical and electrical properties of ZnO devices. A better understanding of the identity, quantity and stability of the defects is required to fabricate high quality devices. Defects are inevitably introduced to semiconductors during their growth processes. The impurities and types of defects introduced to ZnO are different in different crystal growth method. Tuomisto et al. performed a comprehensive study of recovery of point defects induced in vapour phase grown crystals[4]. Chen et al. and Lu et al. carried out positron annihilation studies on hydrothermal grown ZnO crystal[5, 6] . ZnO is a irradiation-hard material[7-9]. With a high electron dosage of $6x10^{17}cm^{-2}$, the vacancy introduction rate is $0.03cm^{-1}$ [10] and the rate is 30 times lower than that of V_{Ga} in GaN [11]. Thus, ZnO devices can be used in high-irradiation environment.

Positron is a powerful non-destructive probe for vacancy type defects such as V_{Zn} and V_{Zn} related defect and complexes[10, 12-15] in ZnO. In the present study, the thermal evolution of defects under both equilibrium (i.e. as-grown) and non-equilibrium conditions (electron-irradiated) were studied by Variable Energy Doppler Broadening Spectroscopy (VEDBS) to obtain the bulk S parameter, positron diffusion length for analysis.

2. Experimental

Pressurized melt grown (MG) ZnO single crystals terminated with Zn face (0001) were purchased from Cermet, inc.. The as-grown samples are n type. The crystals were irradiated by 2MeV electron with a fluence of $6\times10^{17}cm^{-2}$. The current density for the irradiation was $6mA/200cm^{-2}$. The VEDBS measurement was performed using the slow positron beam facility of the University of Hong Kong. The Doppler Broadened gamma photons were detected by high purity Ge detector. The energy resolution of the electronic data acquisition system was 1.3keV at the 514keV photon peak. The Doppler Broadening spectrum was characterized by the valence annihilation S parameter and the core annihilation W parameter. The energy window for the S parameter was 511±0.76keV and that for the W parameter was from 511±3.4keV to 511±6.8keV. Depth profiling was done by measuring the S and W parameter against the different positron incident energy from 0.35 to 22keV. The S-E data was analyzed by the program VEPFIT. Isochronal annealing was carried out for the samples from 100°C to 800°C. The samples were heated in a tube furnace in argon and air ambience for 30min.

3. Results

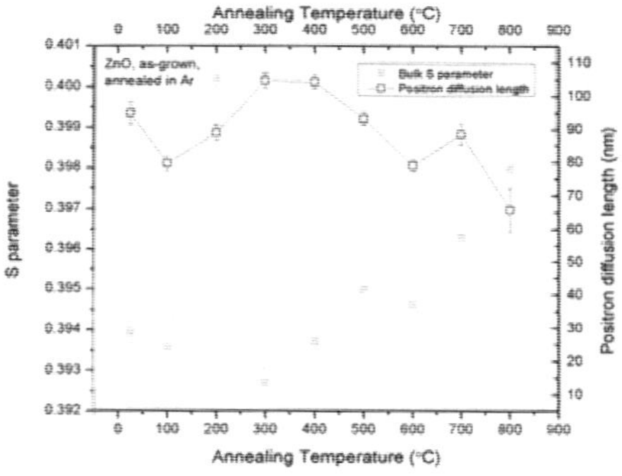

Figure 1. Variation of bulk S parameter and positron diffusion length in as-grown ZnO annealed in argon.

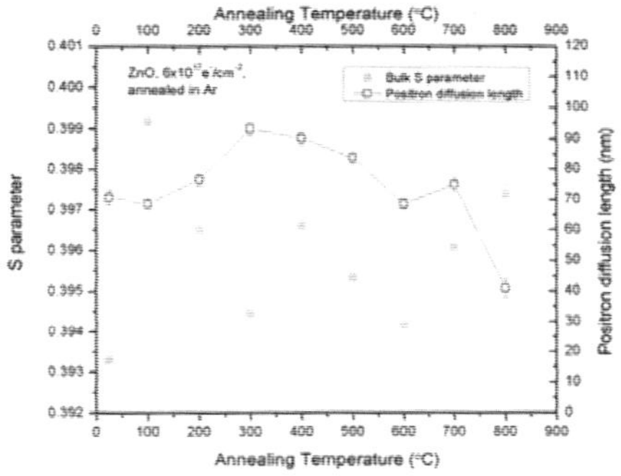

Figure 2. Variation of bulk S parameter and positron diffusion length in electron irradiated ZnO annealed in argon.

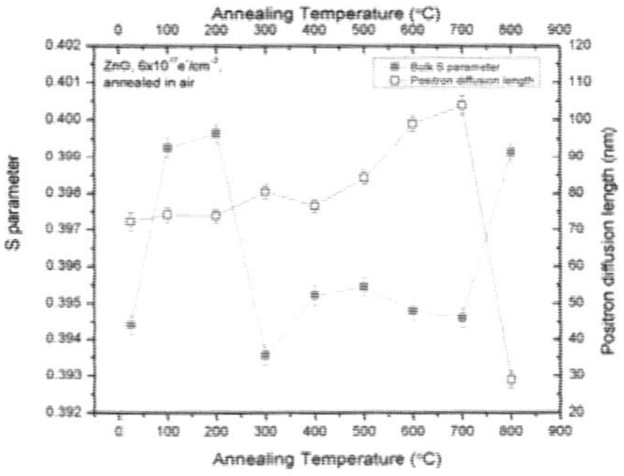

Figure 3. Variation of bulk S parameter and positron diffusion length in electron irradiated ZnO annealed in air.

For the as-grown sample (Fig.1), the S parameter increased sharply from 0.394 at 100°C to 0.400 at 200°C while there was an abnormal slight increase in positron diffusion length (L_+). The change was tentatively attributed to the decrease in positron scattering centre concentration together with an

Journal of Physics: Conference Series **262** (2011) 012059 doi:10.1088/1742-6596/262/1/012059

increase in the positron trapping vacancy-type defect concentration. From 200°C to 300°C, there was a sudden decrease in S parameter from 0.400 to 0.393 with the increase in L_+. This might be due to the decrease of deep positron trap concentration. The S parameter rose from 0.393 at 300°C to 0.395 at 500°C. The increases were accompanied with decreases in L_+ indicating that the positron trapping centre concentration increases. Another S parameter slight decrease from 0.3950 to 0.3946 was observed from 500°C to 600°C with the decrease in L_+. This might be explained by the formation of positron scattering centres. The S parameter peaked at 200°C and 500°C respectively indicating that defect reactions and agglomeration of defects might have taken place in two annealing temperatures. The S parameter reached a minimum at 300°C and 600°C respectively indicating that a recovery process occurred with 2 annealing stages.

For the irradiated sample annealed in argon (Fig. 2), from room temperature to 100°C, the S parameter increased drastically from 0.393 to 0.399 while the L_+ decreased slightly. The anti-correlation showed the increase of vacancy-type concentration. From 100°C to 300°C, the S parameter dropped considerably from 0.399 to 0.394. Together with the increase of L_+, it was revealed that some vacancy-type defects began to anneal out. From 300°C to 400°C, the S parameter rose and peaked at 0.397 with a slight decrease in L_+ indicating that more vacancy-type defects were formed. From 400°C to 600°C, the S parameter decreased from 0.397 to a local minimum of 0.394 and at the same time the L_+ was decreasing steadily. This might implied that the vacancy-type defect concentration decreased but the positron scattering defect concentration increased at the same time. The S parameter peaked at 100°C and 400°C respectively indicating that defect reactions and agglomeration of defects might have taken place at the two annealing temperatures. The S parameter approached a minimum at 300°C and 600°C respectively signifying that the annealing out of positron-trapping defects occurred in 2 stages.

For the irradiated sample annealed in air (Fig. 3), from room temperature to 200°C, the S parameter increased from 0.394 to 0.400 while the L_+ remained unchanged. This indicated the increase of vacancy-type defect concentration. From 200°C to 300°C, the S parameter decreased drastically to a minimum value of 0.394 with a small increase in L_+. This could be explained by the decrease of vacancy-type defect concentration. From 300°C to 500°C, the S parameter rose from 0.394 to 0.395 while the L_+ increases. This showed that secondary vacancy defects were formed and some of the positron scattering centres were annealed out. From 500°C to 700°C, there was a small decrease of S parameter from 0.3955 to 0.3946. L_+ increased in the above temperature range indicating the decrease of vacancy-type defects concentration. In short, the S parameter arrived at a maximum at 200°C and 500°C implying that defect reactions and agglomeration of defects might have taken place at the two temperatures. The S parameter attained a minimum at 300°C and at around 700°C indicating that vacancy-type defects positron-trapping vacancy defects annealed out in 2 stages.

4. Discussions

V_o and/or Zn_i are the possible defects which could act as positron scattering centres [16] since they were the common types of defects which were present in single crystal ZnO[17, 18]. At room temperature, V_o was positively charged [4] and might scatter the positrons by distorting the lattice around its open volumes[16]. The positron binding energy of positron to V_o was very small[19]. Therefore, V_o and/or Zn_i were considered to be the main positron scattering species which could decrease the positron diffusion length in the present study. On the other hand, the rise of S parameter was due to V_{Zn} or V_{Zn} related defects since they were the dominant positron trapping centres in as-grown and irradiated n-type ZnO [10]. Lu et al. and Chen et al. reported the presence of positron defect lifetime 230ps [6] and 231ps [5] respectively in electron irradiated ZnO single crystal. They both concluded that V_{Zn} were introduced to the ZnO single crystals by electron irradiation. Their conclusion were consistent with the theoretical and experimental positron lifetime results by Tuomisto et al. [10]. Therefore the rise and fall in the S parameter in the present study were mainly due to increase and decrease of V_{Zn} and V_{Zn} related defects and complexes concentration.

Tuomisto et al. found that V_{Zn} annealed out of ZnO crystal in 2 annealing stages at about 400K and 600K[4]. Chen et al. observed that V_{Zn} were recovered at 200°C and 700°C from ZnO[5]. At 400°C, additional vacancy-type defects were formed[5]. Our results were different from the above studies.

5. Conclusions

Recovery of V_{Zn} and/or V_{Zn} related defects in as-grown and electron irradiated ZnO crystal occurred with 2 stages at 300°C and at around 600°C to 700°C. Annealing in air or in argon ambience did not affect the recovery temperature.

References

[1] S. J. Pearton, D. P. Norton, R. Frazier, S. Y. Han, C. R. Abernathy and J. M. Zavada, *Iee P-Circ Dev Syst* **152** (4), 312-322 (2005).
[2] U. Ozgur, Y. I. Alivov, C. Liu, A. Teke, M. A. Reshchikov, S. Dogan, V. Avrutin, S. J. Cho and H. Morkoc, *J Appl Phys* **98** (4), 041301 (2005).
[3] D. C. Look, *Mat Sci Eng B-Solid* **80** (1-3), 383-387 (2001).
[4] F. Tuomisto, K. Saarinen, D. C. Look and G. C. Farlow, *Phys Rev B* **72** (8), 085206 (2005).
[5] Z. Q. Chen, S. J. Wang, M. Maekawa, A. Kawasuso, H. Naramoto, X. L. Yuan and T. Sekiguchi, *Phys Rev B* **75** (24), 245206 (2007).
[6] L. W. Lu, C. K. So, C. Y. Zhu, Q. L. Gu, C. J. Li, S. Fung, G. Brauer, W. Anwand, W. Skorupa and C. C. Ling, *Semicond Sci Tech* **23** (9), 095028 (2008).
[7] D. C. Look, D. C. Reynolds, J. W. Hemsky, R. L. Jones and J. R. Sizelove, *Appl Phys Lett* **75** (6), 811-813 (1999).
[8] S. O. Kucheyev, P. N. K. Deenapanray, C. Jagadish, J. S. Williams, M. Yano, K. Koike, S. Sasa, M. Inoue and K. Ogata, *Appl Phys Lett* **81** (18), 3350-3352 (2002).
[9] C. Coskun, D. C. Look, G. C. Farlow and J. R. Sizelove, *Semicond Sci Tech* **19** (6), 752-754 (2004).
[10] F. Tuomisto, V. Ranki, K. Saarinen and D. C. Look, *Phys Rev Lett* **91** (20), 205502 (2003).
[11] J. Oila, V. Ranki, J. Kivioja, K. Saarinen, P. Hautojarvi, J. Likonen, J. M. Baranowski, K. Pakula, T. Suski, M. Leszczynski and I. Grzegory, *Phys Rev B* **63** (4), art. no.-045205 (2001).
[12] G. Brauer, W. Anwand, W. Skorupa, J. Kuriplach, O. Melikhova, C. Moisson, H. von Wenckstern, H. Schmidt, M. Lorenz and M. Grundmann, *Phys Rev B* **74** (4), 045208 (2006).
[13] T. Koida, S. F. Chichibu, A. Uedono, A. Tsukazaki, M. Kawasaki, T. Sota, Y. Segawa and H. Koinuma, *Appl Phys Lett* **82** (4), 532-534 (2003).
[14] S. Brunner, W. Puff, A. G. Balogh and P. Mascher, *Positron Annihilation - Icpa-12* **363-3**, 141-143 (2001).
[15] Z. Q. Chen, M. Maekawa, S. Yamamoto, A. Kawasuso, X. L. Yuan, T. Sekiguchi, R. Suzuki and T. Ohdaira, *Phys Rev B* **69** (3), 035210 (2004).
[16] A. Uedono, T. Koida, A. Tsukazaki, M. Kawasaki, Z. Q. Chen, S. Chichibu and H. Koinuma, *J Appl Phys* **93** (5), 2481-2485 (2003).
[17] K. I. Hagemark, *J Solid State Chem* **16** (3-4), 293-299 (1976).
[18] G. D. Mahan, *J Appl Phys* **54** (7), 3825-3832 (1983).
[19] C. C. Ling, C. K. Cheung, Q. L. Gu, X. M. Dai, S. J. Xu, C. Y. Zhu, J. M. Luo, C. Y. Zhu, K. H. Tam, A. B. Djurisic, C. D. Beling, S. Fung, L. W. Lu, G. Brauer, W. Anwand, W. Skorupa and H. C. Ong, *Appl Surf Sci* **255** (1), 58-62 (2008).

Journal of Physics: Conference Series **262** (2011) 012060 doi:10.1088/1742-6596/262/1/012060

Vacancy evolution in Ni during irradiation at high temperatures studied by in situ positron annihilation spectroscopy

H Tsuchida[1,2,6], **T Iwai**[3], **S Kasai**[2], **H Tanaka**[2], **N Oshima**[4], **R Suzuki**[4], **T Yoshiie**[5] **and A Itoh**[1,2]

[1]Quantum Science and Engineering Center, Kyoto University, Uji, Kyoto 611-0011, Japan

[2]Department of Nuclear Engineering, Kyoto University, Kyoto 606-8501, Japan

[3]Nuclear Professional School, School of Engineering, The University of Tokyo, Ibaraki 319-1188, Japan

[4]Research Institute of Instrumentation Frontier (RIIF), National Institute of Advanced Industrial Science and Technology (AIST), Tsukuba, Ibaraki 305-8568, Japan

[5]Research Reactor Institute, Kyoto University, Kumatori, Osaka 590-0494, Japan

E-mail: tsuchida@nucleng.kyoto-u.ac.jp

Abstract. We present experimental results for in situ study of atomic-vacancy production and its evolution occurring during irradiation by using a slow-positron beam. Thermal stability of the vacancy produced during irradiation is investigated under elevated temperature conditions. An annealed Ni specimen was irradiated with 400 keV He ions at three different temperatures of 296 (RT), 368 and 713 K. Doppler broadenings of positron-annihilation γ-rays were measured and variation of the line-shape parameter S was observed under beam-on (during irradiation) and beam-off (non-irradiation) conditions. Results indicate that variation of the S depends on the specimen temperature, showing that vacancy cluster consisting of about 15 vacancies is formed predominantly via thermal evolution of atomic-sized vacancies under irradiation at high temperatures. We found that formation of the activated vacancies occurs during irradiation, which leads to vacancy clustering.

1. Introduction

Studies of radiation defects in materials have been carried out for many years. Considerable interest has been devoted to study radiation damage (degradation of materials) in radiation environments. Recent investigations focus on an understanding of a characteristic of defects under irradiation to reveal the dynamics underlying defect production, accumulation and evolution. To this end, various apparatus or techniques have been developed for performing in situ observation of irradiation behaviour in materials [1].

Irradiation of energetic ions is known to produce a large number of point defects in a nanometer-sized region near an ion trajectory (collision cascade region). In the cascade region, activated defects are formed, and then they annihilate via interaction with one another, or with various sinks such as

[6] To whom any correspondence should be addressed.

dislocations and a surface. On the other hand, defect clustering occurs via thermal diffusion or migration of the activated defects. In the initial stage of defect clustering, a small-sized vacancy cluster or a latent vacancy cluster, which is invisible in observation of electron microscope, is formed [2]. The formation of latent vacancy cluster plays an important role in processes of void nucleation resulting in void swelling in materials [3].

At present little is known experimentally about the void nucleation because of the experimental difficulty in observation of a small-sized vacancy cluster at nucleation stage during irradiation. Positron annihilation spectroscopy is a powerful technique with a high sensitivity to such atomic-scale open-volume defects or nanovoid. We have recently investigated transient vacancies produced during ion irradiation of non-annealed Ni with positrons [4]. In situ positron annihilation Doppler broadening measurement was performed during ion irradiation at room temperature by using a specially developed experimental system combined with a slow-positron-beam apparatus and a high-energy ion accelerator. The results showed that an increase in the Doppler broadening line-shape parameter *S* occurs during irradiation, implying that vacancy concentration is higher and larger-sized vacancy clusters are formed during irradiation, compared with those of vacancies that have survived after irradiation.

In the present work, we report on thermal stability of vacancies produced during irradiation. A well-annealed Ni was used as a target specimen. The specimen was irradiated with 400 keV He ions at three different temperatures of 296 (RT), 368 and 713 K, and in situ positron annihilation Doppler broadening measurement was performed simultaneously. Temperature dependence of evolution of atomic-scale vacancy defects produced by irradiation is investigated.

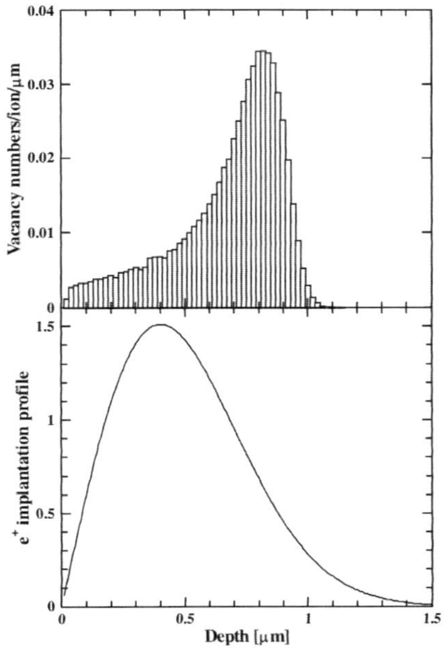

Figure 1. Positron implantation profile for 20 keV positrons in Ni and the number of vacancies produced by one incident ion (400 keV He ion) calculated from TRIM, as a function of the specimen depth.

2. Experimental techniques

The experiments were carried out at High Fluence Irradiation Facility of the University of Tokyo (HIT). The apparatus and experimental procedure were essentially the same as those described in Ref. 5, and only a brief outline is given. A specimen was Ni of 10 μm in thickness that was annealed at 1173 K for an hour in vacuum. The specimen was mounted on a target holder equipped for heating the specimen up to 773 K. Three different specimen-temperature of RT, 368 and 713 K were chosen because mobility of vacancy in Ni differs at each temperature. Temperature of the specimen was measured with a thermocouple, and was monitored continuously during irradiation experiments. Fluctuation of the temperature during the experiments was ±4 K.

Mobility of vacancies in a thermally activated process can be estimated by $v = v_0 \exp(-E_m/kT)$, where v the number of atomic jumps per second, v_0 the lattice vibration frequency (10^{13} s^{-1}), k the Boltzmann constant, T the temperature. The migration energy of single vacancies and divacancies in Ni are E_m=1.25 eV and E_m=0.75 eV, respectively [8]. At the low temperature of 368 K, the value of v is 7.9×10^{-5} jumps/s for single vacancies and 5.5×10^{2} jumps/s for divacancies. This indicates that single vacancy is immobile below 368 K. On the other hand, at the high temperature of 713 K, the value of v is 1.5×10^{4} jumps/s for single vacancies and 5.1×10^{7} jumps/s for divacancies, indicating that both single vacancies and divacancies are mobile.

An irradiation chamber was connected with both slow positron apparatus and high-energy ion accelerator. This experimental system enables to observe in situ vacancy-type defects in materials produced under an ion irradiation environment. Beams of 400 keV He ions were incident on a specimen at 30° to the surface normal, and monoenergetic positron beams were directed parallel to the surface normal. A positron with the incident energy of 20 keV was used so that depth profile of positron implantation overlaps with that of defects produced by the ions. Figure 1 shows depth profile for 20 keV positrons and vacancy defect distribution by 400 keV He ions in Ni, where the defect depth profile by ions was calculated from the TRIM code [6] and the positron implantation profile was calculated using the formula described in Ref. 7.

For the specimen during irradiation and non-irradiation, we performed measurements of the positron annihilation Doppler-broadening spectroscopy. Positron-annihilation γ-rays emitted from the specimen were detected with a high-purity Ge detector (ORTEC GEM-20180-P). A Doppler broadening spectrum of the annihilation γ-rays was characterized by the line-shape parameter S defined as the ratio of counts in the central portion of the annihilation photopeak (ranging from 510.2 keV to 511.8 keV) to the total counts in the peak (from 507.0 keV to 515.0 keV). Overall experimental errors for the S parameter due to a statistical error were ±0.0015. In each run of annihilation spectrum measurements, data-acquisition time was 20 min, and the average fluence of ion beams was $\sim2.3\times10^{14}$ ions/cm^2 (where the average flux was $\sim1.9\times10^{11}$ ions/(cm^2 s)).

3. Results and discussion

Figure 2 shows a typical result for S-parameter variation measured sequentially during non-irradiation (open symbols or odd-numbered runs) and irradiation (closed symbols or even-numbered runs) at temperatures of RT, 368 and 713 K. One can see zigzag variation of S-parameter: values of S observed during irradiation are larger than those measured under non-irradiation condition. This zigzag variation has been also observed in our previous study for non-annealed Ni irradiated with 2.5 MeV C ions at RT [4]. For observations at the low temperature of RT and 368 K, comparing between the value of S for run 1 and that for run 3, the S-parameter changes significantly, indicating that accumulation of surviving or immobile vacancies. On the other hand, such significant change in the S-parameter was not observed at the high temperature of 713 K, suggesting inhibition of defect accumulation.

In order to obtain information on temperature dependence of the variation of S-parameter in more details, the type of defects was studied for the specimens (run 13 shown in Fig. 2) irradiated at a fluence of about 1.0×10^{15} ions/cm^2 by positron lifetime measurements. The measurements were carried out using the high-intensity slow-positron microprobe apparatus at National Institute of Advanced Industrial Science and Technology (AIST) [9]. Time resolution of the spectrometer was typically $\sim$280 ps. Beam spot size of the positron was 200-300 μm in diameter. Although an incident positron energy of 20 keV was used in the measurements, the component of kinetic energy of the positron in parallel direction with the surface normal was about 14 keV. This results from a focusing

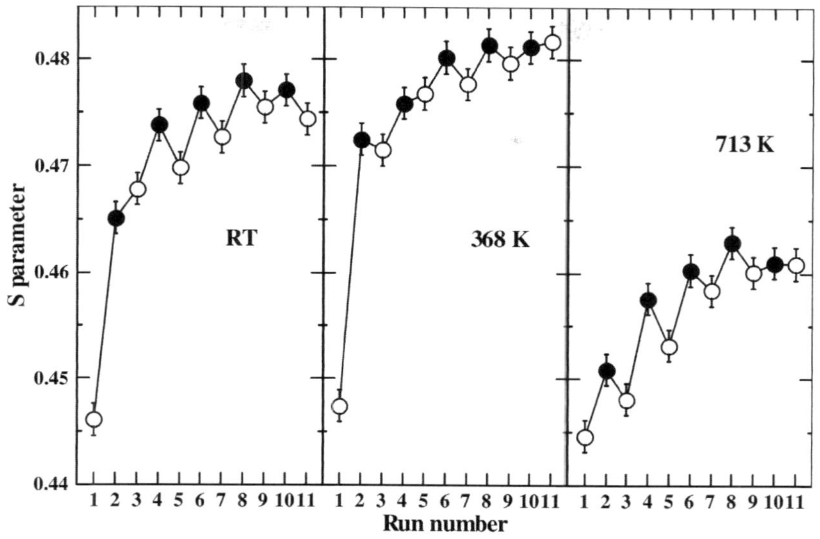

Figure 2. Temperature dependence of *S*-parameter variation measured sequentially during ion irradiation (closed symbols) and non-irradiation (open symbols) for an annealed Ni foil of 10 μm in thickness irradiated by 400 keV He ions.

condition of positron beams. In this case, a mean implantation depth of the positron is about 300 nm. Total counts of more than 1×10^6 were corrected in each lifetime spectrum.

Figure 3 shows experimental results of positron lifetime measurements. Two-component analysis of the lifetime spectra was done. We used the calculated value of positron lifetime taken from Ref. 10 in order to determine the type of vacancies. For the specimen without irradiation, the value of τ_1 corresponds to the bulk lifetime in Ni. The τ_2 component with intensity I_2 of about 18% is attributed to the surface annihilation occurring as a result of back-diffusion of positrons. This is due to that an implantation depth of the positron is short: about 300 nm, as described above. The fraction of positrons diffusing back to the surface, corresponding to the intensity I_2, can be estimated by analysing experimental data of the *S-E* curve (the *S*-parameter as a function of an incident positron energy) with the model developed by Britton et al. [11]. The intensity I_2 was estimated as about 15 %, indicating that the calculation is in good agreement with the experiment. It should be noted that the back-diffusing fraction is significantly affected by the defect concentration, because positrons was trapped at defects before reaching the surface; hence, the back-diffusing fraction reduces significantly for specimen containing defects.

For the specimen irradiated at RT, the value of τ_1 was found to be 207 ps, corresponding to vacancy cluster consisting of about 3 vacancies. The same result was observed for the specimen at 368 K, except for the intensity of the τ_1 component. Result of the τ_2 component (about 300 ps) shows the formation of larger-sized vacancy clusters. The vacancy clusters may be latent vacancy clusters directly formed from collision-cascade damage. The occurrence of the latent vacancy clusters has been observed in the experiment for Ni irradiated with MeV neutrons [2]. On the other hand, the size of vacancies corresponding to the τ_2 component observed at 368 K is larger than that observed at RT. This suggests that vacancy evolution occurs via thermal migration process of vacancy clusters such as divacancies at 368 K.

Remarkably, it is found that for the specimen at the high temperature of 713 K the τ_1 component (115 ps) with intensity of about 60 % corresponds to the bulk lifetime in Ni, and result of the τ_2 component

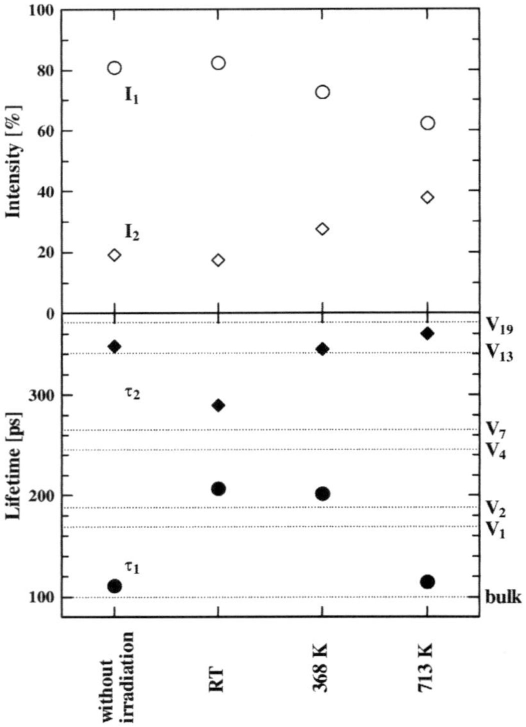

Figure 3. Positron annihilation lifetime (τ_1 (●), τ_2 (◆)) and its intensity (I_1 (○), I_2 (◇)) for the specimens irradiated at a fluence of about 1.0×10^{15} ions/cm^2 (the last run in Fig. 2) at each specimen temperature.

(360 ps) shows formation of a large-sized vacancy cluster consisting of about 15 vacancies. This evidently indicates that atomic vacancies produced during irradiation agglomerate and they develop into a large-sized vacancy cluster.

We consider the increase of *S*-parameter observed during irradiation. The factor may be due to the following process occurring after irradiation. Radiation-induced defects are produced continuously during irradiation (beam-on), and also their evolution (vacancy clustering) takes place simultaneously. After irradiation (beam-off) production of defects completely stops, and then the defect recombination resulting from migration of interstitials occurs mainly because a high-mobility of interstitials. Consequently, a part of the vacancy formed during irradiation is deflated via absorption of a freely migrating interstitial into the vacancy. Therefore, it is likely that the variation of *S*-parameter observed is attributed to a relaxation process occurring between the vacancies produced during irradiation and a freely migration interstitials. This implies that the size of vacancy cluster formed during irradiation is large compared to that survived after irradiation.

4. Conclusions

Using the in situ positron-annihilation Doppler broadening technique, we investigated the vacancy production during irradiation and its thermal evolution for annealed Ni irradiated with 400 keV He ions. We demonstrated that the present in situ observation technique is applicable for the study, in particular of latent vacancy cluster produced in materials under irradiation environments. It was found that (1) vacancy concentration during irradiation is high at each of the temperatures studied compared to that survived after irradiation, (2) stable atomic-scale defects consisting of about 3 vacancies are formed predominantly below the temperatures of 368 K, and (3) agglomeration of the activated vacancies occurs significantly at the high temperature of 713 K, where defect cluster consisting of about 15 vacancies was formed and atomic-sized vacancies annihilate via recombination with a freely migrating interstitial.

Acknowledgements

The authors gratefully acknowledge Takao Omata for his technical supports during the irradiation experiments. We also acknowledge Dr. Komatsu of Hiroshima Institute of Technology for annealing treatment of the specimens and useful discussion. A part of this study was performed under the sponsorship of JNES (Japan Nuclear Energy Safety Organization) open application research project for enhancing the basis of nuclear safety.

References

[1] Birtcher R C, Kirk M A, Furuya K, Lumpkin G R and Ruault M-O 2005 *J. Mater. Res.* **20** 1654
[2] Yohiie T, Hasegawa M, Kojima S, Sato K, Saitoh Y, Yamaguchi S and Kiritani M 1991 *J. Nucl. Mater.* **179-181** 931
[3] Iwai T, Tsuchida H and Awano M 2007 *J. Nucl. Mater.* **367-370** 372
[4] Tsuchida H, Iwai T, Awano M, Kishida M, Katayama I, Jeong S C, Ogawa H, Sakamoto N, Komatsu M and Itoh A 2007 *J. Phys. Condens. Matter* **19** 136205
[5] Iwai T, Ito Y and Koshimizu M 2004 *J. Nucl. Mater.* **329-333** 963
[6] Zieglar J, Biersack J F and Littmark U 1985 *The Transport of Ions in Matter (TRIM) Monte-Carlo calculation* including in Computer program SRIM Version 2003 (Pergamon Press, New York)
[7] Puska M J and Nieminen R M 1994 *Rev. Mod. Phys.* **66** 841
[8] Kiritani M, Konno M, Yoshiie T and Kojima S 1987 *Mater. Sci. Forum* **15-18** 181
[9] Oshima N, Suzuki R, Ohdaira T, Kinomura A, Narumi T, Uedono A and Fujinami M 2009 *Appl. Phys. Lett.* **94** 194104
[10] Ohkubo H, Tang Z, Nagai Y, Hasegawa M, Tawara T and Kiritani M 2003 *Mater. Sci. Eng.* A **350** 95
[11] Britton D T, Rice-Evans P C and Evans J H 1988 *phil. Mag.Lett.* **57** 165

Journal of Physics: Conference Series **262** (2011) 012061 doi:10.1088/1742-6596/262/1/012061

Application of positron annihilation technique to front and backend processes for modern LSI devices

A Uedono[1], S Ishibashi[2], N Oshima[3], T Ohdaira[3] and R Suzuki[3]

[1] Institute of Applied Physics, University of Tsukuba, Tsukuba, Ibaraki 305-8573, Japan
[2] Nanosystem Research Institute (NRI) "RICS", National Institute of Advanced Industrial Science and Technology (AIST), Tsukuba, Ibaraki 305-8568, Japan
[3] Research Institute of Instrumentation Frontier, National Institute of Advanced Industrial Science and Technology (AIST), Tsukuba, Ibaraki 305-8568, Japan
E-mail: uedono@sakura.cc.tsukuba.ac.jp

Abstract. Developments of the semiconductor industry depend on ability to decrease the size of integrated circuits. Significant efforts to introduce new materials and their fabrication processes have been made in order to achieve such trends. However, one faces difficulties in the control of materials properties caused by atomic scale disorder. In the present paper, we used monoenergetic positrons to characterize ultra shallow junction formed on Si substrates and materials buried in LSI circuit structures. It was shown that the positron annihilation technique is a powerful characterization tool for modern Si-technology related materials.

1. Introduction

Developments of the semiconductor industry depend on the ability to decrease the size of integrated circuits [1]. This is often expressed by Moore's Law, and this tendency has resulted in the significant improvements in cost per function, microprocessor throughput, long operating time with batteries, etc. Significant efforts to introduce new materials and their fabrication processes have been made in order to achieve such trends. However, one faces the difficulties in the control of materials properties caused by atomic scale disorder. Positron annihilation has been successfully used to detect defects in Si-technology related materials [2,3]. In the present paper, we used monoenergetic positrons to characterize ultra shallow junction formed on Si substrates [4] and materials buried in LSI circuit structures [5].

Plasma immersion ion implantation has attracted attention for fabricating ultrashallow junctions and conformal doping of the sidewall of fin field-effect transistors, where it is necessary to suppress the short-channel effect and increase the device drive current [1]. Ion bombardment, however, causes the introduction of a large number of defects into the subsurface region. Thus, knowledge about the behavior of defects during annealing processes is crucial. The signal propagation delay and resultant power consumption and cross-talk can be suppressed by integration using low dielectric permittivity (low-*k*) materials and Cu interconnects. The characteristics of those materials are important factors determining the circuit properties, but the patterning and metallization processes could introduce damage and degrade device performance. In the present work, we show that the monoenergetic positrons can characterize the materials buried in the circuits without beam-focusing on integrated device structures.

2. Experiment

The samples investigated were *n*-type Si wafers doped with B using a B_2H_6/He plasma mixture up to a dose of 1×10^{15} cm^{-2} [4]. After the doping, the wafers were annealed by spike rapid thermal annealing (spike RTA) at 1075 °C. The depth profiles of the B concentration were measured by secondary ion mass spectroscopy (SIMS). Before the measurements, step-by-step etching was performed.

The Cu/low-*k* structure was fabricated using the conventional device process [5]. Chemical vapor deposition was used to deposit 800-nm SiO_2 and 250-nm SiOCH films on Si substrates. After the trench formation, the Cu damascene structure was fabricated with Cu thickness of 350 nm. The schematic view of the sample is shown in Fig. 1. The *k* value and density for SiOCH were 2.7 and 1.26 g/cm^3, respectively. A similar pattern structure with SiO_2 films was also fabricated.

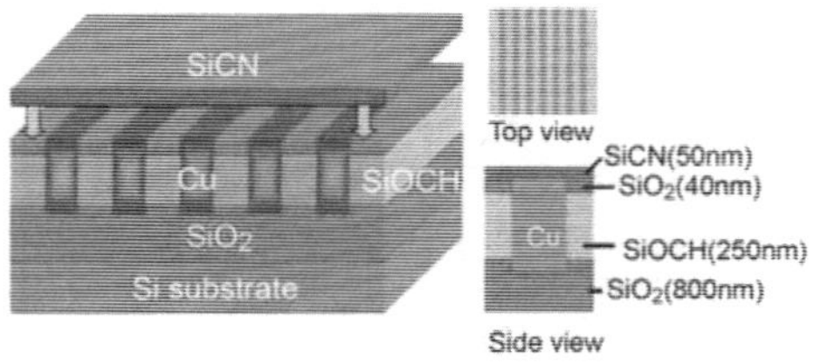

Figure 1. (colour online) Cross-sectional structure of the sample with SiOCH/Cu interconnects. A schematic top and side views are also shown; the line/space widths are 0.27/0.27 μm or 1.08/1.08 μm (pitch pattern = 0.54 or 2.16 μm), so that the area ratio of Cu to SiOCH was 50%

3. Results and discussion

3.1. Vacancy-Boron Complexes in Plasma Immersion Ion-Implanted Si

Figure 2 shows the *S* values of the plasma B-implanted samples before and after RTA as a function of incident positron energy *E*. For the annealed sample, the measurements were made after step-by-step etching (the etching depth was 4 nm). For the as-doped sample, the *S* value at low *E* was larger than that for the unimplanted sample, suggesting the trapping of positrons by vacancy-type defects. For the annealed sample, the *S* value in the subsurface region (0-5 keV) was smaller than the characteristic *S* value obtained for the defect-free Si. The solid curves are fits to the experimental data, and the derived depth distributions of the *S* value are shown in an inset of figure 2. For the annealed sample, the region with the small *S* value is similar to the region of high B concentration, suggesting that the observed decrease in the *S* value could correlate with doped B atoms. Figure 3 shows the *S–E* curves for B-implanted Si and 6H-SiC before and after annealing (conventional ion-implantation). In this figure, the *S* value was normalized to that for the defect-free samples. For Si, the ion energy and dose were 10 keV and 1×10^{13} cm^{-2}, respectively. For 6H-SiC, five fold implantation was used to form a box profile to a depth of 450 nm; the B concentration in the box profile was (1-2)$\times10^{18}$ cm^{-2}. The annealing was performed for 30 min at 1000°C for Si and at 1400°C for 6H-SiC. For the annealed samples, the *S* values in the damaged region were smaller than 1.00 except near the surface. Thus, the results shown in figures 2 and 3 are considered to be typical one for B-implanted Si and SiC. It has been suggested that the vacancy-oxygen pairs or oxygen precipitates in Si may play a certain role in the decrease in the *S* value [6-9]. However, the effect of oxygen in 6H-SiC on the *S* value is unlikely to be large because its defect-free *S* value is small (0.458), which is mainly due to carbon in the matrix [8].

Figure 4 shows the *S–W* relationships for the B-doped samples. The (S,W) values obtained using the projector augmented-wave (PAW) method are also shown [4], where calculations were made for the annihilation of positrons in the delocalized state and in vacancy-type defects such as the divacancy (V_2), its complex with B atoms [$V_{(n+2)}B_n$ (n=0, 2, 4, and 6)], and clusters such as icosahedral B (V_5B_{12}). The value calculated for defect-free Si is located below and to the right of the value for the undoped sample. This difference might be due to several causes such as the limitations of the first-principles calculations applied to Doppler broadening spectra. The (S,W) value of V_4B_2 is close to that for as-doped Si, suggests that the major defect species detected by positrons are the complexes between V_2 and B atoms. The relation between the calculated (S,W) values for V_5B_{12} and defect-free Si was similar to that between the values for the unimplanted and annealed samples. Thus, the decrease in the *S* value for the annealed sample can be attributed to the trapping of positrons by B clusters such as B_{12}. The presence of B_{12} for the present sample was confirmed by measuring soft-x-ray-excited B 1*s* photoelectron spectra [4]. From the calculation, positrons were found to mainly exist outside the cluster and annihilate in the distorted Si matrix around it (figure 5). Because of the negative charges of B_{12} [10], positrons can be weakly captured by the cluster and annihilate in the distorted matrix around the cluster. This trapping mechanism is also likely to be valid for B-implanted 6H-SiC. The results discussed above suggest the possibility of the trapping of positrons by B clusters in Si. The stress introduced by B_{12} could attract interstitial oxygen to the clusters and drive to form boron oxides. Thus, the residual oxygen or oxygen incorporated in Si could couple with B clusters and form complexes such as boron suboxides (B_6O).

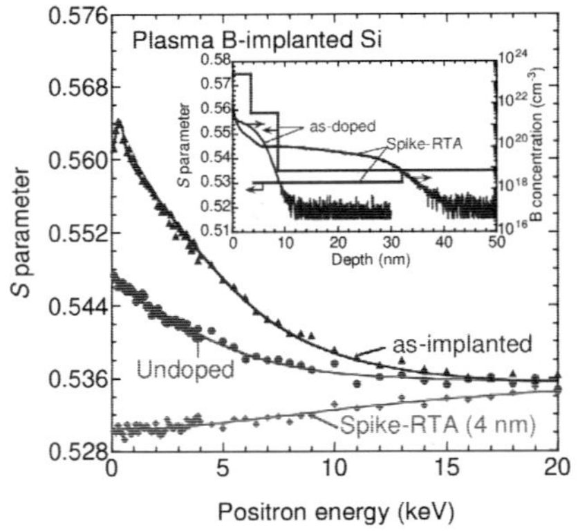

Figure 2. (colour online) *S–E* curves for plasma immersion B-implanted Si before and after spike RTA.

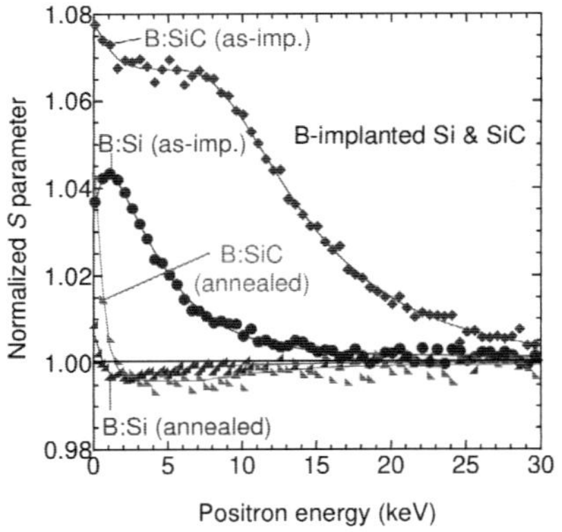

Figure 3. (colour online) *S–E* curves for B-implanted Si and 6H-SiC before and after annealing.

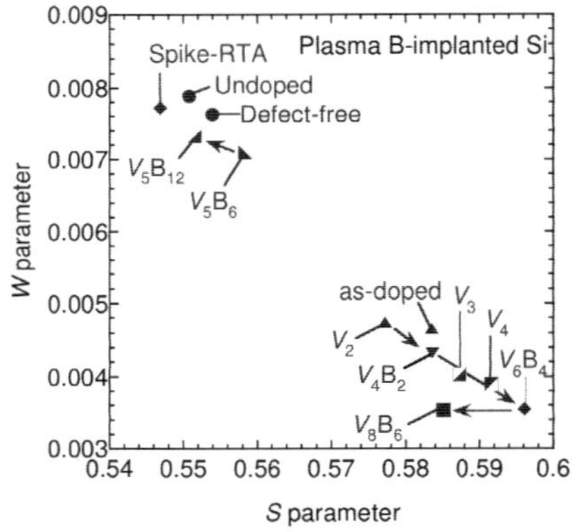

Figure 4. (colour online) *S–W* relationships for B-implanted Si. The values calculated by the PAW method for defect-free Si and various defects were also shown.

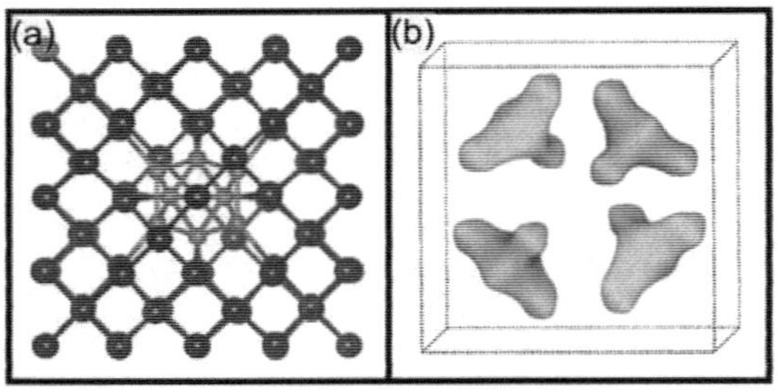

Figure 5. (colour online) Atomic configuration for V_5B_{12} (a) and the positron density distribution around the defects (b).

3.2. Characterization of low-k/Cu damascene structures

Figure 6 shows *S* values of the samples with SiOCH and SiO_2 interlayers. The increase in the *S* value at $E \cong 1$ keV is due to the annihilation of positrons in SiCN. The *S* value started to decrease with increasing *E* because of the annihilation of positrons in SiO_2 and Cu. The increase in the *S* value at E = 2-10 keV was due to the self-annihilation of *p*-Ps in SiOCH. The *S* value above E = 10 keV was mainly attributable to the annihilation of positrons in Si (S = 0.534) and the valley at E = 10 keV was due to the positron annihilation in Cu (0.412). Although the locations of SiOCH and Cu in the cross-sectional view are the same, their contributions to the *S–E* curve appear at different energies. This is because of the difference in their densities and a resultant difference in the corresponding mean implantation depths of positrons. The *S* values for SiOCH were smaller for the samples with the 0.54-μm pattern than those for the samples with the 2.16-μm pattern. A similar decrease in the *S* value was also observed for the samples with a SiO_2 interlayer. The change in those parameters suggests suppression of the Ps formation in SiOCH and SiO_2, and a decrease in the pore size caused by scaling. The probability density function (PDF) of the annihilation rate was shown in figure 7. Three annihilation modes with different positron lifetimes were observed. For the SiOCH-based samples, The shortest, middle, and the long-lived components can be attributed to i) the annihilation of positrons and *p*-Ps, ii) the pick of annihilation of *o*-Ps and iii) the annihilation of *o*-Ps in pores. The mean lifetimes of *o*-Ps trapped by pores were determined to be 7.0 ns and 8.2 ns for the samples with pitch patterns of 0.54 and 2.16 μm, respectively. Using the relationship between the Ps lifetime and pore diameter [11], the mean pore diameters were estimated to be 1.1 and 1.2 nm, suggesting the shrinkage of pore sizes by the scaling. In conclusion, we have shown that the positron annihilation technique is a useful tool for determining point defect behavior in ultra-shallow junctions formed on Si substrates. It was also shown that positron annihilation parameters can be used to trace the variation of pore sizes during the Cu/low-*k* damascene processes.

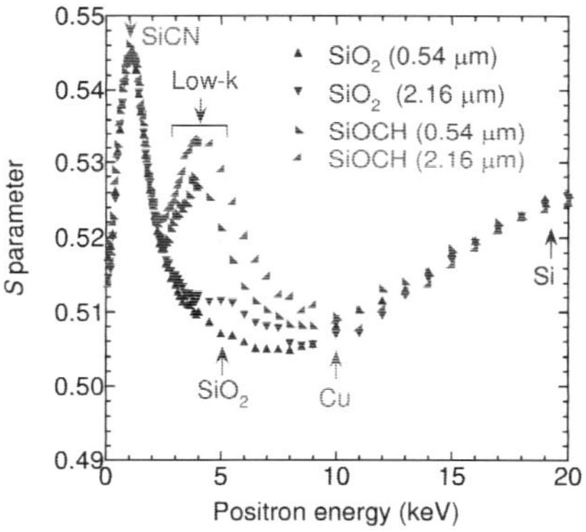

Figure 6. (colour online) *S–E* curves for the samples with SiOCH and SiO_2 interlayers.

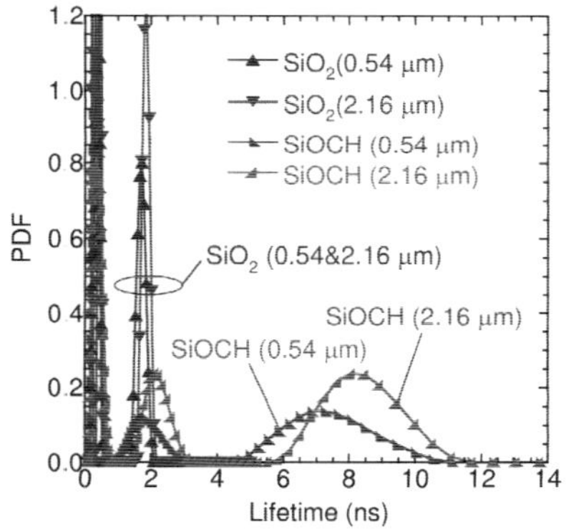

Figure 7. (colour online) PDF of the annihilation rate for the samples with different pitch patterns.

References

[1] *International Technology Roadmap for Semiconductors* 2009 (San Jose: Semiconductor Industry Association)
[2] Schultz P J and Lynn K G 1998 *Rev. Mod. Phys.* **60** 701
[3] Krause-Rehberg R and Leipner H S 1999 *Positron Annihilation in Semiconductors* Solid-State *Sciences, Vol. 127* (Springer: Berlin)
[4] Uedono A, Tsutsui K, Ishibashi S, Watanabe H, Kubota S, Nakagawa Y, Mizuno B, Hattori T and Iwai H 2010 *Jpn. J. Appl. Phys.* **49** 051301
[5] Uedono A, Inoue N, Hayashi Y, Eguchi K, Nakamura T, Hirose Y, Yoshimaru M, Oshima N, Ohdaira T and Suzuki R 2009 *Jpn. J. Appl. Phys.* **48** 120222
[6] Dannefaer S and Kerr D 1986 *J. Appl. Phys.* **60** 1313
[7] Uedono A, Ujihira Y, Ikari A, Haga H and Yoda O 1993 *Hyperfine Interactions* **79** 615
[8] Uedono A, Tanigawa S, Sugiura J and Ogasawara M 1989 *Jpn. J. Appl. Phys.* **28** 1293
[9] Eichler S, Gebauer J, Börner F, Polity A, Krause-Rehberg R, Wendler E, Weber B, Wesch W and Börner H 1997 *Phys. Rev. B* **56** 1393
[10] Yamauchi J, Aoki N and Mizushima I 1997 *Phys. Rev. B* **55** R10245
[11] Dull T L, Frieze W E, Gidley D W, Sun J N and Yee A Y 2001 *J. Phys. Chem. B* **105** 4657

Journal of Physics: Conference Series **262** (2011) 012062 doi:10.1088/1742-6596/262/1/012062

Design optimization of a pulsed positron beam system at IGCAR, Kalpakkam

C Varghese Anto, R Rajaraman, S Abhaya, J Parimala and G Amarendra

Materials Physics Division, Indira Gandhi Centre for Atomic Research,Kalpakkam-603102

E-mail: amar@igcar.gov.in

Abstract. A pulsed positron lifetime system has been designed using RF pulsing scheme. The mechanical design of the ultra high vacuum(UHV) beam line has been made to incorporate the basic requirements of a magnetically guided slow positron beam as well as that of an RF pulsing system consisting of a reflection type grid based chopper, a double gap pre-buncher and a main buncher. The trajectory and the timing characteristics of positrons have been simulated using SIMION code. The amplitude of the chopper and buncher RF voltages has been optimized based on these simulations.

1. Introduction

Positron annihilation spectroscopy is an unique tool for the investigation of various atomistic imperfections especially open volume defects in materials [1]. Among the different positron based techniques, positron lifetime spectroscopy is more powerful in the identification and quantification of vacancy-defects. The inability of combining positron lifetime spectroscopy with the slow positron beam technology has been overcome in the past two decades using various ideas. One method involves the detection of secondary electrons produced upon the implantation of positrons on the surface for the generation of the start signal [2]. Another way is to use an RF pulsing system to generate the start pulse ([3]-[8]) or apply a longitudinal harmonic potential on the beam [9]. Another innovative method is to trap positrons in a field and use a trigger to release the trapped positrons. In this case, the trigger itself can serve the required start signal [10]. When compared to the conventional slow positron beam, pulsed positron beam has the advantage that it enables the study of surfaces, interfaces and thin solid films in a more quantitive way.

Beam based lifetime spectroscopy has been achieved at various labs across the world using the basic techniques listed above ([2]-[10]). Among these, the RF pulsing based system has the best possible time resolution. Any RF pulsing based system consists of two basic elements namely the chopper and the buncher. The principles of these two are borrowed from their use in various ion beam accelerators. The difference between various RF pulsed positron beams already constructed or under construction is in the arrangement, the number and the parameters of these two basic elements. Successful systems consist of one chopper and two bunchers. The way they are arranged are different for example, in one case a chopper is put first followed by two bunchers [4] whereas, in other a pre-buncher is put first followed by a chopper and a second buncher [6].

Two parameters are important here (a)the frequencies at which these pulsing systems work and (b) the RF voltages.

We report here the design of a UHV based pulsed positron beam and the optimization of the RF pulsing parameters of the model system. A system consisting of a grid based chopper and two double gap bunchers, which work at harmonically related frequencies, are utilized here. SIMION code [11] has been utilized to get the trajectory and timing properties of the positrons passing through the model system. The initial results of these simulations are presented in this paper.

2. Design of the UHV system

A pulsed positron beam is basically a combination of a slow positron beam and the RF pulsing system. Therefore the design of a UHV system should cater to the requirements of a slow positron beam as well as that of pulsing system. In the present design we have a source mounting arrangement which is followed by a moderator assembly. A 50 mCi Na^{22} source will be utilized here. The moderator is put on a motion manipulator for moving it to a separate vacuum annealing chamber. We envisage to use a W(100) foil as moderator in transmission geometry. The slow and fast positrons exiting the moderator are separated using a 90° magnetic bend based velocity selector. Beyond this point, we have a housing tube for the RF pulsing system and finally the sample chamber. The above arrangement will be connected to pumping stations, which provides vacuum in the range of 10^{-10} mbar. Two ion pumps with Ti sublimation filaments and a turbo pump are planned towards this. Keeping these basic requirements in picture, a mechanical design was arrived at based on standard vacuum fittings. A schematic view of the system under construction is shown in Fig.1.

The transport of the positrons from the source through the moderator, velocity selector, RF pulsing system and finally to the sample is achieved with the help of a field of 0.01 T. To achieve this field a combination of solenoid and Helmholtz coils are used. Solenoid coil is used in the vertical portion where source and moderator are mounted. Helmholtz coils are used in the horizontal region of the beam to achieve uniform field through the mechanical centre of the beam line without much geometrical hindrance.

The chopper, pre-buncher and the main buncher are housed inside the long horizontal tube in the middle of the beam line. We also have a drift tube and graded accelerator arrangement beyond the main buncher. The sample chamber houses the faraday cage along with the sample. BaF_2 scintillator based photomultiplier tube will be put in one of the ports of the sample chamber for the detection of annihilation 511 keV γ rays.

3. Optimization of RF pulsing parameters

Based on the mechanical design of the UHV beam line, a model of the RF pulsing system has been made using the software package SIMION [11]. SIMION is a finite difference based approach which calculates the 2D/3D electrostatic and scalar magnetic fields solving Laplace equations and then the trajectories of charged particles through it by constant step integration using a highly modified 4th order Runge-Kutta method. SIMION also follows a work-bench protocol where the physical entities making up the system could be placed on a gridded work bench. The electric potential as well as the magnetic field to the various electrodes could then be given and trajectory of the particle can be simulated. We have utilized the user programming capabilities of SIMION using LUA [12] to set static (dc) as well as quasi static (RF) potentials for the electrodes. The magnetic field was set as 0.01T throughout the beam line. The W(100) moderator is biased at 10 V and hence the monoenergetic positrons which pass through the bend have an energy of $\approx$ 13 eV.

The first component of the RF pulsing system is a reflection type three grid chopper [4, 8]. The first grid is grounded whereas the third grid is for pre-acceleration. The second grid is given

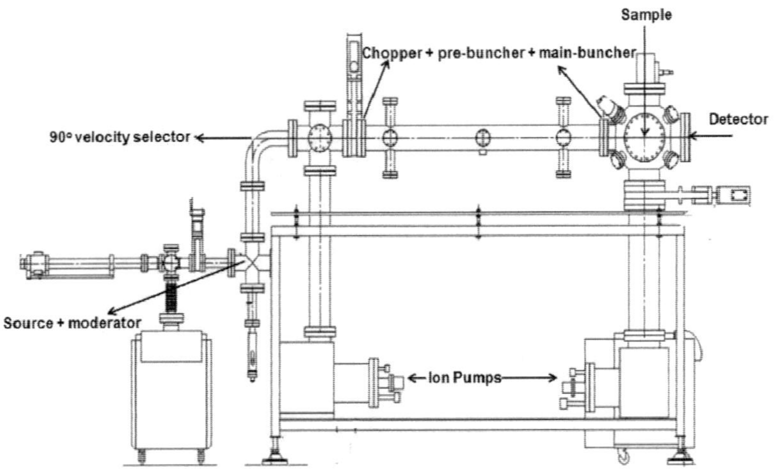

Figure 1. A schematic view of the mechanical design of the UHV pulsed positron beam line. The source chamber is in the vertical portion of the beam line. The long horizontal tube with multiple ports houses the pulsing systems.

a square wave, whose primary purpose is to chop the continuous beam into bunches of positrons. In the present model system, this is achieved by biasing the chopper to a dc voltage of +15 V (to stop the 13 eV positrons) and a square wave of 50 MHz with an amplitude of -15 V . Hence, during 'on-cycle' we have transmittance, whereas during the 'off-cycle' we have reflection.

Following the chopper, we have two double gap bunchers i.e., a pre-buncher followed by a main buncher. The pre-buncher works at the same frequency as that of the chopper. The pre-buncher consists mainly of three tubes separated by a small bunching gap. The middle tube is given a sine wave with required amplitude with respect to the first two tubes. The positrons arrive at the first bunching gap in various time-positions of the negative going slope of the sine wave and hence, get differently velocity-modulated. The length of the middle tube is equal to $\lambda/2$ so that bunching occurs at both gaps. The bunching amplitude is also adjusted so that there is a better time bunching. The purpose of the pre-buncher is also to improve the bunching efficiency as feeding from a lower frequency of 50 MHz of chopper to a buncher at higher frequency results in large loss with fewer positrons in the bunch. The positrons emanating from the pre-buncher are further bunched using the main-buncher, which has similar structure and working principle. The main buncher works at a higher frequency of 200 MHz.

Following the main buncher, there is a drift tube which extends into the sample chamber and a Faraday cage to avoid back scattered positrons interfering with the timing signals generated [5]. The HV is applied to the sample through a graded accelerator consisting of six stages including the faraday cup. The potential is divided among the graded electrodes and the positrons are gradually accelerated from the pre-acceleration value of 250 V to the required HV, resulting in varying implantation depth. All the parameters were adjusted for a sample high voltage of 10 kV. This was done so as to have reasonable drift tube voltages when we moved from small accelerating voltages to the final HV. This also helps to have good time resolutions at the common operating voltages near 5-10 kV rather than at the surface.

The simulation using SIMION was carried out using 1.5×10^5positrons emanating from a source of diameter 6 mm. All the positrons are born at the same time and a time spread at the chopper is introduced by the energy spread in the beginning. For more realistic simulation a

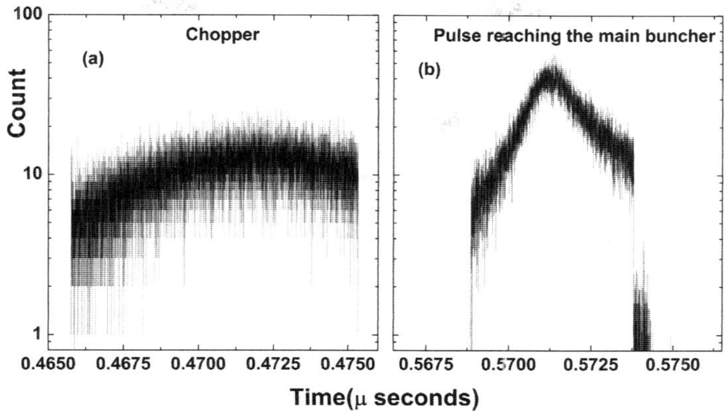

Figure 2. Counts vs Time of arrival for positrons at (a) the chopper and at (b) the main buncher. Positrons having energy between 12 and 14 eV alone are counted at the chopper whereas all positrons are counted at the main buncher.

time spread has to be introduced at the source position also which is currently under progress. The initial energy of positrons has a Gaussian distribution around the central value of 3 eV and has an fwhm of 0.6 eV. Energy was chosen as approximately equal to the work function of W (100) moderator [13]. The FWHM of the positron pulse (in time) at the target is the objective parameter to be minimized. The time of arrival is recorded at multiple planes of interest and at the target position.

For 10 kV acceleration, the drift tube voltage is set at the pre-acceleration voltage of 250 V. The peak voltage combination for the pre-buncher and the main buncher is arrived at by trial and error method. The primary method employed is to under-bunch the positrons at the target using the pre-buncher alone and to adjust the main buncher voltage to achieve the final bunching at the target. In this way, a peak voltage of 7 V is chosen for the pre-buncher where as a peak voltage of 3 V was found suitable for the main buncher.

The result of such a bunching process is shown in Fig.2 and Fig.3. At the chopper, the positrons having an energy between 12 and 14 eV alone are counted whereas at the main buncher all positrons are counted. As can be seen in Fig.2 (a), by the action of 50 MHz square wave at the chopper, we get a 10 ns positron pulse. This square pulse at chopper has been compressed to a pulse of less than 5 ns FWTM at the main buncher after one round of bunching by the pre-buncher, as seen in Fig.2 (b).

By the action of the main buncher, the 5 ns pulse is further compressed at the target as shown in Fig.3. The FWHM of this pulse is 32.4 ps and FWTM is $\leq$ 100 ps. The beam diameter at the target is 6.2 mm. So as to achieve this "pulsing", the energy of the positrons has been modulated at two bunchers. But this should not be very drastic so that the intended use of depth profiling is lost. In the present model, variation in energy at the target for 10 keV is from 9996 eV to 10030 eV. These numbers show that the above model system with the given RF voltages will able to achieve bunching of positrons at the target with minimum energy spread. These simulations were also repeated for different sample HV and appropriate drift tube voltages. The computed FWHM of the positron pulse at the target for all high voltages is below 70 ps.

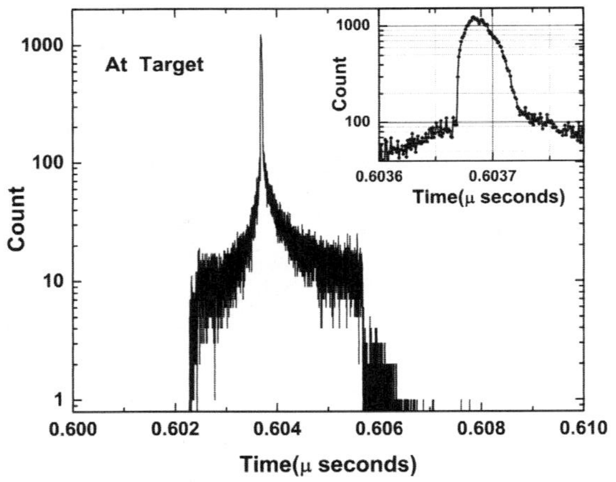

Figure 3. Counts vs Time of arrival for bunched positrons at the target for a target HV of 10 kV. Inset shows the peak region of the pulse.

4. Conclusion

The design of a pulsed positron beam system is described in terms of 90° velocity selector and RF based chopper, pre-buncher and main buncher. Beam parameters are optimized using SIMION simulations. The results show that the present model system is capable of generating positron pulses with FWHM less than 70 ps for all sample high voltages. Based on these results, the parameters for the RF modules were fixed. Currently, the UHV beam line is under construction.

Acknowledgments

One of the authors (C Varghese Anto) wishes to thank Department of Atomic Energy, India for the fellowship. Authors would like to thank Dr. C. S. Sundar and Mr. M. C. Valsakumar for their special interest and encouragement for the project.

REFERENCES

[1] Schultz P J and Lynn K G 1988 *Rev. Mod. Phys.* **60** 701.
[2] Szpala S, Petkov M P and Lynn K G 2002 *Rev. Sci. Instrum.***73** 147.
[3] Schödlbauer D, Sperr P, Kögel G and Triftshäuser W 1988 *Nuclear Instruments and Methods in Physics Research* B **34** 258.
[4] Suzuki R, Kobayashi Y, Mikado T, Ohgaki H, Chiwaki M, Yamazaki T and Tomimasu T 1991 *Japanese Journal of Applied Physics* **30** L 532.
[5] Willutzki P, Störmer J, Kögel G, Sperr P, Britton D T, Steindl R and Triftshäuser W 1994 *Meas. Sci. Technol.* **5** 548.
[6] Fallström K and Laine T 1999 *Applied Surface Science* **149** 44 ; Laakso A, Saarinen K and Hautojärvi P 2001 *Physica* B 308.
[7] Krause-Rehberg R, Sachert S, Brauer G, Rogov A and Noack K 2006 *Applied Surface Science* **252** 3106.
[8] Ohdaira T, Muramatsu M, Suzuki R, Takanawa M, Hashimoto N, Takao K and Kobayashi Y 2007 *Phys. Stat. Sol.*(c) **4** 4020.
[9] Mills Jr. A P 1980 *Applied Physics* **22** 273.
[10] Gilbert S J, Greaves R G and Surko C M 1999 *Applied Physics Letters* **70** 1944.
[11] Dahl D A 2000 *Int. J. Mass Spectrom.(source: Scientific Instrument Services, Inc., Ringoes, NJ, www.simion.com)* **200** 3.
[12] Ierusalimschy R, de Figueiredo L H and Celes W 1996 *Software: Practice and Experience* **26** 635.
[13] Chen D M, Lynn K G, Pareja R and Nielsen B 1985 *Physical Review* B **31** 4123; Amarendra G, Canter K F and Schoepf D C 1996 *J.App. Phys.* **80** 4660.

Journal of Physics: Conference Series **262** (2011) 012063 doi:10.1088/1742-6596/262/1/012063

Hydrogen in red ZnO – defects or lattice expansion

Marc H. Weber and Kelvin G. Lynn

Center for Materials Research, Washington State University, Pullman, WA 99164-2711

m_weber@wsu.edu

Abstract. The effect of hydrogen on ZnO while annealing at 1100 °C with Zn vapour or Ti metal is studied. ZnO turns red only when hydrogen is present in combination with oxygen deficiency. The coloration is reversible at 1100 °C by removing hydrogen either by evacuating or by gettering hydrogen with Ti at low temperature while annealing the ZnO at >950 °C. The latter is a simple method to lower the hydrogen impurity concentration. The annealing data are consistent with an activation energy >1.5 eV to reverse the coloration. Hall data on red ZnO reveal only a shallow donor level around 45 meV, too low for the colour change. Temperature dependent X-ray diffraction reveals a lattice expansion due to the presence of hydrogen which is sufficient to explain the colour change.

1. Introduction

ZnO is a promising wide bandgap semiconducting material with a large number of potential applications. A number of defects both intrinsic and due to impurities pose significant obstacles on the road to commercial applications. As-grown material is of n-type conductivity and no reliable method for p-type doping is known. Hydrogen is an almost omnipresent impurity and known to form a shallow donor when in an interstitial state [1-4]. Other states of hydrogen, some of which are not IR active, have been predicted and observed indirectly [5-7]. A practical method to reduce or even remove hydrogen from ZnO has to be devised in order to minimize the challenges in p-type doping.

ZnO turns red when it is annealed in the presence of Zn vapor at 1100 °C. The red colour has been linked to an excess in Zn [8-12]. Based on the ratio of Zn to O concentration, an excess in Zn is equivalent to a deficiency in oxygen. On an atomic scale, however, the presence of Zn interstitials (Zn_i) is very different from oxygen vacancies (V_O). Stockmann noticed multiple options of accommodating excess Zn [13]. Halliburton *et al.* and Selim *et al.* pointed out that the same red colour was observed after annealing in the presence of phosphorous or titanium rather than zinc vapor, respectively, pointing to V_O [10, 12]. Isolated zinc and oxygen vacancies anneal out below 200 °C and 400 °C, respectively, or merge to clusters which then anneal out below 700 °C [12,14-17]. The red coloration, however, persists suggesting a different defect. The potential contribution of hydrogen, OH ions or water to this effect was never tested and typically presumed insignificant. Here, it is shown that hydrogen in combination with V_O or another form of oxygen deficiency, not interstitial zinc, Zn_i, is responsible for the red colour.

2. Experimental

ZnO from two suppliers and grown by two methods was studied. Hydrothermally grown ZnO was obtained from Tokyo Denpa Co., LTD (TEW) [18] and skull melt grown ZnO was purchased from Cermet Co. [19]. One sample was grown by seeded chemical vapour deposition. All samples were cut

Journal of Physics: Conference Series **262** (2011) 012063 doi:10.1088/1742-6596/262/1/012063

along the c-axis to 10×10 mm2 squares and to 0.5 mm thickness. The supplier polished either on both sides (TEW) or on the Zn-face only (Cermet).

The samples were annealed in high purity low OH quartz ampoules (GE214) in a tube furnace under a variety of ambient conditions. The ampoules containing the samples were connected to an oil free vacuum system, evacuated to ~10^{-7} torr and baked for >12 hrs at 120 °C to remove water vapour. The ampoule-sample assemblies were either sealed off with a hydrogen-oxygen torch for annealing in a closed system, or left connected to the pumping station in an open system to monitor partial pressures with a residual gas analyzer (RGA, Vacuum Generators). A schematic of the annealing system is shown in Figure 1.

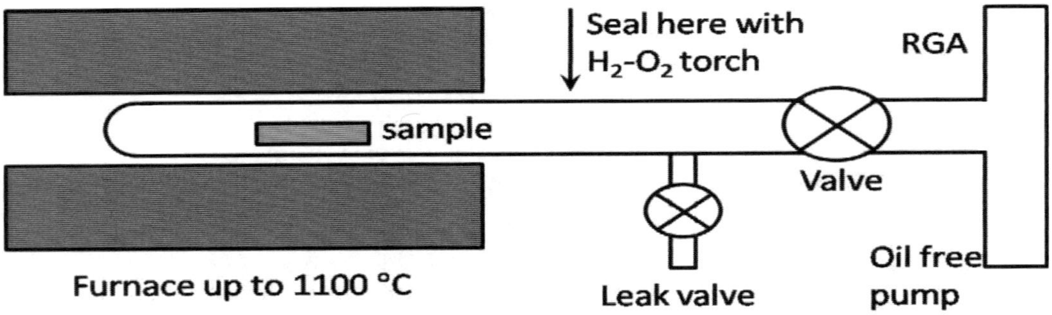

Figure 1: schematic of the experimental apparatus for annealing ZnO under various ambient conditions.

The samples were annealed in the presence of Zn vapour (99.999% pure, 5N Plus Inc.) or Ti metal foil (99.6% pure, Goodfellow Corp.). The hydrogen and water content in the systems was controlled by preheating the ampoules, the Ti foil and the Zn metal pellets in the open system configuration. The evolution of hydrogen as a function of temperature and time was monitored with the RGA. Hydrogen was removed for ZnO by either pumping the evolving gases in the open-system configuration, or by gettering hydrogen with the Ti metal foil in the closed-system configuration. Prior to sealing, oxygen, hydrogen, or deuterium could be introduced into the ampoule through a leak-valve.

Positron annihilation experiments were carried out at the WSU positron beam with an energy range of 0.1 to 70 keV. A high efficiency high resolution (1.5 keV full width half maximum at 511 keV) germanium detector was used for the Doppler broadening experiments. The samples were annealed in situ for 30 minutes at up to 700 °C or in the tube furnace under nitrogen atmosphere above 700 °C. All positron measurements were carried out at room temperature.

3. Results

The release of hydrogen was monitored in the open-system configuration. In addition to the ZnO material [20], several sources of hydrogen may be present. As received Ti and the quartz ampoules may contain significant quantities of hydrogen. Titanium metal is an excellent getter for hydrogen as implemented in Ti-getter vacuum pumps. Ti absorbs hydrogen below ~200 °C and releases it above 400 °C [21]. Ti foils were heated in ampoules to desorb and monitor hydrogen. The H_2 partial pressure rose to 5×10^{-7} torr near 500 °C. The H_2 content in quartz ampoules is not evaluated or listed by manufacturers. Figure 2 shows the results of annealing an empty quartz ampoule twice. Hydrogen diffuses out of the near surface regions of the quartz. The second annealing shows about 10× lower partial pressures. Care must be taken not to heat the ampoule to higher temperatures to prevent further release from deeper in the material. If Zn vapour is present during the annealing, Zn atoms may crack water molecules and generate hydrogen by hydrolysis [22-24].

Journal of Physics: Conference Series **262** (2011) 012063 doi:10.1088/1742-6596/262/1/012063

As received clear ZnO was annealed in a sealed ampoule in the presence of Zn vapour and Ti metal foil without any hydrogen treatment as discussed by Selim *et al.* [12]. In both cases, the ZnO turned reddish at room temperature. When the ampoule and the Ti foil were treated to remove or reduce the hydrogen content, no colour change occurred. ZnO remained clear. The Ti foil was maintained at a temperature sufficiently low to getter hydrogen. The surface roughness of the ZnO increased due to some loss of Zn. This transparent ZnO was then annealed in the presence of Zn vapour in an ampoule treated to lower the hydrogen and water vapour content. The sample remained clear just as with the previous experience with hydrogen depleted Ti. Figure 3 shows photos of the samples at room temperature after annealing. The arrows indicate the annealing step and reversibility. Also, hydrogen depleted ZnO (clear) was annealed in the presence of Ti metal that contained deuterium. This resulted in a yellow sample. Annealing any sample either in oxygen or under vacuum while pumping away any desorbed gases will remove the coloration and result in clear ZnO. Clearly, ZnO turns reddish only when hydrogen is present and oxygen deficient conditions are given. The former or latter alone does not lead to red ZnO.

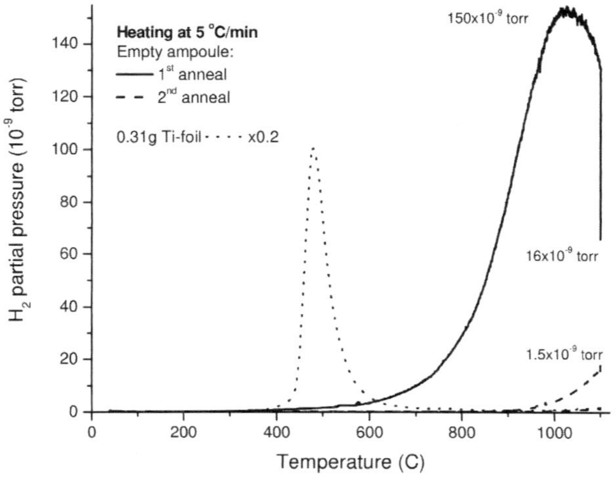

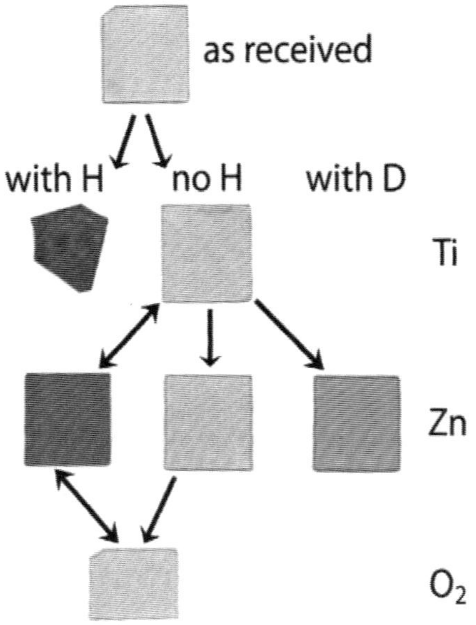

Figure 3: Photos of ZnO samples after annealing at 1100 °C for 30 min in the presence of Ti metal (2nd row) Zn vapour (3rd row) and oxygen (4th row) either in the presence of hydrogen (in the ampoule, ZnO, and/or Ti) (left column) or after pre-treatment to remove or reduce the hydrogen content (middle column). In the right column, Ti was preloaded with some deuterium gas.

Positron annihilation depth profiles can be used to reveal vacancies. Typically, only neutral and negatively charged vacancies can be observed in semiconducting material. Neutral oxygen vacancies are not detectable at room temperature [15, 25]. As demonstrated by Selim *et al.* [12] and Weber *et al.* [26, 27], these oxygen vacancies can be revealed by adding zinc vacancies to the sample and then anneal at up to 700 °C. If oxygen vacancies are present in the original sample, they will become mobile and merge with the added Zn vacancies to form di-vacancies or lager vacancy clusters. In the as-prepared state, clear and red ZnO are indistinguishable and show a small (<3%) increase in S following annealing. When damage including V_{Zn} is added, S rises in the damaged region by 6% and 4.5% for the red and clear ZnO samples respectively, somewhat above the 4% change associated with V_{Zn} [28]. The specific values of saturation trapping of positrons in V_{Zn} remain subject to debate until high quality ZnO similar to Si becomes available. Here, it is relevant only that V_{Zn} and related defects were added and detected. S increases dramatically following annealing at 500 °C in the case of the damaged red ZnO sample only. Pre-existing open volume defects became mobile and merged with the V_{Zn} related vacancies to larger clusters [12]. Oxygen vacancies must be present in the red ZnO. This is shown in Figure 4.

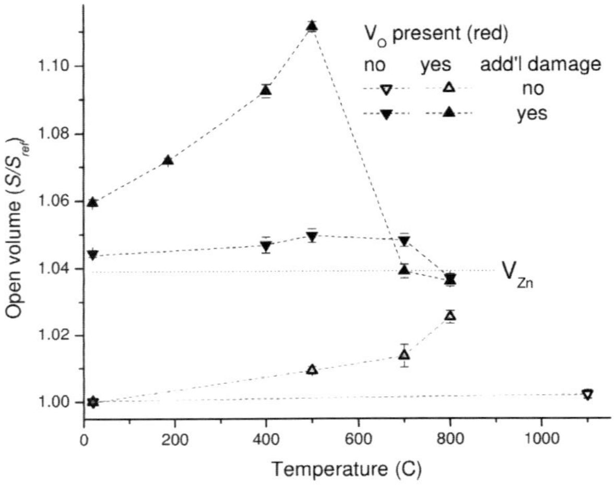

Figure 4: Open volume fraction at 1.2μm implantation depth in ZnO without (downward triangles) and with oxygen vacancies (red) (up-pointing triangles) and with (solid) and without (open) added zinc vacancies (by applying 9 μm polishing grit scratches). S values are shown as a function of annealing temperature. Only the red and damaged ZnO show a dramatic increase in S after up to 700 °C anneals. The value for V_{Zn} was obtained from Ref. [28]. Lines are included to guide the eye.

To detect any hydrogen in red ZnO, a sample was annealed in the open-system configuration at 5 °C/min to increasingly higher temperatures of 700, 850, 950 and 1050 °C. The ampoule was pre-treated to deplete hydrogen and water. The partial pressure of hydrogen was monitored with the RGA. Figure 5 shows the partial pressure of H_2 as a function of temperature. After each annealing step, the sample was cooled to room temperature and the colour noted. The sample became pale after 950 °C and whitish (due to surface roughness) after 1050 °C (see Figure 5). Hydrogen diffuses out around 400 °C without any affect on the sample colour. This is consistent with hydrogen bound at interstitial sites [29]. More hydrogen is released from 800 to 1000 °C and the red coloration disappeared. This hydrogen is more deeply bound at 1.5 eV and consistent with results by Nickel *et al.* [30-33] and calculations by Bang and Chang [34]. In addition to the effusion of hydrogen above 400 °C, Nickel *et al.* observed more molecular hydrogen well above 800 °C and found activation energies of about 1 eV. The former correlates with IR observations of OH vibrations while the latter is not visible to IR [35].

Preliminary low-temperature Hall measurements on a ZnO with hydrogen annealed red sample revealed a shallow donor level at 45±2 meV and no deep levels. This is consistent with observations by Lavrov *et al.* who identified this level with the H_O defect [36]. Clearly, 45 meV are not sufficient to change the colour of ZnO. A different mechanism has to explain the colour change in the absence of a deep level. ZnO is thermochrome, i.e. the bandgap decreases sufficiently with increasing temperature to change the colour of ZnO from transparent at room temperature to dark yellow at ~600 °C. X-ray diffraction data (XRD) show this effect as a small change in the 2Θ diffraction. Preliminary XRD on clear and red ZnO reveal a similar shift in the 2Θ diffraction angles from 72.58° to 72.45° for clear and red ZnO, respectively. Loading of oxygen deficient ZnO with hydrogen exerts pressure on the c-axis and lowers the bandgap by the observed amount with the corresponding change in colour. In order to further verify this, temperature dependent X-ray diffraction measurements are ongoing.

4. Summary

The presence of hydrogen in oxygen deficient ZnO is responsible for the red coloration. A defect involving hydrogen and oxygen vacancies most likely causes H_O to form and the lattice to be expanded sufficiently to lower the bandgap of ZnO. The latter causes the change in colour.

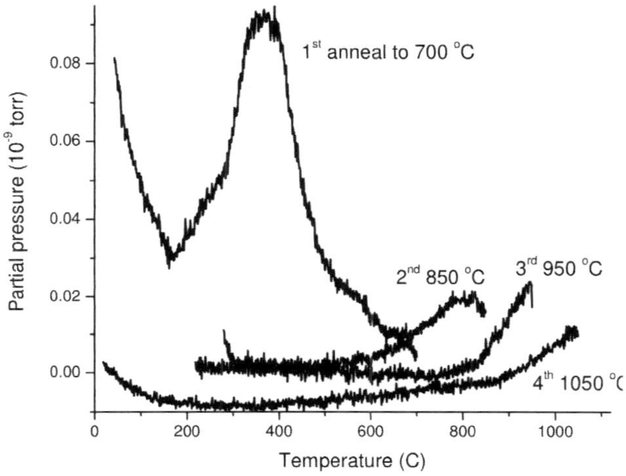

Figure 5: Partial pressure of H_2 as a function of temperature of red ZnO while heating at 5 °C/min to a) 700 °C, b) 850 °C, c) 950 °C, and d) 1050 °C. Heating to between 950 and 1050 °C is required to remove the red colour. Hydrogen diffuses out near 400 C without colour change and at >800 with the colour change. A base pressure of hydrogen after annealing was subtracted.

Acknowledgements

Discussions with Matthew D. McCluskey and prior work by Narendra S. Parmar, Farida Selim, and Christoph Fischer are acknowledged. Some of this work was funded by the US Dept. of Energy under contract number DE-FG02-04ER46103.

References

[1] Mollwo E 1954 *Z. Physik* **138** 478
[2] Thomas D G and Lander J J 1956 *J. Chem. Phys.* **25** 1136
[3] Van de Walle C G 2000 *Phys. Rev. Lett.* **85** 1012
[4] Hofmann D M, Hofstaetter A, Leiter F, Zhou H, Henecker F, Meyer B K, Orlinskii S B, Schmidt J, and Baranov P G 2002 *Phys. Rev. Lett.* **88** 045504
[5] McCluskey M D, Jokela S D and Hlaing Oo W M 2006 *Physica B* **376-377** 690
[6] Lavrov E V 2003 *Physica B* **340-342** 195
[7] Lavrov E V, Börrnert F and Weber J 2005 *Phys. Rev. B* **71** 035205
[8] Ehret W F and Greenstone A 1943 *J. Am. Chem. Soc.* **65** 872
[9] de la Cruz R M, Pareja R, Gonźalez R, Boatner L A and Chen Y 1992 *Phys. Rev. B* **45** 6581
[10] Halliburton L E, Giles N C, Garces N Y, Luo M, Xu C, Bai L and Boatner L A 2005 *Appl. Phys. Lett.* **87** 172108
[11] Jokela S J and McCluskey M D 2007 *Physica B* **401-402** 395-398
[12] Selim F A, Weber M H, Solodovnikov D and Lynn K G 2007 *Phys. Rev. Lett.* **99** 085502
[13] Stöckmann F 1950 *Z. Physik* **127** 563

[14] Chen Z Q, Maekawa M, Kawasuso A, Sakai S and Naramoto H 2006 *Physica B* **376-377** 722
[15] Tuomisto F, Saarinen K, Look D C and Farlow G C 2005 *Phys. Rev. B* **72** 085206
[16] Vlasenko L S and Watkins G D 2005 *Phys. Rev. B* **71** 125210
[17] Chen Z Q, Maekawa M, Kawasuso A, Sakai S and Naramoto H 2006 *J. Appl. Phys.* **99** 093507
[18] Maeda K, Sato M, Niikura I and Fukuda T 2005 *Semicond. Sci. Technol.* **20** S49
[19] Nause J and Nemeth B 2005 *Semicond. Sci. Technol.* **20** S45
[20] Čížek J, Žaludová N, Vlach M, Daniš S, Kuriplach J, Procházka I, Brauer G, Anwand W, Grambole D, Skorupa W, Gemma R, Kirchheim R and Pundt A 2008 *J. Appl. Phys.* **103** 053508
[21] Furuya Y, Takasaki A, Mizuno K and Yoshiie T 2007 *J. Alloys Compd.* **446-447** 447
[22] Bazan J C, Gschaider M E and Alimenti G A 1999 *J. Therm. Anal.* **55** 569
[23] Alimenti G A, Gschaider M E, Bazan J C and Ferreira M L 2004 *J. Colloid Interface Sci.* **276** 24
[24] Melchior T, Piatkowski N and Steinfeld A 2009 *Chem. Eng. Sci.* **64** 1095
[25] Takenaka H and Singh D J 2007 *Phys. Rev. B* **75** 241102(R)
[26] Weber M H, Selim F A, Solodovnikov D and Lynn K G 2008 *Appl. Surf. Sci.* **255** 68-70
[27] Weber M H, Parmar N S, Jones K A and Lynn K G 2010 *J. Electron. Mater.* **39** 573-576
[28] Tuomisto F, Ranki V, Saarinen K and Look D C 2003 *Phys. Rev. Lett.* **91** 205502
[29] Ip K, Overberg M E, Heo Y W, Norton D P, Pearton S J, Stutz C E, Kucheyev S O, Jagadish C, Williams J S, Luo B, Ren F, Look D C and Zavada J M 2003 *Solid State Electron.* **47** 2255-2259
[30] Nickel N H and Fleischer K 2003 *Phys. Rev. Lett.* **90** 197402
[31] Nickel N H and Brendel K 2003 *Phys. Rev. B* **68** 193303
[32] Nickel N H 2006 *Phys. Rev. B* **73** 195204
[33] Nickel N H 2007 *Superlattices and Microstruct.* **42** 3-7
[34] Bang J and Chang K J 2008 *Appl. Phys. Lett.* **92** 132109
[35] Shi G A, Saboktakin M, Stavola M and Pearton S J 2004 *Appl. Phys. Lett.* **85** 5601
[36] Lavrov E V, Herklotz F and Weber J 2009 *Phys. Rev. B* **79**, 165210

Three-dimensional electron-positron momentum distribution of O^{3+}-irradiated 6H SiC using two positron spectroscopy techniques simultaneously

Christopher Williams[1], Larry Burggraf[1], Paul Adamson[2], and James Petrosky[1]

[1]Air Force Institute of Technology, AFIT/ENP, 2950 Hobson Way, Wright-Patterson Air Force Base, OH 45433 USA.

[2]Stockpile Research, Development and Engineering Division, Office of Defense Programs, National Nuclear Security Administration, 1000 Independence Ave SW, Washington, DC 20585 USA.

E-mail: christopher.williams@afit.edu

Abstract. A three-dimensional (3D) positron annihilation spectroscopy system (3DPASS) capable of determining 3D electron-positron (e^--e^+) momentum densities from measurements of deviations from co-linearity and energies of photons from e^--e^+ annihilation events was employed to examine the effects of O-atom defects in 6H SiC. Three-dimensional momentum datasets were determined for 6H SiC irradiated with 24 MeV O^{3+} ions. Angular correlation of annihilation radiation (ACAR) and coincidence Doppler-broadening of annihilation radiation (CDBAR) analyses are presented. In addition, a novel technique is illustrated for analyzing 3D momentum datasets in which the parallel momentum component, $p_{||}$ (obtained from the CDBAR measurement) is selected for annihilation events that possess a particular perpendicular momentum component, $p_{\perp}$ observed in the 2D ACAR spectrum.

1. Introduction

3DPASS measures three-dimensional electron-positron (e^--e^+) momentum distributions by simultaneously measuring correlated deviations from co-linearity and energies of coincident two-photon annihilation events. Data for deviation from co-linearity and for energy are typically collected individually using two common positron annihilation spectroscopy techniques, two-dimensional angular correlation of annihilation radiation (2D ACAR) and two-detector coincident Doppler broadening of annihilation radiation (CDBAR), respectively.[1] Our implementation of 3DPASS, described in [2,3], simultaneously collects a single dataset of correlated energies and positions for coincident annihilation photons using two solid-state double-sided strip detectors (DSSDs). 3DPASS analysis captures the 3D e^--e^+ momentum distribution enabling conservation of total momentum to be used to interpret results.

This research investigated the effect of oxygen-atom defects on the momentum distribution in 6H SiC. 3DPASS measurements were made on 6H SiC samples, undamaged and ion-irradiated by 24-MeV O^{3+}. 2D ACAR and DBAR spectra were extracted from 3DPASS datasets for 6H SiC single crystals that were implanted with oxygen ions. The 3DPASS datasets were further analyzed by extracting DBAR spectra corresponding to particular momentum features in the 2D ACAR spectra, enabling conservation of total momentum to be used to characterize the 3D momentum corresponding to a particular lattice direction.

2. Experiment

Two position-sensitive, high-purity germanium DSSDs coupled with high-speed digital electronics are the heart of our 3DPASS instrument [3]. The digital electronics contain four independent electronics boards, each with two 12-bit ADCs, 8 input channels, and two Alterra Cyclone FPGAs. Output data is stored in two data files: raw data and raw event logs. The raw data are energies recorded in ADC units at each 20 ns clock cycle from single triggers for recordable full-charge events on any of the 32-channels. The raw event log, which is processed to obtain 3DPASS spectra, only stores energy for events that achieve the trigger threshold based on that channel's calibration. Event locations within pixels defined by the orthogonal intersection of front and rear DSSD strips are determined by interpolation based on the ratio of transient charge intensity on strips adjacent to the strip recording a full charge event. [3] Subpixel positions were determined within 0.22 mm standard error by analysis of transient charge intensities on strips adjacent to the strips collecting a full charge event giving angular resolution better than 0.5 mrad for our instrument configuration. DB lineshapes were extracted from the CDBAR energy spectra by selecting events within boundaries defined by equation (1), using $\Delta = 0.3$ keV:

$$E_1 + E_2 = 1022 \text{ keV } \pm \Delta \qquad (1)$$

where E_1 and E_2 are the energies of the annihilation photons detected coincidently in each of the DSSDs and Δ is the width of the diagonal DB stripe extracted from the CDBAR spectrum. An optimization analysis [2] showed that for our system, the value $\Delta = 0.3$ keV minimized fluctuations at the base of the DB spectrum and increased the number of events included in the DB spectrum without degrading the observed DBAR spectrum width. Using only a 106.5 μCi ^{22}Na source, 3DPASS datasets were measured for 6H SiC samples before and after irradiation by oxygen ions. 2D ACAR and CDBAR spectra, as well 3D momentum spectra, were reconstructed from the 3DPASS datasets.

Single-crystal 6H SiC samples procured from CREE Inc. were 254-μm thick research-grade, on-axis, N-type (0001) 6H SiC. O-ion implanted 6H SiC samples were prepared by irradiation with 24 MeV O^{3+} ions at 20 particle-nanoamps at the Sandia National Laboratory's Tandem Ion Beam Facility followed by sample annealing to 1000°C. During irradiation 1.25×10^{14} O^+ ions were implanted at a depth of 10.8 μm, normal to the surface of several 6H SiC. The irradiated 6H SiC samples were dry-etched to ensure the depth of the thin O-ion implanted layer corresponded to the depth of median positron interaction within the 6H SiC samples which was calculated to be 105 mm using a GEANT4 simulation [4-5]. Positrons were directed into the sample on the side opposite to the face that was irradiated with the O ions to reduce the interaction of positrons with vacancy-type defects created during the thermalization of the oxygen ions.

3. Experimental Results

3DPASS datasets contain the three axial components of momentum of the annihilating e^--e^+ pair. In order to fully display the entire 3D momentum, four dimensions are required. Since this is difficult, a convenient approach is to display the momentum distribution for one component, either the component parallel or components in the plane perpendicular to the direction of propagation of the two annihilation photons, for momenta selected in other directions. For example, we selected a momentum feature observed in the ACAR spectra and then displayed the DBAR spectra generated from the parallel component of the momentum of the photons that contributed specifically to that ACAR feature. Momentum anisotropies in the ACAR plane were selected for simplicity and also because the entire range of momentum in the ACAR application, limited to the region associated with mostly valence electrons (± 11 mrad or $\pm 11 \times 10^{-3}$ m_oc (or 11 mm$_o$c) for both directions in the plane), was contained within the DBAR's momentum range (40 mm$_o$c). This 3DPASS capability is demonstrated for 6H SiC samples: virgin and O-ion irradiated.

3DPASS datasets were analyzed to extract conventional 2D ACAR and DBAR spectra as shown in figure 1 for virgin 6H SiC and in figure 2 for O-ion irradiated 6H SiC. Subtle anisotropic momentum features, representing $p_{||}$ momenta, can be extracted from derivatives of the DB lineshape. Most CDBAR momentum features observed for virgin 6H SiC are observed in O-ion irradiated 6H SiC. The exception is the strong feature at 12.0 mm$_o$c which is absent in the O-ion irradiated 6H SiC. Weak momentum features in 6H SiC at about 4, 9, 11 and 15 mm$_o$c are strong in the O-ion irradiated samples, while the converse was observed for features at 0.8 and 1.4 mm_oc. All these momentum features are symmetrical about zero momentum except for the sharp, strong anomalies for O-ion irradiated SiC in which momentum is reduced, e.g. at -1.4 mm_oc in un-annealed, O-ion irradiated SiC. These unsymmetrical features in irradiated SiC we attribute to positron annihilation in the ion tracks produced by O-ion bombardment.

The 3D momentum distribution from the 3DPASS dataset for virgin, un-annealed 6H SiC was analyzed. Events encompassed in a square area surrounding each of the four quadrant anisotropies were extracted from the 2D ACAR spectrum (figure 1a). The data was extracted from the 3DPASS dataset uncorrected for DSSD charge collection strip efficiency, because it was essential that an event recorded in the ACAR spectrum correlated definitively with the same event in the DBAR spectrum; using efficiency corrected data would violate that restriction. The areas of the four quadrant momentum peaks contained a total of 13.4% of the total counts recorded in the entire 2D ACAR spectrum for the virgin 6H SiC.

From the energy difference of the coincident events in the selected ACAR momentum peaks, the parallel component of the momentum distribution was calculated and plotted in the same manner as CDBAR spectra. The 3D momentum lineshape for the selected features was extracted just as was the DB lineshape, using $\Delta = 0.3$ keV for consistency. This 3D momentum lineshape represented the momentum component parallel to the propagation of annihilation photons for virgin, un-annealed 6H SiC for selected peaks in the ACAR spectrum. These peak events have similar momentum component in the plane perpendicular to annihilation photons propagation. The resulting 3D momentum lineshape for the virgin 6H SiC is shown in figure 1b which is dominated by $p_{\|}$ peaks at = 8.1 mm_oc yielding a total, 3D momentum attributed to the anisotropy peaks of 11.4 mm_oc.

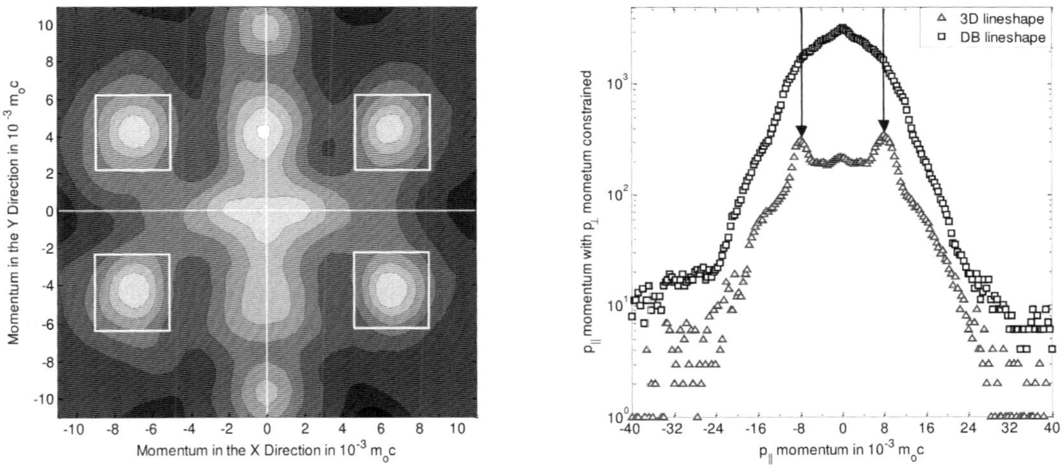

(a) (b)

Figure 1. (a) 2D ACAR spectrum for virgin 6H SiC (b) 3D momentum lineshape virgin 6H SiC illustrating $p_{\|}$ component with $p_\perp$ constrained by the momentum peaks contained in white boxes in 1a.

The two momentum peaks in the 3D momentum lineshape, indicated by the arrows, in the direction parallel to the annihilation photon's propagation corresponded very well to the magnitude of the momentum of the four features that the perpendicular momentum component were constrained to and to the large peaks at ± 2 keV observed in the DB lineshape as discussed in [2]. This indicated that a momentum of ± 8.1 mm_oc existed in both momentum components, in the plane perpendicular and in the direction parallel to the propagation direction of the annihilation photons. The shoulder in the 3D momentum lineshape at approximately ±16 mm_oc did not correspond to any other previously observed feature. 3D momentum lineshapes for the anisotropies centered at (0,4.5 and 0,-4.5) and at (0,10.0) and (0,-10.0) were similarly analyzed, but there was no discernable features above the noise observed in the lineshape, due to the low number of events with momentum corresponding to those anisotropies.

The 3D correlated momentum distribution from the 3DPASS dataset for O-ion irradiated, un-annealed 6H SiC was analyzed. A square area, of the same size (in momentum space) as selected from 2D ACAR for virgin SiC, was extracted surrounding quadrant features in the O-ion irradiated, un-annealed 2D ACAR spectrum. Events contained in the white boxes, shown in figure 2a, for the equivalent momentum peaks approximately centered in the spectrum quadrants were extracted from the 3DPASS dataset uncorrected for the DSSD charge collection strip efficiency. These four momentum peak areas contained 18.7% of the total counts. The resulting 3D momentum lineshape compared to the DB lineshape is shown in figure 2b.

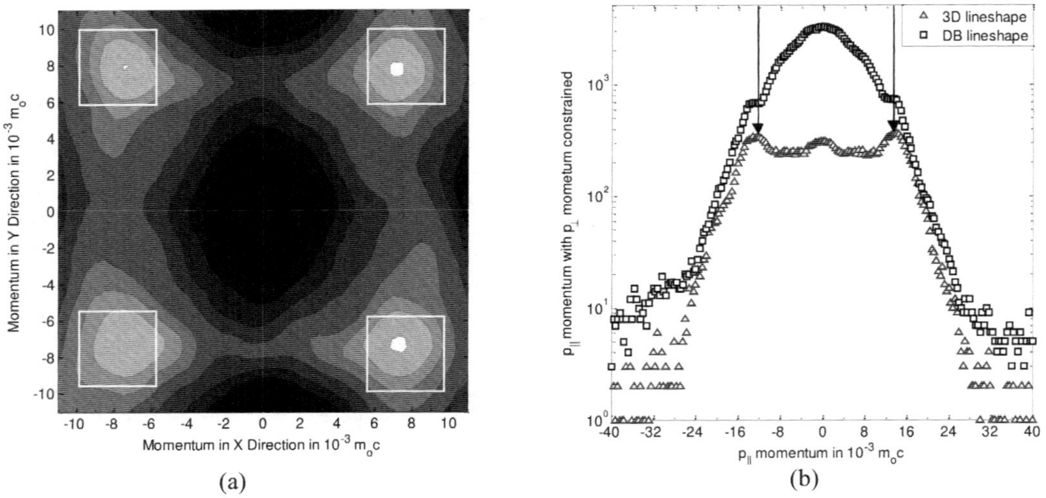

Figure 2. (a) 2D ACAR spectrum for O-ion irradiated, un-annealed 6H SiC (b) 3D momentum lineshape for O-ion irradiated, un-annealed 6H SiC illustrating $p\perp$ momentum

The two momentum peaks in the direction parallel to the annihilation photon's propagation in the 3D momentum lineshape, indicated by the arrows in figure 2b, corresponded very well to the magnitude of the momentum of the two shoulder features at $\pm$ 3 keV observed in the corresponding DB lineshape as discussed in [2]. This not only indicated a momentum of $\pm$ 12 mm$_o$c primarily existed in the component parallel to the direction parallel to the annihilation photons propagation, but also in the momentum in the plane perpendicular to the photon motion. The magnitude of the total momentum of the selected quadrant events was calculated to be 16.0 mm$_o$c from the three dominant momentum components. This illustrates the potential utility of correlating projections of momentum magnitude in simultaneous 2D ACAR and CDBAR measurements which is not possible by analysis of the techniques when applied separately.

4. Conclusions

The large change in the ACAR momentum anisotropies show that the 6H SiC lattice structure at positron annihilation sites was significantly perturbed by the O-ion implantation due to the oxygen atom implantation in interstitial sites. Additionally, the manner in which the anisotropies changed suggests a slight realignment of the symmetry of positron annihilation sites. The change in the positron center of symmetry with O-ion implantation altered the directions of e^{-}-e^{+} momentum projections. The 3DPASS method is capable of directly measuring total momentum corresponding to observed ACAR anisotropies.

Acknowledgements

Views expressed in this article are those of the authors and do not reflect the official policy or position of the Air Force, Department of Defense, or the U. S. Government.

References

[1] Krause-Rehberg R, Leipner H 1999 *Positron Annihilation in Semiconcutors* (Berlin: Springer) 16-20
[2] Williams C, Burggraf L, Adamson P, Petrosky J submitted to *Nuc. Instr. Meth. Phys. Res. A*, 5/ 2010
[3] Williams C, Burggraf L, Adamson P, Petrosky J, Oxley M 2010 *J. Phys Conf. Ser.* **225** 012058
[4] Agostinelli S et al. 2003 *Nuc. Inst. Meth. Phys. Res. A* **506** 250-303
[5] Allison J et al. 2006 *IEEE Trans. Nuc. Sci.* Vol. 53 **1** 270-278

Corrosion of pure aluminium and aluminium alloy: a comparative study using a slow positron beam

Y C Wu[1,2], P H Li[1], X D Xue[1], S J Wang[1], A Kallis[2], P G Coleman[2], T Zhai[3]

[1] Department of Physics, Wuhan University, Wuhan 430072, P.R.China
[2] Department of Physics, University of Bath, Bath BA2 7AY, UK
[3] Chemical and Materials Engineering Department, University of Kentucky, Lexington, KY 40506-0046, USA
E-mail address: ycwu@whu.edu.cn

Abstract. Corrosion-related defects in pure Al and AA 2037 Al alloy have been investigated by positron beam-based Doppler broadening energy spectroscopy. Defect profiles have been analyzed by measuring the *S* parameter as a function of incident positron energy up to 30 keV. When pure Al samples are immersed in 1M NaOH for various times, a significant increase in the *S* parameter near the surface is observed. This implies that the corrosion process involves the creation of defects and nanometer voids. In contrast, a significant decrease in the *S* parameter is observed after the corrosion of water-quenched Al alloy by the same method, which is interpreted as being a result of Cu enrichment near the metal-oxide interface layer.

1. Introduction

Positron annihilation spectroscopy (PAS) has been widely used as a probe for defect studies at the atomic dimension in solids, such as vacancies, vacancy clusters, dislocations, and nanometer-scale voids. The use of variable energy positron beams allows one to probe the depth profile of solids, and is particularly suitable for analysis of surface and near-surface defects [1]. PAS investigations of corrosion-related defects in aluminium have been reported [2, 3]. Significant numbers of vacancy-type (open volume) defects were found near the metal-oxide film interface, and it was concluded that the number of defects and size increased due to corrosion in NaOH or aqueous phosphoric acid solution. Furthermore, these investigations suggested that the cavities were interfacial voids which were revealed by AFM after dissolution of the overlying oxide. Wu et al. reported on corrosion-related defects in iron and AISI 304 and 316 stainless steels studied by Doppler broadening energy spectroscopy (DBES) during corrosion processes [3, 4]. The significant increase in *S*-parameter after corrosion in iron was interpreted as due to the formation of defects and voids. However, a decrease of the *S*-parameter in AISI 304 and 316 stainless steels after corrosion was interpreted as dissolution of the passive oxide film. In this paper, we report the results of the investigation of corrosion-related microstructure and defects during corrosion processes in pure Al and AA 2037 Al alloy using the slow positron beam technique.

2. Experiment

Pure Al foils were purchased from Goodfellow Corporation (Berwyn, PA). Pure Al samples were about 300 μm thick with 99.999% purity, and were cut into 20mmx20mm for DBES measurements. These samples were annealed at 450°C for 30min and served as defect-free reference samples, having only a very small vacancy concentration in the bulk. AA 2037 Al alloys were produced by continuous casting (CC) in the Aleris International factory, USA. The main chemical compositions in the samples are 1.48%Cu, 0.45%Mg and 0.24%Mn. Hot bands were cut into 20mmx20mmx2mm pieces for DBES measurements. The as-quenched samples were annealed at 500°C for 3 hours and then quenched to room temperature in water.

The 1M NaOH immersion solution was prepared from reagent-grade chemicals and deionized water. Corrosion of samples was carried out for different time at room temperature. The DBES experiments were performed using the slow positron beam at the University of Bath. The DBES spectra, recorded using a high-purity Ge solid-state detector, were obtained as a function of positron energy from 0.25 to 30 keV at room temperature. The total count for DBES spectra was 10^6 with a counting rate of 1000 cps near the 511 keV peak.

3. Results and Discussion

3.1. S-energy profiles

Figure 1 shows the *S* parameter as a function of positron incident energy (i.e. *S*-energy profiles) for pure Al and 2037 Al alloy after different dissolution times in 1M NaOH solution. The data points are *S* parameters calculated from individual annihilation spectra measured at particular beam energy. The top *x*-axis is the mean depth of the implanted positrons from the surface at each incident energy as calculated by a well-known equation [1]. The solid lines in the figures are the results of a simulation, to be discussed below. The estimated accuracy of the *S* value is within 0.001. The *S* values in this paper are normalized by dividing by the bulk *S* parameter, which in figure 1(a) is approached as a plateau at energies greater than 15keV. A normalized *S* value larger than one indicates the presence either of a phase having an *S* parameter larger than Al, or of open volume defects.

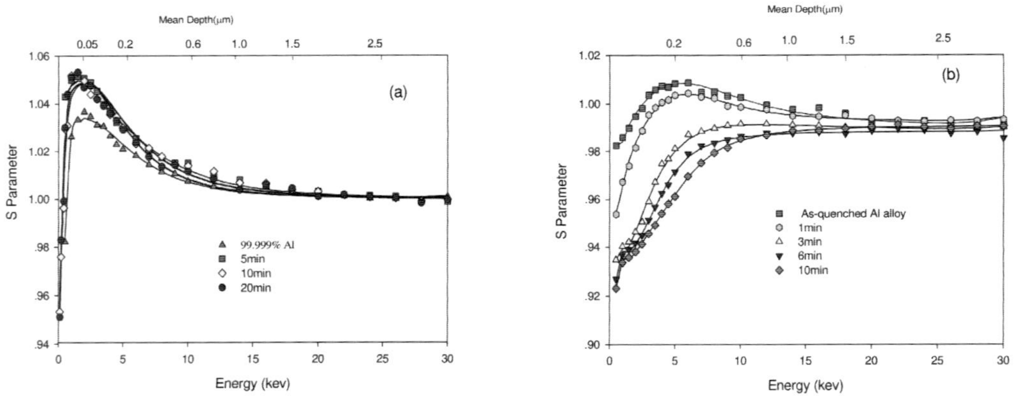

Figure 1 S-energy profiles for pure Al (a) and water-quenched 2037 Al alloy (b) after different dissolution times in 1M NaOH solution. Data points are measured values, and solid lines are results of fitting with simulation.

From the data in figure 1, we observe a different effect between pure Al and Al alloy due to corrosion. When the pure Al samples were immersed in 1MNaOH for 5-20min, a significant increase in the *S* parameter near the surface is observed. The non-monotonic variation of *S* with beam energy indicates that a near-surface defect layer is present having annihilation characteristics different from either the surface or the bulk metal. This general peaked shape of the *S* profiles suggests the presence of open-volume defects such as voids.

However, for water-quenched Al alloy samples, a significant decrease in the *S* parameter is observed after corrosion by the same method. A monotonic variation of *S* with beam energy is observed in NaOH treated samples for times increasing from 3-10min, which will be discussed below.

3.2. S-W plots

Information about the types of defects could be revealed by plotting the *W* parameter as a function of the *S* parameter. The data for all the experiments for pure Al and Al alloy are in figure 2. *S* and *W* values for individual experiments fall on a common locus, which only consists of two straight-line segments connecting three vertices. The low energy (*S,W*) state (0.94, 1.30) represents annihilation of positrons trapped in the oxide film. The other (*S,W*) states in the figure 2 represent defect-free bulk Al (1.00,1.00) or quenched bulk Al alloy (0.99, 1.10), and the open-volume defects near the surface at the metal-oxide interface (*S*=1.06, *W*=0.85). The fairly straight line shown in figure 2 indicates that only two positron states contribute to annihilation – suggesting that only one type of (vacancy-type) defect is present in all the measurements, in the annealed sample as well as NaOH corroded pure Al samples. The curvature commonly seen near the maximum *S* is caused by the relatively small concentrations of this defect, so that all three states contribute to annihilation. Figure 2 also shows that only one type of defect is present in quenched Al alloy without NaOH dissolution, but that an additional effect acted on interfacial layer, which induced in a decrease of *S* parameter after NaOH dissolution.

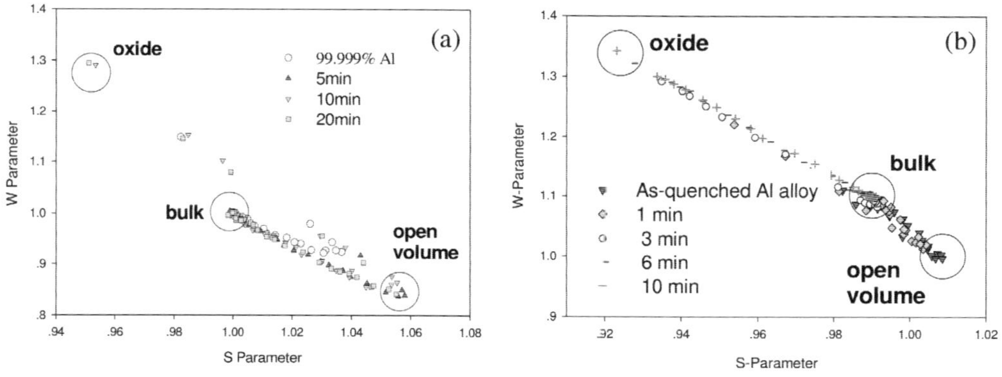

Figure 2 *W*-parameter as a function of the *S* parameter for pure Al (a) and water-quenched 2037 Al alloy (b) after different dissolution times in 1M NaOH solution. (*S* is from Fig 1 data.)

3.3. Fitting of S-energy profiles

The *S*-energy profiles in figure 1 were fitted by solving the positron diffusion-annihilation equation. The measured *S* parameters are fitted using a three-layer model (oxide film, interfacial defect layer and bulk) in the code VEPFIT and the results are plotted as solid lines in figure 1. For pure Al samples, the models used for fitting contained a surface defect layer and a bulk Al layer. Inclusion of a surface oxide layer was found to be unnecessary, as the oxide film thickness is so thin - found to be 5nm by Auger measurements - that the defect layer parameters obtained by the simulation are unaffected by its presence. Thus, the oxide film factors into the model as a surface boundary condition. The only input parameter for the fitting is the bulk Al diffusion length, which is set to 150nm [2]. For water-quenched Al alloy samples, and after dissolution in NaOH for 1min, the VEPFIT fitting model contained a surface defect layer and a bulk Al alloy layer similar to pure Al samples. However, thicker oxide films (about 20nm) are formed after dissolution in NaOH for 3-10min. Thus the VEPFIT models for the alloy contain an oxide film, an interfacial defect layer and the bulk. The fit values of the interfacial defect layer *S* (S_d), positron diffusion length (L_d), and thickness (B_d) are shown in table 1.

Table 1 shows that significant increase in the *S* parameter of the defect layer after dissolution mainly comes from the formation of defects and voids (a size of 0.1nm to several nm) near the surface during corrosion, which is consistent with previous results reported by Hebert et al. [2,3]. In our experiment, the defect layer thickness in pure Al is 114 nm in the annealed sample, compared with about 30 nm after NaOH treatment. The dissolution rate in NaOH solution was measured and found to be 200 nm/min. The initial defect layer would then have dissolved during the 1 min treatment. It is clear that the defect layer after the treatment is dominated by new defects introduced by the immersion. Furthermore, Table 1 also shows that the *S* parameter decreases with dissolution time in 2037 Al alloy, which differs from the increase due to the formation of significant numbers of open volume defects near the metal-oxide film interface induced by NaOH dissolution in pure Al samples. Many experiments [6,7] indicated that Cu has a strong influence on the corrosion of Al alloys, where it is present in matrix regions or as a constituent of secondary phases. In Cu-containing alloys, corrosion is promoted by reduction of oxygen at Cu-rich phases exposed at the alloy surface, with the associated pH rise resulting in alkaline corrosion of the adjacent matrix Cu-rich regions. Corrosion can also be generated from matrix material of low Cu content as a consequence of the growth of an amorphous oxide film. The enrichment of Cu has also been detected following various surface treatments of commercial Al alloys. Further, near-surface accumulation of impurities, including Cu, occurs during alkaline etching of Al [8]. The content of Cu in 2037 Al alloy is only 1.48%, and the annealed Cu has a low *S* value of 0.83 due to presence of 3d-electrons. Therefore, Cu enrichment near the metal-oxide interface layer will induce a decrease of the *S* parameter. The enrichment of Cu increases, and thus the *S* parameter decreases, with dissolution time. The positron diffusion length increases with dissolution time, indicating a progressively lower defect/void concentration. Further investigations are in progress involving AFM, TEM and EDX experiments and will be reported in the future.

99.999% Al			
Time (min)	S_d(nm)	L_d(nm)	B_d(nm)
0	1.034	2.1	114
5	1.048	2.6	38
10	1.045	2.8	32
20	1.049	2.7	29

AA 2037Al alloy			
	S_d(nm)	L_d(nm)	B_d(nm)
0	1.012	31	367
1	1.010	49	367
3	0.996	85	402
6	0.985	87	407
10	0.964	98	408

Table 1 Fitted parameters for the defect layer in pure Al and 2037 Al alloy.

4. Conclusions

When pure Al samples are treated in 1M NaOH for various dissolution times, a significant increase in the *S* parameter near the surface is observed. This implies that the corrosion process involves the creation of defects and nanometer voids. A significant decrease in the *S* parameter is observed after the corrosion of water-quenched Al alloy, which is interpreted as being a result of Cu enrichment near the metal-oxide interface layer

Acknowledgements

This research is supported by the National Natural Science Foundation of China under grants 10975110, 51071111 and the fund of the State Key Laboratory of Solidification Processing in NWPU under grant SKLSP201004.

References

[1] Schultz P J and Lynn K G, 1988 *Rev. Mod. Phys.* **60** 701.
[2] Hebert K R, Gessmann T, Lynn K G andAsoka-Kumar P, 2004 *J. Electrochem. Soc.* **151**, B22.
[3] Wu H, Hebert K R, Gessmann T and Lynn K G, 2002 *J. Electrochem. Soc.* **149** B108.
[4] Wu Y C, Chen Y Q, Wang B, *et al.*, 2006 *Appl. Surf. Sci.*, **252**, 3274.
[5] Wu Y C, Zhang R, Chen H, Li Y, Zhang J and Jean Y C, 2003 *Rad. Phys. Chem.* **68** 599.
[6] Obispo H M, Murr L E and Arrowood R M 2000 *J. Mater. Sci.* **35** 3479.
[7] Cervantes R L, Murr L E and Arrowood R M. 2001 *J. Mater. Sci* **36** 4079.
[8] Huang Renchun, Hebert K R, R, Gessmann T, *et al.* 2004 *J Electrochem Soc* **151** B227

Journal of Physics: Conference Series **262** (2011) 012066 doi:10.1088/1742-6596/262/1/012066

Defect structure of MBE-grown GaCrN diluted magnetic semiconductor films

A Yabuuchi[1], M Maekawa[1], A Kawasuso[1], S Hasegawa[2], Y K Zhou[2] and H Asahi[2]
[1]ASRC, Japan Atomic Energy Agency, 1233, Watanuki, Takasaki, Gunma 370-1292, Japan
[2] ISIR, Osaka University, 8-1, Mihogaoka, Ibaraki, Osaka 567-0047, Japan

E-mail: yabuuchi.atsushi@jaea.go.jp

Abstract. Effects of growth temperature and Si doping on the formation of vacancy defects in molecular beam epitaxy GaCrN films have been studied by positron annihilation spectroscopy. No vacancy defects were detected in the GaCrN films grown at 700°C. In the undoped GaCrN film grown at 540°C, vacancy clusters with sizes of V_8 to V_{12} were responsible for positron trapping. Vacancy clusters were much reduced by Si doping, but complexes related to nitrogen vacancies still survived.

1. Introduction

Chromium-doped GaN is one of the expected GaN-based diluted magnetic semiconductor (DMS) working at room-temperature [1-3]. To obtain the ferromagnetic property, it is important to substitute Ga atoms with high concentration of Cr atoms. However, heavy Cr doping results in the precipitation of CrN secondary phase at normal growth temperature. To avoid the formation of CrN secondary phase, low temperature molecular beam epitaxy (MBE) growth may be a promising manner. In this case, the introduction of vacancy-type defects during MBE growth is anticipated. Recent first-principle calculation studies proposed that the vacancy-type defects in the DMS materials might play an important role for ferromagnetism [4]. Therefore, it is important to investigate the vacancy-type defect structure of the DMS materials. Positron annihilation spectroscopy (PAS) is a capable tool to detect vacancy-type defects. Many PAS studies have been performed on GaN as opto-materials [5]. However, only a few PAS studies are reported about the defect structure of the GaN-based DMS materials [6]. In this study, the defect structure of low-temperature MBE-grown GaCrN films has been probed by a slow positron beam.

2. Experiment

Samples used in this study were GaN-cap(4 nm)/GaCrN(500 nm)/GaN-buffer(40 nm) structures grown by radio frequency (RF) plasma assisted MBE method on commercial GaN templates (2 μm) grown by metal-oxide chemical vapour deposition (MOCVD) methods. The GaN-cap layers and GaN-buffer layers were prepared at 700 °C. The GaCrN (with Cr concentration of 0.7 %) layers were grown at 700 °C or 540 °C with the growth rate of 200 nm/h. The Ga flux, N_2 flow rate and RF plasma power were kept constant at 1.6×10^{-7} Torr, 1.5 sccm and 180 W, respectively. These are referred as HT-GaCrN or LT-GaCrN. In both growth temperatures, Si-doped GaCrN films ([Si]=10^{18} cm^{-3}) were also prepared. The LT- GaCrN samples were eventually heated at 700 °C where the cap layers were grown.

Monoenergetic positron beam with energy (E) of 0.2-30 keV were implanted into the above samples and the Doppler broadening of annihilation radiation (DBAR) measurements were carried out. The DBAR spectra were characterized by S parameter with an energy window of 511$\pm$0.8 keV. All the S parameters were normalized to that obtained for the GaN template. To examine the electron-positron momentum distribution in detail, the coincidence DBAR measurements were also performed. Positron annihilation lifetime measurements were also done with a pulsed positron beam.

3. Calculation of positron annihilation characteristics

To interpret experimental positron lifetime and DBAR data, theoretical calculations were carried out. Positron lifetimes of perfect GaN and some vacancy defects were calculated based on the atomic superposition method developed by Puska et al. [7] with the Borónski–Niemnen enhancement factor [8]. The supercell includes 64 Ga atoms and 64 N atoms, i.e., it is a 4×2×2 wurtzite conventional unit cell. The lattice constants were a=3.18Å and c=5.17Å.

The DBAR spectra were calculated within the local density approximation [7]. The valence electron wavefunctions were calculated with the projector augmented-wave (PAW) method [9] using the ABINIT4.6.4 code [10]. The potentials and projectors were generated using the ATOMPAW code [11]. The valence electron configurations were $3d^{10}4s^{2}4p^{1}$, $2s^{2}2p^{3}$, $3s^{2}3p^{6}3d^{5}4s^{1}$, $3s^{2}3p^{2}$ for Ga, N, Cr and Si atoms, respectively. A supercell including 32 Ga and 32 N atoms with a 2×2×2 conventional wurtzite unit cell was constructed as a GaN crystal. For the calculation of vacancy defects, the lattice relaxations were introduced by the molecular dynamics simulation installed in the code. The cut-off energy of the plane wave basis set was 60 Ryd. The core electron wavefunctions were represented by the Slater function parameterized by Clementi and Roetti [12]. A self-consistent positron wavefunction was calculated based on the two-component density functional theory in order to minimize the energy functional [8]. The Borónski–Nieminen enhancement factor was adopted. The DBAR spectra were obtained by convoluting one-dimensional angular correlation of the annihilation radiation spectra obtained from the momentum density with the Gaussian resolution function having a half width of 3.92×10^{-3} m_0c.

4. Results and discussion

Figure 1 shows S-E curves obtained from the HT-GaCrN samples. It is seen that S parameters in both undoped and Si-doped samples are nearly the same as that of the MOCVD GaN template. This suggests that the concentrations of vacancy defects are under the detection limit. In these samples, CrN precipitations are expected to be formed because of high temperature growth. However, the S-E curves exhibit no clear features arising from CrN precipitations.

Figure 2 shows S-E curves obtained from the LT-GaCrN samples. In the case of the undoped GaCrN, S parameter increases significantly. Although the increase in S parameter is suppressed by Si doping, it is still higher than that for the GaN template. This implies that the size and/or number of vacancy defects are reduced by Si doping, but positrons are still trapped by vacancy defects. It seems conceivable that Si atoms, which are occupying Ga sites, will reduce the size and/or number of defects.

From the annihilation lifetime measurement at E=8 keV (average positron implantation depth is 180 nm), the second lifetime component (τ_2=358 ps and I_2=21 %) was obtained for the undoped LT-GaCrN. Whereas, the positron lifetime for the Si-doped LT-GaCrN did not show clear increase from that of MOCVD GaN template. Figures 3 and 4 shows the DBAR spectra obtained for the undoped and Si-doped LT-GaCrN samples at E=8 keV. To see the detailed spectrum shape, the original spectra were divided by that of the MOCVD GaN template. The DBAR spectrum for the undoped GaCrN represents a typical feature for vacancy defects. That is, an enhancement in the low momentum region ($p<5\times10^{-3}$ m_0c) and a suppression in the high momentum region ($p>10\times10^{-3}$ m_0c). Although similar feature is also observed for the Si-doped GaCrN, an additional bump appears at around $p=10\times10^{-3}$ m_0c. Thus, different types of vacancy defects are responsible for positron trapping in the undoped and Si-doped GaCrN samples.

Table 1 lists the positron lifetimes calculated for a perfect GaN lattice and some vacancy defects in

GaN. Positron lifetime for a nitrogen vacancy (V_N) is 8 ps greater than that for the bulk. Gallium vacancy (V_{Ga}) and divacancy (V_NV_{Ga}) give rise to sufficiently longer lifetimes than the bulk. With increasing the size of vacancy cluster, positron lifetime is prolonged more. The experimental lifetime for the undoped LT-GaCrN (τ_2=358 ps) is comparable to the calculated lifetimes for V_8-V_{12} clusters. As shown in Fig. 3, the experimental DBAR spectrum is also in good agreement with the calculated spectra for V_8 and V_{12} clusters. Thus, in the undoped LT-GaCrN, vacancy defects are generated and agglomerated during the growth and subsequent heating at 700°C for the cap layer growth.

In the case of the Si-doped LT-GaCrN, the observed lifetime is very close to that for the GaN template. This means that at least vacancy clusters are ruled out as the candidates for positron trapping. To identify vacancy defects, the calculated DBAR spectra for a series of small vacancy defects (V_N, V_{Ga}, V_NV_{Ga}, V_N-N_{Ga}, V_{Ga}-Ga_N, V_{Ga}-Cr_N, V_N-Cr_{Ga}, V_{Ga}-Si_N, V_N-Si_{Ga}), that are possible from the geometrical viewpoint, are compared with the experimental spectrum. Consequently, it was found that the experimental spectrum is in better agreement with the calculated spectra for nitrogen vacancy related complexes (V_N-N_{Ga}, V_N-Si_{Ga}) as shown in Fig. 4. The calculated spectrum for nitrogen vacancy (V_N) was closer to the bulk one. Replacement of one Ga atom surrounding a nitrogen vacancy with N or Si provides more space for localizing positrons. Considering the fact that the sample is doped with Si atoms, V_N-Si_{Ga} complexes might be a preferred explanation. Possibly, the occupation of Ga sites by Si atoms reduces the amount of gallium vacancies and residual nitrogen vacancies are combined with Si atoms.

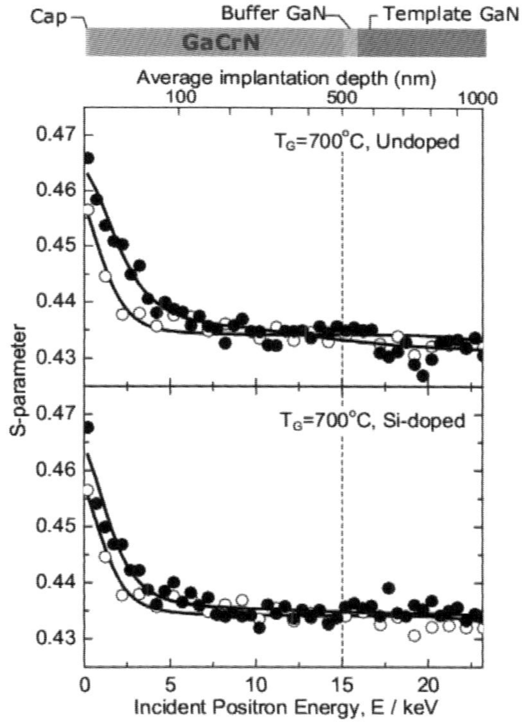

Figure 1. S parameter as a function of incident positron energy obtained from the undoped and Si-doped HT-GaCrN samples (filled circles) and the MOCVD GaN template (open circles). The upper horizontal axis denotes the average implantation depth corresponding to the incident positron energy. The top figure schematically shows the structure of samples.

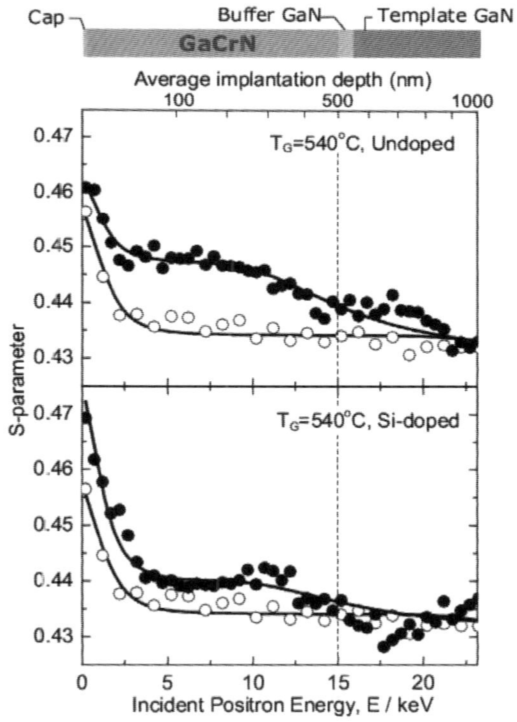

Figure 2. S parameter as a function of incident positron energy obtained from the undoped and Si-doped LT-GaCrN samples (filled circles) and the MOCVD GaN template (open circles). The upper horizontal axis denotes the average implantation depth corresponding to the incident positron energy. The top figure schematically shows the structure of samples.

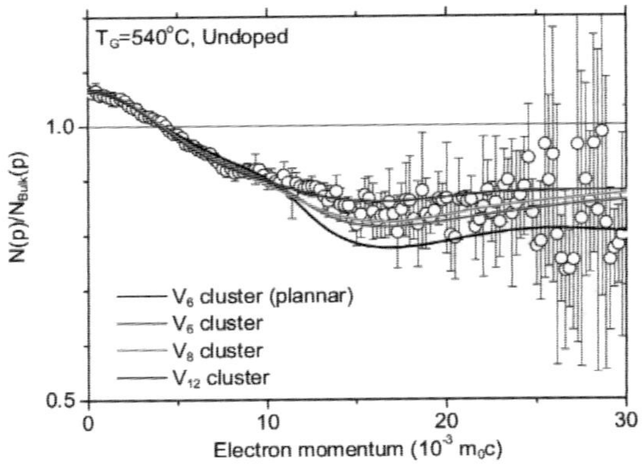

Figure 3. DBAR spectrum obtained for the undoped LT-GaCrN at E=8 keV. The original spectrum is divided by the spectrum of MOCVD GaN. Black, red, green and blue solid lines are the calculated DBAR spectra for V_6 cluster (planar), V_6 cluster, V_8 cluster and V_{12} cluster, respectively.

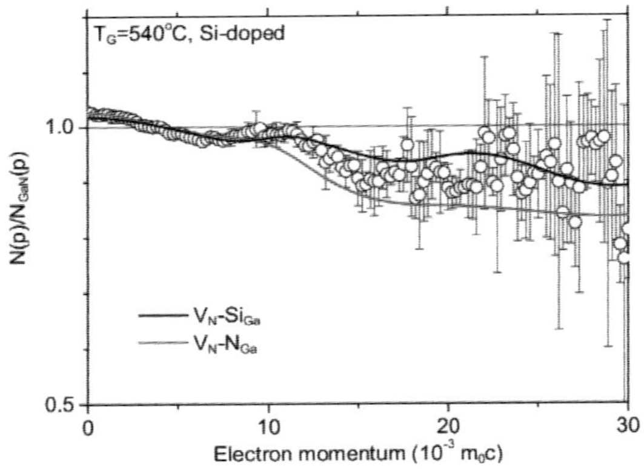

Figure 4. DBAR spectrum obtained for the Si-doped LT-GaCrN at E=8 keV. The original spectrum is divided by the spectrum of MOCVD GaN. Black and red solid lines are the calculated DBAR spectra for V_N-Si_{Ga} and V_N-N_{Ga} complexes, respectively.

Table 1 Positron lifetimes for the perfect GaN lattice and some vacancy defects calculated by the atomic superposition method.

Bulk	V_N	V_{Ga}	V_NV_{Ga}	V_6	V_6 planar	V_8	V_{12}
155 ps	163 ps	204 ps	229 ps	308 ps	311 ps	347 ps	383 ps

5. Conclusion

In this study, vacancy defects in GaCrN grown by low temperature MBE method have been studied by positron annihilation spectroscopy. It was found that no intentional doping results in the formation of large vacancy clusters during growth. Formation of vacancy clusters could be avoided by intentional Si doping, while complexes related to nitrogen vacancies still survive. The further study is needed to elucidate the effect of vacancy defects on the magnetic properties of GaCrN.

References

[1] Hashimoto M, Zhou Y K, Kanamura M and Asahi H 2002 *Solid State Commun.* **122** 37-39.
[2] Hashimoto M, Zhou Y K, Kanamura M, Katayama-Yoshida M and Asahi H 2003 *J. Cryst. Growth* **251** 327-330.
[3] Sato K and Katayama-Yoshida H 2001 *Jpn. J. Appl. Phys.* **40** L485-L487.
[4] Kang J and Chang K J 2007 *J. Appl. Phys.* **102** 083910.
[5] Neugebauer J and de Walle C G V 1996 *Appl. Phys. Lett.* **69** 503-505.
[6] Yang X L, Zhu W X, Wang C D, Fang H, Yu T J, Yang Z J, Zhang G Y, Qin X B, Yu R S and Wang B Y 2009 *Appl. Phys. Lett.* **94** 151907.
[7] Puska M J and Nieminen R M 1994 *Rev. Mod. Phys.* **66** 841–897.
[8] Borónski E and Nieminen R M 1986 *Phys. Rev.* B **34** 3820–3831.
[9] Blöhl P E 1994 *Phys. Rev.* B**50** 17953-17979.
[10] Gonze X, Beuken J -M, Caracas R, Detraux F, Fuchs M, Rignanese G -M, Sindic L, Verstraete M, Zerah G, Jollet F, Torrent M, Roy A, Mikami M, Ghosez Ph, Raty J -Y, and Allan D C 2002 *Comput. Mater. Sci.* **25** 478–492.
[11] Holzwarth N A W, Tackett A R and Matthews G E 2001 *Comput. Phys. Commun.* **135** 329–347.
[12] Clementi E and Roetti C 1974 *At. Data Nucl. Data Tables* **14** 177-478.

Positron microbeam study on vacancy generation caused by stress corrosion crack propagation in austenitic stainless steels

A Yabuuchi, M Maekawa and A Kawasuso
ASRC, Japan Atomic Energy Agency, 1233, Watanuki, Takasaki, Gunma 370-1292, Japan
E-mail: yabuuchi.atsushi@jaea.go.jp

Abstract. A stress-corrosion-cracked Type 304 austenitic stainless steel was investigated by the Doppler broadening of annihilation radiation (DBAR) spectroscopy with a positron microbeam. Clear increase of S parameter was observed over 200-400 μm areas from the stress corrosion cracking (SCC) crack. From the comparison of the DBAR spectra obtained from the stress-corrosion-cracked sample and plastically deformed sample, the increase of S parameter around the SCC crack was attributed to plastic-deformation-induced vacancies.

1. Introduction

Stress corrosion cracking (SCC) is still a serious issue even in the low-carbon austenitic stainless steels which were developed as corrosion resistive materials for nuclear reactors. Recently, a hypothesis of the SCC crack propagation mediated by vacancies near the crack tip stress field is proposed [1-2]. Vacancies in austenitic stainless steels start to migrate at 250 °C [3-5] where a boiling-water reactor is operated. Thus, it is conceivable that vacancies play an important role in the high-temperature water SCC crack propagation mechanism. However, the origin of such vacancies has not yet been fully elucidated. In this study, to investigate the vacancy formation during the SCC crack propagation, an SCC-cracked Type 304 austenitic stainless steel was probed by a positron microbeam.

2. Experiment

All the samples used in this study were prepared from Type 304 austenitic stainless steel foils or plates that were sensitized at 650 °C for 24 hours in vacuum. The sensitized state is defined as a reference. These samples were subjected to the SCC test and plastic deformation test. In the SCC test, a sample ($5\times10\times0.03$ mm^3) having a notch was attached to a tensile holder with a tensile stress of approximately 100 MPa and exposed into boiling $MgCl_2$ water for an accelerated corrosion. The SCC crack started to appear after the corrosion treatment for 6 hours. In the plastic deformation test, a sample with a gauge width of 1 mm, a gauge length of 0.85 mm and a thickness of 50 μm was plastically deformed with a tensile stress up to the nominal strain of 20 %.

The above samples were transferred into a positron microbeam apparatus [6]. The SCC-cracked and deformed samples were scanned by a 20 keV positron microbeam with a diameter of approximately 10 μm with a scanning step of 100 μm and 40 μm, respectively, and DBAR measurements were carried out using a high-purity Ge detector. Accumulated total counts at each measurement were approximately 10^4. To monitor the presence of vacancy defects, the obtained DBAR spectra were characterized by S parameter, which is defined as the normalized peak intensity. All the S parameters were normalized to those obtained for the reference state. To identify the detected defect species, the shapes of DBAR spectra were investigated in detail comparing with theoretical calculation.

3. Calculation of DBAR spectra

To interpret experimental DBAR data, theoretical calculations were carried out within the local density approximation [7]. The valence electron wavefunctions were calculated with the projector augmented-wave (PAW) method [8] using the ABINIT4.6.4 code [9]. The potentials and projectors were generated using the ATOMPAW code [10]. The valence electron configurations were $3s^2 3p^6 3d^6 4s^2$, $3s^2 3p^6 3d^5 4s^1$, $3s^2 3p^6 3d^8 4s^2$ for Fe, Cr and Ni atoms, respectively. A supercell including 23 Fe, 6 Cr and 3 Ni atoms with a 2×2×2 conventional fcc unit cell was constructed as an austenitic stainless steel. The cut-off energy of the plane wave basis set was 40 Ryd. The core electron wavefunctions were represented by the Slater function parameterized by Clementi and Roetti [11]. A self-consistent positron wavefunction was calculated based on the two-component density functional theory in order to minimize the energy functional [12]. The Borónski–Nieminen enhancement factor was adopted. The DBAR spectra were obtained by convoluting one-dimensional angular correlation of the annihilation radiation spectra obtained from the momentum density with the Gaussian resolution function having a half width of 3.92×10^{-3} m_0c.

4. Results and discussion

Figures 1(a) and 1(b) shows optical image and S parameter image, respectively, obtained from the SCC sample. It is seen that S parameter does not increased at the stress free region beside the initial notch. On the contrary, clear increase of S parameter is observed over 200-400 μm areas from the SCC crack. Figure 2(a) and 2(b) shows optical microscope image and S parameter image, respectively, obtained from the deformed sample. The increase of average S parameter in the gauge areas was 2.9 % after the deformation.

Figure 3 shows the experimental DBAR spectrum obtained from the deformed sample and the calculated DBAR spectrum for a monovacancy. To see the detailed spectrum shape, the original spectra were divided point-by-point by the experimental spectrum obtained from the undeformed sample and the calculated spectrum for the perfect lattice of a Type 304 austenitic stainless steel. The calculated spectrum is nicely reproduced the experimental one. Figure 4 shows the DBAR spectrum obtained from the gauge region of the deformed sample and near the crack region of the SCC sample. The original spectra were also divided by the reference spectrum. The DBAR spectrum of the SCC sample is in excellent agreement with that of the deformed sample.

The result shown in Fig. 1(b) suggests that lattice defects were formed near the crack due to the crack propagation. Plastic deformation is considered as an important factor giving rise to the change of S parameter near the crack due to the tensile stress concentrates on the crack tip. Thus, the DBAR spectrum of the deformed sample was measured in order to compare with that of the SCC sample. The spectrum shown in Fig. 3 indicates that the increase of S parameter brought by plastic deformation is caused by a formation of vacancy-type defects. The vacancy concentration introduced by plastic deformation may be given by

$$C_V = \eta \varepsilon^n \tag{1}$$

where η and n are constants and ε is the strain [13]. Assuming $\eta=10^{-2}$ and $n=2$, the vacancy concentration is approximately 4×10^{-4} at $\varepsilon=20$ %. Previous theoretical calculation studies have revealed that positrons are not localized strongly in the pure dislocation and the jog, in addition the positron lifetimes for the pure dislocation and the jog are close to the perfect-lattice value [14,15]. These perceptions imply that all implanted positrons are trapped at vacancy-type defects (e.g., vacancy-impurity complexes, small vacancy clusters or vacancies on the dislocation line). Similar DBAR spectra for the deformed sample and the SCC sample suggest that the plastic-deformation-induced vacancies are formed near the SCC crack. Such vacancies can migrate in the nuclear reactor operation temperature [3-5]. Therefore, it seems to be probable that vacancies migrate and accumulate in grain boundaries of crack tips driven by a stress gradient. The corrosion is another important factor in the SCC. It is also possible that the vacancies observed near the crack were introduced by corrosion. Thus, the effect of corrosion on vacancy formation remains as a matter to be discussed further.

5. Conclusion

In this study, an SCC-cracked Type 304 austenitic stainless steel was probed by a positron microbeam. It was found that lattice defects were formed over 200-400 μm areas from the SCC crack. From the comparison of the DBAR spectra obtained from the SCC sample and the deformed sample, it is concluded that plastic-deformation-induced vacancies are introduced near the SCC crack during crack propagation. Further studies are needed to evaluate the effect of corrosion on vacancy formation in a Type 304 austenitic stainless steel.

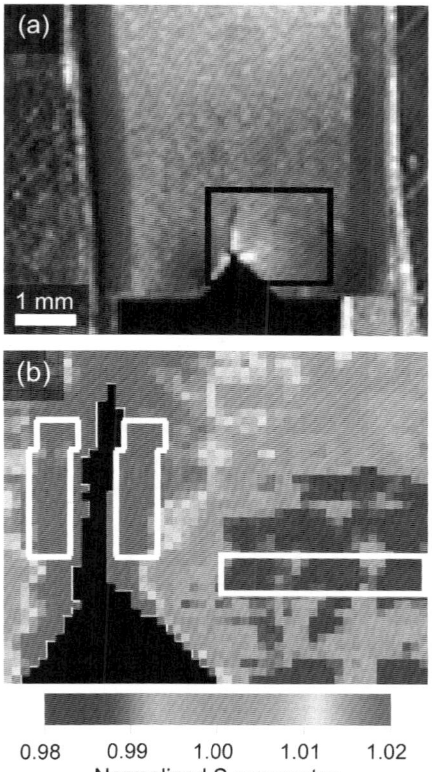

Figure 1. (a) Optical image of the SCC sample. The border in black indicates a scanned area (2.0×1.6 mm^2). (b) S parameter image of the SCC sample. The pixels in black correspond to the SCC crack gap or initial notch region. The borders in white indicate extraction area of the DBAR spectra as the reference and the region where S increased. Each pixel indicates 40×40 μm^2 region.

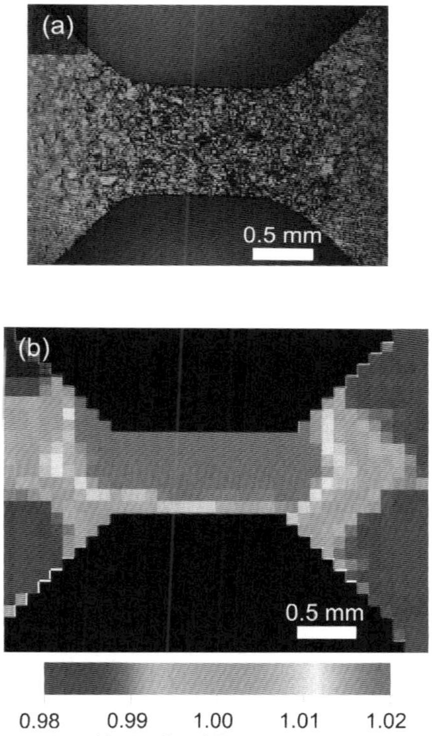

Figure 2. (a) Optical microscope image of the tensile test piece deformed by approximately 20 %. (b) S parameter image obtained from the tensile test piece after the deformation. Each pixel indicates 100×100 μm^2 region.

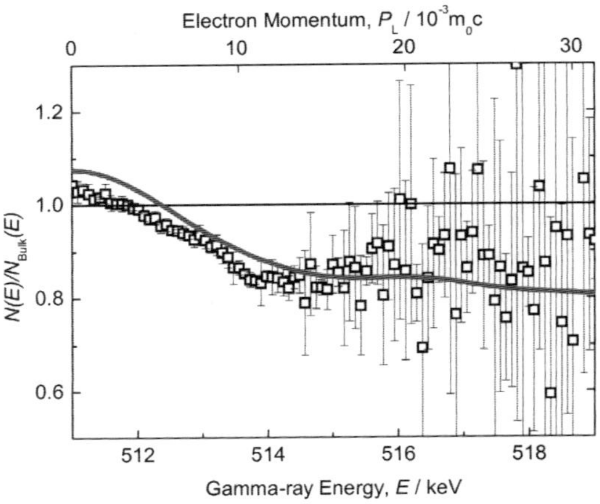

Figure 3. DBAR spectrum obtained from the gauge region of the deformed sample (open square). The original spectrum is divided by the spectrum of the undeformed sample. Solid line is the calculated DBAR spectra for a monovacancy in a Type 304 austenitic stainless steel.

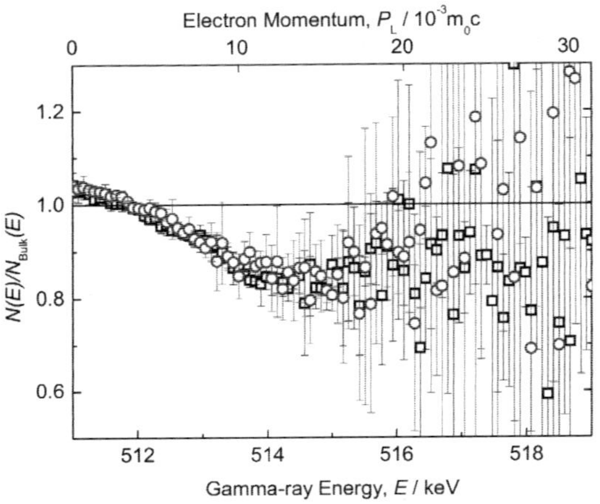

Figure 4. DBAR spectra obtained from the gauge region of the deformed sample (open square) and near the crack region of the SCC sample (open circle). Each original spectrum is divided by the spectrum obtained from the gauge region of the undeformed sample and the region where S unincreased in the SCC sample.

Acknowledgments
The present study was the result of "In-situ analysis of crack tips in stainless steels under tensile stress" entrusted to the Japan Atomic Energy Agency (JAEA) by the Ministry of Education, Culture, Sports, Science and Technology of Japan (MEXT) and was partially supported by the MEXT Grant-in-Aid for Young Scientists (B), No. 22760682, 2010.

References

[1] Staehle R W 2006 "*Proc. Int. Conf. Water Chem. Of Nucl. React. Sys.*," Jeju Is., Korea.
[2] Arioka K, Yamada T and Terachi T 2006 *INSS Journal* **13** 168-178 [in Japanese].
[3] Gil C L, de Lima A P, de Campos N A, Sperr P, Kögel G and Triftshäuser W 1990 *Radiat. Eff. Def. Solids* **112** 111-118.
[4] Dryzek J, Wesseling C, Dryzek E and Cleff B 1994 *Mater. Lett.* **21** 209-214.
[5] Arunkumar J, Abhaya S, Rajarman R, Amarendra G, Nair K G M, Sundar C S and Raj B 2009 *J. Nucl. Mater.* **384** 245-248.
[6] Maekawa M and Kawasuso A 2008 *Appl. Surf. Sci.* **255** 39-41.
[7] Puska M J and Nieminen R M 1994 *Rev. Mod. Phys.* **66** 841–897.
[8] Blöhl P E 1994 *Phys. Rev.* B **50** 17953-17979.
[9] Gonze X, Beuken J -M, Caracas R, Detraux F, Fuchs M, Rignanese G -M, Sindic L, Verstraete M, Zerah G, Jollet F, Torrent M, Roy A, Mikami M, Ghosez Ph, Raty J -Y, and Allan D C 2002 *Comput. Mater. Sci.* **25** 478–492.
[10] Holzwarth N A W, Tackett A R and Matthews G E 2001 *Comput. Phys. Commun.* **135** 329–347.
[11] Clementi E and Roetti C 1974 *At. Data Nucl. Data Tables* **14** 177-478.
[12] Borónski E and Nieminen R M 1986 *Phys. Rev.* B **34** 3820–3831.
[13] Russell B 1963 Phil. Mag. **88** 615-630.
[14] Kamimura Y, Tsutsumi T and Kuramoto E 1995 *Phys. Rev.* B **52** 879-885.
[15] Kamimura Y, Tsutsumi T and Kuramoto E 1997 *J. Phys. Soc. Jpn.* **66** 3090-3096.

Journal of Physics: Conference Series **262** (2011) 012068 doi:10.1088/1742-6596/262/1/012068

Positron beam study of sol-gel prepared ordered mesoporous silica thin films

R S Yu, X B Qin, P Zhang, B Y Wang and L Wei

Key laboratory of Nuclear Analysis Techniques, Institute of High Energy Physics, Chinese Academy of Sciences, No. 19 Yuquan Lu, Beijing 100049, China

E-mail: yursh@ihep.ac.cn (R S Yu), wangboy@ihep.ac.cn (B Y Wang)

Abstract. Several so-gel prepared silica thin films with ordered meso-pores were studied with a variable energy positron beam. Doppler broadening measurements show that in CTAB-templated samples, film porosity is the superior factor upon pore size and affects the final *S* parameter. While for P123 and F127-templated films with similar porosity, as hexagonal pores extend along the direction parallel to film surface, a larger effective pore size could be sensed by positrons, which enhances open volume ratio favourable for Ps formation.

1. Introduction

Mesoporous materials with homogeneous pore size, unique and ordered pore arrangement, as well as high surface area are quite attractive for a wide variety of applications including catalysts, absorbents, reaction vessels, matrices for photochemical species, optical and biomedical applications. Positron annihilation spectroscopy (PAS) has been widely used for the structural characterization of mesopores including pore size, interconnectivity, porosity and open porosity etc [1], but mostly focus on randomly-oriented pores.

In this study, we present positron beam characterization of silica thin films containing ordered mesopores prepared by evaporation induced self-assembly (EISA) method [2,3], with cetyltrimethylammonium bomide (CTAB) and Poly(ethyleneoxide)-Poly(Propylene oxide)-Poly(ethylene oxide) tri-block copolymers (PEO-PPO-PEO) as templates and tetraethylorthosilicate (TEOS) as silica source. Some relationship between the pore structures and annihilation parameters is discussed.

2. Experimental

Two series of mesoporous silica thin films were dip-coated (3 mm/minute) onto Si(100) (n-type, $4\Omega\cdot cm^{-1}$) substrates. The first series were synthesized with CTAB (Alfar-Aesar, molecular weight: 364) as surfactant, and molar ratio for the chemical composition of silica sol is 1 TEOS: 10 H_2O:1.8$\times 10^{-3}$ HCl: 10 EtOH: *x* CTAB, where *x* is 0.1, 0.15, 0.2 and 0.25. The other series of samples were prepared with different surfactants of Pluronic P123 (PEO_{70}-PPO_{20}-PEO_{70}, molecular weight: 5800, Aldrich) and F127 (PEO_{106}-PPO_{70}-PEO_{106}, molecular weight: 12600, Aldrich), molecular ratio of the chemical composition is 1 TEOS: 8.6$\times 10^{-3}$ P123:8.3 H_2O: 2.8$\times 10^{-3}$ HCl: 17.2 EtOH and 1 TEOS: 7.9$\times 10^{-3}$ F127:8.3 H_2O: 2.8$\times 10^{-3}$HCl: 21.4 EtOH, respectively. After aging at room temperature for one day, the samples were calcined by heating at a rate of 1℃/min to 400℃ and holding for four hours to remove the surfactants.

Positron annihilation Doppler broadening measurements were conducted with a magnetically guided variable energy (0-20 keV) positron beam at Beijing intense slow positron facility [4]. And small angle X ray reflectivity (XRR) measurements were performed at Beijing Synchrotron Radiation Facility (BSRF). Film density ρ (unit: g/cm^3) can be deduced by the critical angle θ_c of total external reflection and wavelength λ (1.54 Å) of the X ray following $\theta_c = \lambda(\rho)^{1/2} \times 1.64 \times 10^{-3}$. Suppose the films are composed of only silicon oxide backbones and pores filled with air, film porosity can be estimated by Porosity=$\frac{\rho_0 - \rho}{\rho_0} \times 100\%$, where ρ_0 is the density of a condensed non-porous silica (2.22 g/cm^3). The pore arrangement and size normal to the sample surface can be calculated by $d = 2\pi / \Delta Q_Z$, where $Q_Z = 4\pi \sin\theta / \lambda$, here θ is the angle of adjacent strong Bragg peaks.

3. Results and discussions

3.1. Effect of film porosity on PAS results

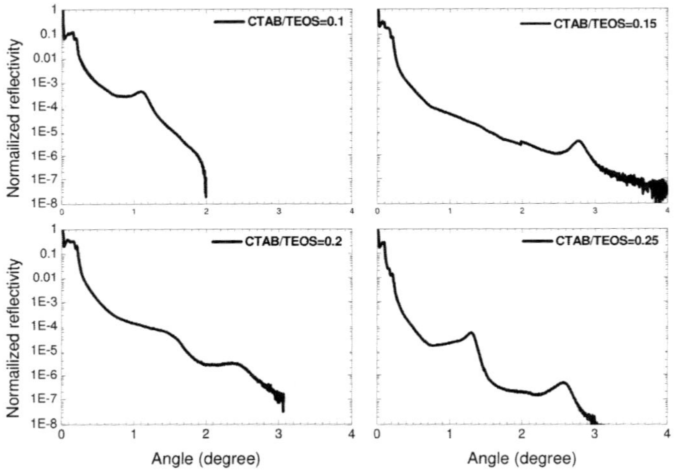

Figure 1. XRR curves of silica thin films synthesized with different CTAB/TEOS molar ratios ranging from 0.1 to 0.25. Porosity of each film was calculated with critical angles appear at very small incident angles below one degree.

XRR results for the first series of samples are illustrated in Figure 1. Film porosity was calculated to be 46%, 47%, 54%, 79% respectively, as the molar ratio of CTAB/TEOS rises from 0.1 to 0.25. Two well resolved Bragg peaks at relatively large angles are seen in the XRR patterns for films with CTAB/TEOS molar ratio of 0.2 and 0.25, indicating the existence of well ordered mesopores in the two samples. For the former sample, the two Bragg peaks are at Q_z= 0.204 Å^{-1} and 0.335 Å^{-1} with *d* spacing of 47.9 Å. For sample with CTAB/TEOS molar ratio of 0.25, the peaks are at Q_z= 0.183 Å^{-1} and 0.365 Å^{-1} with *d* spacing of 34.5 Å. Basically speaking the Ps momentum dispersion decreases, i.e. *S* increases, in a pore with larger size based on uncertainty principle. However, positron annihilation Doppler broadening *S* parameters shown as a function of incident positron energy in Figure 2 clearly indicate that the film with CTAB/TEOS equals to 0.2 (having larger size of pores) has relatively smaller *S* values.

It is reported by Serna *et al.* [5] that the Doppler energy shifts of the free positron and *o*-Ps annihilation are usually close to each other, but much smaller than that related to *p*-Ps, i.e. $S_{e^+} \approx S_{o\text{-}Ps} << S_{p-Ps}$, and as a rule the smaller is Ps formation probability, the smaller the *S* parameter. Reasonably, we infer that it is due to the 25% less of film porosity for the CTAB/Si film with molar ratio of 0.2, Ps yield is reduced here, though the film has a larger *d* spacing value and correspondingly a larger pore size. For the same reason, the smaller characteristic *S* values for silica films shown on Figure 2 at small CTAB/TEOS molar ratios of 0.1 and 0.15 than the other two films is also originated from the low film porosities, as revealed by XRR.

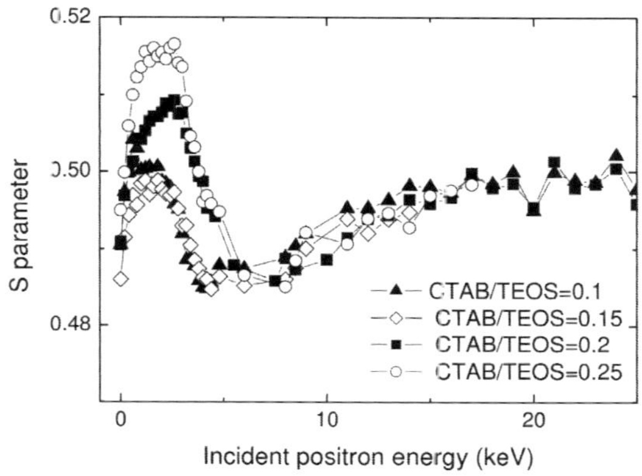

Figure 2. *S-E* curves for silica films prepared with different CTAB/TEOS molar ratio ranging from 0.1 to 0.25. Similar characteristic *S* values were obtained for the two films with lowest CTAB/TEOS molar ratios, because their porosities are quite similar (1% difference).

3.2. Effect of pore shape on PAS results

Figure 3 is the SEM image of two mesoporous silica films with completely different pore shape, i.e. one in hexagonal and the other in cubic. The hexagonal one was prepared with P123 as surfactant and the cubic with F127. XRR results showed that the two films have equivalent film porosity of 30.4% (P123) and 32% (F127), and quite similar pore size of about 5.6 nm (P123) and 5.8 nm (F127) in the direction normal to the film surface. As in our previous discussion, considering the slightly smaller film porosity and pore size for the P123-templated film, it seems that Ps formation probability and positron annihilation *S* values should be lower than that for the F127-templated film. However, results shown in Figure 4 suggest an opposite tendency. Here in this case, pore shape effect plays a dominant role. Since the channel of pores in hexagonal shape extend parallel to the film substrate, while that for the cubic one is in 3D phase as revealed by small angle XRD [2], it could not extend longer and deeper either parallel or normal to the film surface. There can be a larger effective pore size and higher open volume ratio favourable for Ps formation in the former film. Further, it should be noted that pore size deduced from XRR Bragg peaks is only valid in the direction normal to the film surface, while positrons sense the pore size in a three-dimensional way.

4. Conclusion

A series of silica thin films containing ordered mesopores were prepared with sol-gel technique and characterized by a variable energy slow positron beam. Results show that for two CTAB-templated films, the one with a 13.4 Å larger pore size but a much smaller (25%) film porosity corresponds to a reduced positron annihilation Doppler broadening *S* value. Whereas for two P123 and F127-templated mesoporous films, though they have similar porosity and pore size in the direction normal to film surface, the film containing 2D hexagonal pores gives enhanced *S* values owing to channel extension of the pores parallel to the film surface, and consequently a higher open volume ratio favorable for Ps formation.

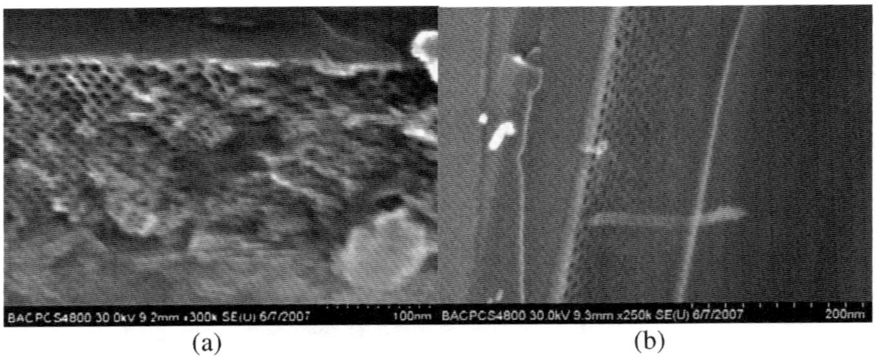

Figure 3. Cross-section view of SEM images for P123 (a) and F127 (b) -templated mesoporous silica thin films.

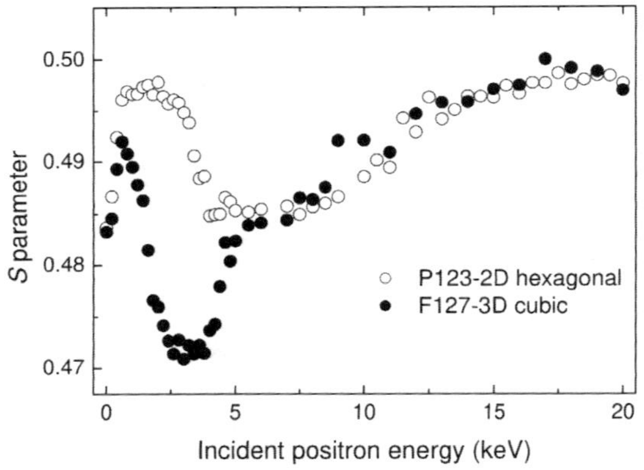

Figure 4. *S-E* curves for silica thin films with 2D hexagonal and 3D cubic pores.

References

[1] Gidley D W, Peng H G and Vallery R S 2006 *Ann. Rev. Mater. Res.* **36** 49
[2] Yu R S, Qin X B, Wang Q Z, Zhang Z, Zhong Y R, Li Z X, Wang B Y, Wei L, Jia Q J and Kurihara T 2009 *Mater. Sci .Forum* **607** 99
[3] Qin X B, Yu R S, Kurihara T, Cao X Z, Wang B Y and Wei L 2009 *Mater. Sci. Forum* **607** 201
[4] Wang B Y, Ma Y Y, Zhang Z, Yu R S and Wang P 2008 *Appl. Surf. Sci.* **255** 119
[5] Serna J, Abbe J Ch and Duplatre G 1986 *Phys. Stat. Sol. (a)* **98** 171

AUTHOR INDEX

AUTHOR INDEX

AUTHOR INDEX

Institute of Physics
Dirac House, Temple Back
Bristol BS1 6BE UK